FDTD Modeling of Metamaterials

Theory and Applications

For a list of recent related Artech House titles, please turn to the back of this book.

FDTD Modeling of Metamaterials

Theory and Applications

Yang Hao
Raj Mittra

ARTECH HOUSE
BOSTON | LONDON
artechhouse.com

Library of Congress Cataloging-in-Publication Data
A catalog record for this book is available from the U.S. Library of Congress.

British Library Cataloguing in Publication Data
A catalogue record for this book is available from the British Library.

ISBN-13: 978-1-59693-160-2

Cover design by Igor Valdman

© 2009 ARTECH HOUSE, INC.
685 Canton Street
Norwood, MA 02062

All rights reserved. Printed and bound in the United States of America. No part of this book may be reproduced or utilized in any form or by any means, electronic or mechanical, including photocopying, recording, or by any information storage and retrieval system, without permission in writing from the publisher.

All terms mentioned in this book that are known to be trademarks or service marks have been appropriately capitalized. Artech House cannot attest to the accuracy of this information. Use of a term in this book should not be regarded as affecting the validity of any trademark or service mark.

10 9 8 7 6 5 4 3 2 1

Contents

Preface	*xi*
Acknowledgments	*xiii*

CHAPTER 1
Introduction — 1

1.1	What Are Electromagnetic Metamaterials?	1
1.2	A Historical Overview of Electromagntic Metamaterials	2
	1.2.1 Artificial Dielectrics	4
	1.2.2 Artificial Magnetic Materials	8
	1.2.3 Bianisotropic Composites	8
	1.2.4 Double-Negative and Indefinite Media	9
	1.2.5 Photonic and Electromagnetic Crystals	11
1.3	Numerical Modeling of Electromagnetic Metamaterials	15
1.4	Layout of the Book	17
	References	18

CHAPTER 2
Fundamentals and Applications of Electromagnetic Bandgap Structures — 25

2.1	Introduction	25
2.2	Bloch's Theorem and the Dispersion Diagram	25
	2.2.1 Translational Symmetry	26
	2.2.2 Bloch's Theorem and Periodic Boundary Condition (PBC)	27
	2.2.3 Brillouin Zone	29
	2.2.4 Dispersion Diagram and EBG	30
2.3	An Overview of Numerical Methods for Modeling EBG Structures	33
	2.3.1 The Generalized Rayleigh's Identity Method and the Korringa-Kohn-Rostoker (KKR) Method	33
	2.3.2 Plane-Wave Expansion Method	35
	2.3.3 The Transfer-Matrix Method	36
	2.3.4 The Finite-Difference Time-Domain (FDTD) Method	39
2.4	An Overview of EBG Applications	41
	2.4.1 In-Phase Reflection	41
	2.4.2 Suppression of Surface Waves	45
	2.4.3 EBGs Operating in Defect Modes	46
	2.4.4 Subwavelength Imaging from the Passband of the EBGs	58

2.5	Summary	61
	References	61

CHAPTER 3
A Brief Introduction to the FDTD Method for Modeling Metamaterials — 67

- 3.1 Introduction — 67
- 3.2 Formulations of the Yee's FDTD Algorithm — 67
  - 3.2.1 Maxwell's Equations — 67
  - 3.2.2 Yee's Orthogonal Mesh — 69
  - 3.2.3 Time Domain Discretization: The Leapfrog Scheme and the Courant Stability Condition (CFL Condition) — 70
- 3.3 Other Spatial Domain Discretization Schemes — 72
  - 3.3.1 Subgridding Mesh — 72
  - 3.3.2 Nonorthogonal Mesh — 75
  - 3.3.3 Hybrid FDTD Meshes — 76
- 3.4 Boundary Conditions — 78
  - 3.4.1 Mur's Absorbing Boundary Conditions (ABCs) — 78
  - 3.4.2 Perfect Matched Layers (PMLs) — 80
  - 3.4.3 Periodic Boundary Condition (PBC) — 81
- 3.5 Bandgap Calculation — 83
  - 3.5.1 Source Excitation — 84
  - 3.5.2 Dispersion Diagram Calculation — 84
  - 3.5.3 Transmission and Reflection Coefficient Calculation — 85
- 3.6 Summary — 87
- References — 88

CHAPTER 4
FDTD Modeling of EBGs and Their Applications — 91

- 4.1 Introduction — 91
- 4.2 FDTD Modeling of Infinite Electromagnetic Bandgap Structures — 91
  - 4.2.1 Physical Model of EBG Structures — 91
  - 4.2.2 Mesh Generation and Simulation Parameters in FDTD Modeling — 93
  - 4.2.3 Simulation Results of Infinite EBGs Using the Conformal and Yee's FDTD — 94
- 4.3 Conformal FDTD Modeling of (Semi-)Finite EBG Structures — 102
  - 4.3.1 FDTD Model and Simulation Results — 102
- 4.4 Design and Modeling of Millimeter-Wave EBG Antennas — 105
  - 4.4.1 Introduction — 105
  - 4.4.2 Design and Modeling of Woodpile EBG — 108
  - 4.4.3 A Millimeter-Wave EBG Antenna Based on a Woodpile Structure — 115
  - 4.4.4 Experimental Results — 117
- 4.5 Conclusions — 121
- References — 121

CHAPTER 5
Left-Handed Metamaterials (LHMs) and Their Applications   123

5.1 Introduction   123
5.2 Effective Medium Theory and Left-Handed Metamaterials   123
    5.2.1 A Composite Medium of Metallic Wires and Split Ring Resonators   124
    5.2.2 Isotropic Three-Dimensional Left-Handed Metamaterials   125
    5.2.3 Left-Handed Metamaterials Using Simple Short Wire Pairs   126
5.3 Applications of Left-Handed Metamaterials   127
    5.3.1 Imaging by a Perfect LHM Lens   127
    5.3.2 Transmission Line Structures of Left-Handed Metamaterials   128
    5.3.3 Directive Electromagnetic Scattering by an Infinite Conducting Cylinder Coated with LHMs   142
    5.3.4 Negative Index Materials (NIM) for Selective Angular Separation of Microwave by Polarization   144
    References   145

CHAPTER 6
Numerical Modeling of Left-Handed Material (LHM) Using a Dispersive FDTD Method   147

6.1 Introduction   147
6.2 The Effective Medium of Left-Handed Materials (LHMs)   148
6.3 Modeling of Left-Handed Metamaterials Using a Dispersive FDTD Method   156
    6.3.1 Two-Dimensional Dispersive FDTD with Auxiliary Differential Equations (ADEs)   156
    6.3.2 Phase Compensation Through Layered LHM Structures   160
    6.3.3 Conjugate Dielectric and Metamaterial Slab as Radomes   161
    6.3.4 Numerical Results   163
6.4 Conclusions   169
    References   169

CHAPTER 7
FDTD Modeling and Figure-of-Merit (FOM) Analysis of Practical Metamaterials   173

7.1 Introduction   173
7.2 EM Response of the Infinite, Doubly Periodic DNG Slab with Plane Wave Illumination   174
    7.2.1 Model Description of the Array Comprising of Split-Ring Resonators and Wires   174
    7.2.2 Scattering Parameters Measurements Obtained from the PBC/FDTD Code   174

	7.2.3 Phase Data Inside the DNG Slab	175
7.3	Retrieval of Effective Material Constitutive Parameters Using the Inversion Approach	182
	7.3.1 Review of the Inversion Approach	182
	7.3.2 Retrieval of the Effective Material Parameters from the Numerical S-Parameters Obtained from FDTD Simulations of Metamaterials	186
	7.3.3 Summary of the Difficulties Encountered Using the Inversion Approach for Effective Medium Characterization	207
7.4	EM Response of a Finite Artificial-DNG Slab with Localized Beam Illumination	208
	7.4.1 Slab with Localized Beam Illumination	209
	7.4.2 FDTD Model	209
	7.4.3 Total Transmission and Reflection Power Under Gaussian Beam Illumination	210
	7.4.4 EM Response of the Artificial-DNG Slab at Normal Incidence with Ey Polarization	213
	7.4.5 EM Response of the Artificial-DNG Slab at Oblique TM_z Incidence Coming from ($\theta = 150°, \phi = 90°$) with Hx Polarization	219
	7.4.6 EM Response of the Artificial-DNG Slab at Oblique TE_z Incidence Coming from $\theta = 150°, \phi = 0°$ with Ey Polarization	223
	7.4.7 EM Response of a Finite Artificial-DNG Slab Excited by Small Dipole	226
7.5	Figure-of-Merit (FOM) Analysis	228
	7.5.1 Loss and Bandwidth of Metamaterials with Different Electrical Sizes and Particle Densities	229
	7.5.2 Figure-of-Merit Analysis by Numerical Experiments	232
7.6	Conclusions	235
	References	236

CHAPTER 8
Accurate FDTD Modeling of a Perfect Lens — 239

8.1	Introduction	239
8.2	Dispersive FDTD Modeling of LHMs with Spatial Averaging at the Boundaries	241
	8.2.1 The (E, D, H, B) Scheme	242
	8.2.2 The (E, J, H, M) Scheme	244
	8.2.3 The Spatial Averaging Methods	245
8.3	Numerical Implementation	250
8.4	Effects of Material Parameters on the Accuracy of Numerical Simulation	255
8.5	Effects of Switching Time	258
8.6	Effects of Transverse Dimensions on Image Quality	260

8.7	Modeling of Subwavelength Imaging	262
8.8	Conclusions	264
	References	264

CHAPTER 9
Spatially Dispersive FDTD Modeling of Wire Medium — 267

9.1	Introduction	267
9.2	Spatial Dispersion in the Wire Medium	269
9.3	Spatially Dispersive FDTD Formulations	270
9.4	Stability and Numerical Dispersion Analysis	274
9.5	Perfectly Matched Layer for Wire Medium Slabs	279
9.6	Numerical Thickness of Wire Medium Slabs	282
9.7	Two-Dimensional FDTD Simulations	286
9.8	Three-Dimensional FDTD Simulations	294
9.9	Experimental Verifications	297
9.10	Internal Imaging by Wire Medium Slabs	299
9.11	Conclusions	303
	References	304

CHAPTER 10
FDTD Modeling of Metamaterials for Optics — 307

10.1	Introduction	307
10.2	Dispersive FDTD Modeling of Silver-Dielectric Layered Structures for Subwavelength Imaging	307
	10.2.1 Introduction	307
	10.2.2 FDTD Modeling of the Silver-Dielectric Layered Structure	310
	10.2.3 Numerical Results and Discussions	311
10.3	A Metamaterial Scanning Near-Field Optical Microscope	316
	10.3.1 Introduction	316
	10.3.2 Theory	317
	10.3.3 Simulation	317
10.4	FDTD Study of Guided Modes in Nanoplasmonic Waveguides	321
	10.4.1 Conformal Dispersive FDTD Method Using Effective Permittivities (EPs)	322
10.5	FDTD Calculation of Dispersion Diagrams	326
	10.5.1 Wave Propagation in Plasmonic Waveguides Formed by Finite Number of Elements	331
10.6	FDTD Modeling of Electromagnetic Cloaking Structures	333
	10.6.1 Dispersive FDTD Modeling of the Cloaking Structure	335
	10.6.2 Numerical Results and Discussion	341
	References	346

CHAPTER 11
Overviews and Final Remarks — 353

11.1	Introduction	353

11.2 Overview of Advantages and Disadvantages of the FDTD
     Method in Modeling Metamaterials   353
11.3 Overview of Metamaterial Applications and Final Remarks   354
     11.3.1 Small Antennas Enclosed by an ENG Shell   357
     11.3.2 Focusing and Superlensing Effects   361
     11.3.3 Performance Enhancement of Planar Antennas   370
     11.3.4 Electromagnetic Cloaks   370
     References   371

List of Abbreviations   373
About the Authors   375
Index   377

Preface

Metamaterials can be generally defined as a class of "artificial" media, possessing extraordinary electromagnetic properties that cannot be found in natural ones. The subject of metamaterials has drawn considerable attention from both the physics and engineering communities worldwide and has received generous support from the funding agencies during recent years. The popularity of this topic has been adequately demonstrated by a rapid surge in the number of publications, special sessions at international conferences, research networks, and launching of new journals on the subject. Metamaterials are periodic electromagnetic structures that are not altogether dissimilar from frequency selective surfaces (FSSs), bianisotropic materials, and optical gratings, all of which have been around in the electromagnetic and optical communities for quite some time. Although there has been much hype recently about the extraordinary performance of devices containing metamaterials, recent studies have indicated that there are a number of fundamental issues, such as high losses and narrow bandwidth characteristics, that must be addressed before these materials can find widespread use in practical applications. Nevertheless, the study of metamaterials has engendered, perhaps for the first time, a widespread interest on the part of physicists, electronic engineers and material scientists, in pursuing collaborative and multidisciplinary efforts, with the common goal of developing an understanding of the fundamental physics of metamaterials, which, in turn, has the potential of achieving new breakthroughs in science and engineering. Research into metamaterials at Queen Mary College, London, was initiated in 2000 and has been supported by several grants from the United Kingdom's Engineering and Physical Science Research Council (EPSRC). A range of computational techniques, including the finite-difference time-domain (FDTD) method, detailed in this book, have been developed for the modeling of metamaterials including electromagnetic bandgap (EBG) structures; left-handed materials (LHMs); artificial dielectrics; plasmonic waveguides; electromagnetic cloaking structures; and, a number of other devices designed for related applications of metamaterials. These computer codes have then been utilized for designing metamaterials and for gaining a physical insight into their electromagnetic characteristics. The FDTD has been widely accepted as one of the most efficient numerical techniques in computational electromagnetics and has been applied to periodic structures including the frequency selective surfaces (FSSs), which have previously found applications mainly as high-performance radomes and spatial filters, but are now finding new applications in metamaterial devices. The Electromagnetic Communication Laboratory of Pennsylvania State University has been engaged in the development of very high-performance computational electromagnetics (CEM) solvers capable of handling upward of 10E+9 unknowns. The GEMS code developed in this lab has played a pivotal role in rigorously analyzing complex electromagnetic structures with multiscale features that often characterize metamaterials.

This book introduces the basics of the FDTD method, especially when it is used to model metamaterials. It shows how to compute the dispersion diagrams, deal with the material dispersion properties, and verify the left-handedness, among other things. Some metamaterials possess unique properties that require special treatments in the numerical code when we analyze them. This book explains how to properly define their material parameters and to characterize the interface of metamaterial slabs and quantify their spatial as well as frequency dispersion characteristics. There has been much recent interest in novel applications of metamaterials to antennas and microwaves and to various devices that have applications in optical engineering. In view of this, the book dedicates an entire chapter solely to this topic. It is shown how these structures can be modeled by using either the effective medium representation or the FDTD code. Though the latter is highly computer-intensive, we have argued that modeling the physical structure numerically and rigorously is the only way to obtain reliable results when attempting to predict the performance of metamaterial devices, because the rigorous results often disagree with those derived by using simplified models based on the effective medium approach. For this reason, we have devoted a substantial amount of space in this book to modeling the problem of the physical structures of metamaterials, instead of using their effective medium representations. In addition, we have analyzed the fundamental limits of metamaterials made from resonant particles, with the hope that the readers will get a true picture of the real-world metamaterials after going through these analyses. We view this book as a complement to a wide array of publications on the FDTD method that have preceded it, and we hope that colleagues in computational electromagnetics will benefit from recent advances in numerical techniques, especially the FDTD, when dealing with the problem of designing metamaterials.

Acknowledgments

First and foremost, we would like to thank our former and current doctoral students for their many contributions to this work. They have been working and contributing diligently for more than five years to enrich and infuse new excitement into the study of FDTD and metamaterials. Without the results of their intensive studies, it would not be possible for us to put together the material for such a comprehensive book. In particular, we would like to thank Dr. Yan Zhao of Queen Mary College, London, for his contributions to the development of a spatially dispersive FDTD (Chapters 8 and 9) and the modeling of plasmonic wave guide and electromagnetic cloaks (Chapter 10); Dr. Wei Song of Queen Mary College for her work on conformal FDTD modeling (Chapters 2 and 3); Dr. Lai-Ching Ma of Penn State University for her work on parameter extractions for left-handed metamaterials (Chapter 7); Yoonjae Lee of Queen Mary College for his work on figure of merit studies on metamaterials (Chapters 5–7); Atiqur Rehman for the development of composite right-left handed transmission lines (CRLH-TL) and their applications to antenna engineering; and Christos Argyropoulos for his input to the modeling of silver lens and electromagnetic cloaks (Chapter 10).

Yang Hao wishes to thank the EPSRC for several related research grants over last five years. Several organizations including the Leverhulme Trust, the Ministry of Defense, and Roke Manor have provided financial support.

We both wish to thank a whole host of our friends and colleagues for many useful discussions. They are Professor Sir John Pendry, Imperial College, London; Professors Peter Clarricoats and Clive Parini; Dr. Pavel Belov, Queen Mary College; Dr. Iain Anderson; Dr. Ian Youngs; Professor John Brown from the Defence Science and Technology Laboratory (DSTL); Dr. Colin Brewitt-Taylor, QinetiQ, United Kingdom; Professor Nader Farahat of Polytechnic University of Puerto Rico; and many other graduate students, postdoctors, and collegues, too numerous to mention individually.

Finally, we wish to thank Professor Haim Cory, Technion, Israel Institute of Technology, for his contribution to this work during his sabbatical leave at Queen Mary College; Professor Julian Evans of University College London and his team; and Dr. Xuesong Lu and Dr. Shoufeng Yang of Queen Mary College for their participation in the fabrication of millimeter-wave metamaterials.

CHAPTER 1
Introduction

1.1 What Are Electromagnetic Metamaterials?

There have been various definitions of *electromagnetic metamaterials* [1–114], where "meta" is a prefix in English meaning "beyond; transcending; more comprehensive." In 2001, Walser [1] from the University of Texas at Austin, coined the term "metamaterial" to refer to artificial composites that "...*achieve material performance beyond the limitations of conventional composites.*" The definition was subsequently expanded by Browning and Wolf of Defense Advanced Research Projects Agency (DARPA) in the context of the DARPA Metamaterials program started also in 2001:

> Metamaterials are a new class of ordered composites that exhibit exceptional properties not readily observed in nature. These properties arise from qualitatively new response functions that are: (1) not observed in the constituent materials and (2) result from the inclusion of artificially fabricated, extrinsic, low dimensional inhomogeneities.

Metamorphose, the European Network of Excellence [2], terms the metamaterials as:

> Artificial electromagnetic (multi-)functional materials engineered to satisfy the prescribed requirements. Superior properties as compared to what can be found in nature are often underlying in the spelling of metamaterial. These new properties emerge due to specific interactions with electromagnetic fields or due to external electrical control. The metamaterials provide a conceptually new range of radio, microwave, and optical technologies.

Sometimes, metamaterials are specifically referred to as a class of artificial materials that have simultaneous negative permittivity and permeability and are also known as *left-handed materials* (LHMs). Present researchers have a tendency to expand the concept of metamaterials so as to make it as broad as possible. The editorial board of IEICE Transactions [3] even questions whether or not artificial materials such as CdS, GaAs, or InGaAs should have been classified as metamaterials. One popular classification of metamaterials is:

> As an ordinary material is made of natural molecules, an artificial material is made of artificial molecules. Due to Maxwell equations' macroscopic property, small particles made of typically metal and dielectric can be considered molecules when put together. The variation of each shape and total alignment makes macroscopically single negative, double negative, or double positive materials.

However, there exists a number of artificial electromagnetic structures, especially at microwave frequencies [e.g., electromagnetic crystals, high-impedance surfaces (HISs), and frequency selective surfaces (FSSs)]. Although these are made of small ordered metallic/dielectric inclusions, they cannot be homogenized by using conventional approaches and described in terms of constitutive parameters such as permittivity and permeability. Smith [4] at Duke University prefers to use the term metamaterials as artificial structures that display properties beyond those available in naturally occurring materials. This definition is a general one, and it may also include artificial dielectrics, artificial magnetics, and bianisotropic materials, which were the subject of extensive research back in the 1960s, 1970s, and 1990s. Most of the concepts in metamaterials originate from solid state physics that deal with the lattice structure of crystals, which is inherently periodic. Indeed periodic structures in nature have fascinating characteristics, which have frequently inspired scientists and engineers alike to think of novel applications of them. Periodic structures have had a long history in electromagnetics dating back to the 1900s, and they can be found as integral parts of microwave filters, traveling wave tubes (TWTs), antenna arrays, leaky wave antennas (LWAs), and FSSs, to name just a few. The role of the periodic structure has been the manipulation of the spectral and spatial spectrum, the selection of spatial harmonics to control the radiation of forward and backward waves, and the control of the phase and group velocities in slow wave structures. Periodic structures are also very popular among optical engineers and are widely used in the design of lasers. For example, a distributed Bragg reflector (DBR) is a structure composed of alternating layers of materials with varying refractive indices, or with periodically varying characteristics, such as height of a dielectric waveguide, which induces a periodic variation in the effective refractive index of the guide. Each layer interface induces a partial reflection of optical waves, at a wavelength for which many reflections undergo a constructive interference, thereby forming a high-quality reflector. The idea has been further extended to two and three dimensions by Yablonovitch [5] and John [6], who have described structures that are now broadly classified as photonic bandgaps (PBGs).

Earlier work of Yablonovitch on PBGs was carried out at microwave frequencies, by using a small dipole in a Fabry-Perot cavity formed by PBGs. They conjectured that such configurations will give a rise to increased directivity of small antennas by focusing their beam [5]. This has engendered new interests in the community of antenna and microwave engineering, and now the so-called electromagnetic bandgap (EBG) structures (in contrast to the PBGs that are their counterparts at optical frequencies) are finding usages in enhancing the performance of antennas and microwave devices.

1.2 A Historical Overview of Electromagntic Metamaterials

Permittivity (ε) and permeability (μ) are two parameters used to characterize the electric and magnetic properties of materials interacting with electromagnetic fields. The *permittivity* is a measure of how much a medium changes to absorb electrical energy when subjected to an electric field [7, 8]. It is defined as a ratio of $\bar{D}$ and $\bar{E}$, where $\bar{D}$ is the electric displacement by the medium and $\bar{E}$ is the electric field strength. The common term dielectric constant is the ratio of permittivity of the material to that of free space ($\varepsilon_0 = 8.85 * 10^{-12}$ F/m). It is also termed as the relative permittivity. *Permeability* is a constant of proportionality that exists between magnetic induction and magnetic field intensity. Free-space permeability (μ_0) is approximately 1.257×10^{-6} H/m. Recently, Ziolkowski [9, 10] has categorized metamaterials by their constitutive parameters as follows (Figure 1.1). Most of the materials in nature have positive permittivity and permeability, and hence, they are referred to as "double-positive (DPS)" media. In contrast, if both of these quantities are negative, they are called "double-negative (DNG)" and are also referred to as LHMs by others. Finally, materials with one negative parameter are named "single-negative (SNG)" and are further classified into two subcategories, namely, "epsilon-negative (ENG)" and "mu-negative (MNG)." Interestingly, natural materials such as cold plasma and silver exhibit negative permittivities at microwave and optical frequencies, respectively, and ferromagnetic materials exhibit a negative permeability behavior in the VHF and UHF regimes. However, to date, no materials that exhibit simultaneous negative permittivity and permeability have been found in nature, and hence, they must be created artificially.

The first comprehensive review of the history of negative refraction and metamaterials was given by Moroz [11]. He indicated that some of metamaterial research started long before Veselago's work and went back to as far as 1905, when Lamb [12] suggested the existence of backward waves, which are associated with

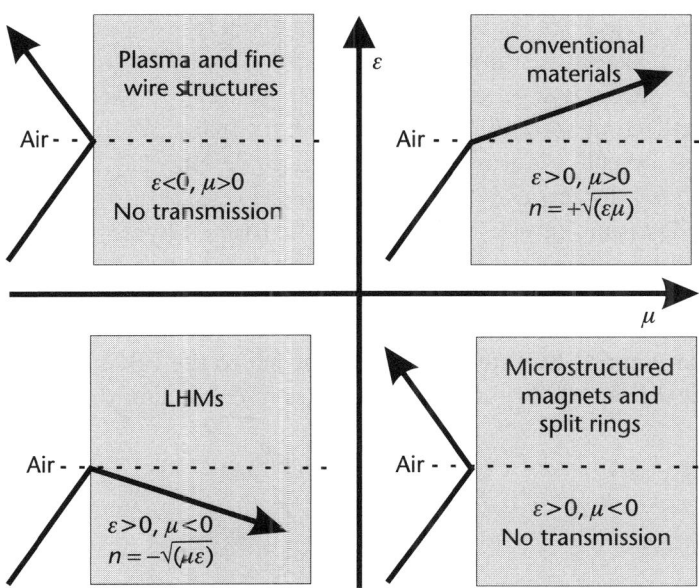

Figure 1.1 A diagram showing the possible domains of electromagnetic materials and wave refraction or reflection directions based on the signs of permittivity and permeability. The arrows represent wave vector directions in each medium. There is wave transmission only when both parameters have the same sign. Waves are refracted positively in conventional materials and negatively in LHMs. (*From:* [10]. © 2001 IEEE. Reprinted with permission.)

waves in which the phase propagates in a direction opposite to that of the energy flow. Moroz's on-line article first appeared in 2003, tracing the roots of research on backward waves, negative refraction, and LHMs. Only recently, Veselago and Narimanov [13], Shivola [14], and Shamonina [15] have linked LHMs, and more generally metamaterials, to the earlier works.

Below we provide a brief historical overview of the metamaterials category as artificial dielectrics, artificial magetics, left-handed metamaterials, high-impedance surfaces, and electromagnetic crystals. Some of those topics will be further discussed in later chapters in the context of numerical modeling and their applications.

1.2.1 Artificial Dielectrics

Artificial dielectrics, the first known metamaterials, usually consist of artificially created "molecules": dielectric or metallic inclusions of a certain shape (see Figure 1.2). These "molecules" can be distributed and oriented in space, either in a regular lattice or in a random manner.

The dimensions of the "molecules" and characteristic distances between neighboring ones are assumed to be very small, as compared to the wavelength. Although the size of a single inclusion is usually much larger than those of real molecules and lattice periods of natural crystals, this allows us to describe the inclusions in terms of material parameters as well as characterize the artificial dielectrics formed with these inclusions macroscopically using the classic Maxwell's equations.

The concept of artificial dielectrics was perhaps first introduced by Kock [16], who used it to design low-weight dielectric lenses at microwave frequencies. The artificial dielectrics also find applications where we need to use a high-permittivity material, such as a titanate with $\varepsilon_r \approx 90$. Since these materials are usually expensive, an alternative is to use metallic inclusions of various shapes with sizes that are small compared to the wavelength to artificially produce a high-permittivity material, which is both low-cost and lightweight [17]. A very interesting example of artificial dielectrics is the wire medium [18–24] (also called "rodded medium"), which has been known since 1950s. It is formed by a regular lattice of conducting

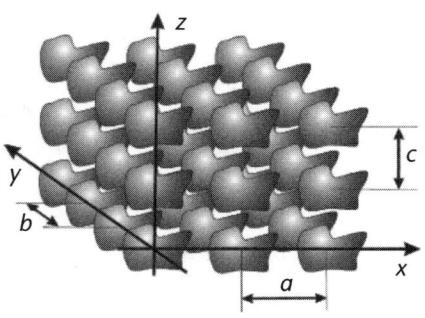

Figure 1.2 The geometry of a generic artificial dielectric.

wires with radii that are small compared to the lattice period (see Figure 1.3). The permittivity of a wire medium has a plasma type frequency dependence: negative below the plasma frequency and positive but smaller than unity above it. This medium is often called an "artificial plasma" since its permittivity has the same

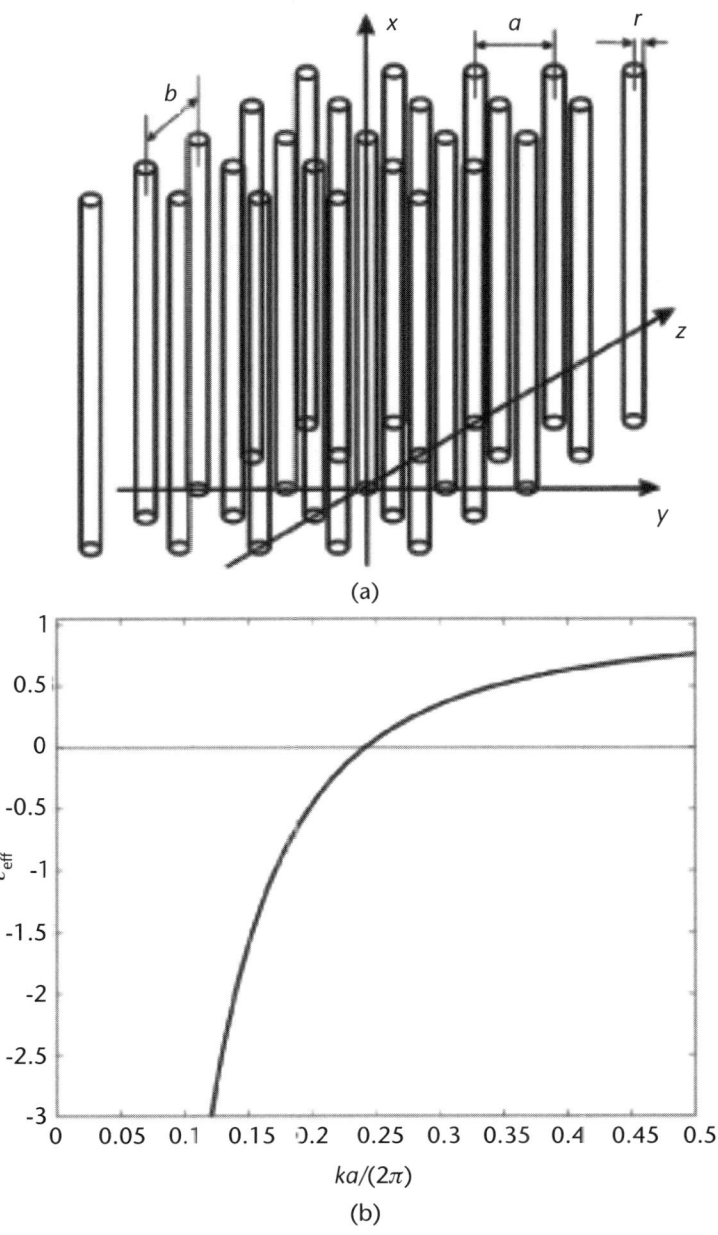

Figure 1.3 (a) The geometry of wire medium: a lattice of parallel conducting thin wires. (b) Permittivity of wire medium demonstrating plasma-like frequency dependent permittivity: negative below the plasma frequency and positive but smaller than the unity above.

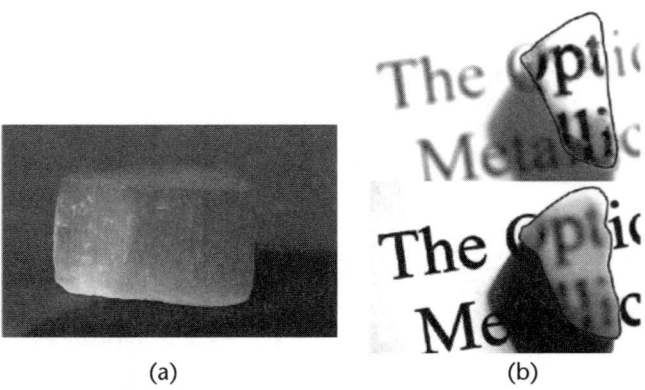

(a) (b)

Figure 1.4 (a) A photo of TV rock, Ulexite (NaCaB$_5$O$_9$·8H$_2$O, hydrated sodium calcium borate hydroxide), a mineral occurring in silky-white rounded crystalline masses or in parallel fibers. (b) The TV rock is placed on a printed surface, and the tiny fibers transmit the image to the top of the rock. This unusual effect is the result of ulexites amazing fiber-optic properties [31, 32].

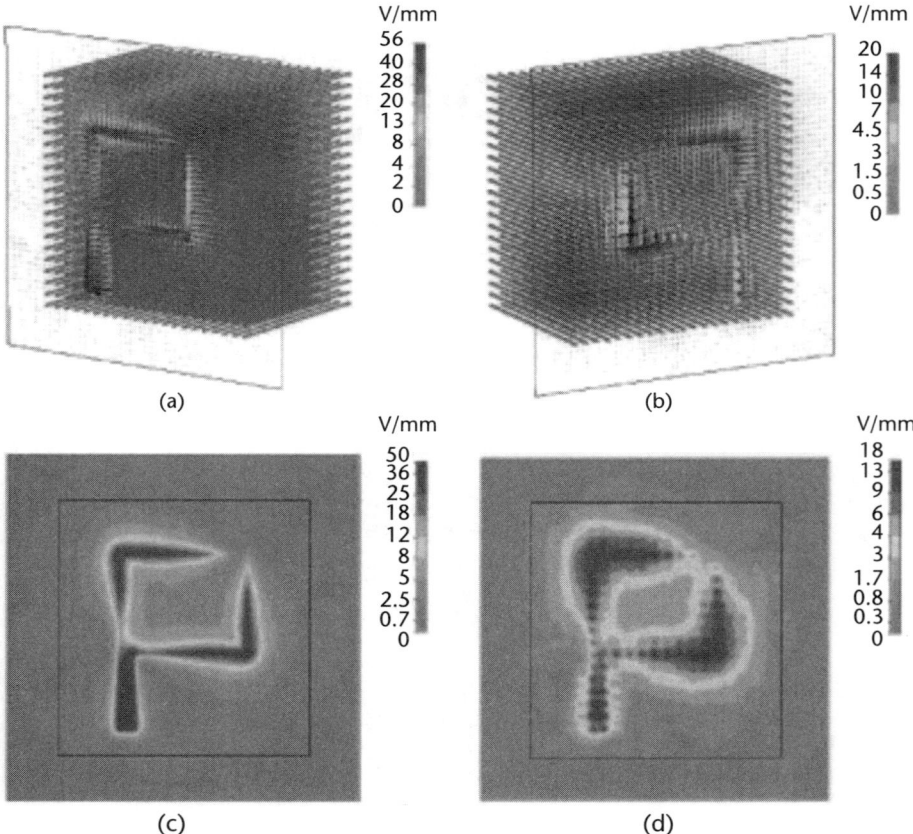

Figure 1.5 Distribution of an electric field and its absolute value: (a) and (c) at a 2.5-mm distance from the front interface and (b) and (d) at a 2.5-mm distance from the back interface [30].

form as that of an ideal (collisionless) electron plasma. Recently, the wire medium has been rediscovered by Pendry [25] and has attracted a great deal of attention from other researchers [26, 27]. This medium possesses unique nonresonant material properties, and its permittivity is negative over a very wide frequency range. This type of wire medium has been used as one of essential components to create LHMs [28] and high-impedance surfaces [29].

In addition, it has been demonstrated that an array of parallel conducting wires [30] can be used as a subwavelength imaging system that effectively acts as a TV Rock–Ulexite, (NaCaB$_5$O$_9$·8H$_2$O, hydrated sodium calcium borate hydroxide), shown in Figure 1.4, a mineral occurring in silky-white rounded crystalline masses and in parallel fibers. It was named after the nineteenth-century German chemist G. L. Ulex who first discovered it.

Figure 1.5 shows the simulation results using the CST microwave studio. A subwavelength distribution of electrical fields at the front interface of the wire medium slab (an analogy to the TV rock at microwave frequencies) [see Figure 1.5(a)] is canalized from the front interface to the back interface and forms an image [see Figure 1.5(b)]. The quality of the image can be clearly seen in Figures 1.5(c, d), where the absolute values of an electrical field in the vicinity of the front and back interfaces are plotted. However, these devices are not true lenses since they do not image the fields in a manner similar to the optical lenses when dealing with objects located at arbitrary distances from the surface of the medium.

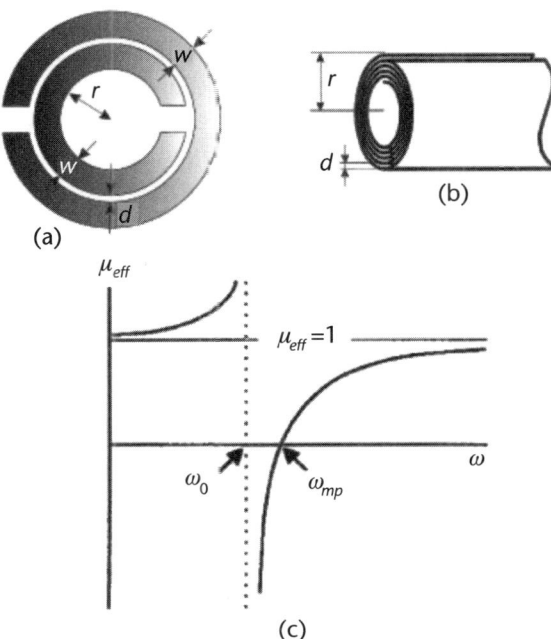

Figure 1.6 Components of artificial magnetic materials: (a) split-ring resonator, (b) Swiss roll, and (c) frequency response of effective permeability.

1.2.2 Artificial Magnetic Materials

Artificial magnetic materials are typically synthesized by using resonant elements, as for example split-ring resonators [see Figure 1.6(a)] or Swiss rolls [see Figure 1.6(b)]. The split-ring resonators are more widely used than the Swiss rolls since they can be manufactured using printed circuit technology. Artificial magnetic materials were known as far back as early the 1950s, and a book by Schelkunoff and Friis [33] provides expressions that can be used to calculate the magnetic flux of a split metallic wire loaded with a small capacitor, formed by a pair of parallel metallic plates. In [34], Schneider used a split metallic tube to produce an NMR probe, and, in [35], Hardy proposed a Swiss roll resonator to achieve magnetic resonance at 200–2,000 MHz. In [36], an artificial magnetic material with frequency-dependent positive permeability has also been synthesized by using double circular ring resonators.

An artificial magnetic material, which is formed by split-ring resonators, possesses negative permeability within a frequency band (bandwidth is typically narrow) near the resonant frequency of the single split-ring resonator; it is widely used to create LHMs [4]. Metallic waveguides filled with such artificial magnetic materials support guided waves at frequencies below the cutoff frequency of hollow waveguides; hence such guides are also referred to as the subwavelength waveguides [38]. This effect can be used for miniaturization of guided wave structures [39], though they do introduce additional losses owing to the presence of high circulating currents in the inclusions when they are resonant.

1.2.3 Bianisotropic Composites

Even when the host medium and the dielectric inclusions in this medium are isotropic on their own—that is, in isolation—the effective medium of the mixture can be anisotropic, and its macroscopic response can vary depending on the polarization of the incident field. A further generalization of this phenomenon is bianisotropy, which is associated with simultaneous presence of both anisotropic and magnetoelectric behavior [14]. For example, lattices of inclusions with of complex shapes (e.g., chiral and omega particles, shown in Figure 1.7), cannot be described in terms of permittivity and permeability corresponding to their behaviour in isolation, because of the electromagnetic coupling between the two sets of inclusions which affects their performance. As a result, the electric field not only

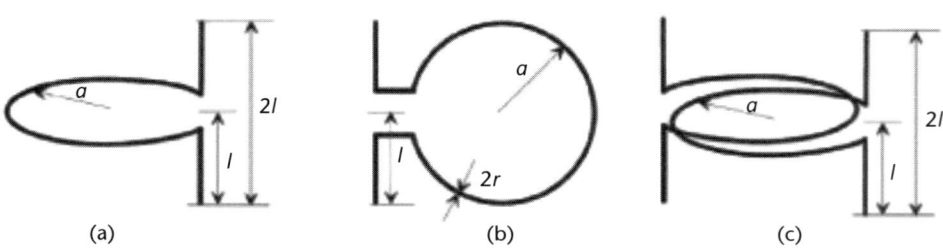

Figure 1.7 Bianisotropic particles: (a) chiral, (b) omega, and (c) double chiral [45].

induces electric but also magnetic polarization in such media (vice versa for the magnetic field), and such materials are referred to as bi-anisotropic media [40–42]. Marques [43] showed that even split-ring resonators possess strong bianisotropic properties, since the dimensions of the two split rings are different. This, in turn, has undesirable effects when we attempt to create a uniaxial artificial magnetic material mentioned in the previous section. However, this problem was mitigated by introducing double split-ring resonators formed by rings with identical sizes [43, 44].

The bianisotropic medium is the most general type of material that can be described in terms of local material parameters. They have been used in many applications, such as the design of radar absorbing materials for stealth technology and polarization transformers. It has been reported that conventional chiral materials can exhibit negative refraction [46–48], and this has opened up novel means of producing artificial materials with negative refraction characterization. In this book, chiral materials will not be covered at any length, since they have been intensively investigated for the past 20 years or so, and there already exists an extensive body of literature covering this topic.

1.2.4 Double-Negative and Indefinite Media

In the existing literature, the DNG media, materials with both negative permittivity ε and permeability μ, have become almost synonymous with metamaterials, though, in reality, the EBGs should share the spotlight just as much with them under the general umbrella of the term metamaterials. In the literature, such materials are also referred to as LHM following the original term proposed by Veselago [49], and sometimes, they are also called backward-wave media [50, 51] and negative index materials [53, 57]. It is worth noting that backward-wave media were first discovered in 1904 by Lamb [54] in mechanical systems, and were revisited by many researchers including Clarricoats in the 1960s [55]. In [55], it was showed that radiation from a slotted circular waveguide partially filled with a dielectric rod (whose material characteristics are not DNG) radiates backward waves that have opposite signs for the group and phase velocities. In addition, negative refraction phenomenon associated with DNG materials can also be observed in photonic crystals and EBG structures (see Section 1.2.5 for details) at the frequencies close to the edges of the bandgaps. This fact was theoretically revealed by Notomi [56], confirmed by others [57], and experimentally verified by Parimi [58] and Sudhakaran [59].

Growing interest in DNG media was incited recently by a seminal paper by Pendry [60], in which he, following Veselago, argued that the resolution of common imaging systems restricted by the so-called diffraction limit can be overcome by the use of a slab of double-negative medium, with $\varepsilon = \mu = -1$. The focusing phenomenon in a perfect lens proposed by Pendry has been attributed to two effects. The propagating modes of a source are focused due to the negative refraction and the evanescent modes experience growing inside the DNG slab, which allows subwavelength details to be restored in the focal plane. Without a doubt, if lossless, wideband and DNG materials become available in practice, they would be a major breakthrough in modern science and engineering and would find a wide variety

Figure 1.8 Realization of DNG material at microwave frequencies [4].

of potential applications such as a subwavelength cavity resonator proposed by Engheta [61] and transparent radomes constructed from conjugate dielectric and DNG slabs [62].

Shelby and Smith [5] recently suggested that a way to realize a uniaxial DNG medium is to use a lattice of metallic wires and resonant magnetic particles, as shown in Figure 1.8. In their design, an array of metallic wires (wire medium as mentioned before) is used to realize negative permittivity, and a lattice of split-ring resonators allows one to create a negative permeability medium. Negative refraction in the Shelby-Smith structure was purportedly proven at microwave frequencies [5, 63, 64], though an alternative explanation of the observed behavior of the scattering characteristics of the slab has recently been offered by Mittra. To date, DNG media based on the Shelby-Smith design have been created for operation in the thetahertz range [61, 62]. Even an isotropic geometry has been proposed [4], though it remains to be proven in practice. The Shelby-Smith structure is not a unique design of DNG media. There are also other possible realizations, such as those presented in [5, 38, 53, 64].

Indefinite media are referred to as artificial materials in which the principle components of the permittivity and permeability tensors have different signs. Such materials were studied in [65, 66], where a variety of effects including negative refraction, backward-wave effect, and near-field focusing were purportedly demonstrated. Anisotropy of the media introduces additional freedom in manipulation of their dispersion and reflection properties, though it is a two-edged sword, since there exist no systematic ways to characterize these media from their reflection and transmission characteristics in terms of permittivity and permeability tensors. Furthermore, it is not at all so straightforward to analyze an antenna metamaterial composite when the latter is anisotropic.

In a 2003 article [67], Holloway speculated that complex structures comprised of split-ring resonators and wires are not necessary for the design of a negative-refraction material, which, in principle, can be realized by using purely dielectric inclusions in a background dielectric medium. This, in turn, has opened up the possibility of synthesizing negative refractive index metamaterials much more simply than has been proposed up to date. Manoz [11] recently expanded on the ideas of Holloway [67] and showed that even a composite of inherently nonmagnetic homogeneous spheres can provide a negative refractive index metamaterial. The absence of inherently magnetic materials is a key difference, which makes it possible to achieve a negative refractive index band even in the deep infrared region [11]. However, it also raises an important and rather fundamental question. Are these effective parameters of a slab of metamaterials containing resonant inclusions simply fictious quantities that are numerically fitted to match reflection and transmission characterization of a planar slab of metamaterial (external characteristics) at a specific incident angle and polarization of the incident waves? In the real-world scenarios, such metamaterials may not bear any resemblance to real homogeneous materials, whose characteristics do not vary with a change in the thickness, incident angle, or polarization. Additionally, real materials are not so highly dispersive as are the resonant structures.

1.2.5 Photonic and Electromagnetic Crystals

Artificial dielectrics are generally designed to operate at wavelengths that are long compared to periods of the lattice In this regime, the inclusions interact primarily quasistatically, and homogenization approaches are applicable. A totally different situation arises in the case of so-called electromagnetic crystals [68, 69] (or photonic crystals at optics [70–74]) when operating at higher frequencies where the interactions between the inclusions play a very significant role. Typically, electromagnetic crystals are periodic structures, operating at wavelengths that are comparable to their period, and one of their inherent features is that they have passbands and stopbands. These stopbands are also referred to as bandgaps, and therefore, these crystals are sometimes referred to as electromagnetic or photonic bandgap structures. The bandgaps are caused by spatial resonances that occur in the crystal, which strongly depend on the direction of propagation of the incident wave (i.e., they are spatially dispersive) [75–77]. Electromagnetic crystals cannot be homogenized in the same manner as ones without resonant inclusions. Nonetheless, one can curve-fit their reflection and transmission characteristics as functions of frequency to define their permittivity and permeability. Whether or not the effective parameter descriptions can be subsequently used to design practical structures remains an open question, which is not yet unequivocally settled.

Incidentally, such photonic or electromagnetic crystal structures do occur in *natural* biological systems, in which nanometer-scale architectures are used to produce striking optical effects: Morpho butterflies use multiple layers of cuticle and air to produce their striking blue color, and some insects use arrays of elements, known as nipple arrays, to reduce reflectivity in their compound eyes as shown in Figure 1.10 [79]. Nireus is found in eastern and central Africa and has dark wings with patches of bright blue-green markings. These markings are unusual in the way they produce their light and color. Optical physicists have studied the scales that make up the brightly colored regions of the creature's wings. These scales contain a pigment that

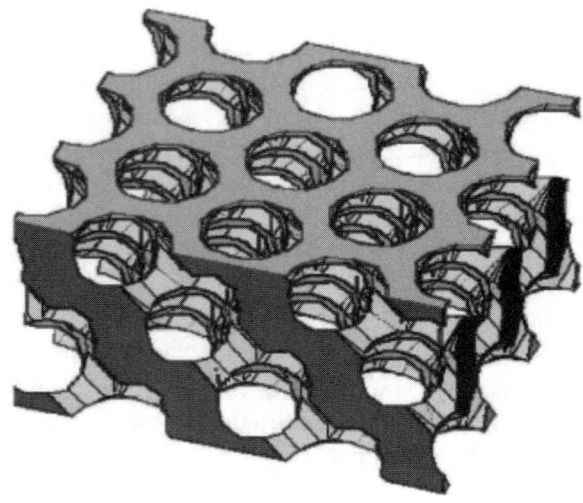

Figure 1.9 The first photonic crystal with a complete bandgap (Yablonovite) [78].

absorbs light at wavelengths around 420 nanometers (roughly sky-blue) and radiates it at 505 nm in the blue-green region where butterfly eyes are particularly sensitive. Indeed, natural photonic structures are providing inspiration for technological applications and might provide us with useful clues for designing them artificially.

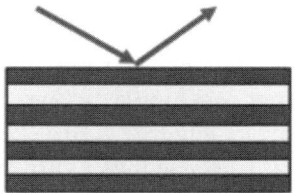

Figure 1.10 The blue-green color on several species of African butterflies is caused by the nanoscale structure of the insects' wings [79].

The interest in photonic crystals arose about 20 years ago, inspired by the investigations described in the pioneering works of Yablonovitch [80] and John [81] in 1987, in which they reported strong localization of photons and inhibited spontaneous emissions due to electromagnetic bandgaps. Ever prior to that, about 15 years earlier, the effect of a photonic bandgap on the spontaneous emission of embedded atoms and molecules had been investigated by Bykov [82, 83]. The first studies and demonstration of a photonic crystal with a complete bandgap (see Figure 1.9) were carried out by Yablonovitch et al. in the early 1990s [78, 84–87]. Since then, photonic and electromagnetic crystals have found numerous applications in FSSs and as components for waveguide and resonators, both in optical and microwave regimes [68, 69–74].

Electromagnetic crystals are also classified as electromagnetic bandgap (EBG) structures and high-impedance surfaces (HISs) at microwave frequencies. A typical transmission coefficient chart of an EBG structure with incidence of plane wave is shown in Figure 1.11(a). The passbands and stopbands correspond to those in Figure 1.11(b) [88]. Some practical applications of EBGs include antenna beam narrowing and shaping using Fabry-Perot like EBG cavity [89–91]; isolation enhancement in diplexer antennas using anisotroptic EBGs [92]; mobile antenna efficiency improvement using Mushroom-like EBGs [93, 94]; EBG horn antennas and arrays at millimeter-wave frequencies [95], and mutual coupling reduction between antenna array elements [96].

Typical HISs are thin composite layers whose reflection coefficient is +1, and hence, the HIS behaves as though it were a magnetic conductor. For this reason, HISs are also referred to as artificial magnetic conductors (AMCs). A well-known example of HIS is the simple mushroom structure [93], as shown in Figure 1.12. The structure is a composite layer that contains a periodic array of metallic patches with metal pins connected to the ground. Its response to the incident electromagnetic wave can be modeled approximately, as that of a resonant parallel LC-circuit. The most important attribute of a HIS is that it interacts with horizontal antennas located close to it in a manner that is constructive in contrast to the interaction with metallic screens that are destructive. Also, in addition to serving as a magnetic wall, the HIS can be designed to suppress surface waves in a certain frequency band. Thus, HISs are prospective candidates for screening of near field and reducing specific absorption rate (SAR), which, in turn, serve to increase the antenna efficiency [94].

An important use of electromagnetic crystals at optics is to control flow of light [71]. The basic idea is to design periodic dielectric structures that have a bandgap in a particular frequency range. By deliberately introducing defects in periodic dielectric rods, it is possible to create waveguiding devices operating at the bandgap frequencies of the periodic structure. Incidentally, such devices, operating at the bandgap frequencies, are not the only option to guide the flow of light. Recently, a new method for guiding electromagnetic waves in structures whose dimensions are below the diffraction limit has been proposed. The structures have been named "plasmonic waveguides," and they operate by virtue of near-field interaction between noble metallic nanoparticles that are closely spaced and can be efficiently excited at their surface plasmon frequency. The underlying principle of their operation relies upon a setup of coupled plasmon modes via dipole

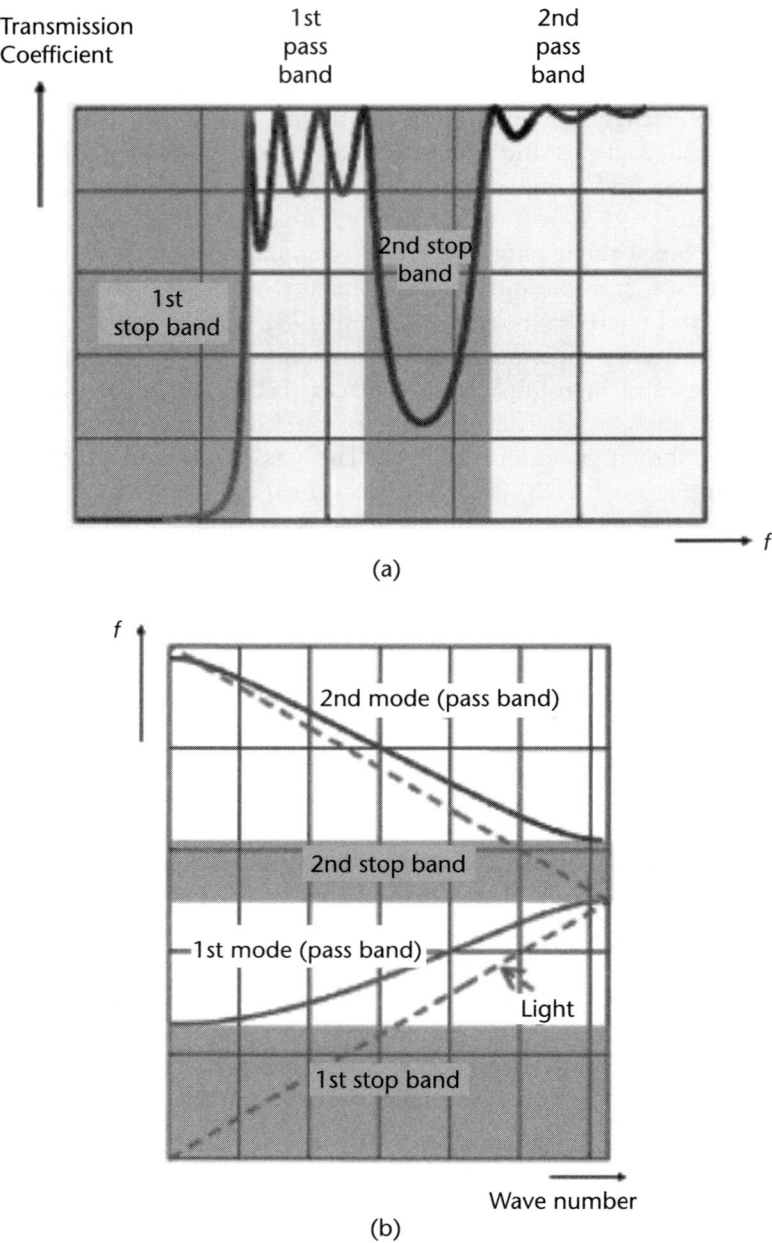

Figure 1.11 (a) Typical EBG structure transmission chart and (b) propagation modes chart.

interactions in the near field that leads to coherent propagation of energy along the array. Analogous structures as waveguides in the microwave regime include periodically arranged metallic cylinders that are useful arrays of fat dipoles and Yagi-Uda antennas, to name just a few.

Figure 1.12 TM surface wave measurement using monopole probe antennas.

1.3 Numerical Modeling of Electromagnetic Metamaterials

Metamaterials have the potential to control and manipulate the wave propagation in a manner that eludes the conventional materials, because of their periodic nature that can either be small-scale or resonant. Their use may enable us to create highly directional antennas, enhance the performance of small antennas by making them appear as though the electrical lengths were much larger, and design highly integrated transceiver systems that can be packaged in a limited space. Although the practical applications of metamaterials are currently limited by their loss characteristics and operational bandwidths, it is very important to develop efficient modeling tools to quantify and represent the characteristics of metamaterials and their behavior, especially when they form an integral part of complex electromagnetic systems.

For more than four decades, the FDTD method has been regarded as a useful electromagnetic modeling tool because of its versatility, and the development of this tool is still ongoing. Researchers all around the world are working on ways to embellish the performance of the FDTD algorithm by developing: new absorbing boundary conditions (ABCs) for novel materials; advanced alternating direction implicit (ADI) algorithms for the unconditionally stable FDTD method; conformal FDTD algorithms based on contour path or effective permittivity for modeling of curved structures; accurate techniques for modeling of thin material layers; parallel implementation of the FDTD algorithm for electrically large problems; and so on. Generally speaking, the FDTD method can be used to solve a wide variety of EM problems pertaining to antennas, electromagnetic compatibility, microwave systems, dosimetry, radar cross section predictions, and others. However, the conventional FDTD algorithm usually cannot handle all possible scenarios. Furthermore, existing commercial software often does not offer the flexibility to modify the algorithms used in their codes. Hence this prevents us from using them for many problems of interest that fall beyond the realm of capabilities of the existing codes. Given this background, this book will focus on improving the conventional FDTD algorithm and enlarging it to handle applications in the area of metamaterial modeling.

There has also been a "divide" between the thought processes of physicists and engineers in the area of metamaterial modeling. Physicists often regarded metamaterials as though they are homogeneous materials, and they frequently assume that these materials satisfy the effective medium theory (EMT). For engineers, losses, operational bandwidth, and the finiteness of metamaterial are all very important. They recognize that "unit-cell" analysis may not be valid for truncated periodic structures. Furthermore, when devices such as antennas are placed close to metamaterials, it becomes highly necessary to account for the interaction between antennas and metamaterials in a numerically rigorous manner. This is particularly important since analytical models exist only for a very small class of electromagnetic metamaterials, consequently, there is often a tendency to attempt to simplify the original problems by making gross approximations to reduce their complexities. However, this often involves sacrificing the accuracy to a point that may make the results totally erroneous and the predictions regarding the performance of the metamaterial devices misleading.

Among metamaterials, electromagnetic crystals are usually numerically studied with the help of methods such as the FDTD; the method of moments (MoM) [68, 69, 72, 74]; Pendry's approach described in [97]; and the Bloch-Floquet method [70, 71], which is based on an expansion of the field into spatial harmonics. There have been a number of attempts to model LHMs mainly based on their EMT models using the FDTD method [98–103]. It appears that the conventional dispersive FDTD method is sufficient to model negative refractions at the interface of LHM and the free space and that it can be used to model the behavior of a planar superlens [98–100]. However, we will demonstrate later in this book that the conventional dispersive FDTD approach suffers from numerical inaccuracies when evanescent waves must be included in the modeling of the structure of interest. In the case of LHM, the evanescent waves may play an important role in subwavelength imaging using a slab of such a material. Earlier FDTD simulations failed to demonstrate the subwavelength imaging property of LHM lenses [98, 99], because they did not accurately model the role of the evanescent waves.

Besides the FDTD method, the pseudospectral time-domain (PSTD) method has been used for the modeling of backward-wave metamaterials [104]. It is claimed [104] that the FDTD method cannot be used to accurately model LHMs due to the numerical artifacts originating from the staggered grid nature of the FDTD meshes. However, it will be demonstrated in this book by comparing transmission coefficients calculated from the FDTD simulation and exact analytical solutions that the FDTD method indeed can be used to accurately characterize the behavior of both the propagating and evanescent waves in LHM slabs with proper field averaging techniques [103, 105].

The FDTD method has also been used to study waveguides formed by several rows of silver nanorods arranged in a hexagonal lattice [106]. Even though the FDTD method has been applied to numerous examples involving the plasmonic structures, its accuracy has not been established as yet, especially when handling complex geometries and metamaterial media. For instance, since the conventional FDTD uses a Cartesian and staggered grid, we often find that it needs to be refined [107–111] when dealing with the curved surfaces, interfaces between different materials, and dispersive media to get accurate results. In fact, modeling dispersive materials with curved surfaces still remains a challenging topic, not only because the algorithm dealing with it is complex, but also because it can become numerically unstable. One approach to addressing this problem is based on the use of effective permittivities (EPs) [112–114] in the Cartesian coordinate system, and to modify the dispersive FDTD scheme in a way such that its stability is not compromised. In this book, we propose a novel conformal dispersive FDTD algorithm that combines the EPs with an auxiliary differential equation (ADE) method [17], then applies the developed method to the problem of modeling plasmonic waveguides formed by an array of circular or elliptically shaped cylinders in the optical frequency regime. The material for the cylinders is silver, which has a negative permittivity, since it behaves like a plasma at these frequencies.

1.4 Layout of the Book

The chapters in this book have been written to illustrate applications of the FDTD to the problem of modeling electromagnetic metamaterials. Chapters 2–5 present the fundamentals of metamaterials and discuss some basic numerical modeling techniques, primarily the FDTD for modeling these materials. Next in Chapters 6–10, we deal with the application of FDTD modeling metamaterials comprised of physical structures and using effective medium models.

Chapter 2 presents the fundamentals of EBG structures, beginning with Maxwell's equations in periodic electromagnetic structures, bandgap theory, dispersion diagrams, spatial harmonics, and phase and group velocities.

Chapter 2 also presents an overview of various computational techniques that are useful for metamaterial modeling, including the plane wave expansion method and the transfer matrix approach. Also, applications of EBGs in both microwave and optical engineering are reviewed.

Chapter 3 deals with the basics of FDTD in the context of metamaterial modeling. It covers the formulation of Yee's FDTD formulations, applications of FDTD to periodic electromagnetic structures, boundary conditions, and excitations in FDTD associated with electromagnetic metamaterials.

Chapter 4 introduces FDTD modeling of EBG structures and their applications to antenna engineering. Calculations of the dispersion diagram of an infinite EBG and the transmission coefficient a (semi-)infinite bandgap structure will be detailed. We shall conduct a case study of designing a millimeter-wave EBG antenna based on the FDTD simulation.

Chapter 5 includes an overview of LHMs and their applications to microwave and antenna engineering. Reviews are focused on LHMs as the context of "materials," although the composite right/left-handed transmission lines (CRLH-TL) are also mentioned.

Chapter 6 presents metamaterial FDTD modeling based on the effective medium theory (EMT). Negative refraction and perfect lenses, two of the characteristics associated with such media, are demonstrated, assuming that the media are ideal in nature, and a discussion of the performance of practical metamaterial slabs is presented in a later chapter. Application of LHMs to the problem of radome is discussed.

Chapter 7 contains an extensive study of FDTD modeling on "real-world" LHMs, and several parameter extraction techniques will be studied for evaluating both electrical and magnetic properties of LHMs. A figure of merit (FoM) study is conducted to relate the bandwidth and loss issues of LHMs to the electrical dimensions of resonant particles.

Chapter 8 addresses a spatial averaging scheme at the boundaries of LHM slabs when modeling the perfect lens.

Chapter 9 presents a novel spatially dispersive FDTD method to illustrate the unique property of uniaxial wire mediums that are capable of mapping the near field akin to an LHM lens, provided that the source and image planes are butted right up against its interface.

Chapter 10 deals with metamaterial FDTD modeling at optics. Specifically the topics such as nanoparticle plasmonic waveguide, subwavelength imaging based on layered silver slabs, and electromagnetic cloaking will be discussed.

Chapter 11 presents an overview of the book and discusses relating to the state of the art in metamaterial modeling and closes with some remarks on future challenges.

References

[1] http://www.ee.duke.edu/~drsmith/neg_ref_home.htm.

[2] http://www.metamorphose-eu.org/.

[3] IECIE, Japan, *IEICE Trans. Electron.*, Vol. E89-C, No. 9, 2006, pp. 1265-1266.

[4] http://www.ee.duke.edu/~drsmith/about_metamaterials.html.

[5] Yablonovitch, E., "Inhibited Spontaneous Emission in Solid-State Physics and Electronics," *Phys. Rev. Lett.*, Vol. 58, No. 20, 1987, pp. 2059-2062.

[6] John, S., "Strong Localization of Photons in Certain Disordered Dielectric Superlattices," *Phys. Rev. Lett.*, Vol. 58, No. 23, 1987, pp. 2486-2489.

[7] Von Hippel, A., and R. Arthur, *Dielectrics and Waves*, Cambridge, MA: MIT Press, 1954.

[8] Balanis, C. A., *Advanced Engineering Electromagnetics*, New York: John Wiley & Sons, 1989, pp. 44-60.

[9] Engheta, N., and R. Ziolkowski, (eds.), *Metamaterials: Physics and Engineering Explorations*, New York: John Wiley & Sons, 2006.

[10] Wiltshire, M. C. K., "Bending of Light in the Wrong Way," *Science*, 292, 2001, pp. 60-61.

[11] www.wave-scattering.com/negative.html.

[12] Lamb, H., "On Group-Velocity," *Proc. London Math. Soc. 1*, 1904, pp. 473-479.

[13] Veselago, V. G., and E. E. Narimanov, "The Left Hand of Brightness: Past, Present and Future of Negative Index Materials," *Nature Materials 5*, 2006, pp. 759-762.

[14] Sihvola, A., "Metamaterials in Electromagnetics," *Metamaterials*, Vol. 1, Issue 1, 2007, pp. 2–11.

[15] Shamonina, E., and L. Solymar, "Metamaterials: How the Subject Started," *Metamaterials*, Vol. 1, Issue 1, 2007, pp. 12–18.

[16] Kock, W., "Metallic delay lenses," *Bell Syst. Tech. J.*, Vol. 27, 1948, pp. 58–82.

[17] Rudge, A. W., et al. (editors), *The Handbook of Antenna Design: Volume II*, London, U.K.: Peter Peregrinus Ltd., 1983.

[18] Brown, J., "Artificial Dielectrics," *Progress in Dielectrics*, Vol. 2, 1960, pp. 195–225.

[19] Brown, J., "Artificial Dielectrics Having Refractive Indices Less Than Unity," *Proc. IEEE*, Vol. 100, No. 62R, 1953, pp. 51–62.

[20] Brown, J., and W. Jackson, "The Properties of Artificial Dielectrics at Centimeter Wavelengths," *Proc. IEEE*, Vol. 102B, No. 1699R, 1955, pp. 11–21.

[21] Seeley, J. S., and J. Brown, "The Use of Artificial Dielectrics in a Beam Scanning Prism," *Proc. IEEE*, Vol. 105C, No. 2735R, 1958, pp. 93–102.

[22] Carne, A., and J. Brown, "Theory of Reflections from the Rodded-Type Artificial Dielectrics," *Proc. IEEE*, Vol. 105C, No. 2742R, 1958, pp. 107–115.

[23] Model, A. M., "Propagation of Plane Electromagnetic Waves in a Space Which Is Filled with Plane Parallel Grids," *Radiotekhnika*, Vol. 10, 1955, pp. 52–57.

[24] Rotman, W., "Plasma Simulations by Artificial Dielectrics and Parallel-Plate Media," *IRE Trans. Ant. Propag.*, Vol. 10, 1962, pp. 82–95.

[25] Pendry, J., et al., "Extremely Low Frequency Plasmons in Metallic Mesostructures," *Phys. Rev. Lett.*, Vol. 76, No. 25, 1996, pp. 4773–4776.

[26] Pokrovsky, A., and A. Efros, "Electrodynamics of Metallic Photonic Crystals and the Problem of Left-Handed Materials," *Phys. Rev. Lett.*, Vol. 89, 2002, p. 093901.

[27] Pokrovsky, A., "Analytical and Numerical Studies of Wire-Mesh Metallic Photonic Crystals," *Phys. Rev. B*, Vol. 69, 2004, p. 195108.

[28] Smith, D. R., et al., "Composite Medium with Simultaneously Negative Permeability and Permittivity," *Phys. Rev. Lett.*, Vol. 84, No. 18, 2000, pp. 4184–4187.

[29] King, R., D. Thiel, and K. Park, "The Synthesis of Surface Reactance Using an Artificial Dielectric," *IEEE Trans. Antennas and Propagat.*, Vol. 31, 1983, pp. 471–476.

[30] Belov, P. A., Y. Hao, and S. Sudhakaran, "Subwavelength Microwave Imaging Using an Array of Parallel Conducting Wires as a Lens," *Phys. Rev. B*, Vol. 73, 2006, p. 033108.

[31] http://www.physlink.com/estore/cart/TVRockUlexite.cfm.

[32] Pendry, J., "Manipulating the Near Field with Metamaterials," *LECPOS*, Santa Barbara, California, July 6, 2002.

[33] Schelkunoff, S., and H. Friis, *Antennas: Theory and Practice*, New York: John Wiley & Sons, 1952.

[34] Schneider, H. J., and P. Dullenkopf, "Slotted Tube Resonator: A New NMR Probe Head at High Observing Frequencies," *Rev. Sci. Instrum.*, Vol. 48, No. 1, 1977, pp. 68–73.

[35] Hardy, W. H., and L. A. Whitehead, "Split-Ring Resonator for Use in Magnetic Resonance from 200-2000 MHz," *Rev. Sci. Instrum.*, Vol. 52, No. 2, 1981, pp. 213–216.

[36] Kostin, M. V., and V. V. Shevchenko, "Artificial Magnetics Based on Double Circular Elements," *Proc. of Bianisotropics '94*, Perigueux, France, 1994, pp. 49–56.

[37] Pendry, J., et al., "Magnetism from Conductors and Enhanced Nonlinear Phenomena," *IEEE Trans. Microwave Theory Tech.*, Vol. 47, No. 11, 1999, pp. 195–225.

[38] Marques, R., et al., "Left-Handed-Media Simulation and Transmission of EM Waves in Subwavelength Split-Ring-Resonator-Loaded Metallic Waveguides," *Phys. Rev. Lett.*, Vol. 89, No. 18, 2002, p. 183901.

[39] Hrabar, S., J. Bartolic, and Z. Sipus, "Waveguide Miniaturization Using Negative Permeability Material," *IEEE Trans. Antennas Propagat.*, Vol. 53, No. 1, 2005, pp. 110–119.

[40] Serdyukov, A., et al., *Electromagnetics of Bianisotropic Materials: Theory and Applications*, Amsterdam: Gordon and Breach Science Publishers, 2001.

[41] Lindell, I., et al., *Electromagnetic Waves in Chiral and Bi-Isotropic Media*, Norwood, MA: Artech House, 1994.

[42] Priou, A., et al., *Advances in Complex Electromagnetic Materials*, NATO ASI Series, Kluwer, 1997.

[43] Marques, R., F. Medina, and R. Rafii-El-Idrissi, "Role of Bianisotropy in Negative Permeability and Left-Handed Metamaterials," *Phys. Rev. B*, Vol. 65, 2002, p. 144440.

[44] Sauviac, B., C. Simovski, and S. Tretyakov, "Double Split-Ring Resonators: Analytical Modeling and Numerical Simulations," *Electromagnetics*, Vol. 24, No. 5, 2004, pp. 317–338.

[45] Tretyakov, S., "Research on Negative Refraction and Backward-Wave Media: A Historical Perspective," *Proc. of EPFL Latsis Symposium 2005, Negative Refraction: Revisiting Electromagnetics from Microwave to Optics*, Lausanne, Switzerland, February 28–March 2, 2005, pp. 30–35.

[46] Tretyakov, S., et al., "Waves and Energy in Chiral Nihility," *J. Electromagnetic Waves Applic.*, Vol. 17, No. 5, 2003, pp. 695–706.

[47] Pendry, J., "A Chiral Route to Negative Refraction," *Science*, Vol. 306, 2004, pp. 1353–1355.

[48] Monzon, C., and D. W. Forester, "Negative Refraction and Focusing of Circularly Polarized Waves in Optically Active Media," *Phys. Rev. Lett.*, Vol. 95, 2005, p. 123904.

[49] Veselago, V., "The Electrodynamics of Substances with Simultaneously Negative Values of ε and μ," *Sov. Phys. Usp.*, Vol. 10, 1968, pp. 509–514.

[50] Lindell, I., et al., "BW Media—Media with Negative Parameters, Capable of Supporting Backward Waves," *Microwave and Optical Technology Letters*, Vol. 31, No. 10, 2001, pp. 129–133.

[51] Ruppin, R., "Surface Polaritons of a Left-Handed Meterial Slab," *J. Phys.: Condens. Matter.*, Vol. 13, 2001, pp. 1811–1819.

[52] Foteinopolou, S., E. Economou, and C. Soukoulis, "Refraction in Media with a Negative Refraction Index," *Phys. Rev. Lett.*, Vol. 90, No. 10, 2003, p. 107402.

[53] Parazzoli, C. G., et al., "Experimental Verification and Simulation of Negative Index of Refraction Using Snell's Law," *Phys. Rev. Lett.*, Vol. 90, 2003, p. 107401.

[54] Lamb, H., "On Group-Velocity," *Proc. London Math. Soc.* 1, 1904, pp. 473–479.

[55] Clarricoats, P. J. B., and B. C. Taylor, "Evanescent and Propagating Modes of Dielectric-Loaded Circular Waveguides," *Proc. IEE* Vol. 111 No. 12, December 1964, pp. 1951–1956.

[56] Notomi, M., "Theory of Light Propagation in Strongly Modulated Photonic Crystals: Refraction Like Behavior in the Vicinity of the Photonic Band Gap," *Phys. Rev. B*, Vol. 62, No. 16, 2000, pp. 10696–10705.

[57] Foteinopolou, S., and C. Soukoulis, "Negative Refraction and Left-Handed Behaviour in Two-Dimensional Photonic Crystals," *Phys. Rev. B*, Vol. 67, 2003, p. 235107.

[58] Parimi, P., et al., "Negative Refraction and Left-Handed Electromagnetism in Microwave Photonic Crystals," *Phys. Rev. Lett.*, Vol. 92, No. 12, 2004, p. 127401.

[59] Sudhakaran, S., Y. Hao, and C. G Parini, "Negative Refraction Phenomenon at Multiple Frequency Bands from Electromagnetic Crystals," *Microwave and Optical Technology Letters*, Vol. 45, 2005, pp. 465–469.

[60] Pendry, J. B., "Negative Refraction Index Makes Perfect Lens," *Phys. Rev. Lett.*, Vol. 85, 2000, pp. 3966–3969.

[61] Engheta, N., "An Idea for Thin, Subwavelength Cavity Resonators Using Metamaterials with Negative Permittivity and Permeability," *Ant. Wireless Propag. Lett.*, Vol. 1, No. 1, 2002, pp. 10–13.

[62] Cory, H., et al., "Use of Conjugate Dielectric and Metamaterial Slabs as Radomes," *IET Proceedings of Microwaves, Antennas & Propagation*, Vol. 1, No. 1, 2007, pp. 137-143.

[63] Shelby, R. A., D. R. Smith, and S. Schultz, "Experimental Verification of a Negative Index of Refraction," *Science*, Vol. 292, 2001, pp. 77-79.

[64] Houck, A., J. Brock, and I. Chuang, "Experimental Verification of Snell's Law in Left Handed Material," *Phys. Rev. Lett.*, Vol. 90, 2003, p. 137401.

[65] Smith, D. R., and D. Schurig, "Electromagnetic Wave Propagation in Media with Indefinite Permittivity and Permeability Tensors," *Phys. Rev. Lett.*, Vol. 90, No. 7, 2003, p. 077405.

[66] Smith, D. R., P. Kolinko, and D. Schurig, "Negative Refraction in Indefinite Media," *J. Opt. Soc. Am. B*, Vol. 21, No. 5, 2004, pp. 1032-1043.

[67] Holloway, C. L., et al., "A Double Negative (DNG) Composite Medium Composed of Magnetodielectric Spherical Particles Embedded in a Matrix," *IEEE Trans. Antennas Propag.*, Vol. 51, 2003, pp. 2596-2603.

[68] Mini special issue on electromagnetic crystal structures, design, synthesis, and applications, *IEEE Trans. Microwave Theory Techniques*, Vol. 47, No. 11, 1999.

[69] Special Issue on Electromagnetic Applications of Photonic Band Gap Materials and Structures, *Progress In Electromagnetic Research, PIER*, Vol. 41, 2003.

[70] Sakoda, K., *Optical Properties of Photonic Crystals*, Springer-Verlag, Berlin, 2005.

[71] Joannopoulos, J., R. Mead, and J. Winn, *Photonic Crystals: Molding the Flow of Light*, Princeton, NJ: Princeton University Press, 1995.

[72] Mini special issue on photonic bandgap structures, *Focus Issue: Photonic Bandgap Calculations, Optics Express*, Vol. 8, No. 3, 2001.

[73] *J. Opt. Soc. Am. B*, Vol. 10, No. 2, 1993, pp. 280-413.

[74] Feature Section on Photonic Crystal Structures and Applications of *IEEE J. Quantum Electron.*, Vol. 38, No. 7, 2002.

[75] Agranovich, V., and V. Ginzburg, *Spatial Dispersion in Crystal Optics and the Theory of Excitons*, New York: Wiley-Interscience, 1966.

[76] Agarwal, G., D. Pattanayak, and E. Wolf, "Electromagnetic Fields in Spatially Dispersive Media," *Phys. Rev. B*, Vol. 10, 1974, pp. 1447-1475.

[77] Birman, J. L., and J. J. Sein, "Optics of Polaritons in Bounded Media," *Phys. Rev. B*, Vol. 6, 1972, pp. 2482-2490.

[78] Yablonovitch, E., T. Gmitter, and K. Leung, "Photonic Band Structure: the Facecentered-Cubic Case Employing Nonspherical Atoms," *Phys. Rev. Lett.*, Vol. 67, No. 17, 1991, pp. 2295-2298.

[79] Vukusic, P., and J. R. Sambles, "Photonic Structures in Biology," *Nature*, Vol. 424, August 14, 2003.

[80] Yablonovitch, E., "Inhibited Spontaneous Emission in Solid-State Physics and Electronics," *Phys. Rev. Lett.*, Vol. 58, No. 20, 1987, pp. 2059-2062.

[81] John, S., "Strong Localization of Photons in Certain Disordered Dielectric Superlattices," *Phys. Rev. Lett.*, Vol. 58, No. 23, 1987, pp. 2486-2489.

[82] Bykov, V. P., "Spontaneous Emission in a Periodic Structure," *Sov. Phys. JETP*, Vol. 35, 1972, pp. 269-273.

[83] Bykov, V. P., "Spontaneous Emission from a Medium with a Band Spectrum," *Sov. J. Quant. Electron*, Vol. 4, 1975, pp. 861-871.

[84] Yablonovitch, E., and T. Gmitter, "Photonic Band Structure: The Face-Centered-Cubic Case,'" *Phys. Rev. Lett.*, Vol. 63, No. 18, 1989, pp. 1950-1953.

[85] Yablonovitch, E., "Photonic Band Structure: The Face-Centered-Cubic Case," *J. Opt. Soc. Am. A*, Vol. 7, No. 9, 1990, pp. 1792-1800.

[86] Yablonovitch, E., et al., "3-Dimensional Photonic Band Structure," *Optical and Quantum Electronics,* Vol. 24, No. 2, 1992, pp. 273-283.

[87] Yablonovitch, E., "Photonic Band-Gap Structures," *J. Opt. Soc. Am. B,* Vol. 10, No. 2, 1993, pp. 283-295.

[88] Kourosh Mahdjoubi, Anne-Claude Tarot, and Ronan Sauleau, "EBG Directive Antennas," *European Antenna Centre of Excellence (ACE) Tutorial on Artificial EBG Surfaces and Metamaterial for Antennas Course,* Gothenburg, Sweden, 2005.

[89] Hao, Y., A. Alomainy, and C. G. Parini, "Antenna Beam Shaping from Offset Defects in UC-EBG Cavities," *Microwave and Optical Technology Letters,* Vol. 43, No. 2, pp. 108-112, 2004.

[90] Boutayeb, H., et al., "Analysis and Design of a Cylindrical EBG-Based Directive Antenna," *IEEE Transactions on Antennas and Propagation,* Volume 54, Issue 1, pp. 211-219, 2006.

[91] Cheype, C., et al., "An Electromagnetic Bandgap Resonator Antenna," *IEEE Transactions on Antennas and Propagation,* Vol. 50, No. 9, pp. 1285-1290, 2002.

[92] Hao, Y., and C. G. Parini, "Isolation Enhancement of Anisotropic UC-PBG Microstrip Diplexer Patch Antenna," *IEEE Antennas and Wireless Propagation Letters,* Vol. 1, 2002.

[93] Zhao, Y., Y. Hao, and C. G. Parini, "Radiation Properties of PIFA on UC-EBG Substrates," *Microwave and Optical Technology Letters,* Vol. 44, No. 1, pp. 21-24, January 5, 2005.

[94] Sievenpiper, D., et al., "High-Impedance Electromagnetic Surfaces with a Forbidden Frequency Band," *IEEE Transactions on Microwave Theory and Techniques,* Vol. 47, No. 11, 1999, pp. 2059-2074.

[95] Weily, A. R., et al. "Linear Array of Woodpile EBG Sectoral Horn Antennas," *IEEE Transactions on Antennas and Propagation,* Vol. 54, No. 8, 2006, pp. 2263-2274.

[96] Yang, F., and Y. Rahmat-Samii, "Microstrip Antennas Integrated with Electromagnetic Band-Gap (EBG) Structures: A Low Mutual Coupling Design for Array Applications," *IEEE Transactions on Antennas and Propagation,* Vol. 51, No. 10, Part 2, 2003, pp. 2936-2946.

[97] Pendry, J., "Calculating Photonic Bandgap Structure," *J. Phys.: Condensed Matter,* Vol. 8, No. 9, 1996, pp. 1085-1108.

[98] Ziolkowski, R. W., and E. Heyman, "Wave Propagation in Media Having Negative Permittivity and Permeability," *Phys. Rev. E,* Vol. 64, 2001, p. 056625.

[99] Loschialpo, P. F., et al., "Electromagnetic Waves Focused by a Negative-Index Planar Lens," *Phys. Rev. E,* Vol. 67, 2003, p. 025602.

[100] Cummer, S. A., "Simulated Causal Subwavelength Focusing by a Negative Refractive Index Slab," *Appl. Phys. Lett.,* Vol. 82, 2003, p. 1503.

[101] Rao, X. S., and C. K. Ong, "Subwavelength Imaging by a Left-Handed Material Superlens," *Phys. Rev. E,* Vol. 68, 2003, p. 067601.

[102] Feise, M. W., and Y. S. Kivshar, "Sub-Wavelength Imaging with a Left-Handed Material Flat Lens," *Phys. Lett. A,* Vol. 334, 2005, pp. 326-323.

[103] Chen, J. J., et al., "Limitation of FDTD in Simulation of a Perfect Lens Imaging System," *Opt. Express,* Vol. 13, 2005, pp. 10840-10845.

[104] Feise, M. W., J. B. Schneider, and P. J. Bevelacqua, "Finite-Difference and Pseudospectral Time-Domain Methods Applied to Backward-Wave Metamaterials," *IEEE Trans. Antennas Propagat.,* Vol. 52, No. 11, 2004, pp. 2955-2962.

[105] Zhao, Y., P. Belov, and Y. Hao, "Accurate Modelling of the Optical Properties of Left-Handed Media Using a Finite-Difference Time-Domain Method," *Phys. Rev. E,* Vol. 75, 2006, p. 037602.

[106] Saj, W. M., "FDTD Simulations of 2D Plasmon Waveguide on Silver Nanorods in Hexagonal Lattice," *Opt. Express,* Vol. 13, No. 13, June 2005, pp. 4818-4827.

[107] Hwang, K.-P., and A. C. Cangellaris, "Effective Permittivities for Second-Order Accurate FDTD Equations at Dielectric Interfaces," *IEEE Microwave Wirel. Compon. Lett.*, Vol. 11, 2001, pp. 158–160.

[108] Hao, Y., and C. J. Railton, "Analyzing Electromagnetic Structures with Curved Boundaries on Cartesian FDTD Meshes," *IEEE Trans. Microwave Theory Tech.*, Vol. 46, January 1998, pp. 82–88.

[109] Luebbers, R., et al., "A Frequency-Dependent Finite-Difference Time-Domain Formulation for Dispersive Materials," *IEEE Trans. Electromagn. Compat.*, Vol. 32, August 1990, pp. 222–227.

[110] Gandhi, O. P., B.-Q. Gao, and J.-Y. Chen, "A Frequency-Dependent Finite-Difference Time-Domain Formulation for General Dispersive Media," *IEEE Trans. Microwave Theory Tech.*, Vol. 41, April 1993, pp. 658–664.

[111] Sullivan, D. M., "Frequency-Dependent FDTD Methods Using Z Transforms," *IEEE Trans. Antennas Propagat.*, Vol. 40, October 1992, pp. 1223–1230.

[112] Kaneda, N., B. Houshmand, and T. Itoh, "FDTD Analysis of Dielectric Resonators with Curved Surfaces," *IEEE Trans Microwave Theory Tech.*, Vol. 45, September 1997, pp. 1645–1649.

[113] Lee, J.-Y., and N.-H Myung, "Locally Tensor Conformal FDTD Method for Modeling Arbitrary Dielectric Surfaces," *Microw. Opt. Tech. Lett.*, Vol. 23, November 1999, pp. 245–249.

[114] Mohammadi, A., and M. Agio, "Contour-Path Effective Permittivities for the Two-Dimensional Finite-Difference Time-Domain Method," *Opt. Express*, Vol. 13, 2005, pp. 10367–10381.

CHAPTER 2
Fundamentals and Applications of Electromagnetic Bandgap Structures

2.1 Introduction

As stated in Chapter 1, periodic structures have been used in both closed metallic and open waveguides long, for instance, in filters and traveling-wave tubes. In the late 1980s, a fully three-dimensional periodic structure, operating at microwave frequencies, was realized by Yablonovitch [1] and his coworkers by mechanically drilling holes into a block of dielectric material. This material prevents the propagation of EM radiation in all three spatial directions; it is transparent at these wavelengths in its solid form. Such artificially engineered structures are generically known as electromagnetic bandgap (EBG) materials [1–96].

The main feature of these materials is the existence of a gap (stopband) in the frequency spectrum of propagating EM waves [2–6]. This bandgap frequency depends on the permittivities of the dielectric inclusions and the background materials, the dimensions of the inclusions/defects, their periodicities and the incidence angle of the electromagnetic waves [7]. This feature leads to a variety of phenomena of both fundamental space [8, 9] and practical way [1, 10] to scientists alike. The use of these artificial materials ranges far and wide and includes antenna substrates, planar filters, and optical waveguides. The EBG technology continues to witness a rapid development, and it is expected that new structures will evolve.

This chapter reviews the basic theory, numerical methods, and applications of the EBG material.

2.2 Bloch's Theorem and the Dispersion Diagram

The role of symmetry is important in an electromagnetic structure when analyzing its behavior. EBG studies are typically based on exploiting interesting symmetry properties, including periodicity in dielectrics of the material, that are amenable to analysis using Bloch's theory. This theory can be used to derive the dispersion diagrams of the EBGs, as will be shown in this section, which will introduce the basic topics pertaining to the study of EBGs, including:

1. Translational symmetry;
2. Bloch's theorem and periodic boundary condition (PBC);
3. Brillouin zones;
4. Dispersion diagram and EBG.

2.2.1 Translational Symmetry

Definition: *A system with translational symmetry [2] is invariant to a translation through a displacement $\vec{d}$.*

Let $\varepsilon(\vec{r})$ be a function defined in a translationally invariant system. Then, when we carry out a translation with an operator $T_{\vec{d}}$, we have:

$$T_{\vec{d}}\left[\varepsilon(\vec{r})\right] = \varepsilon(\vec{r} + \vec{d}) = \varepsilon(\vec{r}) \tag{2.1}$$

A system that has *continuous* translational symmetry is invariant under the $T_{\vec{d}}$'s of any displacement. An example system that has continuous translational symmetry in all three directions is free space $\varepsilon(\vec{r}) = 1$.

Few EBGs have continuous translational symmetry. However, all of them have *discrete translational symmetry*. In other words, they are not translationally invariant under any distance, but only under certain distances—which are multiples of some fixed step length or period. The basic step length is termed as the lattice constant a, while the basic step vector is called the *primitive lattice vector $\vec{a}$*, or the fundamental translation vector $\vec{a}$. So, we have $\varepsilon(\vec{r}) = \varepsilon(\vec{r} + \vec{a})$ and $\varepsilon(\vec{r}) = \varepsilon(\vec{r} + \vec{R})$, where $\vec{R} = s_1\vec{a}_1 + s_2\vec{a}_2 + s_3\vec{a}_3$, and $s_i (i = 1,2,3)$ is an integer. Consequently, these structures can be considered as one basic configuration unit being replicated over and over. This basic unit is referred to as the *unit cell*.

Figures 2.1–2.3 show examples of EBG structures that are periodic in one, two, and all three dimensions, respectively. In Figure 2.1, the dielectric property of the material is repeated in one direction with a period of a. In Figure 2.2, the infinitely long cylindrical rods are periodically loaded in both $\hat{x}$- and $\hat{y}$-directions, both with a distance of a. The material in Figure 2.3 is termed as the woodpile EBG structure and is also referred to as a layer-by-layer photonic crystal in the physics publications [11]. A unit cell of this material is shown in Figure 2.3(b).

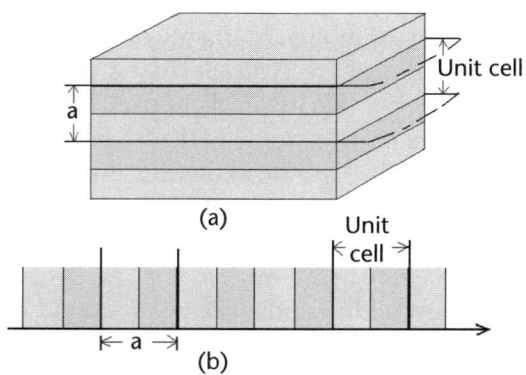

Figure 2.1 A part of a one-dimensional EBG structure. This EBG is made up of infinite planes with two kinds of materials with different dielectric properties, (a is the lattice constance): (a) the three-dimensional view of the EBG and (b) the one-dimensional view of the EBG.

2.2 Bloch's Theorem and the Dispersion Diagram

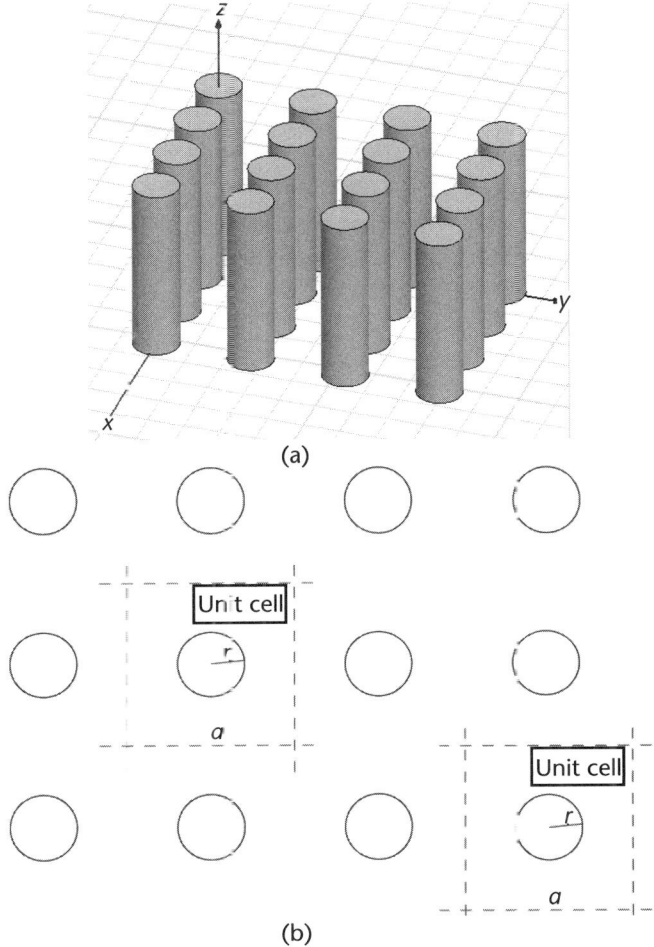

Figure 2.2 (a) A part of the infinitely long two-dimensional EBG structure and (b) unit cell marked on the x-y cut plane of the EBG structure.

2.2.2 Bloch's Theorem and Periodic Boundary Condition (PBC)

Bloch investigated the theory of wave propagation in three-dimensionally periodic media back in 1928, and the same underlying mathematics, however, was also discovered as Floquet's theorem [4] in one dimension in 1883. In his study, Bloch proved that waves in such a medium can propagate without scattering; their behavior is governed by a periodic envelope function multiplied by a plane wave. That is, in an EBG with periodic dielectric function:

$$\varepsilon(\vec{r}) = \varepsilon(\vec{r} + s_1\vec{a}_1 + s_2\vec{a}_2 + s_3\vec{a}_3), s_i = 0, \pm 1, \pm 2, \ldots \quad (2.2)$$

where $\vec{a}_i (i = 1,2,3)$ are three primitive lattice vectors, for a crystal periodic in all three dimensions, the fields can be expressed as plane wave with an envelope (2.3) as follows:

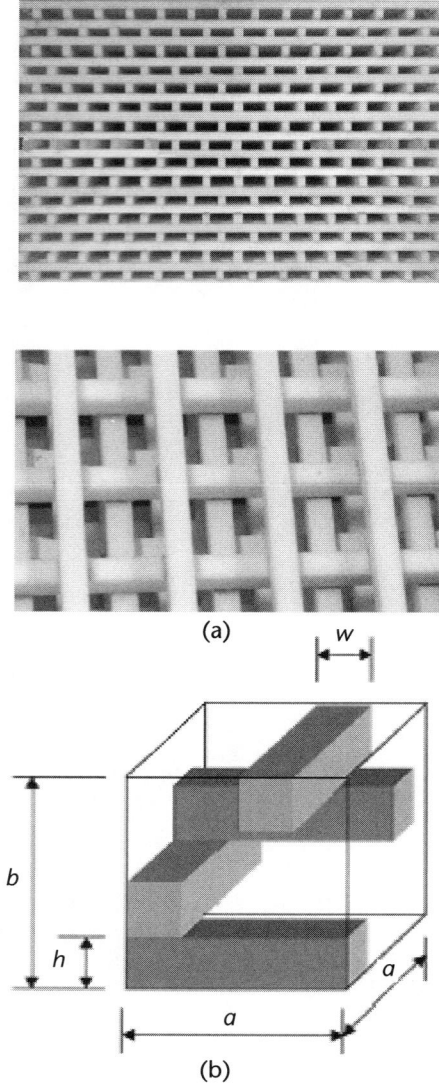

Figure 2.3 (a) Two angles of view of the woodpile EBG structure and (b) unit cell for the woodpile EBG material.

$$\vec{H}(\vec{r}) = e^{i\vec{k}\cdot\vec{r}}\vec{u}(\vec{r}) \qquad (2.3)$$

where $\vec{H}$ is a electric or magnetic field vector and $\vec{u}$ is a periodic envelope function.

The result above is commonly known as Bloch's theorem. In solid-state physics, the form is known as a Bloch state [12], and in mechanics it is referred to as a Floquet mode [13].

Let us consider a simple example as shown in Figure 2.4, which is repetitive in the y-direction, and invariant and infinite in the x-direction. Thus the geometry has continuous translational symmetry in the x-direction, while it possesses

2.2 Bloch's Theorem and the Dispersion Diagram

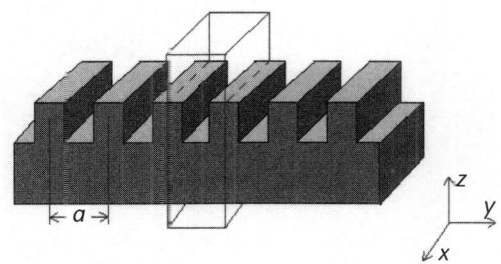

Figure 2.4 A dielectric configuration with discrete translational symmetry. (a is the lattice constant.)

discrete translational symmetry in the y-direction. The unit cell of this geometry is highlighted in Figure 2.4. The primitive lattice vector for this case is given by $\vec{a} = a\hat{y}$.

The fields in this structures are in the Bloch state and can be written as:

$$\vec{H}(\vec{r}) = e^{ik_x x} \cdot e^{ik_y y} \cdot \vec{u}_{k_y}(y,z) \tag{2.4}$$

where $\vec{u}(y,z)$ is a periodic function in y satisfying:

$$\vec{u}(y + s \cdot a, z) = \vec{u}(y, z) \tag{2.5}$$

Equation (2.4) can be interpreted to imply that this state is a plane wave, modulated by a periodic function that arises because of the periodicity of the lattice.

The fields in Bloch's state can be studied using the so-called unit cell approach, in which only elements in one unit cell are investigated, while the elements of the outside cell are related to the one in the unit cell using the following relationship:

$$\begin{aligned}\vec{H}(\vec{r} + la\hat{y}) &= e^{ik_x x} \cdot e^{ik_y(y+la)} \cdot \vec{u}_{k_y}(y + la, z) \\ &= e^{ik_y la} \cdot \left(e^{ik_x x} \cdot e^{ik_y y} \cdot \vec{u}_{k_y}(y,z) \right) \\ &= e^{ik_y la} \vec{H}(\vec{r}) \end{aligned} \tag{2.6}$$

Equation (2.6) is used as the PBC when studying the infinite EBG structures.

2.2.3 Brillouin Zone

A key attribute of the Bloch states is that a Bloch state with a wave vector k_y is identical to the Bloch state whose wave vector is $k_y + mb$, where $b = 2\pi/a$ and m is an integer. That implies that the mode frequencies are also periodic in k_y, i.e. $\omega(k_y) = \omega(k_y + mb)$, and hence, we only need to consider k_y in the range $-\pi/a < k_y \leq \pi/a$. This region that is associated with nonredundant values of k_y is called the Brillouin zone.

Another example shown in Figure 2.5 is made up of cylindrical elements periodically laid out in a square lattice. Figure 2.5(a) shows the physical lattice,

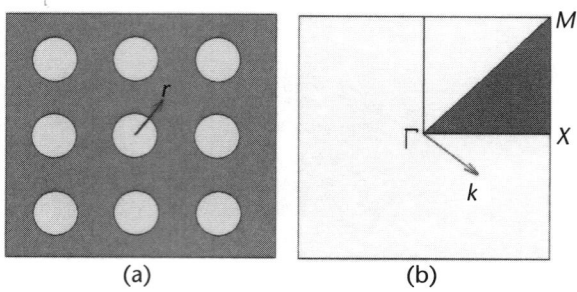

Figure 2.5 (a) The physical lattice of an EBG made using a square lattice. An arbitrary vector *r* is shown. (b) The Brillouin zone of the reciprocal lattice, centered at the origin (Γ). An arbitrary wave vector *k* is shown. The irreducible Brillouin zone is the triangular wedge. The special points at the center, corner, and face are conventionally known as Γ, M and X [2].

while the Brillouine zone of the reciprocal lattice is plotted in Figure 2.5(b). This structure shows not only a discrete translational symmetries along the *x*- and *y*-directions, but also rotation, mirror-reflection, and inversion symmetry. It is stated in [2] that when these symmetries exist in the physical space, the reciprocal space (k-space) shows the same types of symmetry. One consequence of this is that we need only consider the smallest region within the Brillouin zone [see Figure 2.5(b)] for which the k vectors are not related by symmetry and not each and every k-point in the zone. That smallest region is termed as the *irreducible Brillouin zone*, shown in Figure 2.5(b); the rest of the Brillouin zone contains redundant copies of the irreducible zone.

2.2.4 Dispersion Diagram and EBG

With the knowledge of the irreducible Brillouin zone, we can plot the wave numbers of the possible modes against the wave vector $\vec{k}$. Such plots provide the knowledge of the dispersion relationship and the energy flow behaviors in an intuitive way. Figure 2.6 shows two examples of one-dimensional diagrams of one-dimensional EBG structures.

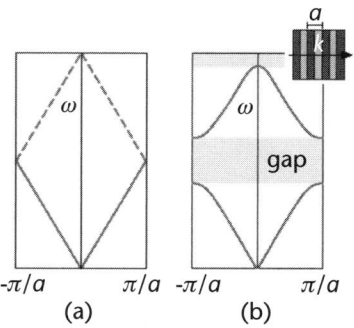

Figure 2.6 (a) Dispersion relation (band diagram), frequency ω versus wave number $\vec{k}$ of a uniform one-dimensional medium. (b) Schematic effect on the bands of a physical periodic dielectric variation (inset), where a gap has been opened by splitting the degeneracy at the $k = \pm\pi/a$. The upper right-hand corner shows the physical lattice and the wave vector. *a* is the lattice constant [4].

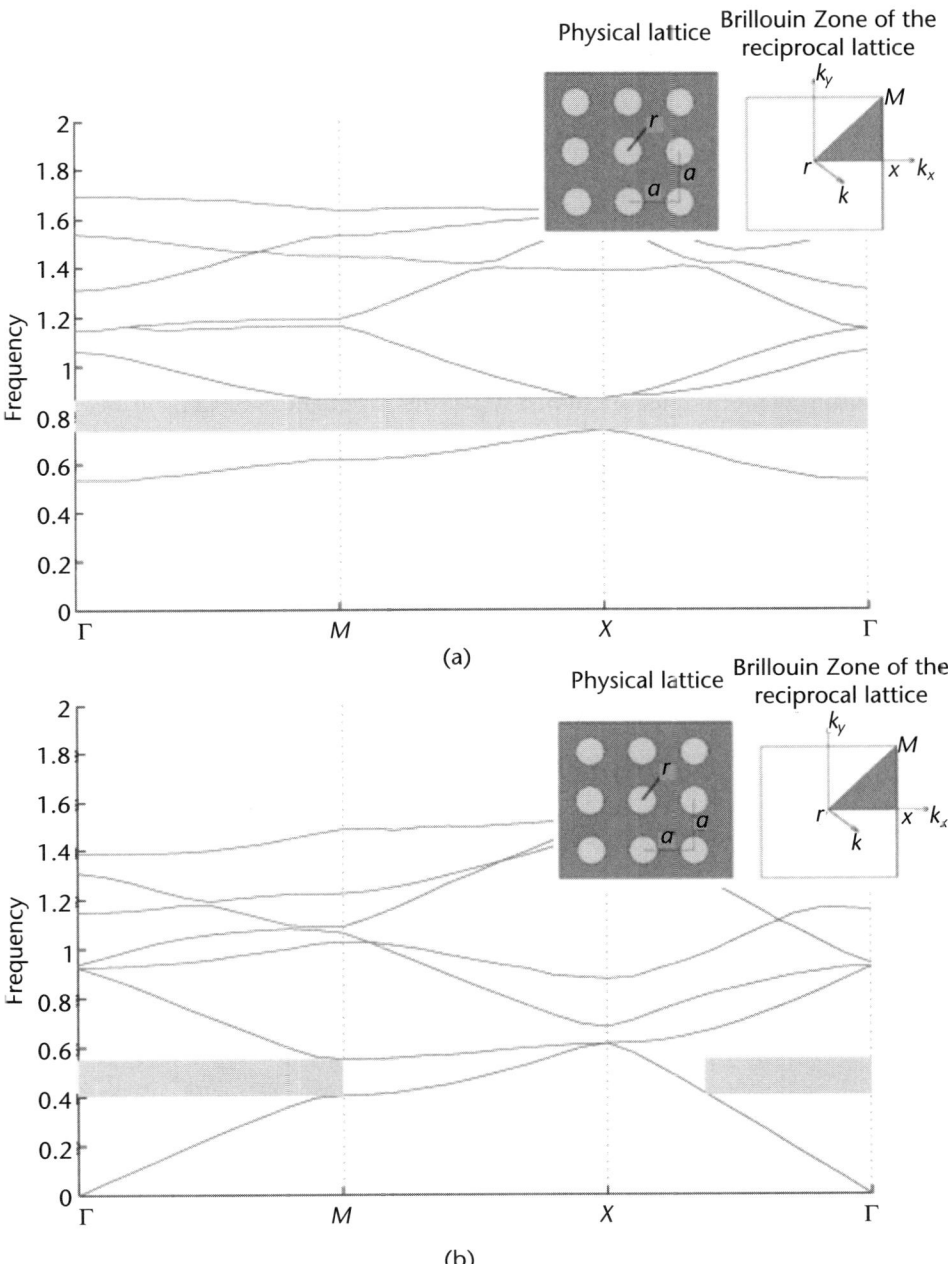

Figure 2.7 (a) A two-dimensional dispersion diagram for a two-dimensional EBG (illustrated in Figure 2.5). (b) A two-dimensional dispersion diagram where an incomplete bandgap can be found. For example, the bandgap from normalized frequency 0.404–0.552 is only valid with k vector from Γ to X, and a subset from M to Γ.

Figure 2.6(a) is the dispersion diagram for a uniform dielectric medium, to which a periodicity of a has been artificially assigned. As we know that the speed of light is reduced by the index of refraction in a uniform medium, the plot is just the light-line given by

$$\omega(\vec{k}) = \frac{c\vec{k}}{\sqrt{\varepsilon}} \qquad (2.7)$$

Because the $\vec{k}$ repeats itself outside the Brillouin zone, the lines fold back into the zone when they reach the edges. The dashed lines show the folding effect of applying Bloch's theorem with an artificial periodicity a.

In Figure 2.6(b), we note that there is a gap in the frequency between the upper and lower branches of the lines. In this frequency region, no mode, regardless of $\vec{k}$, can propagate through the structure. This gap is called an EBG, which can be further classified as:

- *Complete EBG:* A complete electromagnetic bandgap is a range of ω in which there are no propagating (real $\vec{k}$) solutions of Maxwell's equations for any $\vec{k}$, surrounded by propagating states above and below the gap.

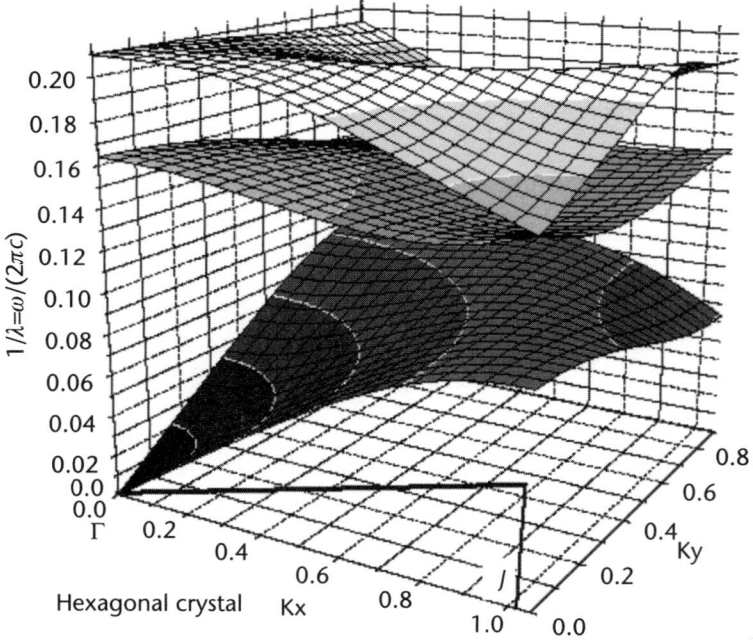

Figure 2.8 A three-dimensional dispersion diagram for a two-dimensional EBG. The EBG is made of circular rods of radius $\rho = 0.6$, with optical index of 2.9, lying in vacuum on a hexagonal lattice with period of 4. The horizontal plane gives the Bloch wave vector k. The vertical axis gives $1/\lambda$. The triangle corresponding to the irreducible Brillouin zone has been drawn in the (kx, ky) plane. The parameters about the EBGs can be found in [14].

- *Incomplete EBG:* This bandgap only exists over a subset of all possible wavevectors, polarizations, and/or symmetries.

Figures 2.7 and 2.8 show two- and three-dimensional dispersion diagrams, respectively, of two two-dimensional EBGs. In this work, only the two-dimensional dispersion diagrams are used as examples for numerical validation.

2.3 An Overview of Numerical Methods for Modeling EBG Structures

Numerical methods are often used to predict the performance of EBGs, either for the purpose of understanding their behavior or for developing new designs. These numerical methods include, the generalized Rayleigh identity method [17–20], the Korringa-Kohn-Rostoker (KKR) approach [21–23], the plane wave expansion (PWE) method [24–26], the transfer matrix method (TMM) [27, 28], and the FDTD method [29]. Among them, the PWE method and the FDTD method are the two that are used most widely. In this section, these numerical methods will be discussed briefly; the details of the FDTD method will be presented in the following chapters.

2.3.1 The Generalized Rayleigh's Identity Method and the Korringa-Kohn-Rostoker (KKR) Method

Nicorovici and McPhedran et al. extended Rayleigh's technique from electrostatic to full electromagnetic problems, for singly-, doubly-, and triply-periodic systems in [18–20], respectively.

Let us now consider EBGs consisting of an array of cylinders or spheres in an isotropic homogeneous dielectric host medium. Denote the fundamental translation vectors of the lattice as $\hat{e}_i (i = 1, 2)$ for a cylinder array or $(i = 1, 2, 3)$ for a spherical array. The wave equations for the electric and magnetic field components decouple from each other and they satisfy the Helmholtz equation:

$$(\nabla^2 + k^2) f(\vec{r}) = 0 \qquad (2.8)$$

where k is the wave vector.

The solution $f(\vec{r})$ must satisfy the quasiperiodicity condition given by

$$f(\vec{r} + \vec{R}_P) = e^{i \vec{k}_i \cdot \vec{R}_P} f(\vec{r}) \qquad \forall p \qquad (2.9)$$

where $\vec{k}_i = (k_i, \theta_i, \varphi_i)$ is the wave vector of the incident radiation, and $\vec{R}_P$ denotes the vector from the origin of the coordinates to the center of the pth cylinder, or sphere, (2.10) and (2.11), respectively, as follows:

$$\vec{R}_P = p_1 \hat{e}_1 + p_2 \hat{e}_2 \equiv (p_1, p_2), \qquad p_i \in \mathbb{Z} \qquad (2.10)$$

$$\vec{R}_P = p_1 \hat{e}_1 + p_2 \hat{e}_2 + p_3 \hat{e}_3 \equiv (p_1, p_2, p_3), \qquad p_i \in \mathbb{Z} \qquad (2.11)$$

The Green's function G, which obeys the inhomogeneous Helmholtz equation in the periodic systems, can be expressed as:

$$(\nabla^2 + k^2)G = -c\sum_p \delta(\vec{r} - \vec{R}_P - \vec{\rho})e^{i\vec{k}_0 \cdot \vec{R}_P} \qquad (2.12)$$

and it satisfies the quasiperiodicity conditions given by

$$G(\vec{r} + \vec{R}_P, \vec{\rho}) = e^{i\vec{k}_0 \cdot \vec{R}_P} G(\vec{r}, \vec{\rho}) \qquad (2.13)$$

$$G(\vec{r}, \vec{\rho} + \vec{R}_P) = e^{-i\vec{k}_0 \cdot \vec{R}_P} G(\vec{r}, \vec{\rho}) \qquad (2.14)$$

where $c = 2\pi$ or 1 for the cylinder and the sphere arrays, respectively.

By introducing the lattice sums, Nicorovici et al. have obtained a representation of the Green's function in terms of a rapidly convergent Neumann series [17–20]. The main advantage of using the lattice sums method is that, for a given problem of electromagnetic scattering by a grating, the coefficients of the series in calculating the Green's function have to be evaluated only once. This appreciably increases the speed of the numerical evaluation of the Green's function at any point in the xy-plane.

If the solution for the periodic lattice is derived via a variation-iteration technique instead of through a direct calculation, the generalized Rayleigh's Identity method becomes the generalization of the KKR method to photonics [20, 30], which was derived by Korringa, Kohn, and Rostoker [21] using different approaches.

The generalized Rayleigh's identity method and the KKR method share some common features. The most attractive feature of these two methods is that the greater part of the work involved in the calculation of the energy bands entails the computation of certain geometrical "structure constants," which need only be calculated once for each type of lattice. This approach leads to a very compact and very rapidly convergent scheme, in comparison to the PWE method, for the case where the potential $V(r)$ is spherically or cylindrically symmetric, within the inscribed spheres of the atomic polyhedra or the cylinders and constant in the space between them. However, the application of the generalized Rayleigh's identity method and the KKR method are limited to modeling structures with spherical or cylindrical symmetries and when $\varepsilon(r)$ is piecewise constant. When the actual potential violates these conditions significantly, the procedure is not suitable (see [23]).

Compared to the generalized Rayleigh's identity method, the KKR method has even limited range of applications. The variational KKR method cannot be used for complex $\varepsilon(r)$. On the other hand, the generalized Rayleigh's identity method is capable of handling problems in which the dielectric constant taking on either finite or infinite values, or imaginary ones. The generalized Rayleigh's identity method can also be applied when the cylinders are comprised of an arbitrary number of circular coaxial shells, filled with materials with different complex dielectric constants. However, because of its variational nature, the KKR method is expected to converge more rapidly within its range of application [30].

2.3.2 Plane-Wave Expansion Method

Comparitively speaking the PWE is much simpler than the generalized Rayleigh's identity and KKR methods, easier to program, and runs substantially faster. In general, the PWE converge more slowly than the two aforementioned methods. However, there are also situations in which the PWE achieves a fairly rapid convergence [23]. Moreover, the PWE method is not limited to spherically or cylindrically symmetric modulation of the potential. Instead, it can readily handle almost all different types of modulations [23, 31].

The PWE method works in the Fourier space. Let us denote

$$\vec{k} = \hat{x}_1 k_1 + \hat{x}_2 k_2 \tag{2.15}$$

as the two-dimensional wave vector and

$$\vec{G}(h) = h_1 \hat{b}_1 + h_2 \hat{b}_2 \tag{2.16}$$

as a reciprocal-lattice vector in the Fourier space. Next we expand the component of the electric field (or the magnetic field) in plane waves as in (2.17). Once the electric field (or the magnetic field) is known, the other EM field vectors can be determined by using:

$$E(\vec{r}) = \sum_{\vec{G}} B_{\vec{G}} e^{i(\vec{k}+\vec{G}) \cdot \vec{r}} \tag{2.17}$$

where $B_{\vec{G}}$'s are the Fourier coefficients.

We also expand the inverse dielectric constance as:

$$\frac{1}{\varepsilon(\vec{r})} = \sum_{\vec{G}} \hat{\kappa}_{\vec{G}} e^{i\vec{G} \cdot \vec{r}} \tag{2.18}$$

in which ε is periodic function in physical space satisfying: $\varepsilon(\vec{r}) = \varepsilon(\vec{r} + \vec{R})$ and $\hat{\kappa}_{\vec{G}}$'s are the Fourier coefficients.

Substituting (2.17) and (2.18) into Maxwell's equations we obtain an eigenvalue equation that reads:

$$\sum_{\vec{G}'} \hat{\kappa}_{\vec{G}-\vec{G}'} |\vec{k} + \vec{G}|^2 B_{\vec{G}'}$$
$$= \frac{\omega^2}{c^2} B_{\vec{G}} \tag{2.19}$$

which can be solved by using a standard eigenvalue solver.

The PWE method is also undergoing a continuous development that will release the method from the restrictions of the complex frequency-dependent dielectric function and enhance the convergence of the PWE approach.

In the original PWE method, the calculation of EBGs comprising components characterized by frequency-dependent, complex dielectric functions presents a more challenging problem than the modeling of purely dielectric materials, since the former requires the solution of a generalized nonlinear eigenvalue problem. This eigenvalue problem can be solved by using a linearization scheme, which requires the diagonalization of an equivalent, enlarged matrix, and hence an increased computational burden. Kuzmiak et al. have reported an alternative PWE approach that incorporates the frequency-dependent dielectric function and reduces the generalized eigenvalue problem to that of diagonalizing of a set of matrices whose size equals the number of plane waves retained in the expansions for the components of the EM field in the system [25]. The PWE method has been further generalized by the above authors to handle complex, frequency-dependent dielectric functions [32].

The PWE method solves the wave equation in the entire region including the inclusion element (region-II) and the embedding background (region-I). However, the dielectric constant has a step discontinuity at the boundary. So the expansion for the dielectric constant in (2.18) becomes poorly convergent, as large basis sets may be required in this case to accurately describe an EBG structure [25]. To avoid this problem, an embedding method is introduced into the PWE method [33] in which the wave equation is only solved in region-I (between the elements), and the region-II (inside the elements) is replaced by an embedding potential defined on the boundary. The convergency of the procedure improves considerably when the embedding method is employed.

Although the PWE method is quite general and can treat inclusion elements of arbitrary shape in EBGs, it exhibits a greater efficiency when the shape of the element has either a spherical or a cylindrical symmetry. Otherwise, it requires a large number of basis sets in the expansion and the method becomes computationally expansive, since the computational time is proportional to the cube of the number of plane waves. Furthermore, when the EBG has a finite size, it is more difficult to expand the parameters in the PWE method.

An interest topic in the context of the EBG study is the behavior of impurity modes associated with the introduction of defects into the EBG structure. While this problem can be tackled by the PWE approach using the supercell method, in which a single defect is placed within each supercell of an artificially periodic system, the calculations require extensive computer time and memory. The numerical method presented below is useful for efficiently calculating the transmission/reflection coefficients of an EBG slab with a finite thickness.

2.3.3 The Transfer-Matrix Method

Pendry and MacKinnon [27] introduced a complementary technique, namely, the transfer-matrix method (TMM), of studying EBG structures in 1992. In the TMM, the total volume of the system is subdivided into small cells and the fields in each cell are coupled to those in the neighboring cells. Then the transfer matrix can be defined by relating the incident fields on one side of the EBG structure to the outgoing fields on the other side.

Writing Maxwell's equations

$$\nabla \times \vec{E} = -\frac{\partial \vec{B}}{\partial t} \qquad (2.20)$$

$$\nabla \times \vec{H} = \frac{\partial \vec{D}}{\partial t} \qquad (2.21)$$

2.3 An Overview of Numerical Methods for Modeling EBG Structures

in $(\vec{k}, \omega)$ space yields:

$$\vec{k} \times \vec{E} = \omega \vec{B} \qquad (2.22)$$

$$\vec{k} \times \vec{H} = -\omega \vec{D} \qquad (2.23)$$

Expressing $\vec{D}$ and $\vec{B}$ in the usual manner, we have

$$\vec{D} = \varepsilon \vec{E} = \varepsilon_0 \varepsilon_r \vec{E} \qquad \vec{B} = \mu \vec{H} = \mu_0 \mu_r \vec{H} \qquad (2.24)$$

Inserting these into (2.22) and (2.23), we can rewrite them as:

$$\begin{bmatrix} \hat{x} & \hat{y} & \hat{z} \\ k_x & k_y & k_z \\ E_x & E_y & E_z \end{bmatrix} = \omega \mu \begin{bmatrix} H_x \hat{x} \\ H_y \hat{y} \\ H_z \hat{z} \end{bmatrix} \qquad (2.25)$$

$$\begin{bmatrix} \hat{x} & \hat{y} & \hat{z} \\ k_x & k_y & k_z \\ H_x & H_y & H_z \end{bmatrix} = -\omega \varepsilon \begin{bmatrix} E_x \hat{x} \\ E_y \hat{y} \\ E_z \hat{z} \end{bmatrix} \qquad (2.26)$$

From (2.25) and (2.26), we can have:

$$\frac{1}{\omega \mu}(k_x E_y - k_y E_x) = H_z \qquad (2.27)$$

$$k_y H_z - k_z H_y = -\omega \varepsilon E_x \qquad (2.28)$$

respectively. Substituting H_z in (2.28), using the expression in (2.27), we get:

$$k_y \left[\frac{1}{\omega \mu}(k_x E_y - k_y E_x) \right] - k_z H_y = -\omega \varepsilon E_x \qquad (2.29)$$

Substituting:

$$H' = \frac{i}{a \omega \varepsilon_0} H \qquad (2.30)$$

we can obtain (2.31).

$$(iak_z)H'_y = \frac{iak_y c^2 \mu_r^{-1}}{a^2 \omega^2}[(iak_x)E_y - (iak_y)E_x] - \varepsilon_r E_x \qquad (2.31)$$

If we use a the simple cubic mesh, we can define the fields by the vectors a, b, c of length a in the $\hat{x}$-, $\hat{y}$-, and $\hat{z}$-directions, respectively. Transforming (2.31) back into real space yields (2.32) below from which the z components of the vectors have been eliminated:

$$H_y'(r+c) = -\varepsilon_r(r+c)E_x(r+c) + H_y'(r)$$
$$-\frac{c^2\mu_r^{-1}(r-b+c)}{a^2\omega^2}[E_y(r+a-b+c) - E_y(r-b+c)$$
$$- E_x(r+c) + E_x(r-b+c)]$$
$$+\frac{c^2\mu_r^{-1}(r+c)}{a^2\omega^2}[E_y(r+a+c) - E_y(r+c)$$
$$- E_x(r+b+c) + E_x(r+c)] \quad (2.32)$$

Similarly, we can derive the other three equations (2.33)–(2.35):

$$H_x'(r+c) = -\varepsilon_r(r+c)E_y(r+c) + H_x'(r)$$
$$-\frac{c^2\mu_r^{-1}(r-a+c)}{a^2\omega^2}[E_y(r+c) - E_y(r-a+c)$$
$$- E_x(r-a+b+c) + E_x(r-a+c)]$$
$$+\frac{c^2\mu_r^{-1}(r+c)}{a^2\omega^2}[E_y(r+a+c) - E_y(r+!c) - E_x(r+b+c) + E_x(r+c)]$$
$$(2.33)$$

$$E_x(r+c) = +\frac{a^2\omega^2}{c^2}\mu_r(r)H_y'(r) + E_x(r)$$
$$+ \varepsilon_r^{-1}(r)[H_y'(r-a) - H_y'(r) - H_x'(r-b) + H_x'(r)]$$
$$- \varepsilon_r^{-1}(r)[H_y'(r) - H_y'(r+a) - H_x'(r+a-b) + H_x'(r+a)] \quad (2.34)$$

$$E_y(r+c) = -\frac{a^2\omega^2}{c^2}\mu_r(r)H_x'(r) + E_y(r)$$
$$+ \varepsilon_r^{-1}(r)[H_y'(r-a) - H_y'(r) - H_x'(r-b) + H_x'(r)]$$
$$- \varepsilon_r^{-1}(r+b)[H_y'(r-a+b) - H_y'(r+b) - H_x'(r) + H_x'(r+b)] \quad (2.35)$$

Equations (2.32) and (2.33) express the H-fields on the next plane in terms of the E residing on the same plane, and the H-fields computed on the previous plane. Equations (2.34) and (2.35) express the E-fields on the next plane of cells in terms of the E- and H-fields on the previous plane. Thus, given the x and y components of the E- and H-fields on one side of a dielectric structure, the x and y components of the E- and H-fields on the other side can be found by integrating through the structure. For a dielectric structure containing $L \times L \times L$ cells, the dimensions of the transfer matrix are $4L^2$.

The TMM has the advantage that the frequency variations of the transmission and attenuation coefficients can be obtained directly from the calculations. This method can also be used for the situations in which the PWE method fails or becomes too time-consuming. In particular, when the permittivity ϵ is frequency dependent, or when ϵ has large imaginary values, Fourier expansion methods are not useful [28]. However, periodic systems with imperfections (defects) can be easily studied by using the TMM method.

The band structure of an infinite periodic system can be calculated using the TMM, but the main advantage of this method is that it enables us to calculate the transmission and reflection coefficients for EM waves incident on a finite-thickness slab of an EBG material as functions of frequencies, provided the material is assumed to be periodic along the directions parallel to the interfaces.

On the other hand, the FDTD method offers greater flexibility when modeling arbitrarily shaped configurations and complex dielectric properties in finite or infinite structures. The computational effort in the FDTD method is proportional to the number of the nodes in the spatial mesh. In the following section, the applications of the FDTD method to the problem of modeling the EBG structures is briefly reviewed. The details of the method and its extensions will be deferred to Chapter 3.

2.3.4 The Finite-Difference Time-Domain (FDTD) Method

The FDTD method is one of the most widely used numerical methods for the solution of electromagnetic problems. It provides us with a simple way to discretize the Maxwell's equations without requiring a complex mathematical formulation, and it does not require any symmetry in the structure being modeled. Furthermore, it computes the solution in the time domain, from which the frequency behavior of the EBGs can be extracted over a wide frequency range. The first FDTD algorithm was introduced by Yee [34] in 1966. Since then, it has witnessed many modifications. Refinements and extensions [35–37] are still evolving even today.

With the computation costs declining steadily, this versatile method is gaining more popularity, and we are witnessing a tremendous amout of FDTD-related research activity. FDTD is finding a wide range of applications in electromagnetics, including photonic structures.

The FDTD approach offers several advantages when used to model EBG structures. It can handle complex geometries associated with EBG structures themselves and their integration with other devices. The FDTD method can deal with a variety of complex materials as well. It can generate the frequency response of an EBG structure over a wide frequency range with a single simulation. By the end of 2007, more than 3,600 papers had been published on the subject of EBG-related research via the FDTD simulation, and three quarters of them were published after 2000. Here we review a small fraction of this research:

- Kesler et al. have reported the antenna design with the use of EBG structures as planar reflectors in 1996 [38]. Field patterns calculated by using the FDTD method were found to be in good agreement with the measurements.
- In 1997, Qian, Radisic, and Itoh reported the investigation of four types of EBG structures as synthesized dielectric materials that possess distinctive stopbands for microstrip lines. Experimental results have been found to agree with the FDTD simulation results [39].

- In 1998, Boroditsky, Coccioli, and Yablonovitch analyzed the dispersion diagrams of a electromagnetic crystal comprised of a perforated dielectric slab and the properties of a microcavity formed by introducing a defect into such a crystal using the FDTD method [40].
- Chutinan and Noda studied a waveguide created by either removing one stripe from a three-dimensional woodpile EBG or filling up or decreasing the sizes of air holes from a two-dimensional EBG slab using the FDTD method in 1999 [41] and 2000 [42].
- Yang and Rahmat-Samii have used the FDTD to analyze the mushroom-like EBGs that find applications as ground planes for low-profile antenna design [43] (2001) [44] (2003) and for lowering the mutual coupling between the elements in an antenna array [45] (2003).
- In 2002 Ozbay et al. presented a study of the localized coupled-cavity modes in two-dimensional dielectric EBG, with the field patterns and the transmission spectra [47].
- Weily and Bird et al. reported the study of a planar resonator antenna based on a woodpile EBG material in 2005 [48] and a linear array of EBG Horn sectoral antennas in 2006 [49].
- In 2007 Kantartzis et al. presented an analysis of DNG metamaterial-based waveguide and antenna devices utilizing the three-dimensional ADI-FDTD method [50].
- Pinto and Obayya reported the study of an EBG cavity using an improved complex-enveloped ADI-FDTD method in 2007 [51].
- There have been a number of attempts to verify the perfect lens concept realized by EBG material [52] using the FDTD method [53–56]. Zhao et al. studied EBGs with material frequency dispersion by means of an auxiliary differential equation (ADE)–based dispersive FDTD methods with averaged permittivity along the material boundaries implemented [57, 58] and with spatial dispersion effects considered [59].

It is worth noting that among the vast applications of FDTD in modeling the EBG-related structures, there are two approaches that are based on the use of nonorthogonal coordinate system:

- The finite difference method developed by Chan et al. [60] (1995) and Pendry et al. [61] (1998) [62] (1999) is often described as an order-N method by its authors. This method has been applied to EBGs with either pure dielectric inclusions [61] or pure metallic inclusions [62]. However, in their present form, they cannot be applied to complex EBGs whose inclusions contain both dielectric and metallic components.
- In 2000, Qiu and He developed an FDTD based on a nonorthogonal coordinate system to study EBGs consisting of a skewed lattice [63, 64]. This method does not rely on any assumptions regarding the dielectric properties of the material to be modeled, and it can tackle complex structures such as those containing a combination of dielectric and metallic components in the EBG cell. However, utilizing a globally uniformed skew grid imposes the staircase approximation when the curved surface is modeled.

2.4 An Overview of EBG Applications

The first attempts of EBG applications in the literature aimed to realize EBG substrates by drilling a periodic pattern of holes in the substrate, or by etching a periodic pattern of circles in the ground plane. However, nowadays, the regularly generated novel ideas expand this rapidly developing scientific area with extremely high rate. An enormous number of designs and a wide range of applications can be found in microwave and radio engineering, optical circuit designs, and optics spectroscopy. The promising effects of EBGs attract researchers from communities other than electromagnetics like acoustics, hydrodynamics, and mechanics [65].

When EBGs are applied to antennas as substrates or high-impedance ground planes [45, 66, 67] or reflector [38, 68–70], their bandgap features are revealed mainly in two ways: the in-phase reflection and the suppression of surface-wave propagation. The in-phase reflection feature leads to low-profile antenna designs [43, 66, 67]. Meanwhile, the feature of surface-wave suppression helps to improve the antenna's performance in ways such as increasing the antenna gain and reducing back radiation [72–75].

There is another well-known characteristic of the EBGs: their ability to support localized electromagnetic modes inside the frequency gap by introducing defects in the periodic lattice; this leads to the development of two important group of applications: the highly-directive antennas [7] and the EBG waveguide. The former group has high directivity due to the limited angular propagation allowed within the EBG material, including EBG resonator/superstrate antenna [14, 48, 76, 77] and EBG cavities [78]. The devices in the second group can efficiently transmit electromagnetic waves, even for 90° bands with a zero radius of curvature [41, 64], including EBG waveguide [79–82], power splitters, directional couplers [83, 84], switches [47], and the EBG filters [85, 86].

There are also applications utilizing the passband of the EBGs, for example, the subwavelength imaging canalization, which is studied extensively by Belov et al. [52, 65, 67, 87, 88] and the references therein.

2.4.1 In-Phase Reflection

A ground plane is used in many antennas, redirecting one-half of the radiation into the opposite direction, improving the antenna gain, and partially shielding objects on the other side. The reflection phase is of special interest when designing the ground plane of the antennas. The reflection phase is defined as the phase of the reflected electric field at the reflecting surface. It is normalized to the phase of the incident electric field at the reflecting surface.

A perfect electric conductor (PEC) has a 180° reflection phase for a normally incident plane wave. That means, in a conventional antenna having a flat metal sheet as a ground plane, if the radiating element is too close to the conductive surface, the 180° out-of-phase reflected waves will tend to cancel the radiation waves, resulting in poor radiation efficiency. This radiation efficiency reduction can also be explained as the image currents generated by a smooth conducting surface tend to cancel the currents in the antenna if the radiating element is too close to the ground plane. This problem is often addressed by including a quarter-wavelength space between the radiating element and the ground plane, so that the reflected wave is in-phase with the radiation wave at the radiating element. However, such a structure then requires a minimum thickness of $\lambda/4$ [3].

The ideal perfect magnetic conductor (PMC) ground plane will have a 0° reflection phase for a normally incident plane wave. However, no natural material has ever been found to realize the magnetic conducting surface [67].

On the other hand, the EBG can be designed to realize a PMC-like surface in a certain frequency band [3]. Ma et al. demonstrated a magnetic surface realized using a two-dimension uniplanar compact electromagnetic bandgap (UC-EBG) structure experimentally and numerically [89]. The UC-EBG has the advantage of ease of fabrication, and most UC-EBGs have been designed by etching a periodic pattern on the ground plane [90]. Except for a stopband over a wide range of

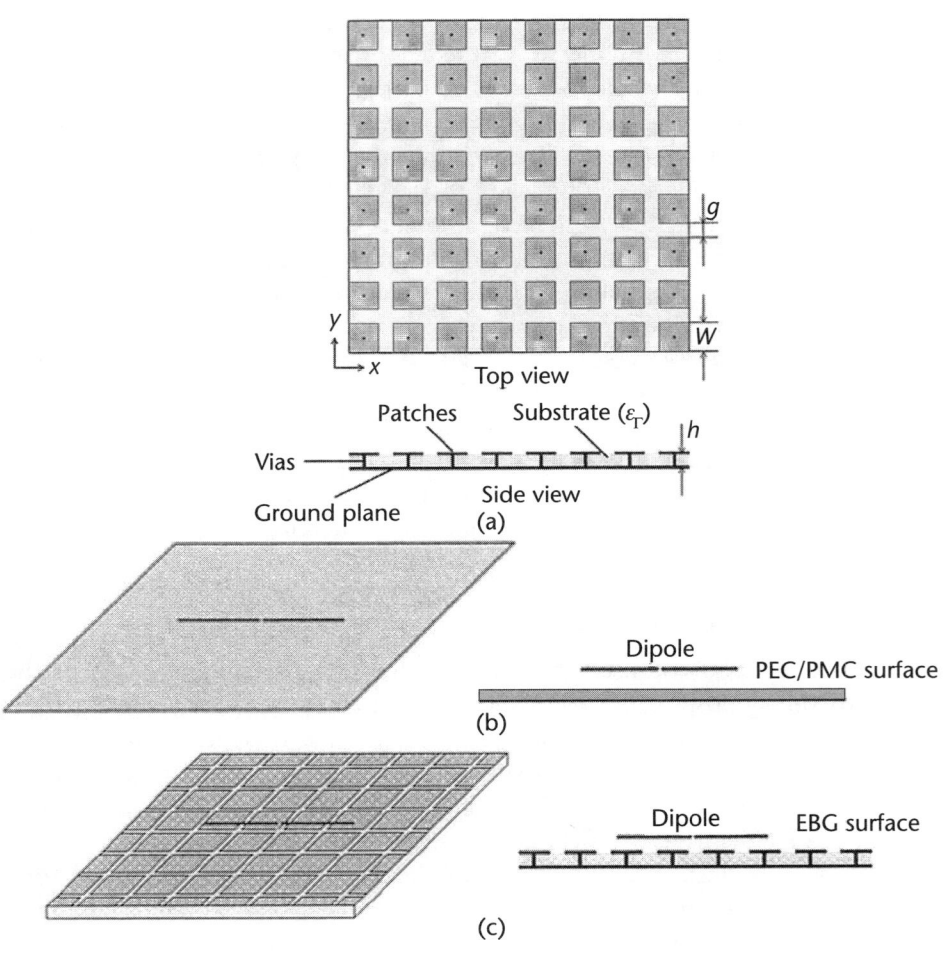

Figure 2.9 (a) Geometry of a mushroom-like EBG structure, which consists of a lattice of metal plates, connected to a solid metal sheet by vertical conducting vias. The EBG has the following parameters: $W = 0.12\lambda_{12GHz}$, $g = 0.02\lambda_{12GHz}$, $h = 0.04\lambda_{12GHz}$, $r = 0.005\lambda_{12GHz}$, $\epsilon_r = 2.20$, where W is the patch width, g is the gap width, h is the substrate thickness, r is the radius of the vias, and ϵ_r is the substrate permittivity. (b) The antenna with the PEC or PMC ground plane. (c) The antenna with the EBG ground plane [44].

2.4 An Overview of EBG Applications

frequency observed, the slow-wave effect is verified when investigating the propagation characteristics of a UC-EBG structure in the passband.

However, EBG surfaces differ from PMC surfaces. The reflection phase of an EBG surface varies continuously from 180° to −180° versus frequency, not only 180° for a PEC surface or 0° for a PMC surface. Fang et al. found through their reflection phase study of a mushroom-type EBG structure that the EBG ground plane requires a reflection phase in the range of 90°± 45° for a low-profile wire antenna to obtain a good return loss.

A finite EBG ground plane with $1\lambda_{12GHz} \times 1\lambda_{12GHz}$ size is used in their analysis. This configuration as shown in Figure 2.9(a) is termed as metallo-dielectric electromagnetic bandgap (MD-EBG) structure and is often referred to as the mushroom-like EBG structure. The height of the dipole over the top surface of the EBG ground plane is $0.02\lambda_{12GHz}$ so the overall height of the dipole antenna from the bottom ground plane of the EBG structure is $0.06\lambda_{12GHz}$. The input impedance is matched to a 50-Ω transmission line.

The return loss of the dipole antenna over the EBG ground plane is compared with those of a dipole antenna over a PEC and PMC ground plane of the same dimension [see Figure 2.9 b, c)]. It is seen from Figure 2.10 that the best return loss of −27 dB is achieved by the dipole antenna over the EBG ground plane. By varying the length of the dipole from $0.26\lambda_{12GHz}$ to $0.60\lambda_{12GHz}$, the return loss

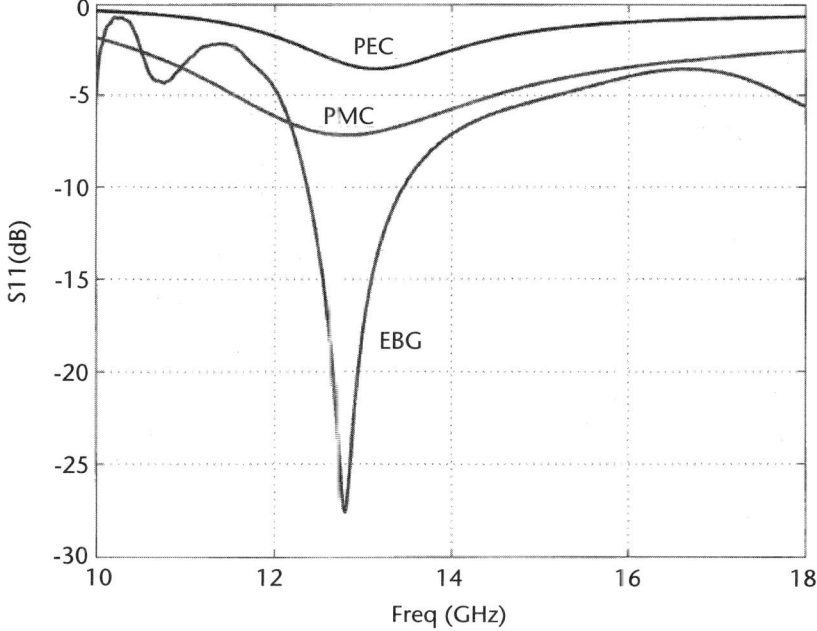

Figure 2.10 FDTD simulated return loss results of the dipole antenna over the EBG, PEC, and PMC ground planes of the same dimension. The dipole length is $0.04\lambda_{12GHz}$ with the PEC and PMC ground plane and the overall antenna height $0.06\lambda_{12GHz}$ [44].

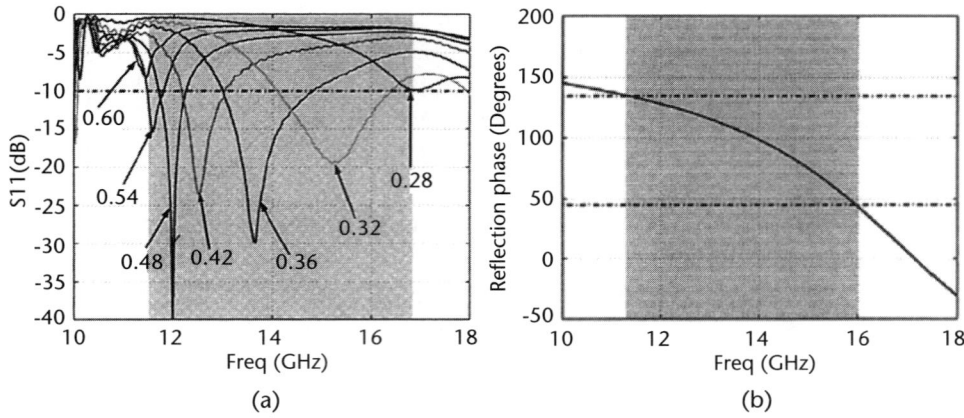

Figure 2.11 The FDTD result of (a) return loss of the dipole with its length varying from $0.26\lambda_{12GHz}$ to $0.60\lambda_{12GHz}$; (b) the reflection phases of the mushroom-like EBG surface versus frequency. The frequency band of the dipole model is 11.5–16.6 GHz according to −10 dB return criteria. The frequency band of the plane wave model is 11.3–16 GHz for $90° \pm 45°$ reflection phase region. [44].

changes; this is plotted in Figure 2.11(a). The frequency band of the dipole model is 11.5–16.6 GHz according to the −10-dB return criteria. It is nearly the same frequency region (11.3–16 GHz) as the reflection phase of the EBG surface varies from $90° + 45°$ to $90° − 45°$ [shown in Figure 2.11(b)].

A dipole antenna over thin grounded high dielectric constant slab can also provide a similar reflection phase curve against frequency and a similar return loss performance. However, with the same dimensions with the EBG ground plane, the dielectric constant of the thin slab needs to be increased to $\epsilon_r = 20$. More to the point, the EBG antenna also demonstrates a higher gain and lower back lobe in terms of return loss due to the suppression of surface waves, which will be discussed in the next section. More detailed discussion of this EBG antenna can be found in [44].

As demonstrated in the above example, with the in-phase reflection, the radiation element of the antennas can be put very close to the EBG structure. In this way many low-profile antenna designs can be realized. A low-profile cavity backed

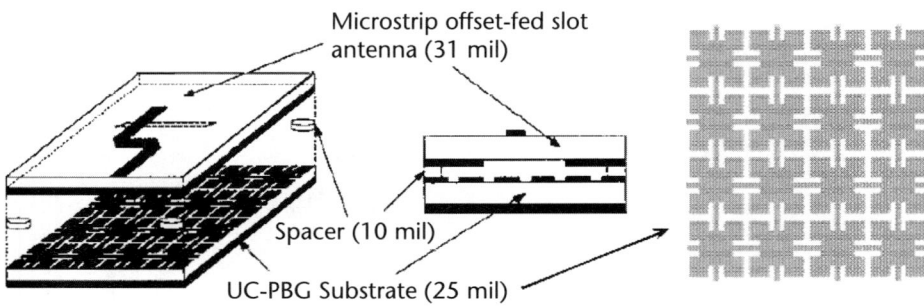

Figure 2.12 Schematic cross-section of the proposed slot antenna loaded with UC-PBG reflector, and the top view of the UC-PBG [66].

2.4 An Overview of EBG Applications

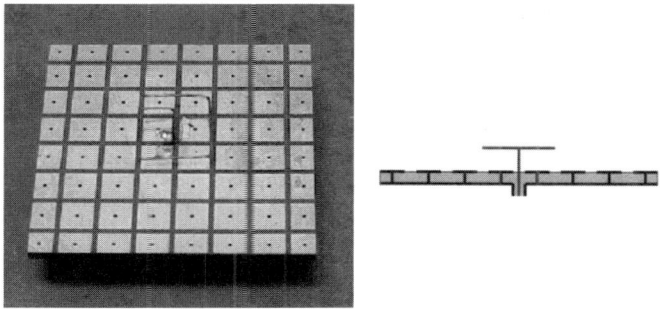

Figure 2.13 Configuration of a square curl antenna over an EBG surface. The size of the ground plane is $1\lambda \times 1\lambda$ where λ is the free-space wavelength at working frequency 1.57 GHz. The low-profile curl antenna is with height of 0.06λ [43].

slot antenna on a uniplanar UC-PBG structure was proposed in [66] (Figure 2.12). The cavity depth of the proposed antenna (35 mil) is 16 times thinner than that of a conventional $\lambda/4$ wavelength cavity slot antenna (559 mil). Similarly, a low-profile circularly polarized patch antenna was proposed in [43] (Figure 2.13). The EBG ground plane size can be as small as $0.82\lambda \times 0.82\lambda$. The overall height of the EBG structure in conjunction with the antenna proposed can be 0.1λ (λ corresponds to the working frequency in free space).

2.4.2 Suppression of Surface Waves

Another property of metals is that they support surface waves. These are propagating electromagnetic waves that are bound to the interface of two dissimilar materials (e.g., metal and free space). At microwave frequencies, they are nothing more than the normal currents that occur on any electric conductor [3]. The surface waves decay exponentially into the surrounding materials and will not couple to external plane waves if the metal surface is smooth and flat [91]. However, the presence of bends, discontinuities, or surface textures will result in the surface waves radiating vertically [3].

On a finite ground plane, surface waves propagate until they reach an edge or corner, where they can radiate into free space. More to the point, if multiple antennas share the same ground plane, surface currents can cause unwanted mutual coupling.

By incorporating a special texture on a conducting surface, it is possible to alter its EM properties [71]. In the circumstance where the period of the surface texture is much smaller than the wavelength, the structure can be described using an effective medium model, and its qualities can be summarized and expressed by the surface impedance [91]. A smooth conducting sheet has low surface impedance, but with a specially designed geometry, a textured surface can have high surface impedance.

The surface impedance of the textured metal surface can be characterized by an equivalent parallel resonant LC circuit. At low frequencies, it is inductive and supports transverse magnetic (TM) waves. At high frequencies it is capacitive, and supports transverse electric (TE) waves. Near the LC resonance frequency, the surface impedance is very high. In this region, waves are not bound to the surface; instead, they radiate readily into the surrounding space.

With the suppression of surface wave, the radiation pattern of the antenna can be improved by the EBG ground plane. Figure 2.14 shows a higher gain and lower back lobe than the dipole antenna over thin grounded high dielectric constant ($\epsilon_r = 20$) slab with the same dimension. Its application has also been found in the design of mobile antennas [46].

2.4.3 EBGs Operating in Defect Modes

Although the EBGs are structures based on periodicity, by introducing defects in the periodic lattice, enormous interesting applications have been generated because of

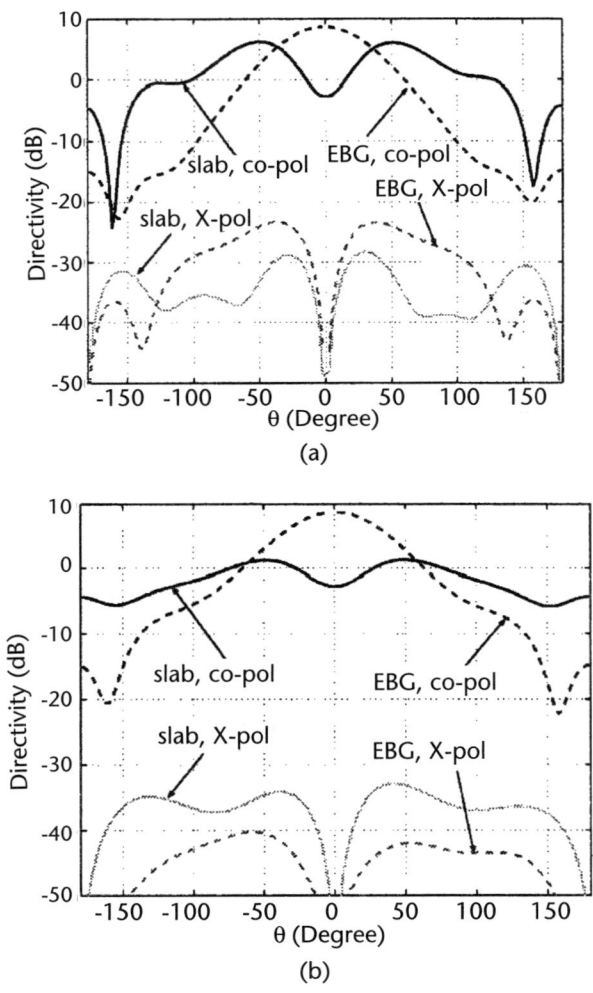

Figure 2.14 Radiation pattern comparison of dipoles near the thin grounded high dielectric constant slab and the EBG surface: (a) E-plane pattern and (b) H-plane pattern. The patterns are calculated at the resonant frequency of 13.6 GHz. Since the high dielectric constant substrate is used in the grounded slab and strong surface waves are excited, the dipole on the slab shows a lower gain and higher back lobe [44].

2.4 An Overview of EBG Applications

the ability of the defect EBG to support localized electromagnetic modes inside the frequency gap. Defects can be realized in many ways, including removing segments of elements from an EBG, changing the shape or EM properties of segments of elements, or replacing a segment of EBG with other materials. In this section, examples of EBG superstrates, EBG waveguides, EBG splitters and couplers, and EBG tunable filters will be presented.

2.4.3.1 EBG Superstrates

As EBGs can be placed under an antenna as a substrate to miniaturize antennas and reduce backward radiation, they can also be applied above the antenna to enhance antenna performance on the directivity of the antenna [14–16, 76, 94–96].

Figure 2.15 shows an example of a high directive EBG resonator antenna utilizing a frequency-selective surface (FSS) superstrate designed by Lee et al. [95]. FSSs are chosen in Lee's design because they are easier to fabricate using the etching processes and they help to achieve a more compact EBG antenna design, especially in terms of the antenna thickness. Figure 2.16 shows the field distribution of the EBG antenna, Figure 2.17 demonstrates a substantial enhancement on the directivity of the antenna by utilizing the EBG superstrate.

Enoch et al. reported in [14] a device that radiates energy in a very narrow angular range, based on a two-dimensional EBG that is made of circular rods of radius $r = 0.6$ mm, with optical index $n_r = 2.9$, lying in vacuum. The rods are arranged on a triangular/hexagonal lattice with period $a = 4$ mm (distance between

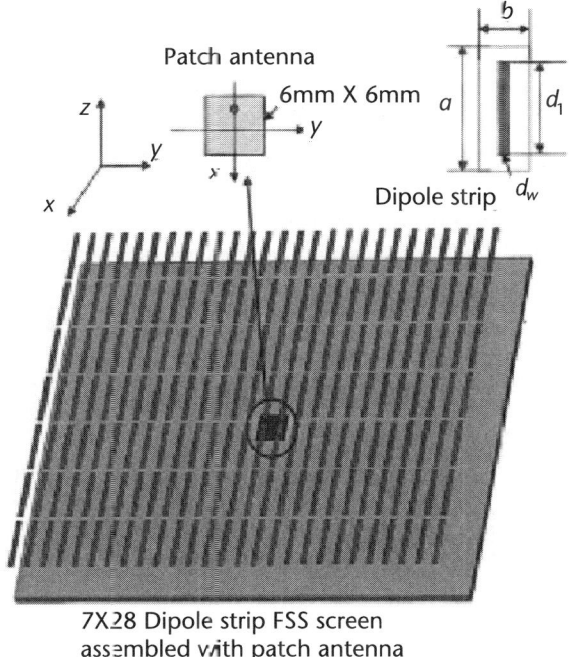

Figure 2.15 Geometry of a patch antenna with a strip dipole FSS superstrate [95].

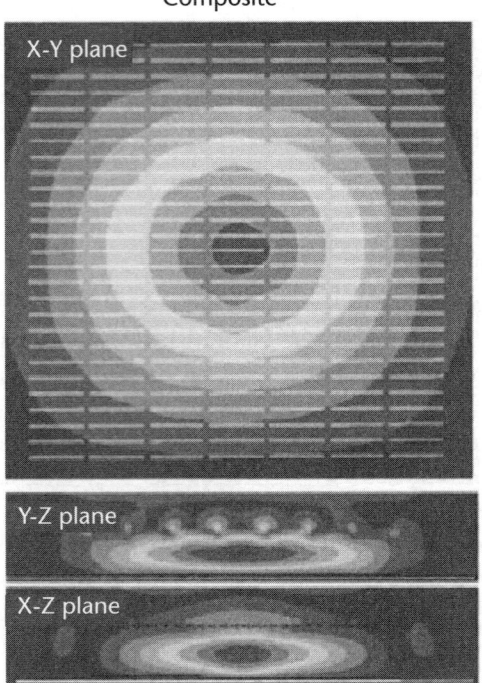

Figure 2.16 Field distribution of the EBG antenna [95].

the centers of the rods). The dispersion diagram of this EBG is presented in Figure 2.18.

When the cell of this original EBG is expanded in the y direction (i.e., the vertical spacing between two grids is enlarged from $\sqrt{3}a/2 \approx 3.46$ mm to 3.9 mm),

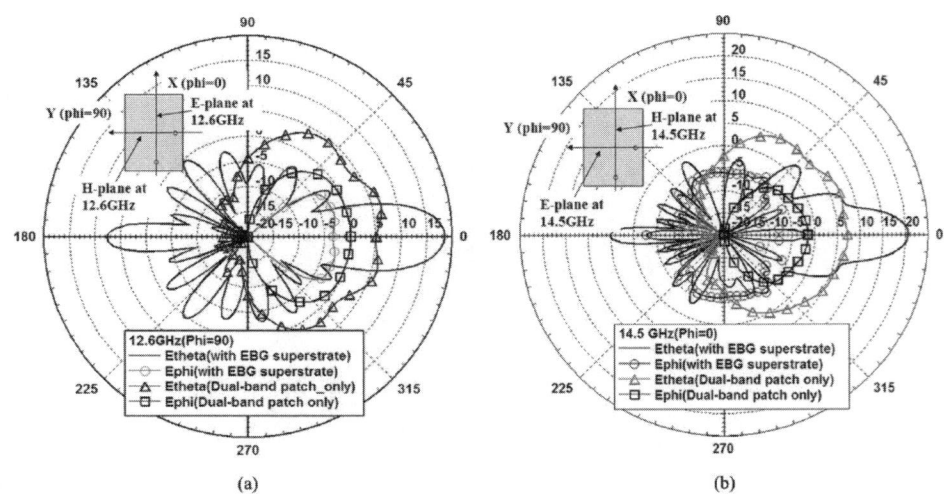

Figure 2.17 Comparison of the radiation patterns of the FSS antenna composite and the patch antenna only. The directivity of the antenna is substantially enhanced [95].

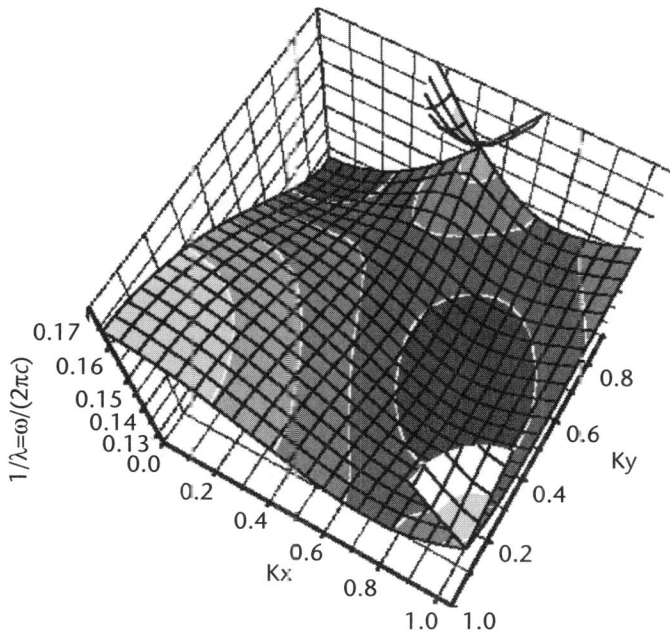

Figure 2.18 Three-dimensional dispersion diagram of the two-dimensional EBG with circular rods lying in vacuum in a triangular lattice [14].

a new EBG is formed with dispersion diagram shown in Figure 2.19. At the wavelength $\lambda = 7.93$ mm corresponding to the horizontal plane at the bottom of Figure 2.19, the constant-frequency dispersion diagram of the expanded EBG reduces to a small curve (see Figure 2.20). As a consequence, any source embedded in a slice of this expanded EBG will radiate only with a small range of wavevector $\vec{k}$.

By placing the original (unexpanded) EBG, which exhibits bandgap at the same wavelength, backwards radiation is eliminated. In this way, a device with field radiated in a narrow angular range using almost any excitation is achieved, as seen in Figure 2.21. The half-power beamwidth is equal to $\pm 4.0°$ (see Figure 2.22), in comparison to the $\pm 4.5°$ achievable by an aperture having the same width illuminated by a field with constant amplitude and phase. By increasing the lateral size of the structure and sacrificing the range of possible working wavelengths, an even narrower radiation pattern can be obtained.

2.4.3.2 EBG Waveguides, Splitters, and Couplers

The efficient guiding and bending of light by integrated photonic devices are important to designing optical circuits for technological and optical computing applications. Conventional dielectric or metallic waveguides have large scattering losses when sharp bends are introduced. However, EBG studies enable an efficient way of guiding waves even for sharp bends [41, 80–82].

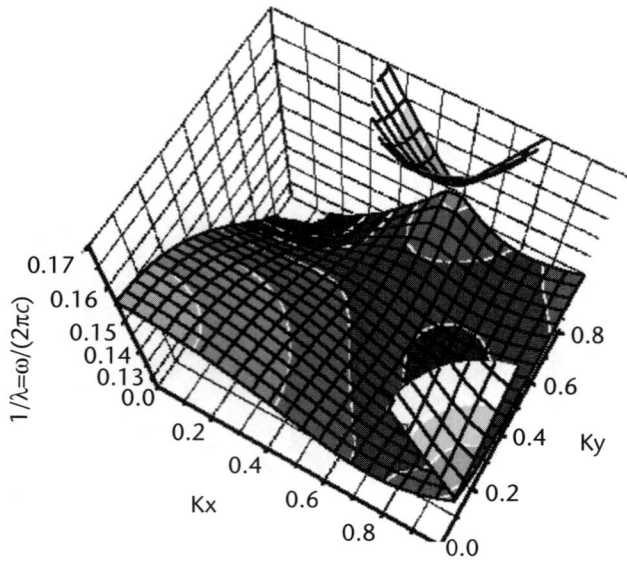

Figure 2.19 The dispersion diagram of the expanded EBG [14].

Ozbay et al. demonstrated a zigzag coupled cavity waveguide (CCW) formed by removing consecutive rods from a two-dimensional EBG with rods loaded in free space in triangular lattice [shown in Figure 2.23(a)]. A defect band is observed

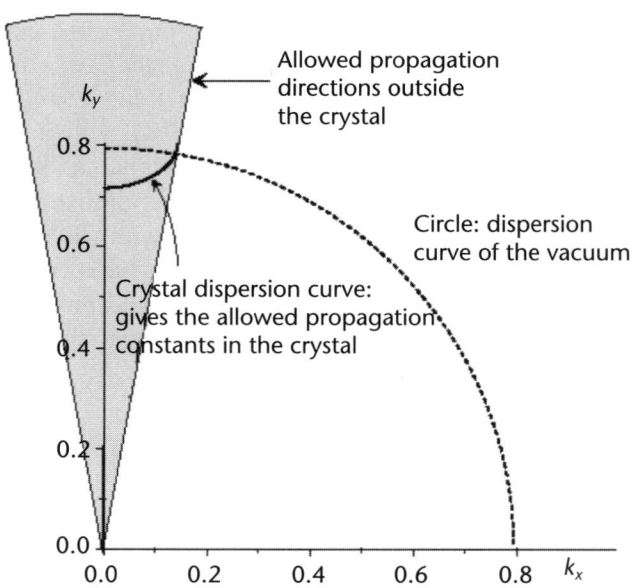

Figure 2.20 Constant-frequency dispersion diagram of the expanded EBG for $\lambda = 7.93$ mm [14].

2.4 An Overview of EBG Applications

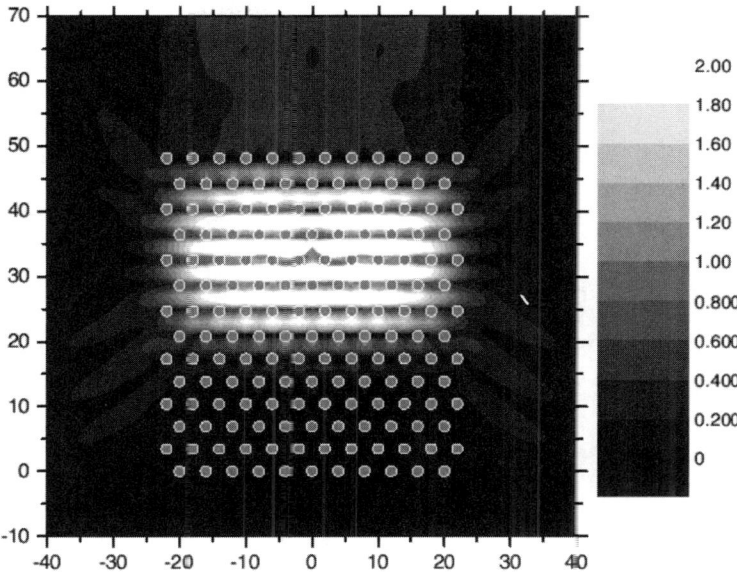

Figure 2.21 Total field modulus radiated by the structure excited by the wire source at $\lambda = 7.93$ mm [14].

between $0.857\omega_0$ to $0.949\omega_0$. Complete transmission is seen for certain frequencies within the defect band. Since the defect band shows sharp band edges compared to the EBG edges, it is suggested in [47] that this property can be used to construct photonic switches by changing the position of the defect band.

A Y-shaped splitter (shown in Figure 2.24) is also presented in [47] in order to demonstrate the splitting of EM power. The splitter consists of one-input CCW and two-output CCWs. The input and output waveguide ports contain five and six coupled cavities, respectively. As can be seen in Figure 2.24(a), the propagating mode inside the input CCW splits equally into two output CCW ports for all frequencies within the defect band. The electric field distribution inside the splitter is computed for frequency $\omega = 0.916\omega_0$ and is shown in Figure 2.24(b).

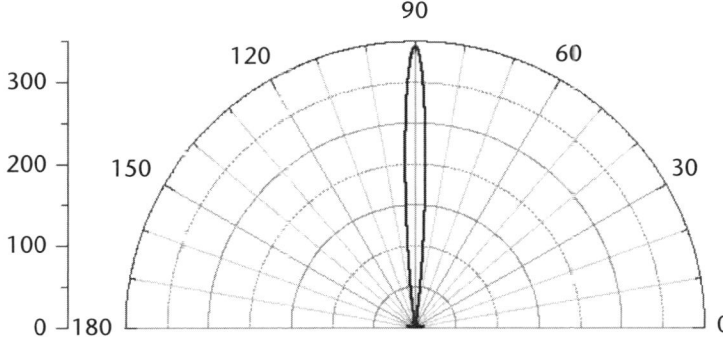

Figure 2.22 Polar emission diagram for the structure excited by the wire source at $\lambda = 7.93$ mm [14].

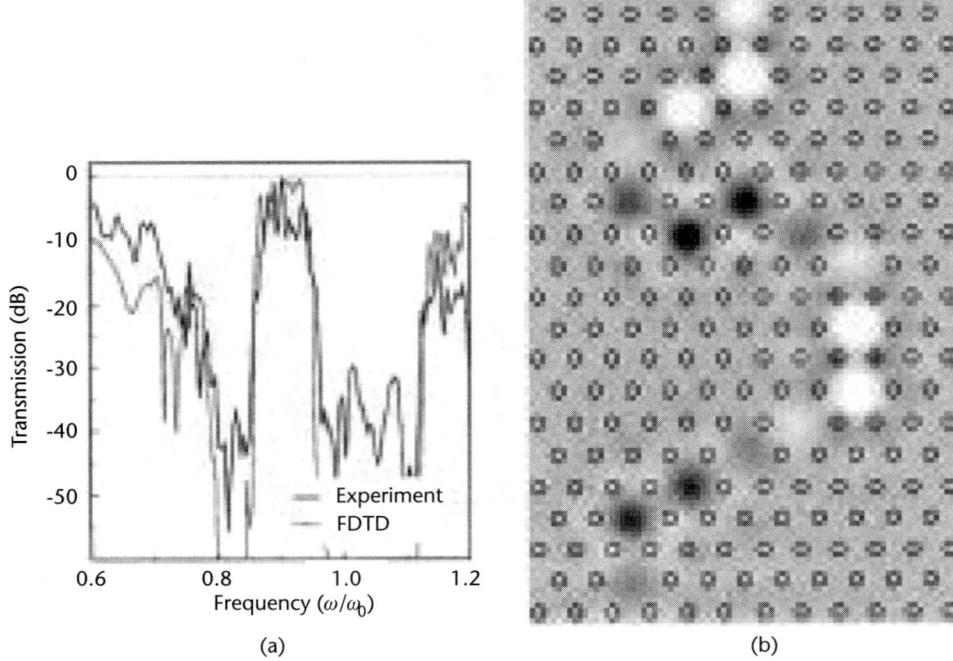

Figure 2.23 (a) Measured (solid line) and calculated (dotted line) transmission spectra of a zigzag CCW waveguide that contains 16 cavities. (b) Calculated field distribution of the zigzag CCW waveguide. (*From:* [47]. © 2002 IEEE. Reprinted with permission.)

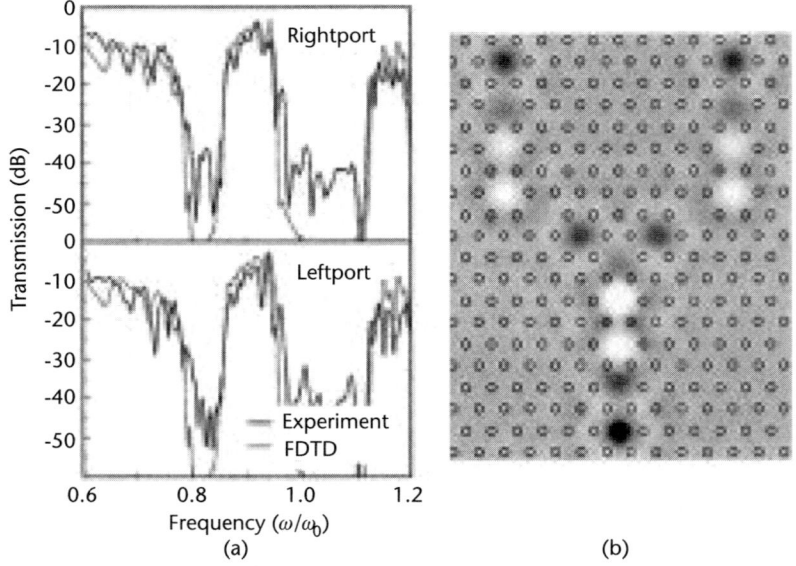

Figure 2.24 (a) Measured (solid line) and calculated (dotted line) transmission spectra of a Y-shaped coupled-cavity based splitter. (b) Calculated power distribution inside the input and output waveguide channels of the splitter for frequency $\omega = 0.916\omega_0$. (*From:* [47]. © 2002 IEEE. Reprinted with permission.)

2.4 An Overview of EBG Applications

If a single rod is placed to the left side of the junction of the Y-splitter, as is shown in Figure 2.25(b), the splitter structure becomes a photonic switch. Because the symmetry of the Y-shaped structure is broken, the power at each output waveguide port is drastically changed. In this way, the amount of power flow into the output ports can be regulated.

EBG waveguides can also be realized based on the three-dimensional layer-by-layer dielectric EBG structures [41, 82]. Figure 2.26(a) shows Bayindir et al.'s experimental setup. The woodpile EBG constructing the CCWs consists of square-shaped alumina rods having a refractive index 3.1 at the microwave frequencies. The dimension of each rod is 0.32 cm × 0.32 cm × 15.25 cm. The offset between the rods is 1.12 cm. The bandgap of the EBG extends from 10.6 to 12.8 GHz. The defect is formed by removing a single rod from a unit cell of the EBG crystal. The electric-field vector of the incident EM field was parallel to the rods of the defect lines. When the defect exists in adjacent unit cells, a very high transmission of the EM wave was observed within a frequency range inside the bandgap of the otherwise perfect EBG, which is hereinafter referred to as the waveguide band. Nearly a complete transmission was observed within the waveguide band for a straight EBG waveguide [Figure 2.26(b)] and greater than 90% transmission for an EBG waveguide with a 40° bend [Figure 2.26(c)].

2.4.3.3 EBG Tunable Filters

A variety of filtering devices, based on the two-dimensional dielectric and metallic EBGs containing nematic liquid crystal materials as defect elements or layers, has been explored by Kosmidou et al. [86]. The defects states originating from the

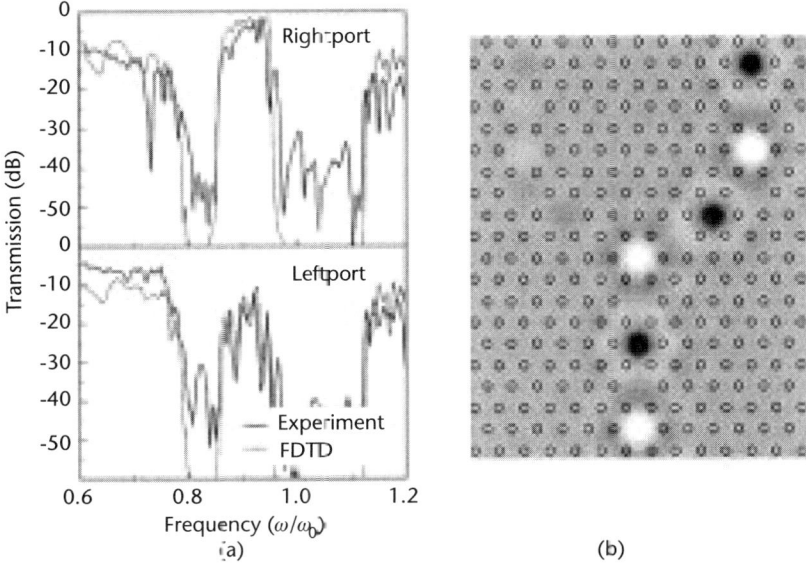

Figure 2.25 (a) Measured (solid line) and calculated (dotted line) transmission spectra of a coupled-cavity switching structure. (b) Calculated field pattern clearly indicates that most of the power is coupled to the right port. (*From:* [47]. © 2002 IEEE. Reprinted with permission.)

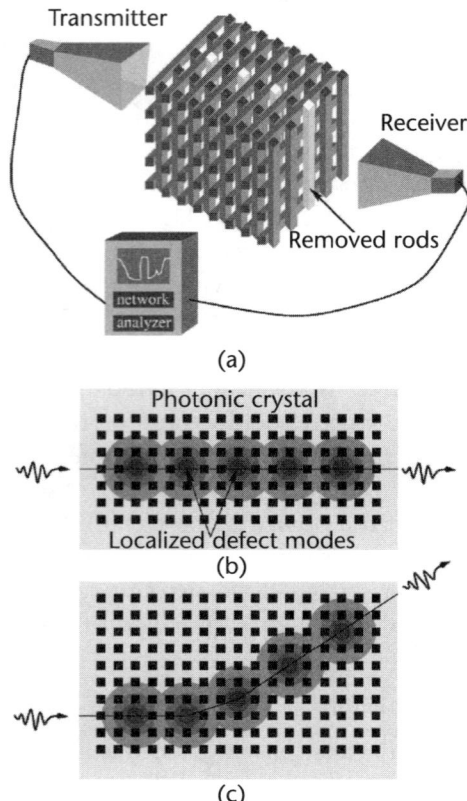

Figure 2.26 (a) Experimental setup for measuring the transmission-amplitude and transmission-phase spectra of the coupled cavity waveguides (CCWs) in 3-D electromagnetic crystals. (b) A mechanism to guide light through localized defect modes in a woodpile EBGs. (c) Bending of the EM waves around sharp corners [82].

liquid crystal impurities are reviewed by FDTD simulation to be tunable by the application of a local static electric field. Narrow mode linewidths, almost 0.2 nm, and tuning ranges in the order of tens of nanometers, covering in some cases both the C- and L-bands, can be achieved with low operating voltages (0–4V). So the proposed devices are rendered suitable to operate as a spectral filter in modern optical communication systems.

Depicted in Figure 2.27 is the defect EBG filter with air voids filled with liquid crystal as defect elements. It is based on a electromagnetic crystal consisting of a triangular lattice of infinitely long air cylinders embedded in silicon. The radius of the circular cross-section of the air rods is set to $0.3a$, where a is the lattice constant. The relative permittivity of Si is considered to be $\epsilon_r = 11.4$. The defect air voids at the center of the device are filled with E7, a typical nematic liquid crystal material characterized by ordinary and extraordinary refractive indexes equal to 1.49 and 1.66, respectively. The defect area is surrounded by five periods of the EBG cells, whereas the lateral width (in the $\hat{y}$-direction) of the device is presumed to be infinite. The optical axis direction inside the liquid crystal, lying

2.4 An Overview of EBG Applications

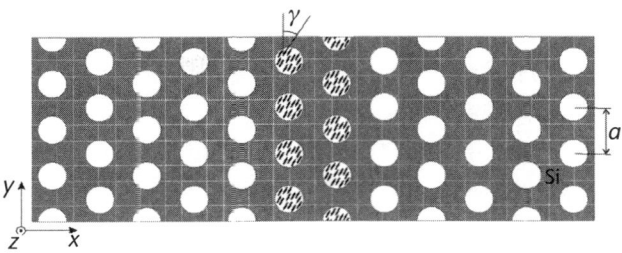

Figure 2.27 Dielectric EBG structure having air voids filled with liquid crystal as defect elements [86].

in the x-y plane, is defined by the tilt angle γ, which can be altered by means of applying a local static electric field.

Figures 2.28 and 2.29 show how defect modes can be tuned by changing the number of defect void rows and by tuning the tilt angle γ in the defect cylinder rows.

In Figure 2.30, it can be seen that the discrete defect cylinders are replaced by an E7 layer interposed between two blocks of the EBGs. Between the EBG and the E7 layer are two thin films of indium tin oxide (ITO) with the related refractive index equal to 1.9 and thickness 0.2a, on which electrods are attached to provide a local static voltage across the liquid crystal slab. Figure 2.31 shows how the tilt

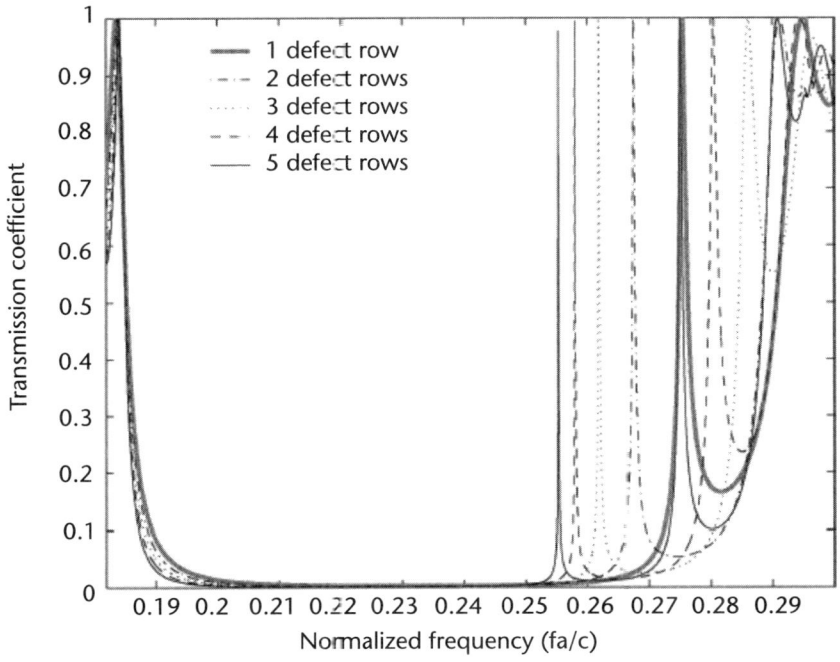

Figure 2.28 Transmission coefficient for various numbers of defect cylinder rows when $\gamma = 45°$ [86].

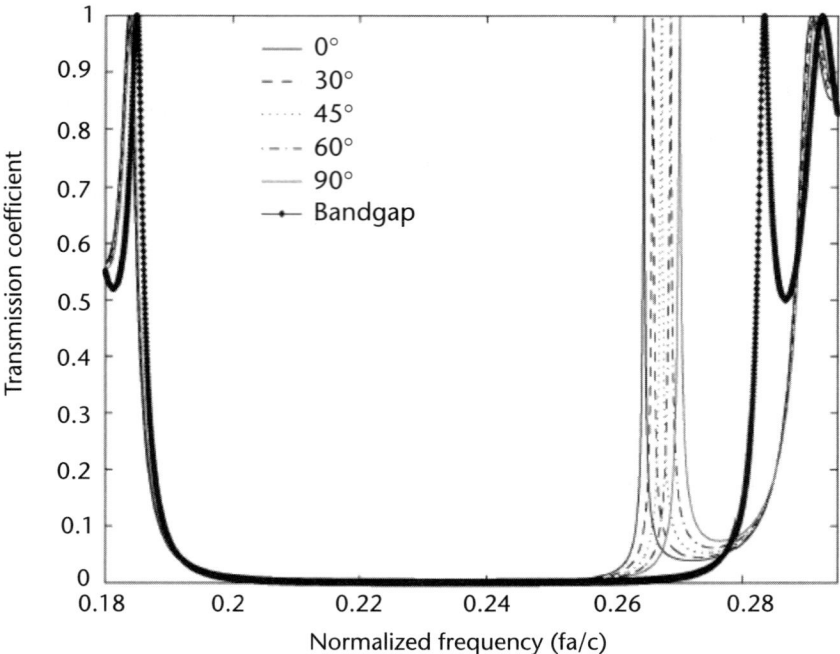

Figure 2.29 Transmission coefficient for various values of the tilt angle γ in the case of two defect cylinder rows [86].

angle γ in the E7 slab varies according to the spatial position and to the different static voltage applied. By altering the optical axis orientation inside of the defect slab, the thickness of E7 layer L_c, or the distance L_s, the effective permittivity will change, which in turn causes the change of the position of the defect modes. Figure 2.32 demonstrates how the position of the defect modes can be tuned by tuning the applied voltage, assuming $L_c = 3a$ and $L_s = 0.567a$.

Another two-dimensional EBG tunable filter is based on EBGs with infinitely long metallic rods loaded in a background material with low refractive index ($n_r = 1.32$) in a square lattice. The radius of each cylinder cross-section is set to $0.2a$ as illustrated in Figure 2.33(a), each with four periods of the EBG cells in

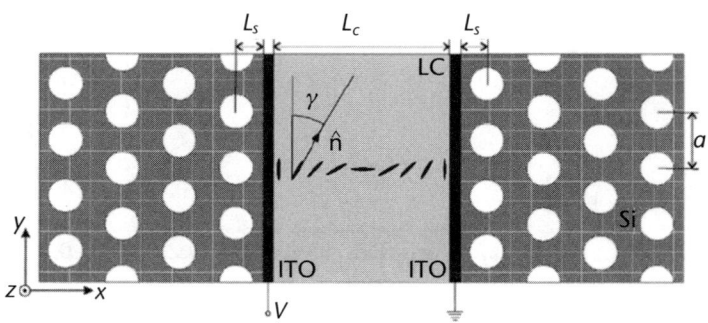

Figure 2.30 Dielectric EBG structure with a liquid crystal defect layer [86].

2.4 An Overview of EBG Applications

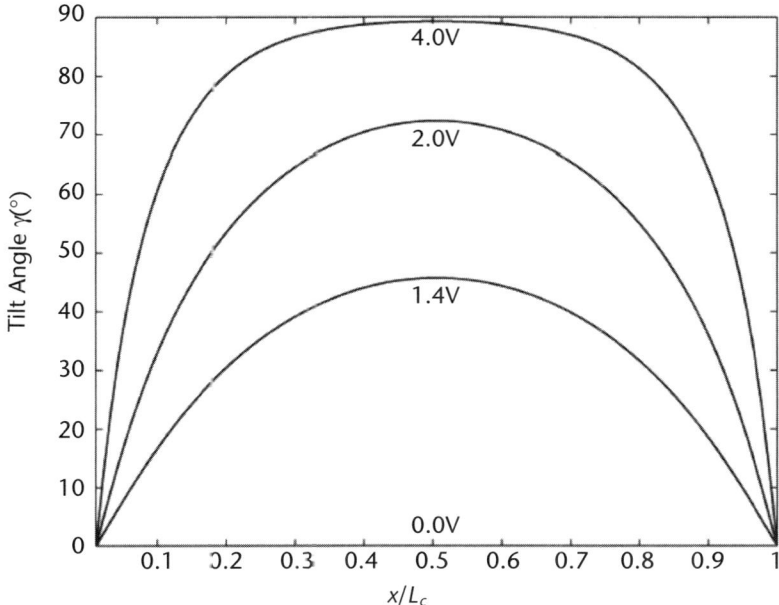

Figure 2.31 Director orientation profile across the liquid crystal defect layer for different values of the applied voltage [86].

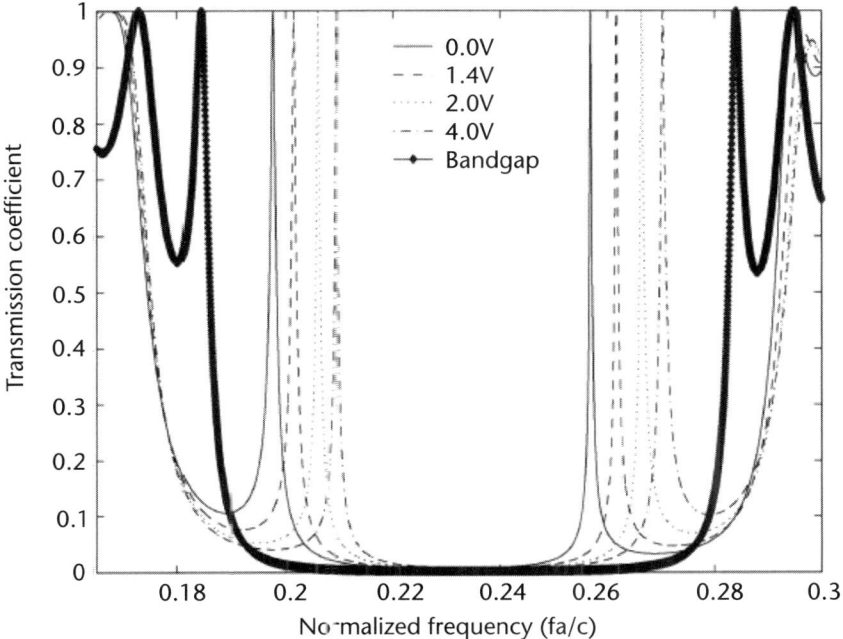

Figure 2.32 Transmission coefficient versus normalized frequency for various values of the applied voltage when $L_c = 3a$, $L_s = 0.567a$ [86].

the $\hat{x}$-direction, and with infinite length assumed in the $\hat{y}$-direction. Figure 2.33(b) shows the tuning effect of the defect modes by changing the applied voltage, assuming $L_c = 4a$ and $L_s = a$ [86].

2.4.4 Subwavelength Imaging from the Passband of the EBGs

Resolution of common imaging systems is restricted by the so-called diffraction limit. This effect limits the minimum diameter d of spot of light formed at the focus of a lens, given as:

$$d = 1.22\lambda \frac{f}{a} \tag{2.36}$$

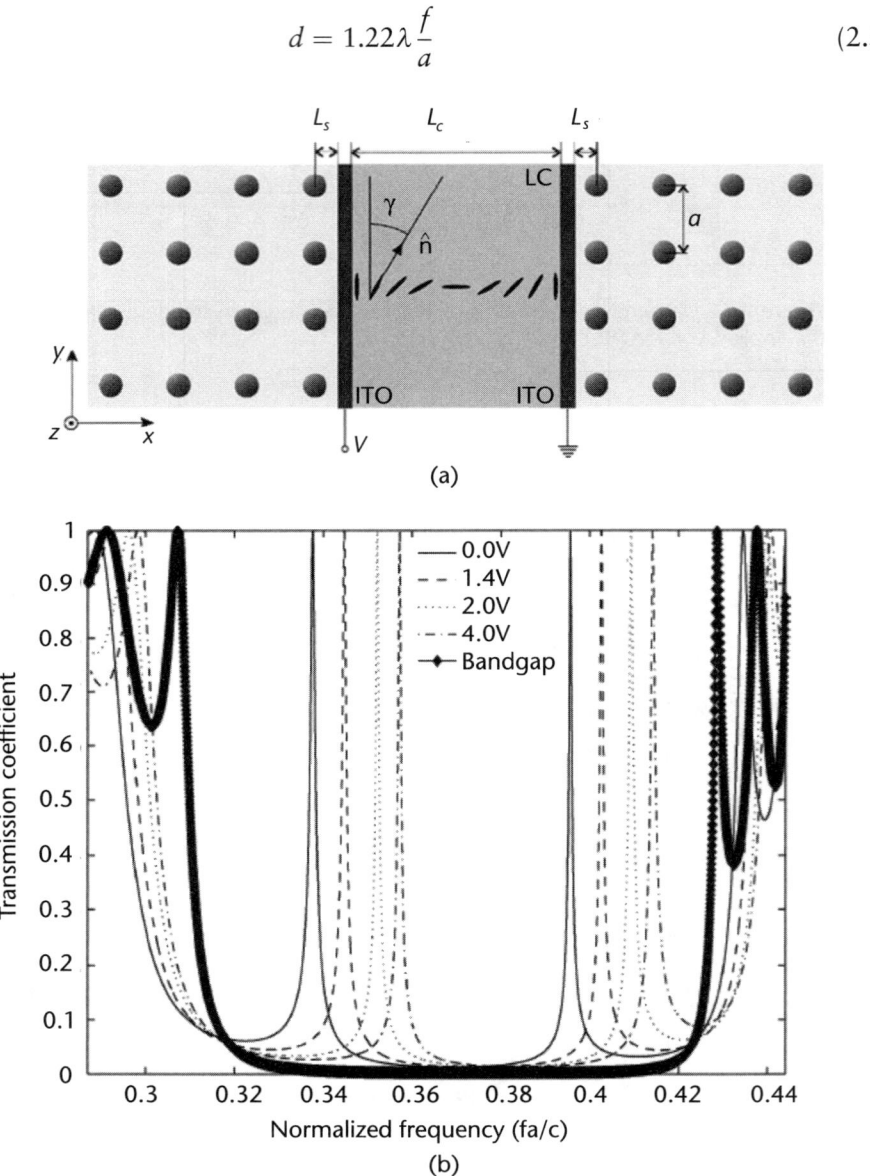

Figure 2.33 (a) Metallic EBG structure with a liquid crystal defect layer. (b) Transmission coefficient versus normalized frequency for various values of the applied voltage when $L_c = 4a$, $L_s = a$ [86].

2.4 An Overview of EBG Applications

where λ is the wavelength of the light, f is the focal length of the lens, and a is the diameter of the beam of light, or (if the beam is filling the lens) the diameter of the lens. As a result, even if one could fabricate an imperfection-free optical system, there is still a limit to the resolution of an image created by the conventional optical lens. In order to overcome the diffraction limit, an artificial material (EBG) with electromagnetic properties that is dramatically different from the materials occurring in nature was proposed as a candidate for perfect lens and theoretical possibility of subwavelength imaging was demonstrated by Pendry in his seminal work [52]. Belov et al. experimentally demonstrated a possibility to channel the near-field distribution of a line source with subwavelength details through an EBG crystal. A channeled intensity maximum having a radius of $\lambda/10$ has been achieved by the use of an electrically dense lattice of capacitively loaded wires [65, 88].

Figure 2.34 shows the experimental implementations of the EBGs composed of capacitively loaded wire medium (CLWM) and the EBG lens. Figure 2.35 shows isofrequency contours for the frequency region $ka = 0.43 \sim 0.47$. The isofrequency

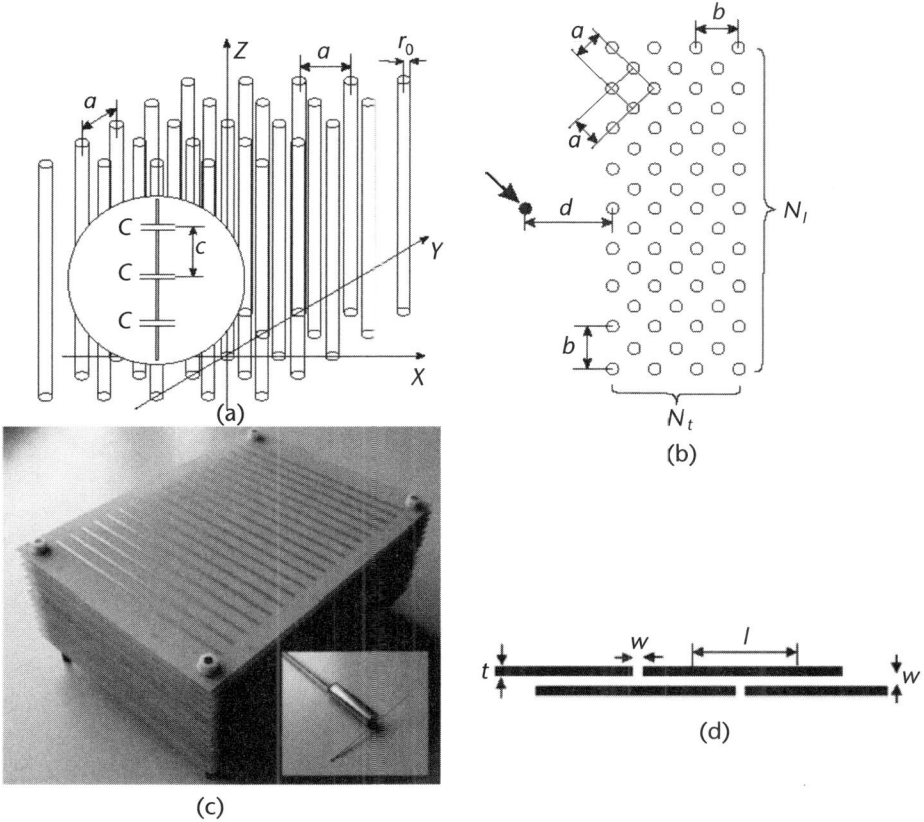

Figure 2.34 (a) A schematic illustration of the EBG structure composed of CLWM. (b) A schematic illustration of the lens formed by the EBG (CLWM). (c) The implemented CLWM EBG sample and the probe used in the measurements. (d) A schematic illustration of the loaded wires (a piece of it) [88].

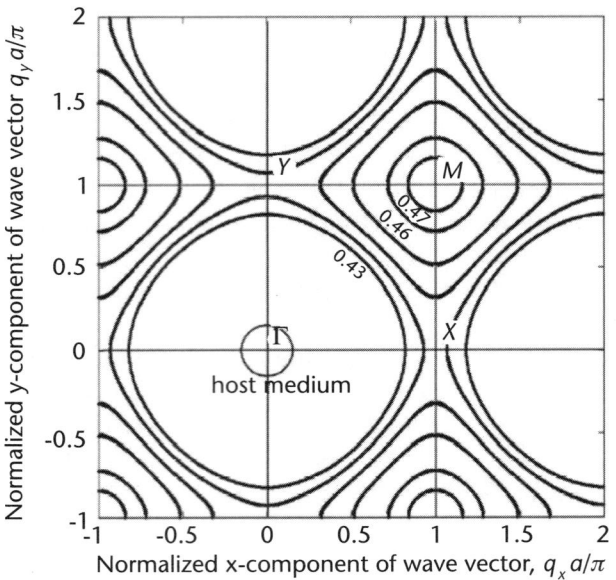

Figure 2.35 Isofrequency contours for the CLWM. The numbers correspond to values of normalized frequency ka [88].

contour of the host material for $ka = 0.46$ is shown as the small circle around Γ point. The part of the isofrequency contour for the EBG corresponding to $ka = 0.46$, and located within the first Brillouin Zone, is practically flat. This part is perpendicular to the diagonal of the first Brillouin Zone. Thus, in order to achieve channeling regime, the interfaces of the slab are oriented orthogonally to the (11)-direction as shown in Figure 2.34(b).

Figure 2.36 depicts the simulated amplitude and intensity distribution of a line source working on $ka = 0.46$ excited near the interface of the CLWM slab. A clear channel through the slab can be observed together with a bright spot having a radius of $\lambda/6$ (determined from the intensity distribution at level $max(intensity)/2$)

Figure 2.36 Simulated distribution of electric field (a) amplitude and (b) intensity for the sub-wavelength lens formed by the CLWM operating in the canalization regime. It shows that when a point source is presented near one interface of the CLWM slab, an image near the back interface is observed with the radius of the focal spot approximately $\lambda/6$ [88].

behind the slab. An experimental verification of subwavelength imaging using CLWM slabs [shown in Figure 2.34(c)] demonstrated an impressive resolution of $\lambda/10$. This lens can be designed thick, since the required tunnel thickness is not related with the distance to the source. The application of this CLWM lens being used in the near-field microscopy in the optical range is suggested in [65], when the needle of a microscope used as a probe can be located physically far from the tested source.

2.5 Summary

The periodic structures are presently one of the most rapidly advancing sectors in the electromagnetic arena. This chapter reviews the basic theory of EBG structures, the numerical methods that are popular in modeling EBGs, and examples from the vase applications of the EBGs.

References

[1] E. Yablonovitch, "Inhibited Spontaneous Emission in Solid-State Physics and Electronics," *Phys. Rev. Lett.*, Vol. 58, 2059–2062, 1987.

[2] J. D. Joannopoulos, R. D. Meac and J. N. Winn, *Photonic Crystals: Molding the Flow of Light*, Princeton University Press, Princeton, NJ, 1995.

[3] D. Sievenpiper, L. Zhang, R. F. J. Broas, N. G. Alexopolus, and E. Yablonovitch, "High-impedance electromagnetic surfaces with a forbidden frequency band," *IEEE Trans. Microwave Theory Tech.*, Vol. 47, 2059–2074, 1999.

[4] S. G. Johnson and J. D. Joannopoulos, "Introduction to Photonic Crystals: Bloch's Theorem, Band Diagrams and Gaps (But No Defects)," tutorial, ab-initio.mit.edu/photons/tutorial/photonic-intro.pdf, 2003.

[5] S. Germani, L. Minelli, M. Bozzi, and L. Perregrini, and P. de Maagt, "Modeling of MD-EBG by the Mom/BI-REM Method," *Proc. 27th ESA Antenna Workshop on Innovative Periodic Antennas*, Proceedings ESA WPP-222, 2004

[6] W. Axmann, P. Kuchment, and Leonid Kunyansky, "Asymptotic Methods for Thin High-Contrast Two-Dimensional PBG Materials," *J. Lightwave Tech.* Vol. 17, No. 11, 1996–2007, 1999.

[7] M. Thevenot, C. Cheype, A. Reineix, and B. Jecko, "Directive photonic-bandgap antennas," *IEEE Transactions on Microwave Theory and Techniques*, Vol. 47, Issue 11, 2115–2122, 1999.

[8] G. Kuriz and A. Z. Genack, "Suppression of Molecular Interactions in Periodic Dielectric Structures," *Phys. Rev Lett.*, Vol. 61, 2269–2271, 1988.

[9] S. John and J. Wang, "Quantum electrodynamics near a photonic band gap: Photon bound states and dressed atoms," *Phys. Rev. Lett.*, Vol. 64, 2418–2421, 1990.

[10] R. D. Meade, K. D. Brommer, A. M. Rappe, and J. D. Joannopoulos, "Photonic bound states in periodic dielectric materials," *Phys. Rev. B*, Vol. 44, Issue 24, 13772–13774, 1991.

[11] C.-T. Hwang and R.-B. Wu, "Treating Late-Time Instability of Hybrid Finite-Element/Finite-Difference Time Domain Method," *IEEE Trans. Antennas Propagat.*, Vol. 47, No. 2, 227–232, 1999.

[12] C. Kittel, *Introduction to Solid State Physics*, 6th ed., New York: Wiley, 1986.

[13] J. Mathews and R. Walker, *Mathematical Methods of Physics*, W.A. Benjamin, New York, 1964.

[14] S. Enoch, G. Tayeb, and D. Maystre, "Dispersion diagrams of Bloch modes applied to the design of directive sources," *Progress in Electromagnetic Research, Special Issue on Electromagnetic Applications of PBG Materials and Structures,* PIER, Vol. 41, 61–81, 2003.

[15] Y. Hao, A. Alomainy, and C. G. Parini, "Antenna Beam Shaping from Offset Defects in UC-EBG Cavities," *Microwave and Optical Technology Letters,* 43 (2): 108–112 October 20, 2004.

[16] Y. Lee, X. Lu, Y. Hao, S. Yang, R. Ubic, J. R. G. Evans, and C. G. Parini, "A Directive Millimetrewave Antenna Based on Freeformed Woodpile EBG Structure," *Electronics Letters,* 43 (4): 195–196 February 15, 2007.

[17] N. A. Nicorovici, R. C. McPhedran, and L. C. Botten, "Photonic band gaps for arrays of perfectly conducting cylinders," *Phys. Rev. E,* Vol. 52, No. 1, 1135–1145, 1995.

[18] N. A. Nicorovici and R. C. McPhedran, "Lattice sums for off-axis electromagnetic scattering by grating," *Phys. Rev. E,* Vol. 50, No. 4, 3143–3160, 1994.

[19] S. K. Chin, N. A. Nicorovici, and R. C. McPhedran, "Green's function and lattice sums for electromagnetic scattering by a square array of cylinders," *Phys. Rev. E,* Vol. 49, No. 5, 4590–4606, 1994.

[20] N. A. Nicorovici, and R. C. McPhedran, "Propagation of electromagnetic waves in periodic lattices of spheres: Green's function and lattice sums," *Phys. Rev. E,* Vol. 51, No. 1, 690–702, 1995.

[21] W. Kohn and N. Rostoker, "Solution of the Schrodinger Equation in Periodic Lattices with an Application to Metallic Lithium," *Phys. Rev.,* Vol. 94, 1111–1120, 1954.

[22] K. M. Leung and Y. Qiu, "Multiple-scattering calculation of the two-dimensional photonic band structure," *Phys. Rev. B,* Vol. 48, 7767–7771, 1993.

[23] K. M. Leung and Y. F. Liu, "Photon band structures: The plane-wave method," *Phys. Rev. B,* Vol. 41, 10188–10190, 1990.

[24] Z. Zhang and S. Satpathy, "Electromagnetic wave propagation in periodic structures: Bloch wave solution of Maxwell's equations," *Phys. Rev. Lett.,* Vol. 65, Issue. 21, 2650–2653, 1990.

[25] V. Kuzmiak, A. A. Maradudin, and F. Pincemin, "Photonic band structures of two-dimensional systems containing metallic components," *Phys. Rev. B,* Vol. 50, 16835–16844, 1994.

[26] M. Plihal and A. A. Maradudin, "Photonic band structure of two-dimensional systems: The triangular lattice," *Phys. Rev. B,* Vol. 44, 8565–8571, 1991.

[27] J. B. Pendry and A. MacKinnon, "Calculation of photon dispersion relations," *Phys. Rev. Lett.,* Vol. 69, 2772–2775, 1992.

[28] M. Sigalas, C. M. Soukoulis, E. N. Economou, C. T. Chan, and K. M. Ho, "Photonic band gaps and defects in two dimensions: Studies of the transmission coefficient," *Phys. Rev. B,* Vol. 48, Issue 19, 14121–14126, 1993.

[29] A. Taflove and S. C. Hagness, *Computational Electrodynamics: The Finite-Difference Time-Domain Method,* 2nd ed., Norwood, MA: Artech House, 2000.

[30] A. Moroz, "Inward and outward integral equations and the KKR method for photons," *J. Phys.: Condens. Matter,* Vol. 6, 171–182, 1994.

[31] K. Sakoda, N. Kawai, T. Ito, A. Chutinan, S. Noda, T. Mitsuyu, and K. Hirao, "Photonic bands of metallic systems. I. Principle of calculation and accuracy," *Phys. Rev. B,* Vol. 64, 045116–045123, 2001.

[32] V. Kuzmiak and A. A. Maradudin, "Photonic band structures of one- and two-dimensional periodic systems with metallic components in the presence of dissipation," *Phys. Rev. B,* Vol. 55, 7427–7444, 1997.

[33] R. Kemp and J. E. Inglesfield, "Embedding approach for rapid convergence of plane waves in photonic calculations," *Phys. Rev. B,* Vol. 65, 115103–115112, 2002.

[34] K. S. Yee, "Numerical solution of initial boundary value problems involving Maxwells equations in isotropic media," *IEEE Trans. Antennas Propagat.*, Vol. AP-14, No. 3, 302–307, 1966.

[35] R. Holland, "Finite-Difference solution of Maxwell's equations in generalized nonorthogonal coordinates," *IEEE Trans. on Nuclear Science*, NS-30, 4589–4591, 1983.

[36] C. J. Railton, D. L. Paul, I. J. Craddock, and G. S. Hilton, "The treatment of geometrically small structures in FDTD by the modification of assigned material parameters," *IEEE Transactions on Antennas and Propagation*, Vol. 53, Issue 12, 4129–4136, 2005.

[37] F. Zheng, Z. Chen, and J. Zhang, "Toward the development of a three-dimensional unconditionally stable finite-difference time-domain method," *IEEE Transactions on Microwave Theory and Techniques*, Vol. 48, No. 9, 1550–1558, 2000.

[38] M. P. Kesler, J. G. Maloney, B. L. Shirley, and G. S. Smith, "Antenna design with the use of photonic band-gap materials as all-dielectric planar reflectors," *Microwave and Optical Technology Letters*, Vol. 11, Issue 4, 169–174, 1996.

[39] Y. Qian, V. Radisic, and T. Itoh, "Simulation and experiment of photonic band-gap structures for microstrip circuits," *Microwave Conference Proceedings, APMC '97*, Vol. 2, 585–588, 1997.

[40] M. Boroditsky, R. Coccioli, and E. Yablonovitch, "Analysis of photonic crystals for light-emitting diodes using the finite-difference time domain technique," *Proceedings of SPIE*, Vol. 3283, 184–190, 1998.

[41] A. Chutinan and S. Noda, "Highly confined waveguides and waveguide bends in three-dimensional photonic crystal," *Applied Physics Letters*, Vol. 75, No. 24, 3739–3741, 1999.

[42] A. Chutinan and S. Noda, "Waveguides and waveguide bends in two-dimensional photonic crystal slabs," *Phys. Rev. B*, 62, Issue 7, 4488–4492, 2000.

[43] F. Yang and Y. Rahmat-Samii, "A low-profile circularly polarized curl antenna over an electromagnetic bandgap (EBG) surface," *Microwave and Optical Technology Letters*, Vol. 31, No. 4, 264–267, 2001.

[44] F. Yang and Y. Rahmat-Samii, "Reflection phase characterizations of the EBG ground plane for low profile wire antenna applications," *IEEE Transactions on Antennas and Propagation*, Vol. 51, Issue 10, Part 1, 2691–2703, 2003.

[45] F. Yang and Y. Rahmat-Samii, "Microstrip antennas integrated with electromagnetic band-gap (EBG) structures: a low mutual coupling design for array applications," *IEEE Transactions on Antennas and Propagation*, Vol. 51, Issue 10, 2936–2946, 2003.

[46] Y. Zhao, Y. Hao, and C. G. Parini, "Radiation Properties of PIFA on UC-EBG substrates," *Microwave and Optical Technology Letters*, 44 (1): 21–24, January 5, 2005.

[47] E. Ozbay, M. Bayindir, I. Bulu, and E. Cubukcu, "Investigation of localized coupled-cavity modes in two-dimensional photonic bandgap structures," *IEEE Journal of Quantum Electronics*, Vol. 38, Issue 7, 837–843, 2002.

[48] A. R. Weily, L. Horvath, K. P. Esselle, B. C. Sanders, and T. S. Bird, "A planar resonator antenna based on a woodpile EBG material," *IEEE Transactions on Antennas and Propagation*, Vol. 53, Issue 1, Part 1, 216–223, 2005.

[49] A. R. Weily, K. P. Esselle, T. S. Bird, and B. C. Sanders, "Linear array of woodpile EBG sectoral horn antennas," *IEEE Transactions on Antennas and Propagation*, Vol. 54, Issue 8, 2263–2274, 2006.

[50] N. V. Kantartzis, D. L. Sounas, C. S. Antonopoulos, and T. D. Tsiboukis, "A Wideband ADI-FDTD Algorithm for the Design of Double Negative Metamaterial-Based Waveguides and Antenna Substrates," *IEEE Transactions on Magnetics*, Vol. 43, Issue 4, 1329–1332, 2007.

[51] D. Pinto and S. S. A. Obayya, "Improved Complex-Envelope Alternating-Direction-Implicit Finite-Difference-Time-Domain Method for Photonic-Bandgap Cavities," *Journal of Lightwave Technology*, Vol. 25, Issue 1, 440–447, 2007.

[52] J. Pendry, "Negative refraction index makes perfect lens," *Phys. Rev. Lett.*, Vol. 85, 3966–3969, 2000.

[53] S. A. Cummer, "Simulated causal subwavelength focusing by a negative refractive index slab," *Appl. Phys. Lett.*, Vol. 82, 1503–1505, 2003.

[54] M. W. Feise and Y. S. Kivshar, "Sub-wavelength imaging with a left-handed material flat lens," *Phys. Lett. A*, Vol. 334, pp. 326–330, 2005.

[55] X. S. Rao and C. K. Ong, "Subwavelength imaging by a left-handed material superlens," *Phys. Rev. E*, Vol. 68, 067601 (1–3), 2003.

[56] J. J. Chen, T. M. Grzegorczyk, B. I. Wu, and J. A. Kong, "Limitation of FDTD in simulation of a perfect lens imaging system," *Opt. Express*, Vol. 13, 10840–10845, 2005.

[57] Y. Zhao, P. Belov, and Y. Hao, "Improvement of Numerical Accuracy in FDTD Modeling of Left-Handed Metamaterials," *Metamaterials for Microwave and (Sub) Millimetrewave Applications: Electromagnetic Bandgap and Double Negative Design, Structures, Devices and Experimental Validation*, 2006. The Institution of Engineering and Technology Seminar, September 2006, 153–157.

[58] Y. Zhao, P. Belov and Y. Hao, "Accurate modeling of left-handed metamaterials using a finite-difference time-domain method with spatial averaging at the boundaries," *Journal of Optics, A, Pure and Applied Optics*, Vol. 9, S468–S475, 2007.

[59] Y. Zhao, P. Belov, and Y. Hao, "Spatially dispersive finite-difference time-domain analysis of sub-wavelength imaging by the wire medium slabs," *Optics Express*, Vol. 14, No. 12, 5154–5167, 2006.

[60] C. T. Chan, Q. L. Yu, and K. M. Ho, "Order-N spectral method for electromagnetic waves," *Phys. Rev. B*, 51, 16635–16642, 1995.

[61] A. J. Ward and J. B. Pendry, "Calculating photonic Greens functions using a nonorthogonal finite-difference time-domain method," *Phys. Rev. B*, 58, 7252–7259, 1998.

[62] J. Arriaga, A. J. Ward, and J. B. Pendry, "Order-N photonic band structures for metals and other dispersive materials," *Phys. Rev. B*, 59, 1874–1877, 1999.

[63] M. Qiu and S. He, "A nonorthogonal finite-difference time-domain method for computing the band structure of a two-dimensional photonic crystal with dielectric and metallic inclusions," *J. Appl. Phys.*, Vol. 87, Issue No. 12, 8268–8275, 2002.

[64] M. Qiu and S. He, "Guided modes in a two-dimensional metallic photonic crystal waveguide," *Phys. Lett. A*, Vol. 76, 425–429, 2000.

[65] P. Belov, "Analytical Modeling of Metamaterials and a New Principle of Sub-Wavelength Imaging," *Dissertation for the degree of Doctor of Science in Technology*, 2006, http://lib.tkk.fi/Diss/2006/isbn9512283786/isbn9512283786.pdf.

[66] J. Y. Park, C. C. Chang, Y. Qian, and T. Itoh, "An improved low-profile cavity-backed slot antenna loaded with 2D UC-PBG reflector," *Proc. IEEE AP-S Dig.*, Vol. 4, 194–197, 2001.

[67] Z. Li and Y. Rahmat-Samii, "PBG, PMC and PEC surface for antenna applications: A comparative study," *2000 IEEE AP-S Dig.*, Vol. 2, 674–677, 2000.

[68] G. S. Smith, M. P. Kesler, and J. G. Maloney, "Dipole antennas used with all-dielectric, woodpile photonic band-gap reflectors: gain, field patterns, and input impedance," *Microwave Opt. Technol. Lett.*, Vol. 21, No. 3, 191–196, 1999.

[69] M. Thèvenot, A. Reineix, and B. Jecko, "A dielectric photonic parabolic reflector," *Microwave Opt. Technol. Lett.*, Vol. 21, No. 6, 411C414, 1999.

[70] D. F. Sievenpiper, J. H. Schaffner, H. J. Song, R. Y. Loo, and G. Tangonan, "Two-dimensional beam steering using an electrically tunable impedance surface," *IEEE Trans. Antennas Propag.*, Vol. 51, No. 10, 2713–2722, 2003.

2.5 Summary

[71] Y. Hao and C. G. Parini, "Microstrip Antennas on Various UC-PBG Substrates," *IEICE Trans. Electron,* Vol. E86-C, No. 8, pp. 1536–1541, 2003.

[72] F. Yang and Y. Rahmat-Samii, "Mutual coupling reduction of microstrip antennas using electromagnetic band-gap structure," *Proc. IEEE AP-S Dig.,* Vol. 2, 478–481, 2001.

[73] M. Rahman and M. Stuchly, "Wide-band microstrip patch antenna with planar PBG structure," *Proc. IEEE AP-S Dig.,* Vol. 2, 486–489, 2001.

[74] R. Coccioli, F. R. Yang, K. P. Ma, and T. Itoh, "Aperture-coupled patch antenna on UC-PBG substrate," *IEEE Trans. Microwave Theory Tech.,* Vol. 47, 2123–2130, 1999.

[75] S. Sharma and L. Shafai, "Enhanced performance of an aperture-coupled rectangular microstrip antenna on a simplified unipolar compact photonic bandgap (UCPBG) structure," *Proc. IEEE AP-S Dig.,* Vol. 2, 498–501, 2001.

[76] D. H. Lee, Y. J. Lee, J. Yeo, R. Mittra, and W.S. Park, "Design of novel thin frequency selective surface superstrates for dual-band directivity enhancement," *IET Microwaves, Antennas and Propagation,* Vol. 1, Issue 1, 248–254, 2007.

[77] A. Pirhadi, F. Keshmiri, M. Hakkak, and R. B. Karimzadeh, "Design of Compact Dual Band High Directive Electromagnetic Bandgap (EBG) Resonator Antenna Using Artificial Magnetic Conductor," *IEEE Transactions on Antennas and Propagation,* Vol. 55, Issue 6, Part 2, 1682–1690, 2007.

[78] S. Foteinopoulou and C. M. Soukoulis, "Theoretical investigation of one-dimensional cavities in two-dimensional photonic crystals," *IEEE Journal of Quantum Electronics,* Vol. 38, Issue 7, 844–849, 2002.

[79] M. Loncar, J. Vuckovic, and A. Scherer, "Methods for controlling positions of guided modes of photonic-crystal waveguides," *J. Opt. Soc. Amer. B.,* Vol. 18, 1362–1368, 2001.

[80] S. Kawashima, M. Okano, M. Imada, and S. Noda, "Design of compound-defect waveguides in three-dimensional photonic crystals," *Optics Express,* Vol. 14, Issue 13, 6303–6307, 2006.

[81] A. Mekis, J. C. Chen, I. Kurland, S. Fan, P.R. Villeneuve, and J. D. Joannopoulos, "High Transmission Through Sharp Bends in Photonic Crystal Waveguides," *Phys. Rev. Lett.,* Vol. 77, Issue 18, 3787–3790, 1996.

[82] M. Bayindir, B. Temelkuran, and E. Ozbay, "Propagation of photons by hopping: A waveguiding mechanism through localized coupled cavities in three-dimensional photonic crystals," *Phys. Rev. B,* Vol. 61, Issue 18, R11855–R11858, 2000.

[83] J. Yonekura, M. Ikeda, and T. Baba, "Analysis of finite 2D photonic crystals of columns and lightwave devices using the scattering matrix method," *Journal of Lightwave Technology,* Vol. 17, Issue 8, 1500–1508, 1999.

[84] A. Martinez, F. Cuesta, and J. Marti, "Ultrashort 2-D photonic crystal directional couplers photonics," *IEEE Technology Letters,* Vol. 15, Issue 5, 694–696, 2003.

[85] Y. J. Lee, J. Yeo, R. Mittra, and W. S. Park, "Application of electromagnetic bandgap (EBG) superstrates with controllable defects for a class of patch antennas as spatial angular filters," *IEEE Transactions on Antennas and Propagation,* Vol. 53, Issue 1, Part 1, 224–235, 2005.

[86] E. P. Kosmidou, E. E. Kriezis, and T. D. Tsiboukis, "Analysis of tunable photonic crystal devices comprising liquid crystal materials as defects," *IEEE Journal of Quantum Electronics,* Vol. 41, Issue 5, 657–665, 2005.

[87] P. A. Belov, Y. Hao, and S. Sudhakaran, "Subwavelength microwave imaging using an array of parallel conducting wires as a lens," *Phys. Rev. B.,* Vol. 73, 033108 (4 pages), 2006.

[88] P. Ikonen, P. A. Belov, C. R. Simovski, and S. I. Maslovski, "Experimental demonstration of subwavelength field channeling at microwave frequencies using a capacitively loaded wire medium," *Phys. Rev. B,* Vol. 73, 073102 (1–4), 2006.

[89] K.-P. Ma, K. Hirose, F.-R. Yang, Y. Qian, and T. Itoh, "Realisation of magnetic conducting surface using novel photonic bandgap structure," *Electronics Letters,* Vol. 34, Issue 21, 2041–2042, 1998.

[90] C. Caloz, C. C. Chang, and T. Itoh, "A novel anisotropic uniplanar compact photonic band-gap (UC-PBG) ground plane," *31st European Microwave Conference,* Vol. 2, 29–32, London, 2001.

[91] D. Sievenpiper, L. Zhang, R. Broas, and E. Yablonovitch, "High-Impedance Electromagnetic Surfaces with Forbidden Bands at Radio and Microwave Frequencies," *SPIE Conference on Terahertz and Giagahertz Photonics,* Denver, CO, 1999.

[92] S. Kitson, W. Barnes, and J. Sambles, "Full Photonic Band Gap for Surface Modes in the Visible," *Phys. Rev. Lett.,* Vol. 77, 2670–2673, 1996.

[93] P. S. Kildal, "Artificially Soft and Hard Sufaces in Electromagnetics," *IEEE Trans. Ant. Prop.,* Vol. 38, 1537–1544, 1990.

[94] C. Cheype, C. Serier, M. Thevenot, T. Monediere, A. Reineix, and B. Jecko, "An electromagnetic bandgap resonator antenna," *IEEE Transactions on Antennas and Propagation,* Vol. 50, Issue 9, 1285–1290, 2002.

[95] Y. J. Lee, J. Yeo, R. Mittra, and W. S. Park, "Design of a high-directivity Electromagnetic Band Gap (EBG) resonator antenna using a frequency-selective surface (FSS) superstrate," *Microwave and Optical Tech. Lett.,* Vol. 43, No. 6, 462–467, 2004.

[96] Y. J. Lee, J. Yeo, R. Mittra, and W.S. Park, "Thin frequency selective surface (FSS) superstrate with different periodicities for dual-band directivity enhancement," *IEEE International Workshop on Antenna Technology: Small Antennas and Novel Metamaterials. IWAT 2005,* March 7–9, 2005, 375–378.

CHAPTER 3
A Brief Introduction to the FDTD Method for Modeling Metamaterials

3.1 Introduction

The FDTD method [1] has been proven to be one of the most effective numerical methods in the study of metamaterials. As a direct solution to the Maxwell's equations, FDTD offers a simple yet straightforward way to model complex periodic structures. Since it is a time domain solver, it is convenient for dealing with the characteristics of metamaterials over a wide frequency band.

The foundation of FDTD was laid down by Yee in 1966 [2]. Yee chose a geometric relationship for the spatial sampling of the vector components of the electric and magnetic fields that enables representing both the differential and integral forms of Maxwell's equations in a robust manner. The FDTD algorithm as proposed by Yee in his original paper is second-order accurate in both time and space. Furthermore, in this algorithm, the numerical dispersion effects can be kept small by using a cell size that is sufficiently small in comparison to the wavelength, say on the order of $\lambda/20$ or $\lambda/30$. This chapter begins with a brief review of the FDTD fundamentals, and then Chapter 4 discusses its application to the problem of modeling the EBG structure.

3.2 Formulations of the Yee's FDTD Algorithm

The Yee algorithm simultaneously deals with both electric and magnetic fields in time and space using the coupled form of Maxwell's curl equations, rather than by solving the wave equation for either the electric field (or the magnetic field) alone.

3.2.1 Maxwell's Equations

Consider a region of space that has no electric or magnetic current sources, but may have materials that absorb electric or magnetic field energy. The time-dependent Maxwell's equations are given in differential and integral forms by:

$$\frac{\partial \vec{B}}{\partial t} = -\nabla \times \vec{E} - \sigma^* \vec{H} \tag{3.1}$$

$$\frac{\partial \vec{D}}{\partial t} = \nabla \times \vec{H} - \sigma \vec{E} \tag{3.2}$$

$$\nabla \cdot \vec{D} = 0 \tag{3.3}$$

$$\nabla \cdot \vec{B} = 0 \tag{3.4}$$

and

$$\frac{\partial}{\partial t}\iint_A \vec{B} \cdot d\vec{A} = -\oint_l \vec{E} \cdot d\vec{l} - \iint_A \sigma^* \vec{H} \cdot d\vec{A} \tag{3.5}$$

$$\frac{\partial}{\partial t}\iint_A \vec{D} \cdot d\vec{A} = \oint_l \vec{H} \cdot d\vec{l} - \iint_A \sigma \vec{E} \cdot d\vec{A} \tag{3.6}$$

$$\oiint_A \vec{D} \cdot d\vec{A} = 0 \tag{3.7}$$

$$\oiint_A \vec{B} \cdot d\vec{A} = 0 \tag{3.8}$$

where

σ^*: Equivalent magnetic loss (ohms/meter);
σ: Electric conductivity (siemens/meter);
$\vec{E}$: Electric field, also called the electric flux density (volt/meter);
$\vec{H}$: Magnetic field strength (ampere/meter);
$\vec{D}$: Electric displacement field (coulomb/meter^2);
$\vec{B}$: Magnetic field, also called the magnetic flux density (tesla, or volt-seconds/meter^2);
A: Surface area (meter^2).

In linear, isotropic, nondispersive materials, $\vec{D}$ and $\vec{B}$ are simply related to $\vec{E}$ and $\vec{H}$ as follows:

$$\vec{D} = \varepsilon \vec{E} = \varepsilon_0 \varepsilon_r \vec{E} \quad and \quad \vec{B} = \mu \vec{H} = \mu_0 \mu_r \vec{H} \tag{3.9}$$

where ε and μ are the medium permittivity and permeability, ε_0 and μ_0 are the permittivity and permeability of free space, and ε_r and μ_r are the relative permittivity and permeability.

Substituting (3.9) into (3.1) and (3.2), we obtain the Maxwell's curl equations in linear, isotropic, nondispersive materials that read:

$$\frac{\partial \vec{H}}{\partial t} = -\frac{1}{\mu}\nabla \times \vec{E} - \frac{1}{\mu}\sigma^* \vec{H} \tag{3.10}$$

$$\frac{\partial \vec{E}}{\partial t} = \frac{1}{\varepsilon}\nabla \times \vec{H} - \frac{1}{\varepsilon}\sigma \vec{E} \tag{3.11}$$

3.2 Formulations of the Yee's FDTD Algorithm

Expansion of the vector components of the curl operators of (3.10) and (3.11) yields the following system of six coupled scalar equations under Cartesian coordinate:

$$\frac{\partial H_x}{\partial t} = -\frac{1}{\mu}\left[\frac{\partial E_z}{\partial y} - \frac{\partial E_y}{\partial z}\right] - \frac{\sigma^*}{\mu}H_x \qquad (3.12)$$

$$\frac{\partial H_y}{\partial t} = -\frac{1}{\mu}\left[\frac{\partial E_x}{\partial z} - \frac{\partial E_z}{\partial x}\right] - \frac{\sigma^*}{\mu}H_y \qquad (3.13)$$

$$\frac{\partial H_z}{\partial t} = -\frac{1}{\mu}\left[\frac{\partial E_y}{\partial x} - \frac{\partial E_x}{\partial y}\right] - \frac{\sigma^*}{\mu}H_z \qquad (3.14)$$

$$\frac{\partial E_x}{\partial t} = \frac{1}{\varepsilon}\left[\frac{\partial H_z}{\partial y} - \frac{\partial H_y}{\partial z}\right] - \frac{\sigma}{\varepsilon}E_x \qquad (3.15)$$

$$\frac{\partial E_y}{\partial t} = \frac{1}{\varepsilon}\left[\frac{\partial H_x}{\partial z} - \frac{\partial H_z}{\partial x}\right] - \frac{\sigma}{\varepsilon}E_y \qquad (3.16)$$

$$\frac{\partial E_z}{\partial t} = \frac{1}{\varepsilon}\left[\frac{\partial H_y}{\partial x} - \frac{\partial H_x}{\partial y}\right] - \frac{\sigma}{\varepsilon}E_z \qquad (3.17)$$

The system of six coupled partial differential equations (3.12)–(3.17) forms the basis of the FDTD numerical algorithm for modeling electromagnetic wave interactions with arbitrary three-dimensional objects.

Yee's FDTD scheme discretizes Maxwell's curl equations by approximating the time and space first-order partial derivatives with central differences, and then solving the resulting equations by using a leapfrog scheme.

3.2.2 Yee's Orthogonal Mesh

The Yee's algorithm positions its $\vec{E}$ and $\vec{H}$ components at the centers of the grid lines and surfaces such that each $\vec{E}$ component is surrounded by four $\vec{H}$ components, and vice versa. This provides an elegant yet simple picture of three-dimensional space being filled by interlinked arrays of Faraday's law and Ampere's law contours. Thus, it is possible to identify the $\vec{E}$ components associated with the displacement current flux linking with the $\vec{H}$ loops and, correspondingly, the $\vec{H}$ components associated with the magnetic flux are linked with the $\vec{E}$ loops, as shown in Figure 3.1.

Utilizing Yee's spatial gridding scheme, the partial spatial derivatives in (3.12)–(3.17) can be approximated by a central difference in space such as a sample equation here:

$$\frac{\partial E_y}{\partial z}\Big|_{(i+\frac{1}{2},j,k)} \approx \frac{E_y(i+\frac{1}{2},j,k+\frac{1}{2}) - E_y(i+\frac{1}{2},j,k-\frac{1}{2})}{\Delta z} \qquad (3.18)$$

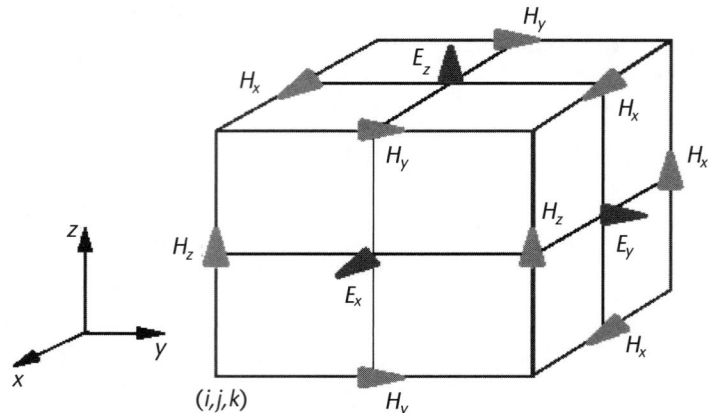

Figure 3.1 Yee's spatial grid.

Consequently (3.12) becomes:

$$\frac{\partial H_x(i+\frac{1}{2},j,k)}{\partial t} + \frac{1}{\mu(i+\frac{1}{2},j,k)} \cdot \sigma^*\left(i+\frac{1}{2},j,k\right) H_x\left(i+\frac{1}{2},j,k\right) = -\frac{1}{\mu(i+\frac{1}{2},j,k)}$$
$$\cdot \left[\frac{E_z(i+\frac{1}{2},j+\frac{1}{2},k)-E_z(i+\frac{1}{2},j-\frac{1}{2},k)}{\Delta y} - \frac{E_y(i+\frac{1}{2},j,k+\frac{1}{2})-E_y(i+\frac{1}{2},j,k-\frac{1}{2})}{\Delta z}\right]$$

(3.19)

3.2.3 Time Domain Discretization: The Leapfrog Scheme and the Courant Stability Condition (CFL Condition)

Yee's algorithm also utilizes central differencing in time for the $\vec{E}$ and $\vec{H}$ components and then solves them by using a leapfrog scheme as shown in Figure 3.2. All of the $\vec{E}$ components in the modeled space are computed and stored in memory by using the previously computed values of $\vec{E}$ and the newly updated $\vec{H}$ field data. At the next step, $\vec{H}$ is recomputed based on the previously obtained $\vec{H}$ and the newly obtained $\vec{E}$. This process continues until the time-stepping is terminated.

A central difference approximation is applied to (3.19) as follows:

$$\left.\frac{\partial H_x}{\partial t}\right|^{(n\Delta t)} \approx \frac{H_x^{(n\Delta t + \frac{1}{2}\Delta t)} - H_x^{(n\Delta t - \frac{1}{2}\Delta t)}}{\Delta t} \quad (3.20)$$

$$H_x^{(n\Delta t)} \approx \frac{H_x^{(n\Delta t + \frac{1}{2}\Delta t)} + H_x^{(n\Delta t - \frac{1}{2}\Delta t)}}{2} \quad (3.21)$$

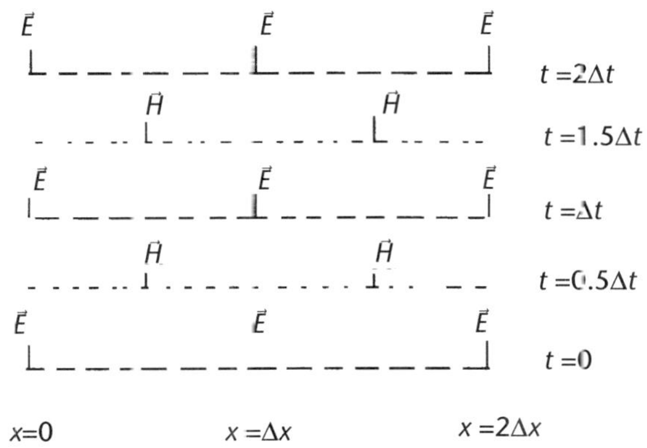

3.2 Formulations of the Yee's FDTD Algorithm

Figure 3.2 Leapfrog scheme—the temporal scheme of the FDTD method.

Equation (3.19) thus becomes a discretized equation, (3.22), which can be solved conveniently on the computer.

$$H_x\left(i+\frac{1}{2},j,k\right)^{(n\Delta t+\frac{1}{2}\Delta t)} = \frac{\frac{1}{\Delta t}-\frac{\sigma^*}{2\mu}}{\frac{1}{\Delta t}+\frac{\sigma^*}{2\mu}} H_x\left(i+\frac{1}{2},j,k\right)^{(n\Delta t-\frac{1}{2}\Delta t)}$$
$$-\frac{1}{(\frac{1}{\Delta t}+\frac{\sigma^*}{2\mu})\cdot\mu\Delta y}\left[E_z\left(i+\frac{1}{2},j+\frac{1}{2},k\right)^{(n\Delta t)} - E_z\left(i+\frac{1}{2},j-\frac{1}{2},k\right)^{(n\Delta t)}\right]$$
$$+\frac{1}{(\frac{1}{\Delta t}+\frac{\sigma^*}{2\mu})\cdot\mu\Delta z}\left[E_y\left(i+\frac{1}{2},j,k+\frac{1}{2}\right)^{(n\Delta t)} - E_y\left(i+\frac{1}{2},j,k-\frac{1}{2}\right)^{(n\Delta t)}\right]$$

(3.22)

Numerical stability of the Yee algorithm requires that we set an upper bound on the time step (Δt) that is determined by the spatial increments Δx, Δy, and Δz in accordance with the Courant-Friedrich-Levy (CFL) stability condition. In three dimensions this condition is given by

$$\Delta t \leq \Delta t_{max} = \frac{1}{c\sqrt{\frac{1}{\Delta x^2}+\frac{1}{\Delta y^2}+\frac{1}{\Delta z^2}}} \qquad (3.23)$$

In a cubic grid (where $\Delta x = \Delta y = \Delta z = \Delta$), (3.23) can be expressed as

$$\Delta t \leq \Delta t_{max} = \frac{\Delta}{c\sqrt{3}} \qquad (\Delta x = \Delta y = \Delta z = \Delta) \qquad (3.24)$$

Enforcement of this upper bound on Δt guarantees the stability of the algorithm, which is essential to guarantee its robustness when applied to a wide variety of electromagnetic wave modeling problems. However, there are applications of FDTD modeling that find the CFL stability bound too restrictive. For example, when simulating structures with fine-scale geometries, the cell size Δ needs to be much less than the shortest wavelength λ_{min}. So for a fixed total time of simulation T, which is determined by the nature of the configuration being modeled, choosing a time step Δt limited by CFL can cause the total number of time steps N_{sim} required to become very large, given by:

$$N_{sim} = \frac{T_{sim}}{\Delta t} \tag{3.25}$$

Since relaxing the CFL condition in the conventional FDTD is not a viable option, the simulation of structures with fine details can become highly computer-intensive, and even prohibitive in terms of simulation time.

Many attempts have been made to relax or even remove the CFL stability constraint. Some early works involved an application of the alternating-direction-implicit (ADI) technique [3], which is unconditionally stable. In the first attempt of the implementation of ADI in the FDTD, which dates back to 1984 [3], the finite-difference operator was factored into three implicit operators in the three coordinate directions, namely x, y, and z. However, it was difficult to prove the numerical stability of this scheme at that time [1]. In 1999, a 2-D FDTD algorithm that was not restricted by the Courant stability condition was proposed for a 2D-TE wave [4], and the ADI method was again implemented in this algorithm. The resulting FDTD formulation was found to be unconditionally stable [4, 5]. As a consequence, it was possible to remove the CFL constraint on the FDTD algorithm and the selection of the time step was only governed by the accuracy desired [5].

3.3 Other Spatial Domain Discretization Schemes

Since the FDTD method is a grid-based algorithm, mesh generation plays a very important role in its implementation. A properly defined mesh helps reduce numerical errors and increases computational efficiency.

An orthogonal, uniform, and Cartesian meshing scheme is the most simple and straightforward one to implement and is most commonly used in FDTD modeling. An orthogonal FDTD grid generally introduces minimal numerical errors [6], and hence, a boundary-orthogonal mesh is preferred even in a conformal mesh generation scheme. However, a staircase approximation of curved structures often introduces numerical inaccuracy (for instance, numerical dispersion) [7].

A variety of mesh generation schemes have been developed in the context of FDTD, and they have led to several modifications of the original Yee FDTD scheme. Additionally, subgridding, nonorthogonal meshing, and the use of hybrid meshes have been employed in a wide range of applications, such as investigation of large structures with fine details and the modeling of objects comprised of curved or oblique surfaces.

3.3.1 Subgridding Mesh

To ensure numerical accuracy it is necessary to choose the cell size (the spatial increment Δx) in FDTD discretization to be much smaller (typically less than $\lambda/10$) in the frequency range of interest. Consequently, the simulation of an electrically

3.3 Other Spatial Domain Discretization Schemes

large object with locally fine structures using overall an fine mesh (small Δx) and a small Δt owning to the small cell size is computationally costly in the FDTD modeling.

One approach to alleviating this problem is to introduce a subgridding mesh scheme. The basic idea followed in this scheme entails dividing the computational volume into subregions and simulating them with variable step sizes. A coarse grid is used in a large volume, and fine meshes are applied only for regions containing objects with fine features or in the vicinity of discontinuities. To minimize numerical reflections from an abrupt transition between very coarse and very fine meshes regions, a slow taper in the cell size by a factor of 2 or 3 and a sequence of subgrids can be employed when necessary [8].

To explain the field updating procedure we refer to Figure 3.3. The fields inside the coarse and fine mesh subregions, shown in Figure 3.3, are calculated by using the conventional FDTD equations, (3.26) and (3.27), respectively. The time increments in each subregion can either be chosen in accordance with the smallest spatial increment, or can be related to the spatial increments in each subregion. On the coarse-fine grid boundary, an interpolation is utilized to calculate the tangential

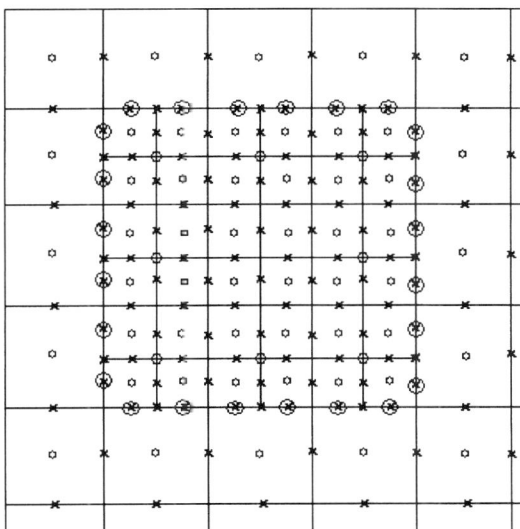

x Electric field points
o Magnetic field points
⊗ Electric field values obtained by spatial and time interpolations
⊙ Initial values for the magnetic field obtained by spatial and time average of the four neighbors

Figure 3.3 A cross-section of a computational domain meshed according to the subgridding algorithm. Positions where the field quantities are calculated are shown. Since the spatial increment in the fine mesh is only half that of the coarse grid, the time increment for the fine mesh domain is equal to half of that in the coarse domain [12].

electric fields and the boundary layer magnetic fields in the fine mesh region. For the subgridding FDTD, various mesh and interpolation schemes can be found in [8–11].

The subgridding mesh scheme requires less computer memory and therefore, expands the problem-solving capability of the FDTD method. When properly implemented, the subgridding algorithm has been found to exhibit good numerical stability [12].

$$H_x^{n+\frac{1}{2}}(i,j,k) = H_x^{n-\frac{1}{2}}(i,j,k)$$
$$- \frac{\Delta t}{\mu} \left[\frac{E_z^n(i,j,k) - E_z^n(i,j-1,k)}{\Delta y} - \frac{E_y^n(i,j,k) - E_y^n(i,j,k-1)}{\Delta z} \right]$$

$$H_y^{n+\frac{1}{2}}(i,j,k) = H_y^{n-\frac{1}{2}}(i,j,k)$$
$$- \frac{\Delta t}{\mu} \left[\frac{E_x^n(i,j,k) - E_x^n(i,j,k-1)}{\Delta z} - \frac{E_z^n(i,j,k) - E_z^n(i-1,j,k)}{\Delta x} \right]$$

$$H_z^{n+\frac{1}{2}}(i,j,k) = H_z^{n-\frac{1}{2}}(i,j,k)$$
$$- \frac{\Delta t}{\mu} \left[\frac{E_y^n(i,j,k) - E_y^n(i-1,j,k)}{\Delta x} - \frac{E_x^n(i,j,k) - E_x^n(i,j-1,k)}{\Delta y} \right]$$
(3.26)

$$E_x^{n+1}(i,j,k) = E_x^n(i,j,k)$$
$$+ \frac{\Delta t}{\varepsilon} \left[\frac{H_z^{n+\frac{1}{2}}(i,j+1,k) - H_z^{n+\frac{1}{2}}(i,j,k)}{\Delta y} - \frac{H_y^{n+\frac{1}{2}}(i,j,k+1) - H_y^{n+\frac{1}{2}}(i,j,k)}{\Delta z} \right]$$

$$E_y^{n+1}(i,j,k) = E_y^n(i,j,k)$$
$$+ \frac{\Delta t}{\varepsilon} \left[\frac{H_x^{n+\frac{1}{2}}(i,j,k+1) - H_x^{n+\frac{1}{2}}(i,j,k)}{\Delta z} - \frac{H_z^{n+\frac{1}{2}}(i+1,j,k) - H_z^{n+\frac{1}{2}}(i,j,k)}{\Delta x} \right]$$

$$E_z^{n+1}(i,j,k) = E_z^n(i,j,k)$$
$$+ \frac{\Delta t}{\varepsilon} \left[\frac{H_y^{n+\frac{1}{2}}(i+1,j,k) - H_y^{n+\frac{1}{2}}(i,j,k)}{\Delta x} - \frac{H_x^{n+\frac{1}{2}}(i,j+1,k) - H_x^{n+\frac{1}{2}}(i,j,k)}{\Delta y} \right]$$
(3.27)

Equations (3.28) and (3.29) show examples of how the fields at the coarse-fine grid boundary are updated using the neighboring averaging scheme and employing a time interval calculated based on the smallest sell size in the fine mesh region (see Figure 3.4).

$$E_f(2,1,1) = \frac{1}{4} E_c(1,1) + \frac{3}{4} E_c(2,1) \qquad (3.28)$$

3.3 Other Spatial Domain Discretization Schemes

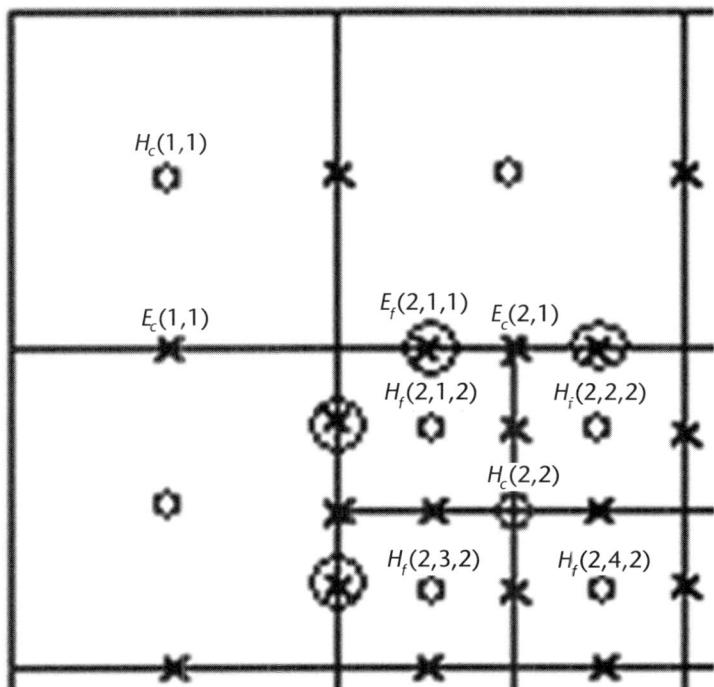

Figure 3.4 Enlarged view of the top-left corner of Figure 3.3. In the coarse-fine grid boundary, the electric field in fine mesh [$E_f(2,1,1)$] is updated by electric field in coarse mesh region (E_c) using the neighboring averaging equation (3.28); and the magnetic field in coarse mesh [$H_c(2,2)$] is updated by magnetic field in the fine mesh region (H_f) using the neighboring averaging equation (3.29).

$$H_c(2,2) = \frac{1}{4}H_f(2,1,2) + \frac{1}{4}H_f(2,2,2) + \frac{1}{4}H_f(2,3,2) + \frac{1}{4}H_f(2,4,2) \qquad (3.29)$$

In (3.28) and (3.29), (E_f and H_f) and (E_c and H_c) denote the fields in the fine and coarse mesh subregions, respectively.

3.3.2 Nonorthogonal Mesh

Many real-world electromagnetic problems are characterized by geometries with curved boundaries, or surfaces that are tilted relative to a Cartesian grid. Approximating such an oblique or a curved surface by using a staircased mesh usually requires a very fine mesh, which in turn, dictates the use of very small time step in the FDTD algorithm, with the resulting increase in the computation time.

Since Maxwell's equations are vector equations that can be implemented in any coordinate system, they can be expressed in the nonorthogonal coordinate system as shown in [13]. In 1983, Holland incorporated Maxwell's equations in nonorthogonal coordinate systems into the FDTD method and developed a nonorthogonal FDTD (NFDTD) algorithm, based on general nonorthogonal grids [3]. In this scheme, oblique surfaces or curved structures are meshed conformally

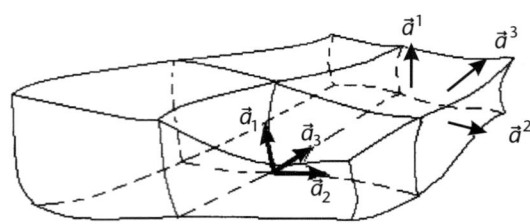

Figure 3.5 A part of a three-dimensional nonorthogonal mesh showing the covariant and contravariant vectors.

and more accurately yet with a coarser mesh. Since then, the NFDTD method has been refined by many researchers, including Yee [14], Lee [15], Mittra [16], Jurgens [17, 18], Railton [19], Hao [20] and Douvalis [21].

In nonorthogonal coordinate systems, an arbitrary vector can be expressed as a linear combination of two types of components according to two bases, namely the covariant and contravariant components of this vector. In the FDTD modeling of an EM problem, the covariant component relates physically to the flow of the vector along the contour of an arbitrary surface, while the contravariant component represents the flux density of this vector passing through this surface (Figure 3.5).

However, compared with the conventional Cartesian FDTD method, the global curvilinear FDTD must store many additional metric tensors, which are essential parameters in the NFDTD scheme and are calculated from the spatial increment of each grid. While the contravariant components are updated in a way similar to that in Yee's scheme, the covariant components must be computed from the contravariant ones by using two additional projection equations containing the metric tensors.

As a result, the global curvilinear FDTD method is computationally more intensive than the conventional Cartesian FDTD method. However, to alleviate this problem, it is possible to use a hybrid meshing scheme [i.e., the local distorted NFDTD (LD-NFDTD) [19, 20] algorithm].

3.3.3 Hybrid FDTD Meshes

Any mesh generation scheme has its own advantages and disadvantages. However, it is possible to combine different schemes to devise an efficient and accurate FDTD grid. Brief descriptions of several hybrid mesh generation schemes are presented as follows:

- *Hybrid conformal and orthogonal grid:* This type of mesh can be devised by employing conformal cells only at and near the curved boundaries within an underlying Cartesian coordinate system [22]. In other words, the curvilinear meshes are used in the immediate vicinity of the curved boundary, while the vast majority of the mesh away from the curved boundary can be

3.3 Other Spatial Domain Discretization Schemes

rectangular/square. The so-called LNFDTD scheme, implemented on such a grid, achieves improved accuracy as well as versatility compared to the conventional Yee's Cartesian FDTD method, but without compromising the computational efficiency [20].

- *Conformal grid employing a subgridded meshes:* The subgridding in the space domain can be applied to combine with the conformal grid leading to a subgridding NFDTD method. The computational efficiency is expected to be improved in comparison with a subgridding scheme based on orthogonal meshes, or a pure NFDTD scheme. Besides the subgridding in space, a time subgridding scheme can also be employed in the NFDTD method and is reported to be helpful in reducing the late time instability in the NFDTD method [22].

- *Conformal grids with triangular meshes:* In [23], Schuhmann et al. observed that degenerate cells in a NFDTD mesh are responsible for introducing local field errors. They not only lead to irregular convergence behavior but also contribute to the late time instability. To overcome this problem, it has been suggested that the NFDTD be combined with a triangular meshing scheme. Figure 3.6 demonstrates the staircase approximation, the nonorthogonal

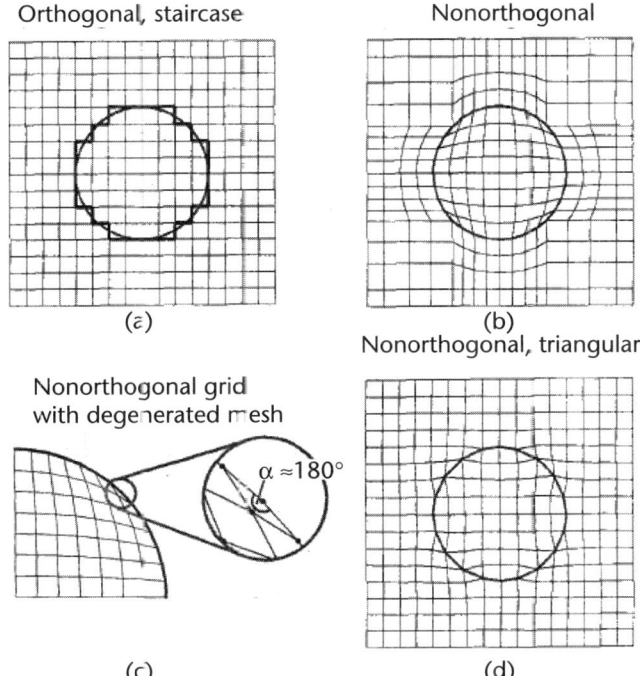

Figure 3.6 Meshes of the cross-section of a cylindrical cavity: (a) the staircase approximation; (b) the nonorthogonal mesh; (c) details of the degenerated cell in the nonorthogonal mesh, which is most responsible for the numerical error and late time instability; and (d) the triangular fillings NFDTD [23].

mesh, the degenerated cell in the nonorthogonal mesh, and the proposed triangular fillings NFDTD, respectively, when meshing a 2-D cylindrical cavity. This scheme is regarded as a more flexible scheme and may considerably improve the efficiency and accuracy of the NFDTD [23].

3.4 Boundary Conditions

In numerical modeling, many geometries of interest are defined in "open" regions where the spatial domain of the computed field is unbounded in one or more coordinate directions. Since the data storage in a computer is limited by the size of memory, it is not possible to handle an open region problem directly. To mitigate this problem, an absorbing boundary condition (ABC) is used to truncate the computational domain that is designed to suppress spurious reflections of the outgoing waves to an acceptable level.

The ABCs can be divided into two different categories: those derived from differential equations and others based on the use of absorbing materials. The most widely used ABC in the first group is the one derived by Engquist and Majda [24] with the discretization given by Mur [25]. It is based on an approximation of the outgoing wave equation being expressed using a Taylor approximation. In contrast to this, material-based ABCs are realized by surrounding the computational domain with a lossy material that dampens the outgoing fields. In this group, the perfectly matched layer (PML) technique [26–31], which was put forward by Berenger in 1994, exhibits an accuracy level that is significantly better than most other ABCs. Consequently, it is widely used in the FDTD simulations.

An ideal metamaterial structure, which has an infinite periodicity, does not exist in the real world, because it is necessarily truncated and hence finite. However, it is always interesting and useful to study the characteristics of an infinite metamaterial, which provides considerable insight into its applicability for the problem at hand, even when it deals with a truncated metamaterial. The implementation of the periodic boundary condition (PBC) is a tool that enables the modeling of an infinite metamaterial.

3.4.1 Mur's Absorbing Boundary Conditions (ABCs)

Engquist and Majda derived a theory for one-way wave equations that describes wave propagation only in specified directions. For example, consider the two-dimensional wave equation in Cartesian coordinates:

$$\frac{\partial^2 U}{\partial x^2} + \frac{\partial^2 U}{\partial y^2} - \frac{1}{c^2}\frac{\partial^2 U}{\partial t^2} = 0 \tag{3.30}$$

where U is a scalar field component and c is the phase velocity of the wave. We define the partial differential operator as:

$$L = \frac{\partial^2}{\partial x^2} + \frac{\partial^2}{\partial y^2} - \frac{1}{c^2}\frac{\partial^2}{\partial t^2} = D_x^2 + D_y^2 - \frac{1}{c^2}D_t^2 \tag{3.31}$$

3.4 Boundary Conditions

and use it to write the wave equation as:

$$LU = L^-L^+U = 0 \tag{3.32}$$

where L^- and L^- are the factors of the wave operator L, defined as:

$$L^- \equiv D_x - \frac{D_t}{c}\sqrt{1-S^2} \quad \text{and} \quad L^+ \equiv D_x + \frac{D_t}{c}\sqrt{1-S^2} \tag{3.33}$$

with

$$S = \frac{D_y}{D_t/c} \tag{3.34}$$

Waves that satisfy the left operator equation, namely

$$L^-U = 0 \tag{3.35}$$

only propagate towards the $-x$-direction and ideally, will not bounce back into the computational domain.

Usually, S given in (3.34) is very small, and hence, a Taylor series can be used to approximate the square-root function in (3.33) by two-term expansion as follows:

$$\sqrt{1-S^2} \cong 1 - \frac{1}{2}S^2 \tag{3.36}$$

Substituting (3.36) into (3.33) we obtain:

$$L^- = D_x - \frac{D_t}{c}\left[1 - \frac{1}{2}\left(\frac{cD_y}{D_t}\right)^2\right] = D_x - \frac{D_t}{c} + \frac{cD_y^2}{2D_t} \tag{3.37}$$

Then substituting (3.37) into (3.35), multiplying by D_t we get a second-order accurate ABC at the boundary $x = 0$, given by:

$$D_t L^- U = \frac{\partial^2 U}{\partial x \partial t} - \frac{1}{c}\frac{\partial^2 U}{\partial t^2} + \frac{c}{2}\frac{\partial^2 U}{\partial y^2} = 0 \tag{3.38}$$

Mur used a simple central-difference scheme to interpret (3.38) in the Yee's space (with spatial increments Δx and Δy) and the time (with time step Δt) domain. For example, in the second-order ABC, the mixed x and t derivative is written as:

$$\frac{\partial^2 U|_{1/2,j}^n}{\partial x \partial t} = \frac{1}{2\Delta t}\left[\left(\frac{U|_{1,j}^{n+1} - U|_{0,j}^{n+1}}{\Delta x}\right) - \left(\frac{U|_{1,j}^{n-1} - U|_{0,j}^{n-1}}{\Delta x}\right)\right] \tag{3.39}$$

The discretized version of the tangential field under discretization at the boundary (e.g., $U|_{0,j}^{n+1}$) is calculated as follows:

$$U|_{0,j}^{n+1} = -U|_{1,j}^{n+1} + \frac{c\Delta t - \Delta}{c\Delta t + \Delta}\left(U|_{1,j}^{n+1} + U|_{0,j}^{n-1}\right) + \frac{2\Delta}{c\Delta t + \Delta}\left(U|_{1,j}^n + U|_{0,j}^n\right)$$
$$+ \frac{(c\Delta t)^2}{2\Delta(c\Delta t + \Delta)}\left(U|_{0,j+1}^n - 2U|_{0,j}^n + U|_{0,j-1}^n\right)$$
$$+ \frac{(c\Delta t)^2}{2\Delta(c\Delta t + \Delta)}\left(U|_{1,j+1}^n - 2U|_{1,j}^n + U|_{1,j-1}^n\right) \tag{3.40}$$

where we have assumed, for the sake of simplicity, $\Delta x = \Delta y = \Delta$. Remove the y-directive term, and the first-order Mur's ABC at $x = 0$ boundary is obtained:

$$U|_{0,j}^{n+1} = -U|_{1,j}^{n+1} + \frac{c\Delta t - \Delta}{c\Delta t + \Delta}\left(U|_{1,j}^{n+1} + U|_{0,j}^{n-1}\right) + \frac{2\Delta}{c\Delta t + \Delta}\left(U|_{1,j}^n + U|_{0,j}^n\right) \quad (3.41)$$

3.4.2 Perfect Matched Layers (PMLs)

In the perfect matched layer (PML) [26, 27], each component of the electromagnetic field is split into two parts. In the Cartesian coordinates, the six field components yield 12 subcomponents, denoted by E_{xy}, E_{xz}, E_{yx}, E_{yz}, E_{zx}, E_{zy}, H_{xy}, H_{xz}, H_{yx}, H_{yz}, H_{zx}, and H_{zy}. Using the above components, Maxwell's equations are replaced by the following 12 equations,

$$\varepsilon\frac{\partial E_{xy}}{\partial t} + \sigma_y E_{xy} = \frac{\partial H_{zx} + H_{zy}}{\partial y} \quad (3.42)$$

$$\varepsilon\frac{\partial E_{xz}}{\partial t} + \sigma_z E_{xz} = -\frac{\partial H_{yz} + H_{yx}}{\partial z} \quad (3.43)$$

$$\varepsilon\frac{\partial E_{yz}}{\partial t} + \sigma_z E_{yz} = \frac{\partial H_{xy} + H_{xz}}{\partial z} \quad (3.44)$$

$$\varepsilon\frac{\partial E_{yx}}{\partial t} + \sigma_z E_{yx} = -\frac{\partial H_{zx} + H_{zy}}{\partial x} \quad (3.45)$$

$$\varepsilon\frac{\partial E_{zx}}{\partial t} + \sigma_z E_{zx} = \frac{\partial H_{yz} + H_{yx}}{\partial x} \quad (3.46)$$

$$\varepsilon\frac{\partial E_{zy}}{\partial t} + \sigma_z E_{zy} = -\frac{\partial H_{xy} + H_{xz}}{\partial y} \quad (3.47)$$

$$\mu\frac{\partial H_{xy}}{\partial t} + \sigma_x^* H_{xy} = -\frac{\partial E_{zx} + E_{zy}}{\partial y} \quad (3.48)$$

$$\mu\frac{\partial H_{xz}}{\partial t} + \sigma_y^* H_{xz} = \frac{\partial E_{yz} + E_{yx}}{\partial z} \quad (3.49)$$

$$\mu\frac{\partial H_{yz}}{\partial t} + \sigma_x^* H_{yz} = -\frac{\partial E_{xy} + E_{xz}}{\partial z} \quad (3.50)$$

$$\mu\frac{\partial H_{yx}}{\partial t} + \sigma_y^* H_{yx} = \frac{\partial E_{zx} + E_{zy}}{\partial x} \quad (3.51)$$

$$\mu\frac{\partial H_{zx}}{\partial t} + \sigma_x^* H_{zx} = -\frac{\partial E_{yz} + E_{yx}}{\partial x} \quad (3.52)$$

3.4 Boundary Conditions

$$\mu \frac{\partial H_{zy}}{\partial t} + \sigma_y^* H_{zy} = \frac{\partial E_{xy} + E_{xz}}{\partial y} \tag{3.53}$$

where the parameters $(\sigma_x, \sigma_y, \sigma_z, \sigma_x^*, \sigma_y^*, \sigma_z^*)$ are homogeneous electric and magnetic conductivities. Applying the central difference approximation to the temporal and spatial partial differential operator, it yields the relevant FDTD equations incorporating the PML absorbing boundary conditions. For example, (3.42) becomes

$$E_{xy}^{n+\frac{1}{2}}(i,j,k) = \frac{\frac{1}{\Delta t} - \frac{\sigma_y}{2\varepsilon}}{\frac{1}{\Delta t} + \frac{\sigma_y}{2\varepsilon}} E_{xy}^{n-\frac{1}{2}}(i,j,k) + \frac{1}{\left(\frac{1}{\Delta t} + \frac{\sigma_y}{2\varepsilon}\right) \cdot \varepsilon \Delta y}$$
$$\cdot \left[H_{zx}^n(i,j+1,k) - H_{zx}^n(i,j,k) + H_{zy}^n(i,j+1,k) - H_{zy}^n(i,j,k) \right] \tag{3.54}$$

It has been shown that for any propagating plane wave normally incident at the interface $\hat{a}$ ($\hat{a}, \hat{b}, \hat{c} = \hat{x}, \hat{y}, \hat{z}$) lying between PML media that have the same ε and μ, the wave will be transmitted into and in between the PML layers with no reflection if the transverse conductivities $\sigma_b, \sigma_b^*, \sigma_c, \sigma_c^*$ are equal and all the pairs of conductivities $(\sigma_x, \sigma_x^*), (\sigma_y, \sigma_y^*), (\sigma_z, \sigma_z^*)$ satisfy the matching impedance condition $(\sigma/\varepsilon = \sigma^*/\mu)$.

This approach is based on the splitting of the field components into two subcomponents. Sacks et al. [28] and Gedney [29], among others, were able to formulate the PML types of ABC based on a Maxwellian formulation that removed the need to split the field. Veihl and Mittra [30] have presented a slightly different formualtion that also utilized the unsplit form of the PMLs. Understandably, implementation of the unsplit field PML to the FDTD method is computationally more efficient than its split counterpart. Furthermore, though not discussed here, it can be extended to nonorthogonal and unstructured grids.

3.4.3 Periodic Boundary Condition (PBC)

As is mentioned earlier, the waves or the fields are in Bloch's state in an infinite periodic structure and can be studied by using the unit cell approach, in which only elements in one unit cell are modeled and the fields in the adjacent unit cells are expressed explicitly using the PBC [31, 32]:

$$\vec{F}(\vec{r} + l_1 \vec{a}_1 + l_2 \vec{a}_2 + l_3 \vec{a}_3) = e^{i \vec{k} \cdot l_1 \vec{a}_1 + l_2 \vec{a}_2 + l_3 \vec{a}_3} \vec{F}(\vec{r}) \tag{3.55}$$

where $\vec{a}_i (i = 1, 2, 3)$ is the component of the lattice constant vector $\vec{a} = \sum_{i=1}^{3} \vec{a}_i$ along the three direction of periodicity, $l_i (i = 1, 2, 3)$ is any integer and k is the wave vector. F can be either the EM field components or the contravariant EM fluxes.

Equation (3.55) is an important boundary condition in metamaterial modeling because it provides an efficient numerical approach for analyzing an infinite periodic structure. When the electromagnetic field components of the cells on the boundary of the FDTD computational domain are updated, we can utilize the PBCs to express the field components in the cell outside the computational domain, in terms of the field components in the cells inside the computational domain. In addition, we introduce a phase shift calculated from the dimension of the unit cell in the updating procedure. For example, (3.56) shows how H_x at the boundary $z = 1$ is updated by using the PBCs when $E_y(i,j,0)$ is not available in the FDTD domain. Equation (3.57) shows the updating equation for E_x at boundary $z = max_z$, where max_z denotes the maximum number in the $\hat{z}$ direction in the FDTD domain.

$$H_x^{n+\frac{1}{2}}(i,j,1) = H_x^{n-\frac{1}{2}}(i,j,1)$$
$$- \frac{\Delta t}{\mu} \left[\frac{E_z^n(i,j,1) - E_z^n(i,j-1,1)}{\Delta y} - \frac{E_y^n(i,j,1) - E_y^n(i,j,max_z) \cdot e^{-ik \cdot max_z \Delta z}}{\Delta z} \right]$$

(3.56)

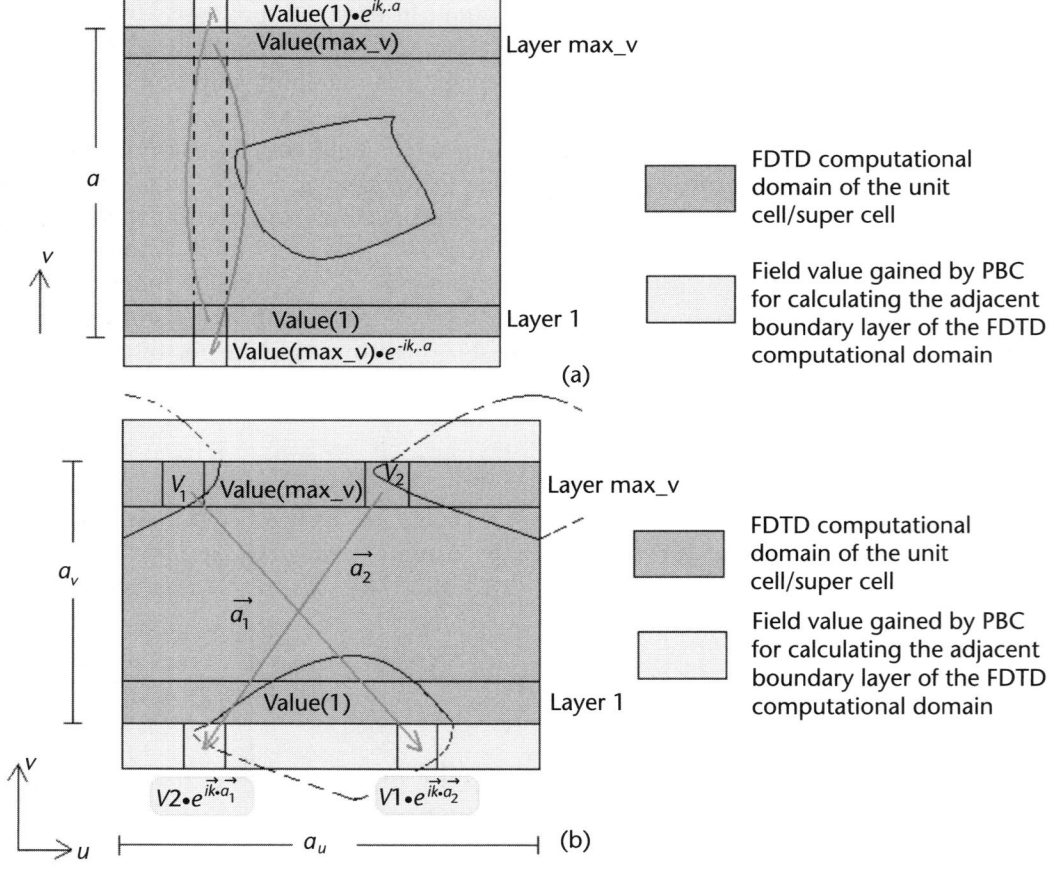

Figure 3.7 PBCs when calculating the infinite EBGs. The inclusion of the EBGs (marked by the solid black line) can be of any shape. The FDTD computational domain is limited to one unit cell/super cell. To calculate the field of the boundary layer of the computational domain (layer 1 or layer max_v in the graph), fields at the adjacent unit cell/super cells are needed but they are outside the computational domain. However, they can be expressed using the field value within the domain (layer max_v or layer 1) applying (3.55). (a) PBCs in EBGs with rectangular lattice. (b) PBCs in EBGs with triangular lattice.

$$E_x^{n+\frac{1}{2}}(i,j,max_z) = E_x^{n-\frac{1}{2}}(i,j,max_z)$$
$$-\frac{\Delta t}{\varepsilon}\left[\frac{H_z^n(i,j+1,max_z)-H_z^n(i,j,max_z)}{\Delta y} - \frac{H_y^n(i,j,1)\cdot e^{ik\cdot max_z \Delta z} - H_y^n(i,j,max_z)}{\Delta z}\right]$$
(3.57)

The other equations for the field components at all the boundaries can be updated in a similar manner. Figure 3.7(a, b) illustrate the interpolations of PBCs in the EBGs with square and triangular lattices, respectively.

3.5 Bandgap Calculation

Some of the EBG characteristics can be obtained in an intuitive way from the results of the FDTD modeling of bandgap structures. For example, the passband and stopband behaviors and the transmission/reflection coefficients can be obtained readily from the time-domain field response. However, it becomes necessary to carry out some postprocessing (see Figure 3.8) to derive the dispersion relationship (bandgap characteristics) of an EBG structure. These will be further discussed later.

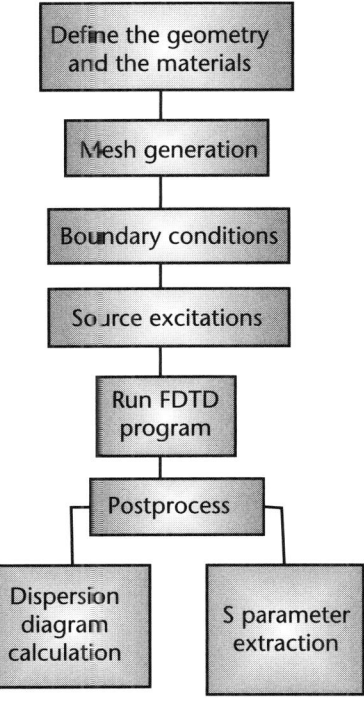

Figure 3.8 The FDTD procedure in modeling EBG structures.

3.5.1 Source Excitation

A modulated Gaussian pulse (also termed as the Gabor pulse) with the following form is typically used in the simulation of EBG structures:

$$S(t) = e^{-\frac{t^2}{2\sigma^2}} \cos(2\pi\xi t + \phi) \qquad (3.58)$$

where σ controls the (effective) time width of the pulse and, hence the bandwidth of the source [33]. Also, ξ and ϕ are the frequency and phase, respectively, of the single frequency wave that modulates the Gaussian pulse. Figure 3.9 shows a typical modulated Gaussian waveform with $\sigma = 8 \times 10^{-6}$ s, $\xi = 100$ kHz, and $\phi = 0$ rad.

A modulated Gaussian pulse is chosen for the excitation because it offers good time and frequency resolutions. In contrast to a truncated sine pulse, the pulse energy of the modulated Gaussian pulse is more concentrated near the center of the pulse in the time as well as the frequency spectrum. We can define a desired form of a modulated Gaussian pulse tailored for the desired application by controlling the time width σ.

3.5.2 Dispersion Diagram Calculation

The dispersion diagram is very useful for studying the bandgap characteristics of an infinite EBGs. As discussed in Chapter 2, the dispersion diagram is a plot of the possible modes as functions of the wave vector in the irreducible Brillouin zone. As mentioned earlier, by using the PBCs, the infinite EBGs can be modeled by using only one cell, or only a group of cells if there are defects in the EBGs. The cell and the group of cells will be referred to in the following sections as the unit cell and the super cell, respectively [34]. The $\hat{k}$ vector from the irreducible Brillouin zone is used to set up the PBCs.

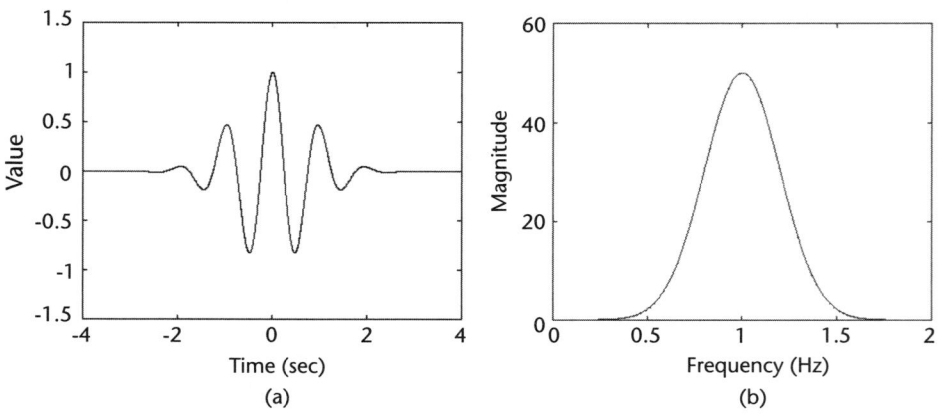

Figure 3.9 The modulated Gaussian pulse: (a) the shape of the modulated Gaussian pulse in the time domain and (b) the magnitude of its Fourier transform [33].

3.5.2.1 The Unit Cell Approach

After a mesh is set up for the unit cell, one can randomly choose a few points in the mesh as the source points [32] and several points as the probing field (or monitoring) points. The latter should be evenly spaced and be sufficiently dense to capture the nuances of the possible modal distributions of the fields that would be generated.

A modulated Gaussian pulse is applied at the source points to excite the electromagnetic (EM) modes of the EBG over a wide range of frequencies. As the field evolves during the FDTD simulation, only the true transmission modes would survive, while the pseudotransmission modes would decay [35]. As we proceed with the FDTD simulation, we record the temporal responses at the probing points at each time step until the time-domain signature stabilized sufficiently. The temporal signature should be long enough to achieve the desired frequency resolution after the Fourier transformation, but not so long as to run into instabilities (especially the late time instabilities in the NFDTD algorithm), or to prolong the computation effort unnecessarily.

The next step is to Fourier-transform the temporal signatures to obtain the frequency spectra that typically exhibit peaks at certain frequency values, indicating the existence of the transmission modes, or the eigenmodes supported by the EBGs, corresponding to the wave vector $\hat{k}$. Plotting these frequency values against the wave vector $\hat{k}$ yields the dispersion diagram of the EBGs.

3.5.2.2 The Super-Cell Approach

The super-cell approach is very similar to the unit cell approach we described above, although the modeling domain now consists of defect cell(s) normally embedded within regular unit cells. PBCs are used to terminate the modeling domain under the assumption that the structure is infinite. In the direction in which the defects appear periodically, the PBCs are imposed one period (of the defects) away from each other. If the defects spoil the periodicity of the structure, we need to create it by including a sufficient number of regular cells that nest it, thus creating a supercell. Since the PBCs artificially introduce the periodicity, the number of EBG unit cells should be sufficiently large to isolate the EM modes from the spurious defects in the neighboring super cells. On the other hand this unit cell layer should be as small as possible in order to maximize the computational efficiency. This number is often determined experientially [35] and is normally more than 10. Figure 3.10 shows the super cell used by Chutinan et al. when they model the waveguide created by filling in one column of the air holes in the EBGs [34].

3.5.3 Transmission and Reflection Coefficient Calculation

If the bandgap structure is not infinite in the direction of its periodicity, then the transmission coefficient is helpful when we wish to find the bandgap of the structure of interest.

Let us consider a two-dimensional EBG with four arrays of cylindrical rods in the $\hat{x}$-direction and infinitely loaded rods in the $\hat{y}$-direction in free space. Since the rods are infinite in the $\hat{y}$-direction, the PBC can still be used to terminate the

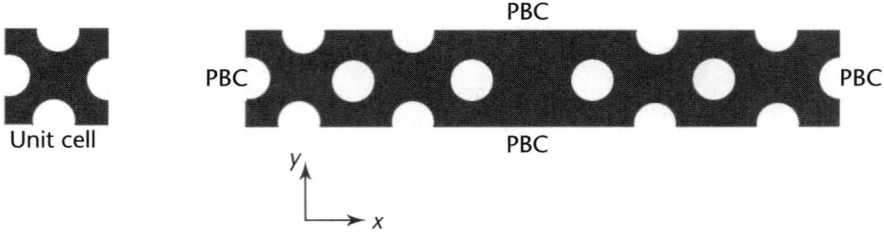

Figure 3.10 The super cell of the waveguide created by filling up one column of the air holes in the EBGs [34]. In the y-direction, the defects are periodic with period of one unit cell, so one unit cell is used in between the PBCs. In the x-direction, five unit cells are used between the PBCs to isolate the modes from the neighboring spurious defects.

computational domain in this direction. In the $\hat{x}$-direction, in which there exist four arrays of rods, an ABC is used to terminate the computational domain at a suitable distance away from the scattering structure. By combining PBCs with ABCs in this way, the computational efficiency of modeling the finite EBGs can be enhanced. The boundary condition set-up used for this purpose is shown in Figure 3.11.

Let a plane wave traveling in the $\hat{x}$-direction with the time signature of a modulated Gaussian pulse be excited from a line source at one end of the structure. Temporal signatures of two lines of probes located at the two ends of the structures (see Figure 3.11) are recorded and analyzed to calculate the transmission and reflection coefficients, respectively. For probe set 1, the Fourier transformation can be applied directly to the time domain signal. The averaged frequency spectrum along the probe line shows the transmission coefficient as a function of frequency from which the bandgap can be readily found. The procedure for computing the reflection coefficient is similar, but the signal is recorded for this calculation after the first pulse has passed.

If the structure is finite in size, then the entire computational domain should be terminated with ABCs in all directions. A modulated Gaussian pulse is launched at one side of the EBGs (either a point or a line source) and the probes are located at

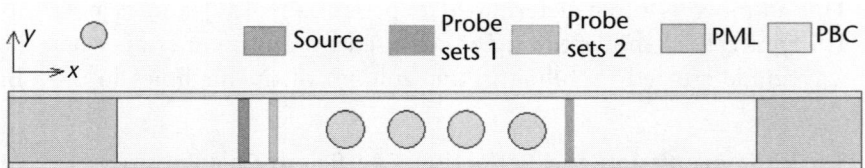

Figure 3.11 Numerical model for a two-dimensional EBG structure of semi-finite size. There are four arrays of cylindrical rods in the x-direction and infinitely loaded rods in the y-direction. So the computation domain is terminated by the PBCs in the y-direction and by ABCs in the x-direction. A plane wave source in form of modulated Gaussian pulse is defined at one side of the EBGs. The responses at the other side of the EBGs are collected as probe set 1 for calculating the transmission coefficient (S_{21}). The responses at the same side of the EBGs are collected as probe set 2 for calculating the reflection coefficient (S_{11}).

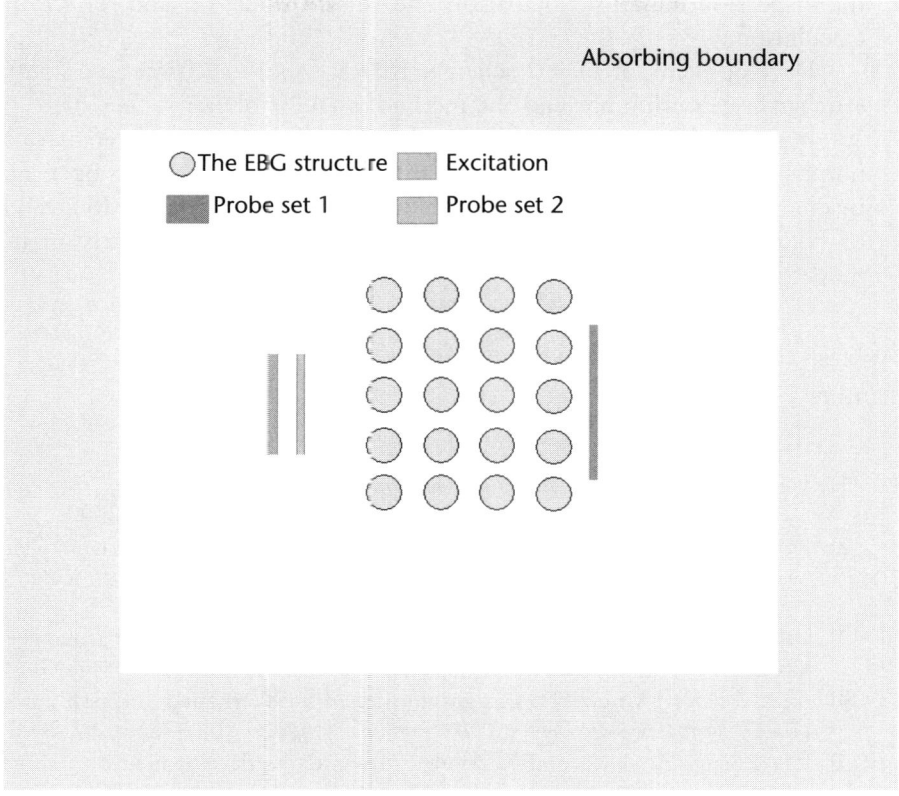

Figure 3.12 Numerical model for an EBG structure of finite size. The computational domain is terminated by the ABCs in all the directions. A modulated Gaussian pulse is excited at one side of the EBGs. The responses at probe set 1 and probe set 2 with the existence of the EBG are calibrated by those without EBG in the calculation of the transmission coefficient (S_{21}) and the reflection coefficient (S_{11}).

the other side. An additional simulation is needed in this case with the EBG structure replaced by free space. Then the bandgap can be found with the transmission coefficient from the EBGs by using the free space case as a reference (Figure 3.12).

3.6 Summary

The FDTD method is widely used because it is simple to implement numerically. It provides a flexible means for directly solving Maxwell's time-dependent curl equations by using finite differences to discretize them. It can be used to solve various types of electromagnetic problems, including the anisotropic or nonlinear problems. This chapter briefly reviewed the fundamentals of the FDTD method, including Yee's spatial and temporal grid, the updating formulation, and two important boundary conditions—the ABCs and the PBCs. The techniques specifically tailored for calculating the EBG-related parameters are also presented, including the dispersion diagram calculation and the transmission and reflection coefficient calculation.

Developments of FDTD schemes that are more accurate and computationally efficient are continuing, and the method is still evolving.

It is worth noting that FDTD algorithm has recently been parallelized by a number of workers, including Yu and Mittra [36], and this has open up new vistas by enabling us to solve problems with very large degrees of freedom (DOF), upward of billions of unknowns, such as those encountered frequently in metamaterial modelings.

References

[1] A. Taflove and S.C. Hagness, *Computational Electrodynamics: The Finite-Difference Time-Domain Method*, 2nd ed., Norwood, MA: Artech House, 2000.

[2] K.S. Yee, "Numerical solution of initial boundary value problems involving Maxwell's equations in isotropic media," *IEEE Trans. Antennas Propagat.*, Vol. AP-14, No. 3, 302–307, 1966.

[3] R. Holland, "Finite-difference solution of Maxwell's equations in generalized nonorthogonal coordinates," *IEEE Trans. on Nuclear Science*, NS-30, 4589–4591, 1983.

[4] T. Namiki, "A new FDTD algorithm based on alternating-direction implicit method," *IEEE Trans. on Microwave Theory and Techniques*, Vol. 47, No. 10, 2003–2007, 1999.

[5] F. Zheng, Z. Chen, and J. Zhang, "Toward the development of a three-dimensional unconditionally stable finite-difference time-domain method," *IEEE Trans. on Microwave Theory and Techniques*, Vol. 48, No. 9, 1550–1558, 2000.

[6] J.F. Thompson, Z.U.A. Ward, and C.W. Mastin, *Numerical Grid Generation: Foundations and Applications*, New York: Elsevier North-Holland, Inc., 1985.

[7] A.C. Cangellaris and D.B. Wright, "Analysis of the numerical error caused by the stair-stepped approximation of a conducting boundary in FDTD simulations of electromagnetic phenomena," *IEEE Trans. Antennas Propagat.*, Vol. 39, 1518–1525, 1991.

[8] M. Okoniewski, E. Okoniewska, and M.A. Stuchly, "Three-dimensional subgridding algorithm for FDTD," *IEEE Transactions on Antennas and Propagation*, Vol. 45, Issue 3, 422–429, 1997.

[9] I.S. Kim and W.J.R. Hoefer, "A local mesh refinement algorithm for the time domain-finite difference method using Maxwell's curl equations," *IEEE Trans. on Microwave Theory and Techniques*, Vol. 38, Issue 6, 812–815, 1990.

[10] K.M. Krishnaiah and C.J. Railton, "Passive equivalent circuit of FDTD: an application to subgridding," *Electronics Letters*, Vol. 33, Issue 15, 1277–1278, 1997.

[11] D.T. Prescott and N.V. Shuley, "A method for incorporating different sized cells into the finite-difference time-domain analysis technique," *IEEE Microwave and Guided Wave Letters*, (see also *IEEE Microwave and Wireless Components Letters*), Vol 2, Issue 11, 434–436, 1992.

[12] S.S. Zivanovic, K.S. Yee, and K.K. Mei, "A subgridding method for the time-domain Finite-Difference Method to Solve Maxwell's Equations," *IEEE Transactions on Microwave Theory and Techniques*, Vol. 39, No. 3, 471–479, 1991.

[13] J. Stratton, *Electromagnetic Theory*, New York: McGraw-Hill, 1941.

[14] K.S. Yee, J.S. Chen, and A.H. Chang, "Conformal finite-difference time-domain (FD-TD) with overlapping grid," *IEEE Trans. Antennas Propagat.*, Vol. 40, 1068–1075, 1992.

[15] J.F. Lee, R. Palandech, and R. Mittra, "Modeling three-dimensional discontinuities in waveguides using nonorthogonal FDTD algorithm," *IEEE Transactions on Microwave Theory and Techniques*, Vol. 40, No. 2, 346-352, 1992.

[16] P.H. Harms, J.-F. Lee, and R. Mittra, "A study of the nonorthogonal FDTD method versus the conventional FDTD technique for computing resonant frequencies of cylindrical cavities," *IEEE Transactions on Microwave Theory and Techniques*, Vol. 40, Issue 4, 741-746, 1992.

[17] T.G. Jurgens, A. Taflove, K. Umashankar, and T.G. Moore, "Finite-difference time-domain modeling of curved surfaces," *IEEE Transactions on Antennas and Propagation*, Vol 40, Issue 4, 357-366, 1992

[18] T.G. Jurgens and A. Taflove, "Three-dimensional contour FDTD modeling of scattering from single and multiple bodies," *IEEE Transactions on Antennas and Propagation*, Vol. 41, Issue 12, 1703-1708, 1993.

[19] H. Yang and C.J. Railton, "Efficient and accurate FDTD algorithm for the treatment of curved material boundaries," *Microwaves, Antennas and Propagation, IEE Proceedings*, Vol. 144, Issue 5, 382-388, 1997.

[20] Y. Hao and C.J. Railton, "Analyzing electromagnetic structures with curved boundaries on Cartesian FDTD meshes," *IEEE Transactions on Microwave Theory and Techniques*, Vol. 46, No. 1, 82-86. 1998.

[21] V. Douvalis, Y. Hao, and C.G. Parini, "Reduction of late time instabilities of the finite difference time domain method in curvilinear coordinates," *Computation in Electromagnetics. CEM 2002. The Fourth International Conference on (Ref. No. 2002/063)*, 8-11, 2002.

[22] V. Douvalis, Y. Hao, and C.G. Parini, "Reduction of late time instabilities of the local-distorted nonorthogonal finite difference time domain method," *IEEE Antennas and Propagation Society International Symposium*, Vol. 3, 628-631, 2002.

[23] R. Schuhmann and T. Weiland, "FDTD on Nonorthogonal Grid with Triangular Fillings," *IEEE Transactions on Magnetics*, Vol. 35, No. 3, 1470-1473, 1999.

[24] B. Engquist and A. Majda, "Absorbing boundary conditions for the numerical simulations of waves," *Math. Comp.*, Vol. 31, 629-651, 1977.

[25] G. Mur, "Absorbing boundary conditions for the finite difference qpproximation of the time-domain electromagnetic-field equations," *IEEE Trans. Electromagn. Compat.*, Vol. 23, 377-382, 1981.

[26] J.P. Bérenger, "A Perfect Matched Layer for the Absorption of Electromagnetic Waves," *J. Comput. Phys.*, Vol. 114, 185-200, 1994.

[27] J.P. Bérenger, "Three-Dimensional Perfect Matched Layer for the Absorption of Electromagnetic Waves," *J. Comput. Phys.*, Vol. 127, 363-379, 1995.

[28] Z.S. Sacks, D.M. Kingsland, R. Lee, and J.-F. Lee, "A perfectly matched anisotropic absorber for use as an absorbing boundary condition," *IEEE Trans. Antennas Propagat.*, Vol. 43, no. 12, 1460-1463, 1995.

[29] S.D. Gedney, "An anisotropic perfectly matched layer-absorbing medium for the truncation of FDTD lattices," *IEEE Trans. Antennas Propagat.*, Vol. 44, 1630-1639, 1996.

[30] J.C. Veihl and R. Mittra, "An efficient implementation of Berengers perfectly matched layer (PML) for finite-difference time-domain mesh truncation," *IEEE Microwave Guided Wave Lett.*, Vol. 6, 94-96, 1996.

[31] L. Zhao and A.C. Cangellaris, "GT-PML: Generalized theory of perfectly matched layers and its application to the reflectionless truncation of finite-difference time-domain grids," *IEEE Trans. Microwave Theory Tech.*, Vol. 44, 2555-2563, 1996.

[32] C.T. Chan, Q.L. Yu, and K.M. Ho, "Order-N spectral method for electromagnetic waves," *Phys. Rev. B, 51*, 16635-16642, 1995.

[33] J-C. Hong, K.H. Sun, and Y.Y. Kim, "The matching pursuit approach based on the modulated Gaussian pulse for efficient guided-wave damage inspection," *Smart Mater. Struct.*, Vol. 14, 548–560, 2005.

[34] A. Chutinan and S. Noda, "Waveguides and waveguide bends in two-dimensional photonic crystal slabs," *Phys. Rev. B*, Vol. 62, Issue 7, 4488–4492, 2000.

[35] M. Qiu and S. He, "Guided modes in a two-dimensional metallic photonic crystal waveguide," *Phys. Lett. A*, Vol. 76, 425–429, 2000.

[36] W. Yu, Y. Liu, T. Su, and R. Mittra, "A robust parallel conformal finite difference time domain processing package using the MPI library," *IEEE Antennas Propag. Mag.*, Vol. 47, No. 3, pp. 39–59, June 2005.

CHAPTER 4
FDTD Modeling of EBGs and Their Applications

4.1 Introduction

In this chapter, some examples of FDTD modeling of electromagnetic bandgap (EBG) structures and their applications will be presented. In addition, the procedure for computing the dispersion diagram of an infinite EBG structure, as well as deriving the transmission coefficient of a (semi-)infinite bandgap structure will be detailed. Since most of existing EBGs either have curved unit cells, or rhombic lattices, modeling such structures often requires a special treatment because the conventional Yee's FDTD is not only inefficient; it may also introduce numerical artifacts when dealing with curved boundaries by using staircase approximations. To address this issue, a conformal FDTD algorithm, namely NFDTD, will be applied to model the aforementioned structures. A comparison of the numerical efficiency and accuracy of Yee and the conformal FDTD algorithms will be carried out.

Finally, a design of a millimeter-wave EBG antenna based on the FDTD simulation will be presented, and the results will be compared with the measured data for the same antenna at 95 GHz.

4.2 FDTD Modeling of Infinite Electromagnetic Bandgap Structures

In this section, we will carry out a study of 2-D infinite EBG structures with curved inclusions by using both the conformal and Yee's FDTD schemes. We will discuss the requirements of spatial resolution, computer memory, processing time, efficiency, and accuracy of the conformal method by using the unit cell approach discussed in Chapter 3.

4.2.1 Physical Model of EBG Structures

An array of metallic rods in free space periodically arranged in square and triangular/rhombic lattices will now be modeled by using both the NFDTD and the Yee's schemes. The structures are assumed to be infinite in the $\hat{x}$- and the $\hat{y}$-directions both, with a lattice constant (period) equal to a, and they are infinitely long in the $\hat{z}$-direction. Each rod is made from copper, whose conductivity is approximately $\sigma = 5.8 \times 10^7$ S/m at microwave frequencies. The ratio of the radius r to the lattice constant a is chosen to be $r/a = 0.2$. Numerical simulations will be performed to determine the dispersion diagrams for both the transverse electric (TE) and transverse magnetic (TM) polarizations, for EBGs with square (Figure 4.1), as well as triangular/rhombic lattices (Figure 4.2).

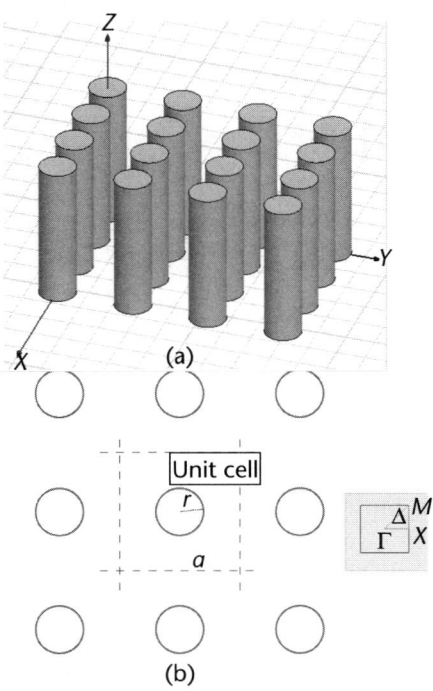

Figure 4.1 (a) A three-dimensional view of EBG with square lattices (radius r, spacing a, and $r/a = 0.2$). (b) The x-y plane cut out of the EBGs.

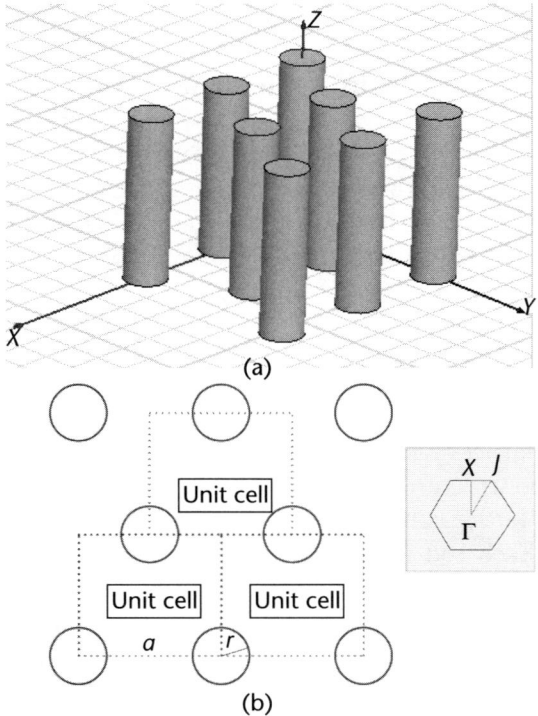

Figure 4.2 (a) A three-dimensional view with triangular/rhombic (radius r, spacing a, and $r/a = 0.2$). (b) The x-y plane cut out of the EBGs.

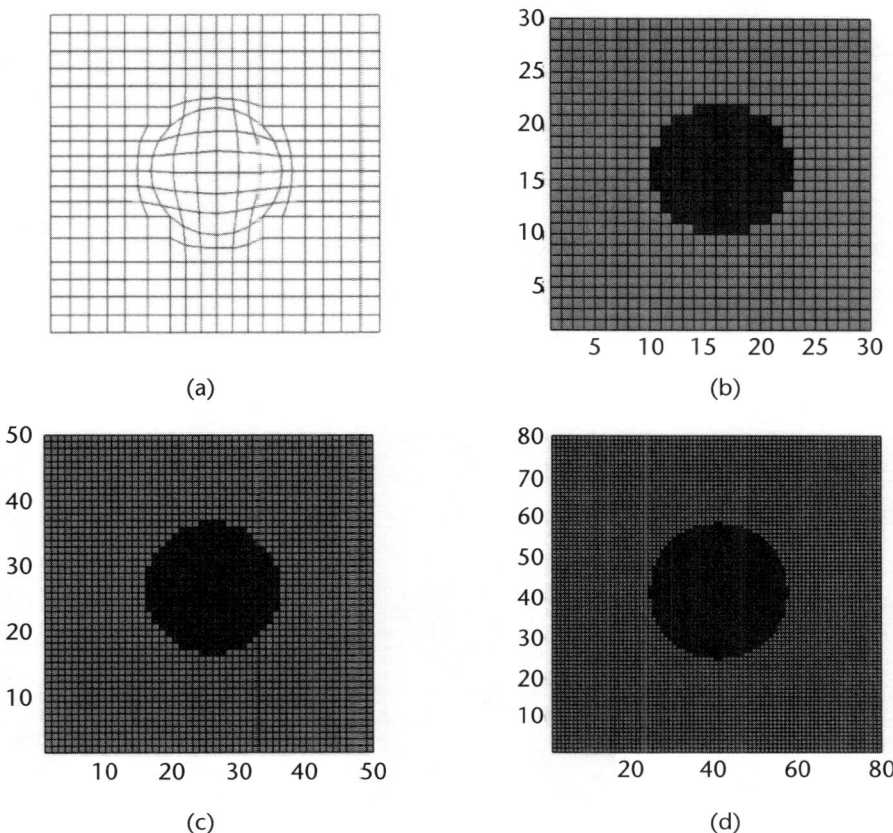

Figure 4.3 Examples of the mesh schemes for a unit cell from the square lattice in the NFCTD and Yee's FDTD modeling: (a) (18 × 18) NFDTD cells, (b) (30 × 30) FDTD cells, (c) (50 × 50) FDTD cells, and (d) (80 × 80) FDTD cells.

4.2.2 Mesh Generation and Simulation Parameters in FDTD Modeling

We discretize the unit cells in square and triangular/rhombic lattices [see Figures 4.1(b) and 4.2(b)] using both Cartesian and structured conformal grids, with various cell sizes; examples are shown in Figures 4.3 and 4.4.

Since the metallic rods in the EBGs are periodically arranged, and because the structure is infinite in both the $\hat{x}$- and $\hat{y}$-directions, we can model it simply by using their unit cells to which we impose the PBCs. Other parameters used in the FDTD simulation include the excitation source, probe settings, definition of computational domain and the boundary conditions, all of which are illustrated in Figure 4.5.

A modulated Gaussian pulse is used to provide a wide-band excitation at different positions inside the computation domain comprising a unit element of the EBGs. The location of the excitation source, as well as of the field probes, is often chosen randomly, though they are required to excite and detect all possible resonant modes inside the structure. We can apply the fast Fourier transformation

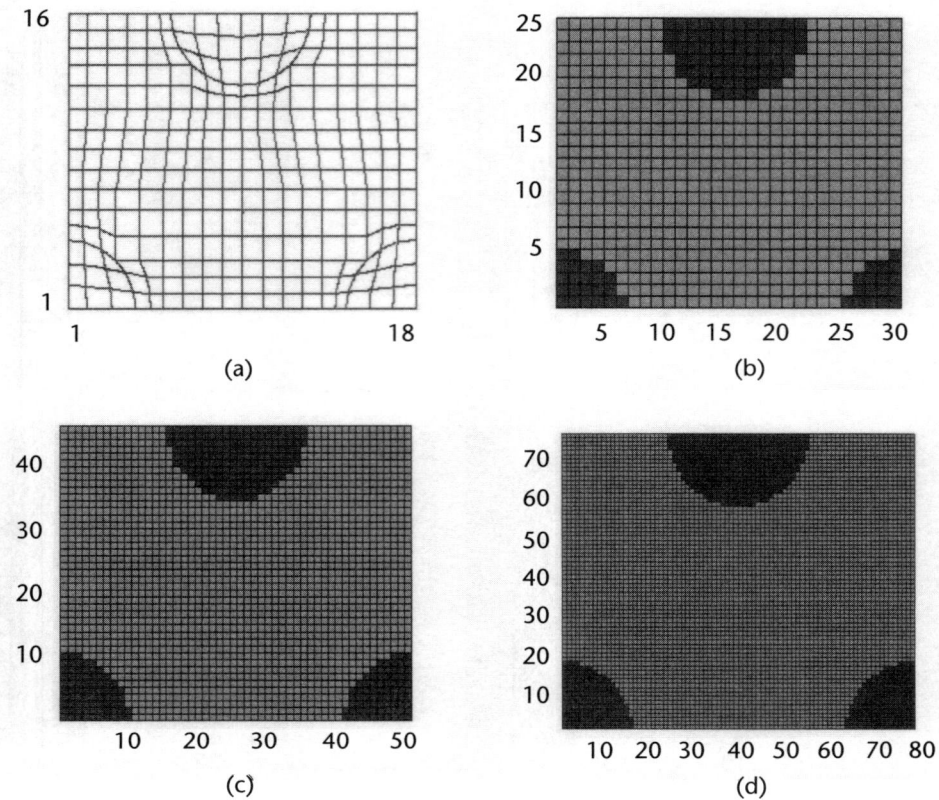

Figure 4.4 Examples of the mesh schemes for an unit cell from the triangular lattice in the NFDTD and Yee's FDTD modeling: (a) (18 × 15) NFDTD cells, (b) (30 × 26) FDTD cells, (c) (50 × 42) FDTD cells, and (d) (80 × 69) FDTD cells.

(FFT) to the temporal signatures collected from different probes, to derive the frequency spectra response, from which we can extract the dispersion diagram of the EBG for various angles of excitation.

4.2.3 Simulation Results of Infinite EBGs Using the Conformal and Yee's FDTD

In 1983, Holland developed a NFDTD algorithm and opened up the possibility for a more general, efficient and accurate numerical time domain method for modeling curved boundaries [1]. In this method, the FDTD technique is no longer restricted to an orthogonal Cartesian grid, and a generalized curvilinear coordinate system is used instead. As a consequence, an arbitrary structure with a curved boundary, or a canted surface, can be meshed conformally, without employing the staircase approximation as is the case in the Yee's algorithm. Since the NFDTD method was proposed, it has been successfully used to analyze the optical dielectric waveguide, the dielectric-loaded resonant cavity, microstrip discontinuities, and periodic structures at oblique incidence.

4.2 FDTD Modeling of Infinite Electromagnetic Bandgap Structures

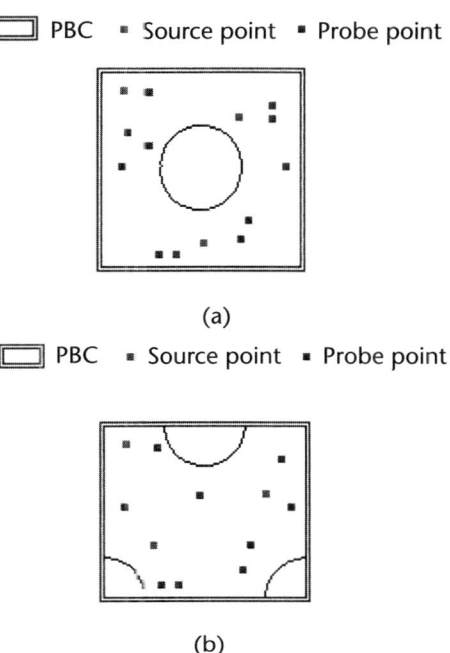

Figure 4.5 Modeling schemes of the unit cell approach for EBG structures: (a) for a square lattice of Figure 4.1 and (b) for a triangular/rhombic lattice of Figure 4.2.

In a nonorthogonal coordinate system, the Maxwell's curl equations in a source-free and loss-free medium, can be written as [1–3]:

$$-\mu \frac{\partial h^1}{\partial t} = \frac{1}{\sqrt{g}} \left(\frac{\partial e_3}{\partial \vec{u}_2} - \frac{\partial e_2}{\partial \vec{u}_3} \right)$$

$$-\mu \frac{\partial h^2}{\partial t} = \frac{1}{\sqrt{g}} \left(\frac{\partial e_1}{\partial \vec{u}_3} - \frac{\partial e_3}{\partial \vec{u}_1} \right)$$

$$-\mu \frac{\partial h^3}{\partial t} = \frac{1}{\sqrt{g}} \left(\frac{\partial e_2}{\partial \vec{u}_1} - \frac{\partial e_1}{\partial \vec{u}_2} \right)$$

$$\varepsilon \frac{\partial e^1}{\partial t} = \frac{1}{\sqrt{g}} \left(\frac{\partial h_3}{\partial \vec{u}_2} - \frac{\partial h_2}{\partial \vec{u}_3} \right)$$

$$\varepsilon \frac{\partial e^2}{\partial t} = \frac{1}{\sqrt{g}} \left(\frac{\partial h_1}{\partial \vec{u}_3} - \frac{\partial h_3}{\partial \vec{u}_1} \right)$$

$$\varepsilon \frac{\partial e^3}{\partial t} = \frac{1}{\sqrt{g}} \left(\frac{\partial h_2}{\partial \vec{u}_1} - \frac{\partial h_1}{\partial \vec{u}_2} \right) \tag{4.1}$$

By following procedure similar to the one in the Cartesian FDTD scheme, we can discretize (4.1) using the a central-difference approximation, as follows:

$$h^{1(n+\frac{1}{2})}\left(i+\frac{1}{2},j,k\right) = h^{1(n-\frac{1}{2})}\left(i+\frac{1}{2},j,k\right)$$

$$-\frac{\Delta t}{\mu\left(i+\frac{1}{2},j,k\right)\sqrt{g\left(i+\frac{1}{2},j,k\right)}}$$

$$\cdot \left(e_3^{(n)}\left(i+\frac{1}{2},j+\frac{1}{2},k\right) - e_3^{(n)}\left(i+\frac{1}{2},j-\frac{1}{2},k\right) \right)$$

$$+\frac{\Delta t}{\mu\left(i+\frac{1}{2},j,k\right)\sqrt{g\left(i+\frac{1}{2},j,k\right)}}$$

$$\cdot \left(e_2^{(n)}\left(i+\frac{1}{2},j,k+\frac{1}{2}\right) - e_2^{(n)}\left(i+\frac{1}{2},j,k-\frac{1}{2}\right) \right)$$

$$h^{2(n+\frac{1}{2})}\left(i,j+\frac{1}{2},k\right) = h^{2(n-\frac{1}{2})}\left(i,j+\frac{1}{2},k\right)$$

$$-\frac{\Delta t}{\mu\left(i,j+\frac{1}{2},k\right)\sqrt{g\left(i,j+\frac{1}{2},k\right)}}$$

$$\cdot \left(e_1^{(n)}\left(i,j+\frac{1}{2},k+\frac{1}{2}\right) - e_1^{(n)}\left(i,j+\frac{1}{2},k-\frac{1}{2}\right) \right)$$

$$+\frac{\Delta t}{\mu\left(i,j+\frac{1}{2},k\right)\sqrt{g\left(i,j+\frac{1}{2},k\right)}}$$

$$\cdot \left(e_3^{(n)}\left(i+\frac{1}{2},j+\frac{1}{2},k\right) - e_3^{(n)}\left(i-\frac{1}{2},j+\frac{1}{2},k\right) \right)$$

$$h^{3(n+\frac{1}{2})}\left(i,j,k+\frac{1}{2}\right) = h^{3(n-\frac{1}{2})}\left(i,j,k+\frac{1}{2}\right)$$

$$-\frac{\Delta t}{\mu\left(i,j,k+\frac{1}{2}\right)\sqrt{g\left(i,j,k+\frac{1}{2}\right)}}$$

$$\cdot \left(e_2^{(n)}\left(i+\frac{1}{2},j,k+\frac{1}{2}\right) - e_2^{(n)}\left(i-\frac{1}{2},j,k+\frac{1}{2}\right) \right)$$

$$+\frac{\Delta t}{\mu\left(i,j,k+\frac{1}{2}\right)\sqrt{g\left(i,j,k+\frac{1}{2}\right)}}$$

$$\cdot \left(e_1^{(n)}\left(i,j+\frac{1}{2},k+\frac{1}{2}\right) - e_1^{(n)}\left(i,j-\frac{1}{2},k+\frac{1}{2}\right) \right)$$

4.2 FDTD Modeling of Infinite Electromagnetic Bandgap Structures

$$e^{1\,(n+1)}\left(i,j+\frac{1}{2},k+\frac{1}{2}\right) = e^{1\,(n)}\left(i,j+\frac{1}{2},k+\frac{1}{2}\right)$$

$$+ \frac{\Delta t}{\varepsilon\left(i,j+\frac{1}{2},k+\frac{1}{2}\right)\sqrt{g\left(i,j+\frac{1}{2},k+\frac{1}{2}\right)}}$$

$$\cdot \left(h_3^{(n+\frac{1}{2})}\left(i,j+1,k+\frac{1}{2}\right) - h_3^{(n+\frac{1}{2})}\left(i,j,k+\frac{1}{2}\right)\right)$$

$$- \frac{\Delta t}{\varepsilon\left(i,j+\frac{1}{2},k+\frac{1}{2}\right)\sqrt{g\left(i,j+\frac{1}{2},k+\frac{1}{2}\right)}}$$

$$\cdot \left(h_2^{(n+\frac{1}{2})}\left(i,j+\frac{1}{2},k+1\right) - h_2^{(n+\frac{1}{2})}\left(i,j+\frac{1}{2},k\right)\right)$$

$$e^{2\,(n+1)}\left(i+\frac{1}{2},j,k+\frac{1}{2}\right) = e^{2\,(n)}\left(i+\frac{1}{2},j,k+\frac{1}{2}\right)$$

$$+ \frac{\Delta t}{\varepsilon\left(i+\frac{1}{2},j,k+\frac{1}{2}\right)\sqrt{g\left(i+\frac{1}{2},j,k+\frac{1}{2}\right)}}$$

$$\cdot \left(h_1^{(n+\frac{1}{2})}\left(i+\frac{1}{2},j,k+1\right) - h_1^{(n+\frac{1}{2})}\left(i+\frac{1}{2},j,k\right)\right)$$

$$- \frac{\Delta t}{\varepsilon\left(i-\frac{1}{2},j,k+\frac{1}{2}\right)\sqrt{g\left(i+\frac{1}{2},j,k+\frac{1}{2}\right)}}$$

$$\cdot \left(h_3^{(n+\frac{1}{2})}\left(i+1,j,k+\frac{1}{2}\right) - h_3^{(n+\frac{1}{2})}\left(i,j,k+\frac{1}{2}\right)\right)$$

$$e^{3\,(n+1)}\left(i+\frac{1}{2},j+\frac{1}{2},k\right) = e^{3\,(n)}\left(i+\frac{1}{2},j+\frac{1}{2},k\right)$$

$$+ \frac{\Delta t}{\varepsilon\left(i+\frac{1}{2},j+\frac{1}{2},k\right)\sqrt{g\left(i+\frac{1}{2},j+\frac{1}{2},k\right)}}$$

$$\cdot \left(h_2^{(n-\frac{1}{2})}\left(i+1,j+\frac{1}{2},k\right) - h_2^{(n+\frac{1}{2})}\left(i,j+\frac{1}{2},k\right)\right)$$

$$- \frac{\Delta t}{\varepsilon\left(i+\frac{1}{2},j+\frac{1}{2},k\right)\sqrt{g\left(i+\frac{1}{2},j+\frac{1}{2},k\right)}}$$

$$\cdot \left(h_1^{(n+\frac{1}{2})}\left(i+\frac{1}{2},j+1,k\right) - h_1^{(n+\frac{1}{2})}\left(i+\frac{1}{2},j,k\right)\right)$$

(4.2)

To derive an explicit time-marching nonorthogonal FDTD scheme, we need to compute the covariant field components (e_i/h_i) that must be calculated using their

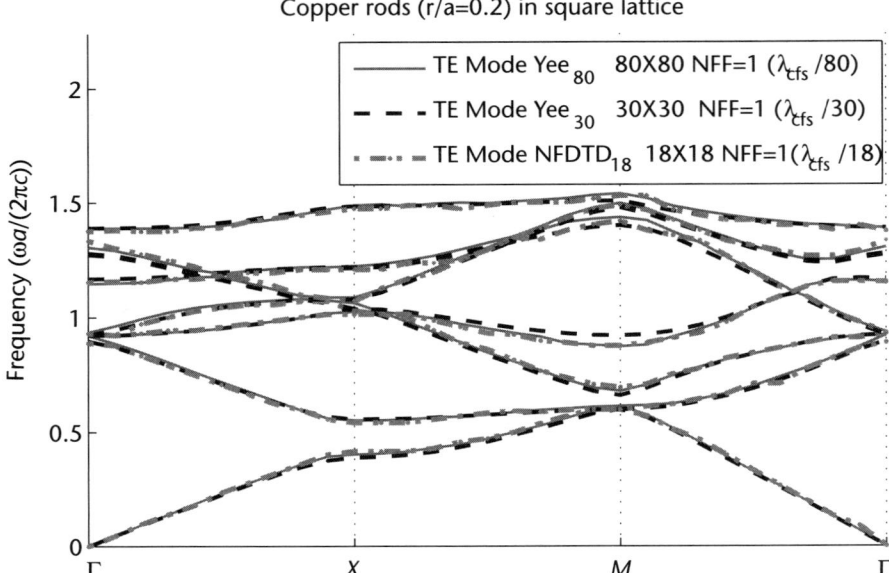

Figure 4.6 The dispersion diagram of the first few TE modes, for the copper rods in free space in square lattice. The Yee's FDTD and the NFDTD result with the low spatial resolution is plotted to compare with the high spatial resolution result. It can be seen that the maximum disagreement appears at M point. (NFF is the normalized central feeding frequency parameter of the modulated Gaussian pulse. It is a frequency parameter in the modulated Gaussian pulse.)

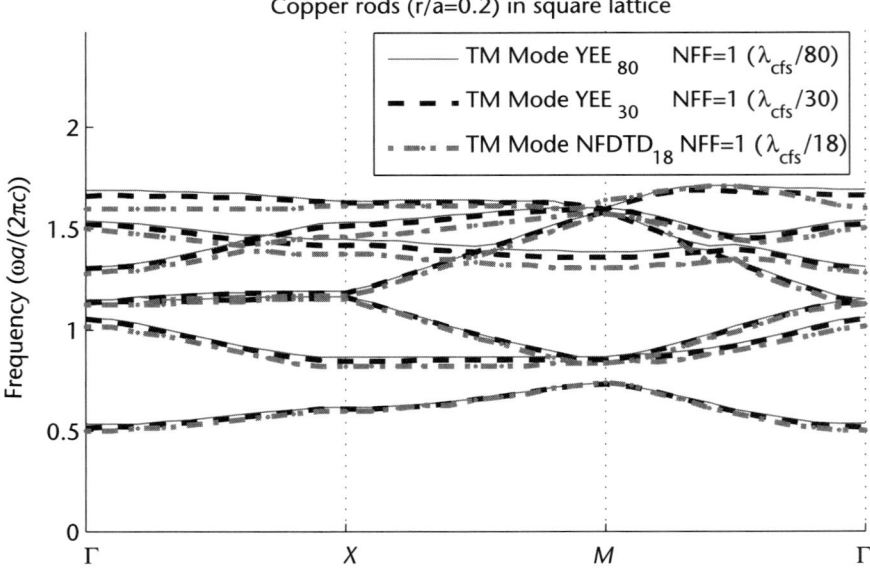

Figure 4.7 The dispersion diagram of the first few TM modes, from the copper rods in free space in square lattice. The Yee's FDTD simulation result with high spatial resolution is plotted as a reference result. The Yee's FDTD and the NFDTD result with the low spatial resolution is plotted to compare with the high spatial resolution result. It can be seen that the maximum disagreement appears at M point.

4.2 FDTD Modeling of Infinite Electromagnetic Bandgap Structures

dual, namely the contravariant field components (e^i/h^i); and thus, in turn, we require additional projection equations (4.3). Since the points where the covariant field values are sampled should be physically collocated with the contravariant ones, we need an interpolation scheme (i.e., a neighbor averaging projection scheme) of the type given here:

$$
\begin{aligned}
v^1(i,j,k) = &\ g_{11}(i,j,k)v^1(i,j,k) \\
&+ \frac{g_{12}(i,j,k)}{4}\Big(v^2(i-1,j,k) + v^2(i-1,j+1,k) \\
&+ v^2(i,j,k) + v^2(i,j+1,k)\Big) \\
&+ \frac{g_{13}(i,j,k)}{4}\Big(v^3(i-1,j,k) + v^3(i-1,j,k+1) \\
&+ v^3(i,j,k) + v^3(i,j,k+1)\Big)
\end{aligned}
\tag{4.3}
$$

where v stands for either e or h. The above implies that the contravariant components v^1, v^2, and v^3 can be obtained from (4.2).

We have computed the dispersion diagrams of the EBG (shown in Figures 4.3 and 4.4) by using the conformal FDTD scheme described earlier, and compare them with those obtained by using the conventional Yee's algorithm. The results are shown in Figures 4.6–4.9. The simulated results for calculating the dispersion diagrams are convergent provided that the spatial resolution of the FDTD scheme

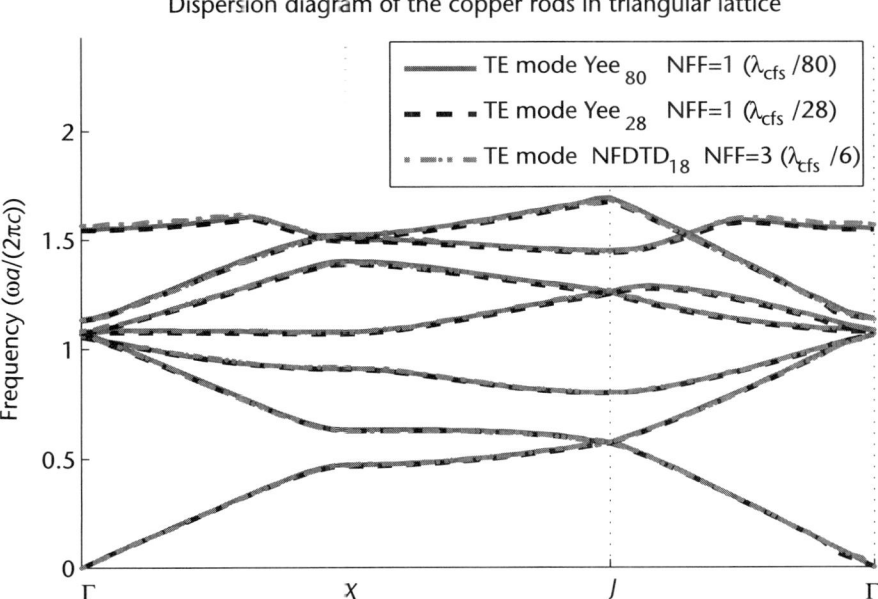

Figure 4.8 The dispersion diagram of the first few TE modes, from the copper rods in free space in triangular/rhombic lattice.

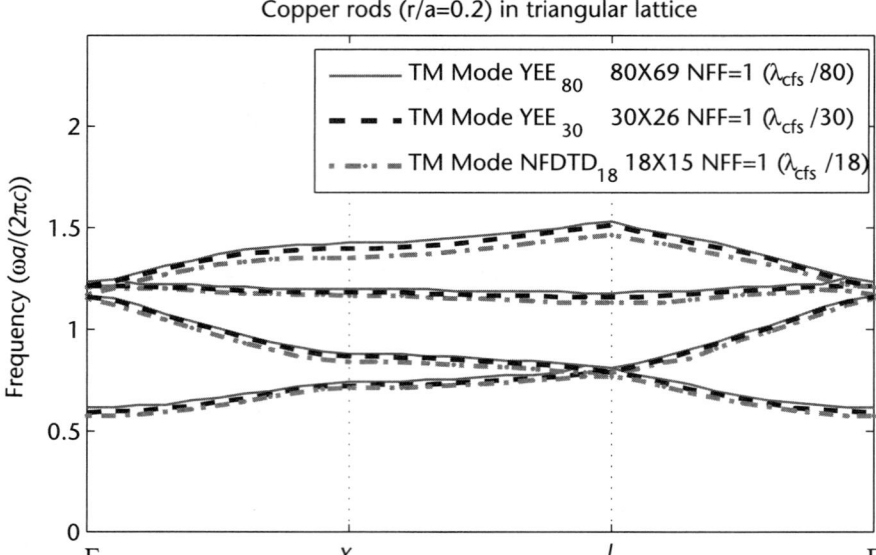

Figure 4.9 The dispersion diagram of the first few TM modes, from the copper rods in free space in triangular/rhombic lattice. The Yee's FDTD and the NFDTD result with the low spatial resolution is plotted to compare with the high spatial resolution result.

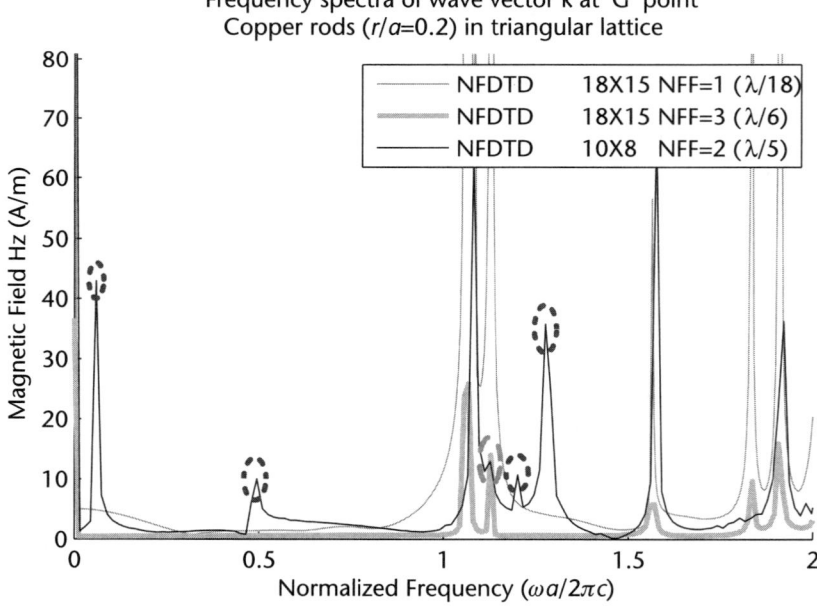

Figure 4.10 The TE mode frequency spectra of the wave vector $\hat{k}$ at $\vec{\Gamma}$ point, for the modeling of the cylindrical copper rods in triangular lattice. The positions of peaks in the two lines agree well, indicating that all the modes founded by the low spatial resolution ($dx_{\text{NFDTD}} = \frac{\lambda_{cfs}}{6}$) are all genuine. The comparison of these three lines indicates that $dx_{\text{NFDTD}} = \frac{\lambda_{cfs}}{6}$ is the minimum spatial resolution required by NFDTD for this modeling.

is adequate. We observe that the maximum errors in the results obtained by using Yee's FDTD with a low spatial resolution occur near $\vec{M}$ and $\vec{\Gamma}$ points in the dispersion diagram. We also note that the NFDTD scheme offers a significant advantage over the conventional scheme in that it allows us to use larger cell sizes in the simulation and, hence, improves the efficiency of numerical modeling scheme, as is evident from Figure 4.10). In the following figures, we denote "cfs" as the center frequency of excitation source.

Figure 4.11 shows that spurious modes arise at low frequencies when Yee's algorithm is employed to model the EBG using the discretization $dx = \frac{\lambda_{cfs}}{26}$. Numerical studies show that for this structure, a maximum cell size of $\frac{\lambda_{cfs}}{28}$ is needed to accurately predict the dispersion diagram when Yee's scheme is used.

Although the choice of cell size largely depends on the physical configuration of the EBGs being modeled, for the same structure, it is seen from Figure 4.12 that the conformal FDTD algorithm always requires fewer cells than Yee's FDTD. Furthermore, Yee's FDTD requires us to use considerably smaller cells, especially when computing the high-order modes. Consequently, substantial computer resources (memory and CPU time) are required, and these facts make it impossible to simulate electrically large finite EBG structures using Yee's algorithm.

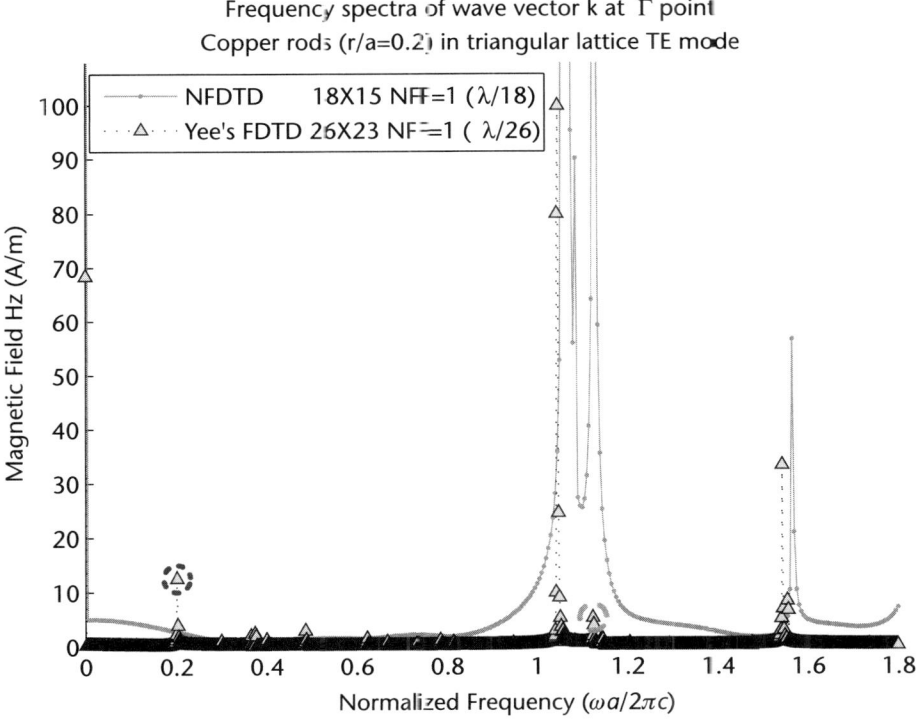

Figure 4.11 The TE mode frequency spectra of the wave vector $\hat{k}$ at Γ point of the cylindrical copper rods in triangular lattice. Yee's FDTD results in a spatial resolution of $dx = \frac{\lambda_{cfs}}{26}$ that provides a result with spurious energy in the low-frequency band, which is strong enough to compete with a genuine mode. The NFDTD results in a spatial resolution of $dx = \frac{\lambda_{fs}}{18}$, which provides all the genuine modes used as a reference.

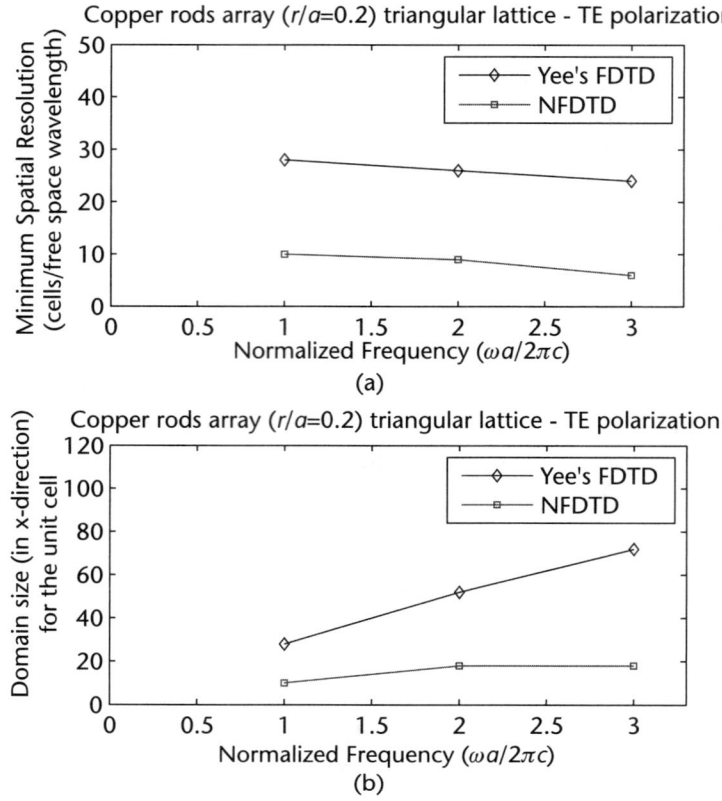

Figure 4.12 (a) The minimum spatial resolution and (b) mesh size required for Yee's FDTD and NFDTD algorithms for the triangular lattice TE mode unit cell simulation.

4.3 Conformal FDTD Modeling of (Semi-)Finite EBG Structures

In this section, we will study a semifinite two-dimension EBG structure by using the conformal FDTD and the conventional Yee algorithms. The bandgaps for the EBG structure are computed by evaluating the transmission coefficients and the FDTD simulation results are compared with those from the transfer-matrix method (TMM) [4].

The EBG structure considered herein is shown in Figure 4.13. It consists of an array of cylindrical dielectric scatterers, with a radius $r = 0.48$ cm and a dielectric constant $\varepsilon_r = 9$. They form a square lattice with a lattice constant (period) $a = 1.27$ cm. These rods are infinitely arranged in the $\hat{x}$-direction. There are eight layers of scatterers in the $\hat{y}$-direction. The rods are infinitely long in the $\hat{z}$-direction. The ratio of the radius r to the lattice constant a is set to be $r/a = 0.378$, and numerical simulations are performed for both the TE and TM polarization.

4.3.1 FDTD Model and Simulation Results

The computational domain for modeling the EBG, shown in Figure 4.13, can be set up as follows: only one row of dielectric rods need be considered, since the

4.3 Conformal FDTD Modeling of (Semi-)Finite EBG Structures

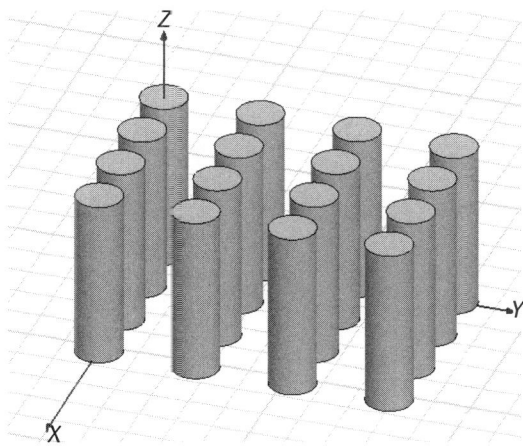

Figure 4.13 The dielectric cylinders with radius $r = 0.48$ cm, and dielectric constant $\varepsilon_r = 9$, surrounded by air, forming a square lattice with a lattice constant $a = 1.27$ cm, infinite in the z-direction; infinite in the x-direction; finite (8 elements) in the y-direction.

EBG can be truncated by applying the PBCs and a perfect matched layer (PML). We use a wideband Gaussian plane wave source, which covers the frequency range of $0 \sim 20$ GHz to simulate the structure, and the probes are located at each side of the EBGs in order to calculate the transmission coefficient and investigate the existence of bandgaps. The FDTD model is illustrated in Figure 4.14.

The FDTD grid of one unit cell is displayed in Figures 4.15 and 4.16, which show a nonuniform distribution of the cell sizes used to simulate the EBG.

The temporal responses at the positions of two probes are plotted in Figures 4.17–4.19.

The transmission coefficients versus frequency for the TE polarization, derived from the NFDTD simulation, are plotted in Figure 4.20. It can be seen that the simulation results converge provided that at least 12×12 cells per unit cell are employed. Two bandgaps, in the vicinity of 11 GHz and 15.5 GHz, are clearly seen in Figure 4.20. The first stopband is not visible because of small EBG dimensions

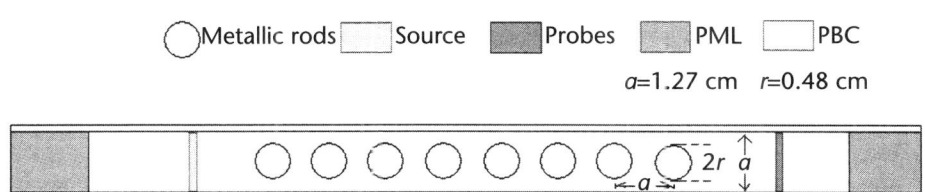

Figure 4.14 The cut plane of the EBG structure consists of eight infinite dielectric cylindrical rods array in square lattice, modeled by the NFDTD method. The rods are assumed to be infinitely long, so a two-dimension model is applied. In the direction in which the eight elements are aligned, PML is used to terminate the computational domain. In the direction in which the array is infinitely arranged, the unit cell approach is applied with PBCs used to terminate the computational domain.

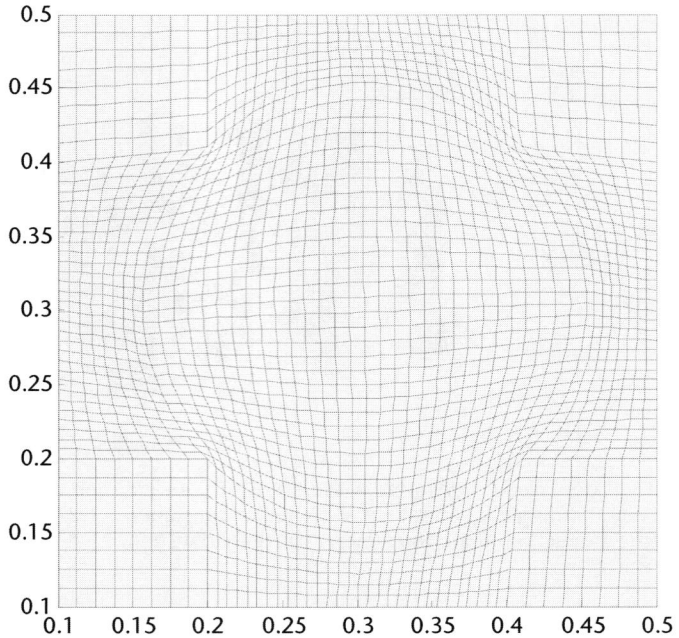

Figure 4.15 The conformal mesh of the unit cell of the structure described in Figure 4.13 in the x-y cut plane with spatial resolution of 48 × 48 per unit cell.

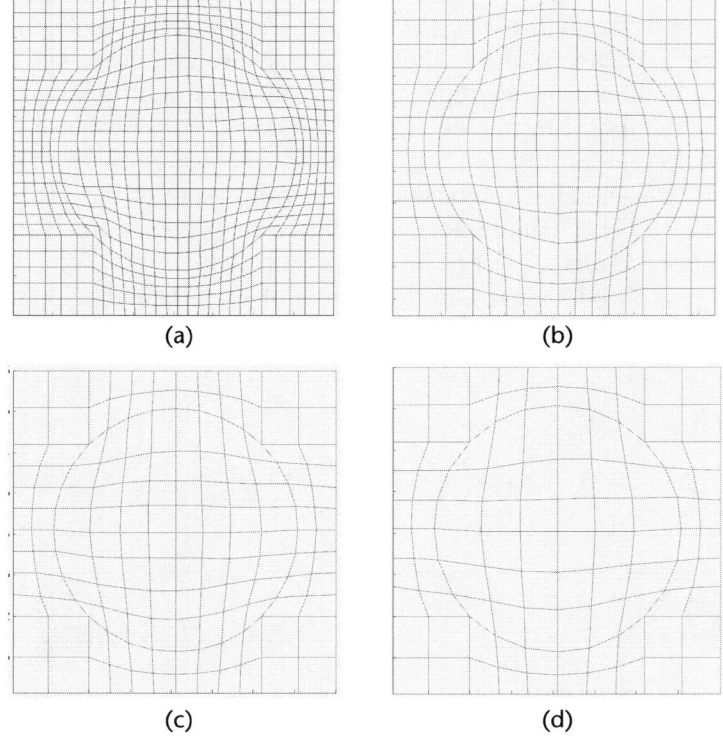

Figure 4.16 The conformal mesh of the unit cell of structure (Figure 4.13) in the x-y cut plane with different spatial resolutions: (a) 26 × 26; (b) 16 × 16; (c) 12 × 12; and (d) 10 × 10 per unit cell.

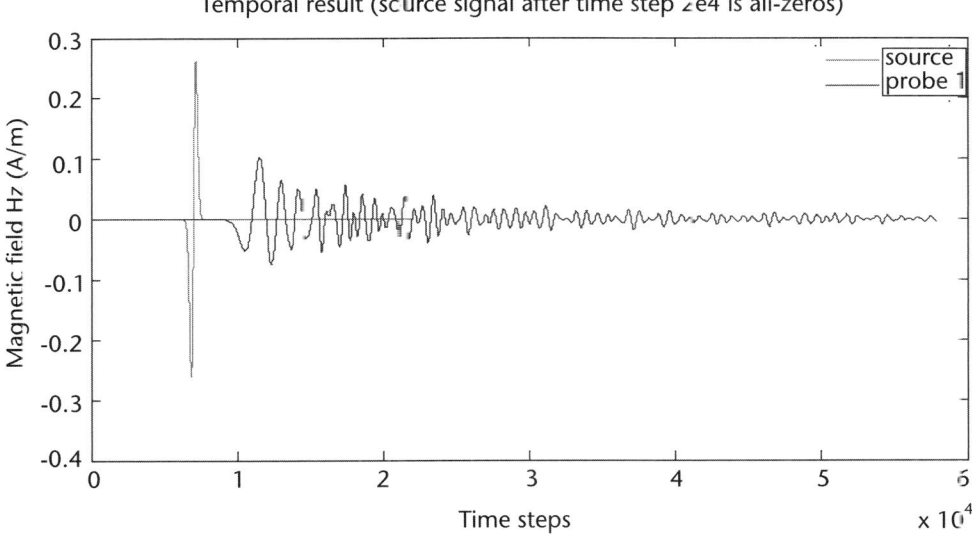

Figure 4.17 The NFDTD temporal results (using mesh Figure 4.15) after a Gaussian pulse plane wave excitation at one side of the EBG slab.

compared to the wavelength of operation. Results from the conformal FDTD simulation show very good agreement with those from the TMM (Figure 4.21), as well as with Yee's FDTD results, provided the cell sizes are sufficiently small (see Figure 4.22). Good agreement is also seen for the TM mode (see Figures 4.23 and 4.24).

4.4 Design and Modeling of Millimeter-Wave EBG Antennas

4.4.1 Introduction

Millimeter-wave systems are becoming increasingly important in many scientific, civil, and military applications because they have potentially wider bandwidths useful for transmitting large amounts of data and achieving an enhanced resolution in radar imaging. Recently, it has been demonstrated by various groups that novel metamaterials and devices for microwave and millimeter-wave frequencies can be realized by using EBG structures. EBG structures, also known as photonic bandgap structures (PBGs) [5,6] in optics, are now finding numerous applications in microwave and millimeter-wave devices [7,8]. In general, EBG structures are comprised of dielectric or metallic elements, arranged in a periodic grating that exhibits forbidden frequency bands (bandgaps). The full potential of EBG structures can be realized by using the full three-dimensional (3-D) bandgap. Thus fabrication of 3-D EBG structures is of significant importance. The woodpile structure shown in Figure 4.25, also called a layer-by-layer structure, consists of stacked diffraction gratings, in which the dielectric rods in adjacent layers are arranged such that they are orthogonal to each other. This structure possesses face-centered-tetragonal symmetries and provides a full 3-D bandgap in which wave propagations are

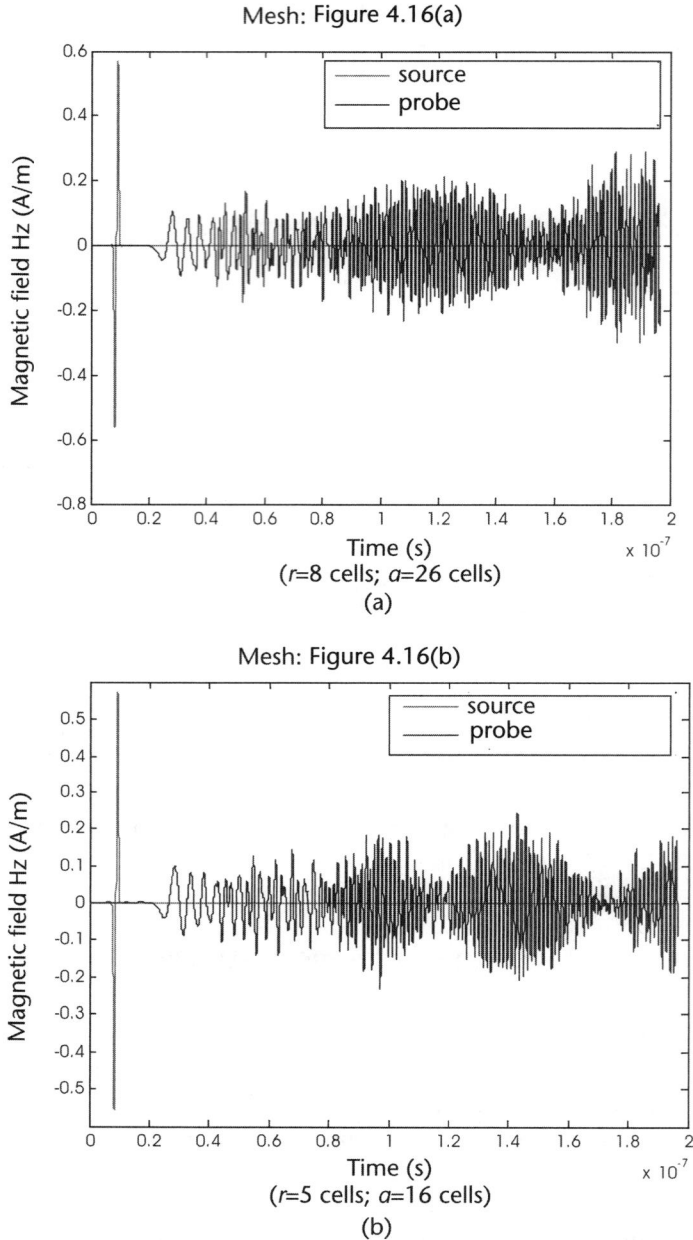

Figure 4.18 The NFDTD temporal results with different spatial resolutions after a modulated Gaussian pulse plane wave excitation at one side of the EBG slab: (a) mesh Figure 4.16(a) is used and (b) mesh Figure 4.16(b) is used.

prohibited in any spacial direction. Such a periodic structure can be easily fabricated for microwave applications by using columns of individually machined dielectric materials with preferred dimensions. However, conventional machining is difficult at best at millimeter-wave frequencies, because of small dimensions of the structures ranging from 50 to 500 μm. Various sophisticated micro-fabrication

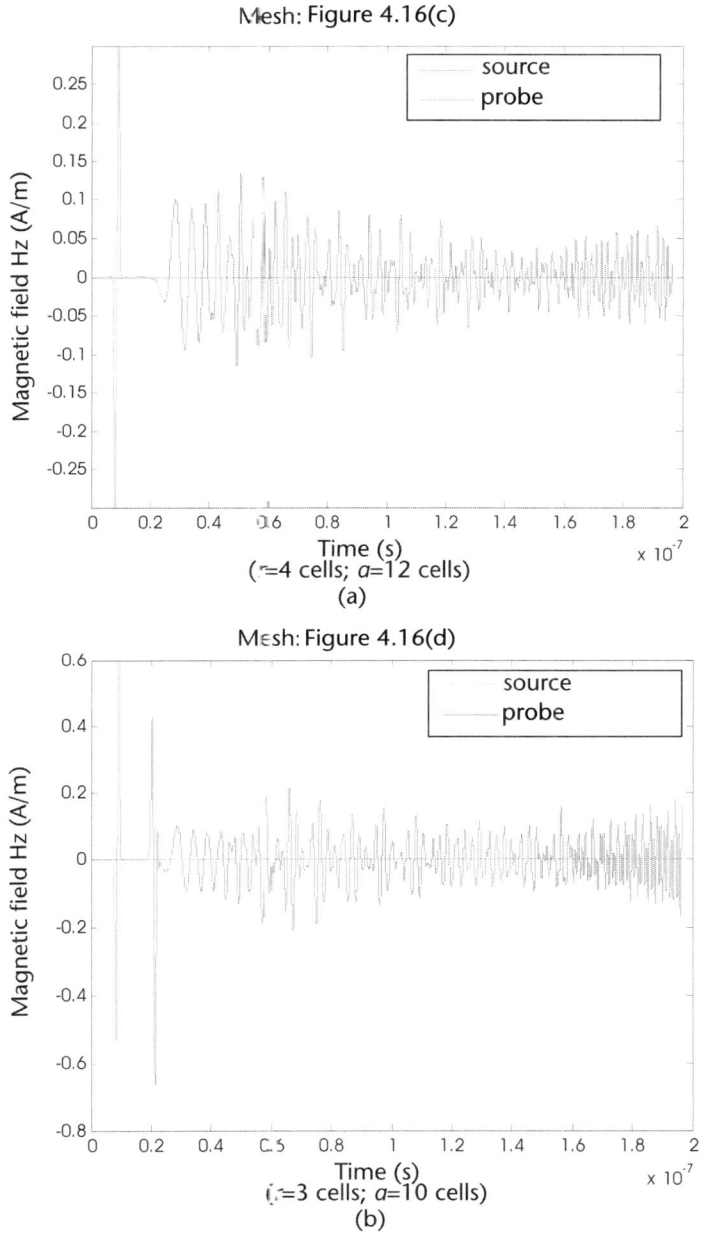

Figure 4.19 The NFDTD temporal results with different spatial resolutions after a modulated Gaussian pulse plane wave excitation at one side of the EBG slab: (a) mesh Figure 4.16(c) is used and (b) mesh Figure 4.16(d) is used.

techniques, such as silicon lithography and wafer fusion are available for microstructures, but those are more appropriate for terahertz frequency and photonic wavelengths, and it would be costly to fabricate 3-D structures with large number of layers for applications in the W-band (75–111 GHz) range. In this chapter, we present a design and FDTD modeling of 3-D EBG materials for millimeter-wave

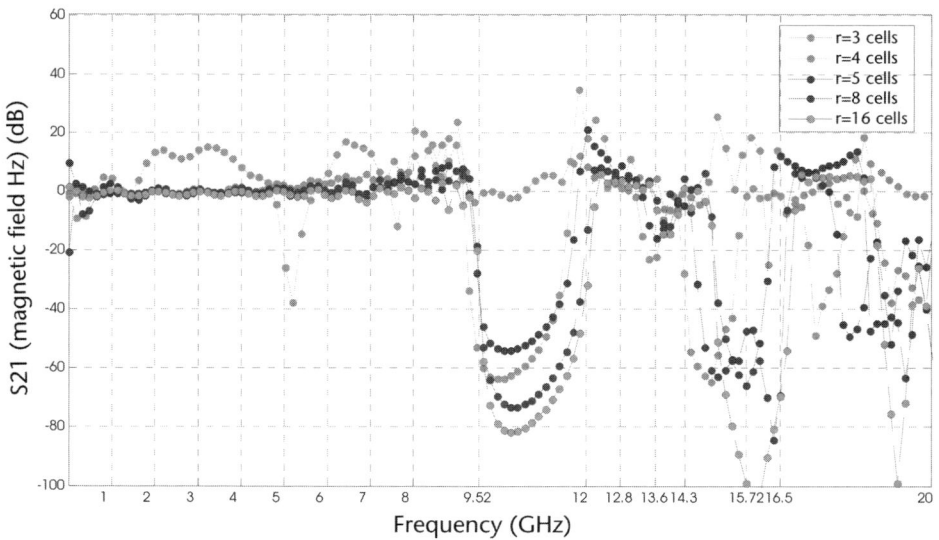

Figure 4.20 The transmission coefficients (TE mode) of the dielectric rods (Figure 4.13) calculated using the NFDTD method. *r* is the radius of the cylinder.

applications based on extrusion freeforming of ceramic material [9]. The proposed fabrication method can also be extended to submillimeter waves and proven useful for the construction of cylindrical and spherical EBGs.

4.4.2 Design and Modeling of Woodpile EBG

In order to design the woodpile structure for millimeter-wave frequencies (93–97 GHz), we first determine the EBG dimension, namely the width w of dielectric

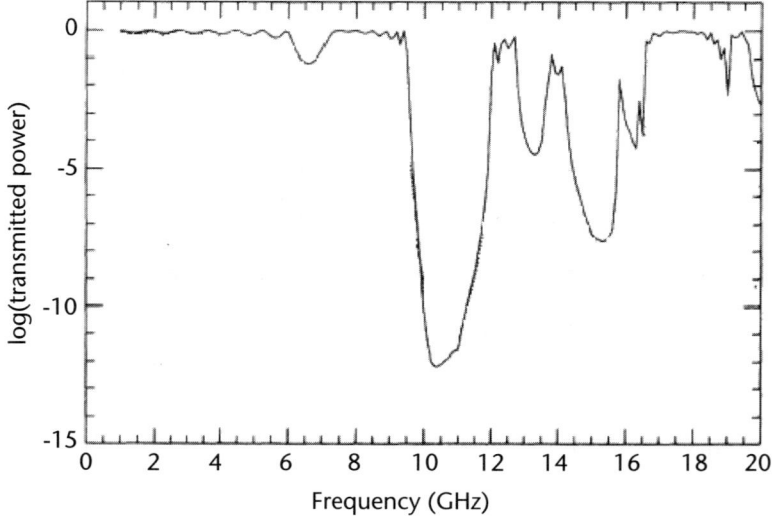

Figure 4.21 The theoretical transmission coefficients (TE mode) of the dielectric rods (Figure 4.13) predicted by the transfer-matrix method [4].

4.4 Design and Modeling of Millimeter-Wave EBG Antennas

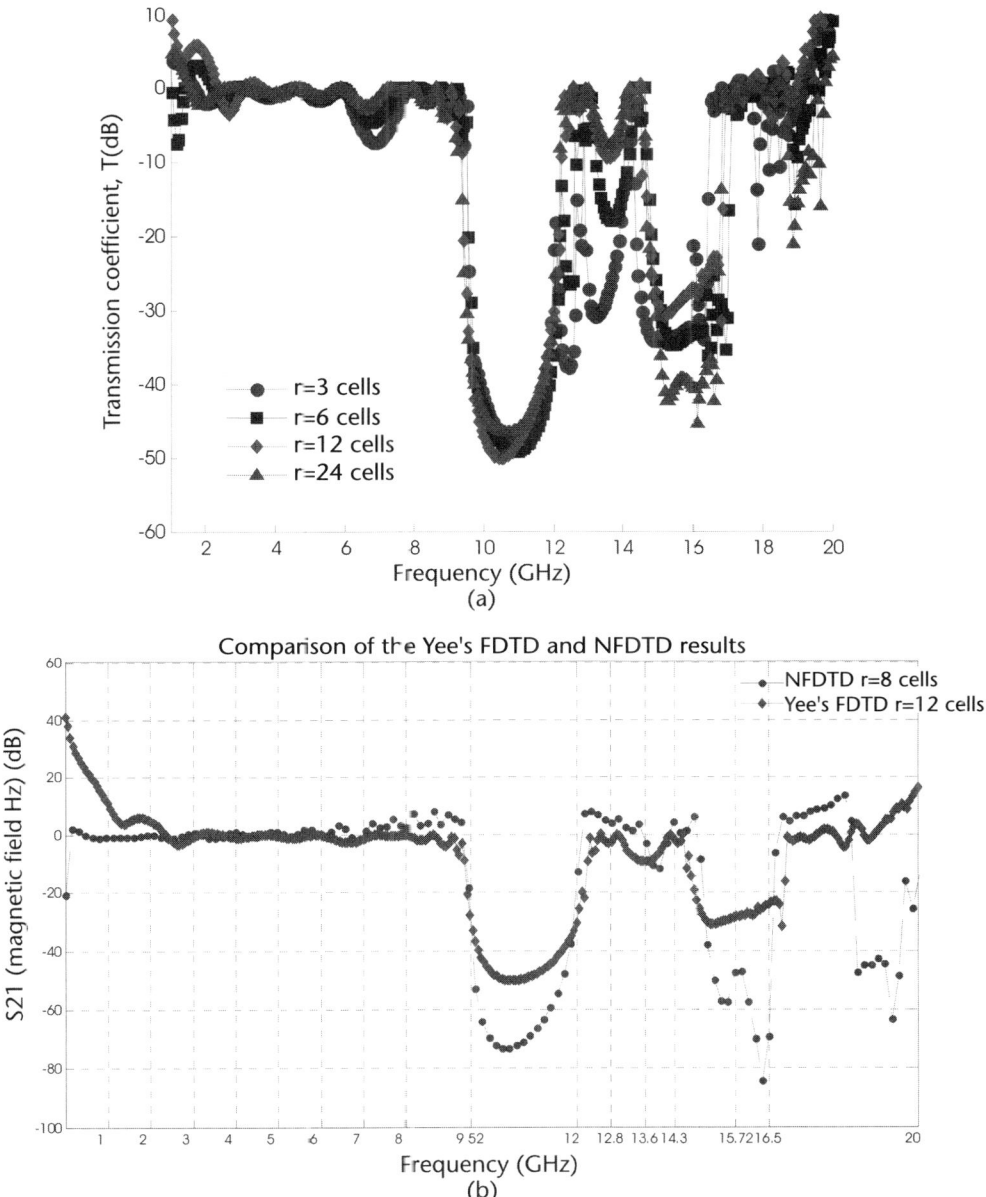

Figure 4.22 The transmission coefficients (TE mode) of the EBGs consisting of dielectric rods (Figure 4.13) (a) Yee's FDTD simulation results and (b) the comparison of NFDTD and Yee's FDTD results with similar spatial resolutions.

rods and their period a based on the calculation of dispersion diagram. There are several different methods available to compute the dispersion diagram of the periodic structures. These include the plane wave expansion method [10,11], multiple scattering theory [12], transfer matrix method [13], and finite difference method [14]. In this work, we use the FDTD method [15] for computing the dispersion diagrams of periodic structures. One of the important advantages of

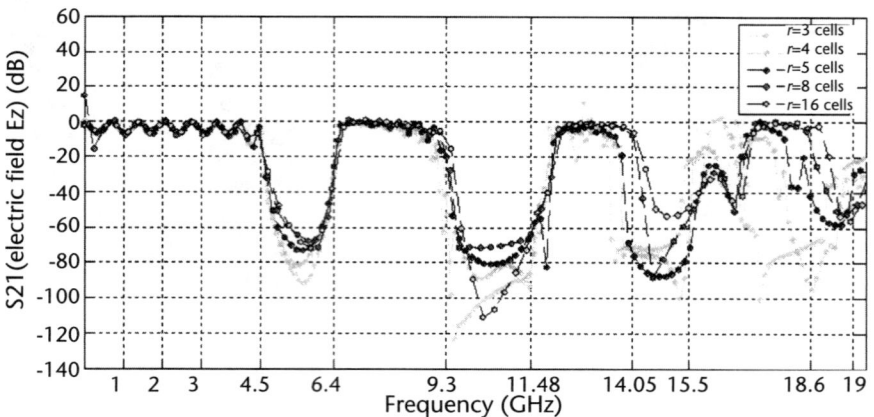

Figure 4.23 The transmission coefficients (TM mode) of the dielectric rods (Figure 4.13) calculated using the NFDTD method with different spatial resolutions. *r* is the radius of the cylinder.

the FDTD method is that it is capable of handling inhomogeneous materials in 3-D forms. It is also possible to simulate periodic structures in which the materials are dispersive. We use the standard 3-D FDTD algorithm and apply a PBC to compute the dispersion diagrams, as described in Section 3.5.

Since the structure is periodic, we choose the unit cell of the lattice as the computation domain. The unit cell of the woodpile lattice is shown in Figure 4.26 and the Brillouin zone of the woodpile is shown in Figure 4.27.

Since the field distribution in the computation domain should satisfy the Bloch condition, the field can be written in the following form:

$$E(\vec{r}) = e^{j\vec{k}\cdot\vec{r}} u_e(\vec{r}) \tag{4.4}$$

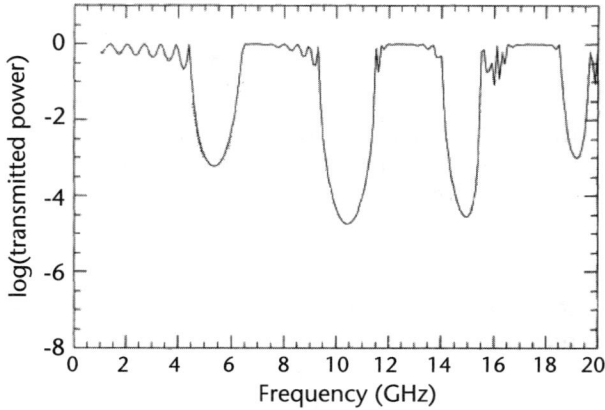

Figure 4.24 The theoretical transmission coefficients (TM mode) of the dielectric rods (Figure 4.13) predicted by the transfer-matrix method [4].

4.4 Design and Modeling of Millimeter-Wave EBG Antennas   111

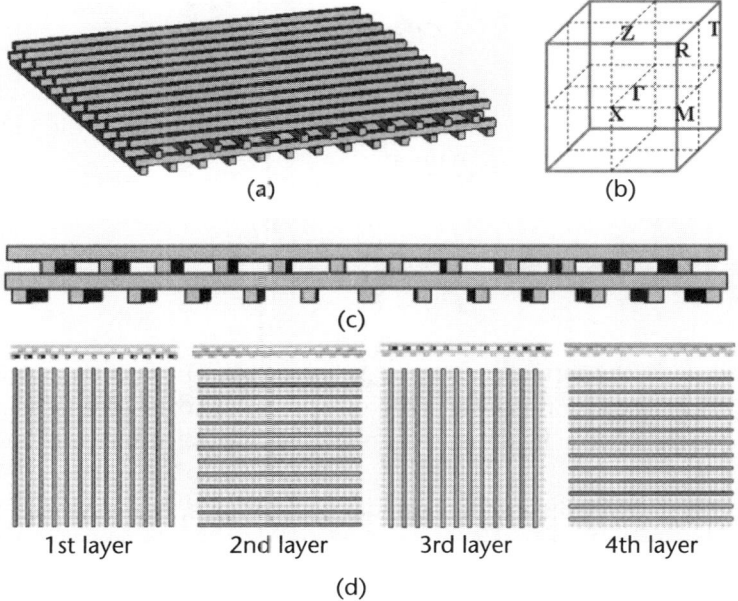

Figure 4.25 (a) Geometry of the woodpile structure, (b) Brillouin zone defined for the unit cell, (c) side view of the woodpile, and (d) top view of each layer of the woodpile.

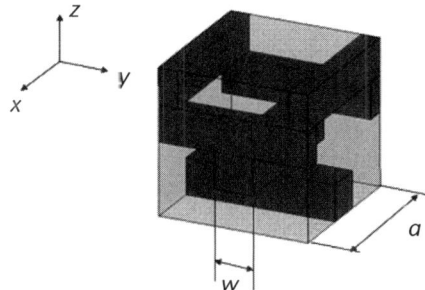

Figure 4.26 Unit cell geometry of the woodpile.

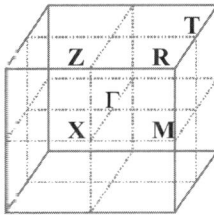

Figure 4.27 Brillouin zone of the woodpile EBG.

$$H(\vec{r}) = e^{j\vec{k}\cdot\vec{r}} u_h(\vec{r}) \qquad (4.5)$$

where $u_e(\vec{r})$ and $u_h(\vec{r})$ are periodic functions on the lattice [i.e., $u_e(\vec{r}) = u_e(\vec{r}+\vec{R})$ and $u_h(\vec{r}) = u_h(\vec{r}+\vec{R})$ for all lattice vectors]. Therefore, the boundary conditions for the computational domain are:

$$E(\vec{r}+\vec{R}) = e^{j\vec{k}\cdot\vec{R}} E(\vec{r}) \qquad (4.6)$$

$$H(\vec{r}+\vec{R}) = e^{j\vec{k}\cdot\vec{R}} H(\vec{r}) \qquad (4.7)$$

where $\vec{R}$ is the lattice vector, as is introduced in Section 3.4.3. Equations (4.6) and (4.7) suggest that the fields on the boundaries of the computation domain are related to the fields at the corresponding locations on the other side of the boundary.

Once we have set the boundary conditions, we vary the magnitude of the k-vector component ($|\vec{k}_i|$) defined along each lattice vector with magnitude (a_i) from 0 to π/a_i, with finite intervals, and excite the computational domain with a wide-band source.

The location of the source can be anywhere in the background medium (the host medium that contains the woodpile rods and air is assumed in this work) in the computational domain. After a certain number of time steps, we record the temporal field values inside the unit cell and search for the maxima in the spectrum, corresponding to the propagating modes in the EBGs for each **k**, via the use of the FFT. Since we are only interested in the spectral distribution of the eigen-frequencies, we take the FFT of the time domain output from $t = 0$ to the end of the time step. In our simulation, the computational domain was discretized by using $20 \times 20 \times 20$ cells along the x-, y-, and z-directions, respectively. The time domain fields have been measured at 20 different locations inside the domain.

The spectral distribution of the each time domain data is then summed up to obtain the results shown in Figure 4.28(a). The peaks indicate the eigen-frequencies that satisfy the boundary condition.

In order to obtain the dispersion diagram, we vary the k-values according to the direction defined in the Brillouin zone and carry out the FDTD simulations for each value of **k**. From the spectral distribution obtained for each **k**, we extract the location of the peaks. Finally, we plot the peak locations versus the normalized frequency versus **k** values for different directions, as shown in Figure 4.28.

Figure 4.28(b) shows the computed dispersion diagram of the woodpile structure with $w/a = 0.25$, where w is the width of the dielectric rod, and a is the period of the square lattice. The dielectric constant of the rod is 9.6. The above figure clearly shows that a complete bandgap exists between the normalized frequencies of 0.44 and 0.5. We also note that the widest bandgap exists along the Γ-Z direction.

The design parameters for 95 GHz have been found as $w = 0.41$ mm and $a = 1.7$ mm with $\varepsilon_r = 9.6$ for the permittivity of the dielectric material. The transmission property was then characterized along the horizontal direction using the FDTD simulation, for the case of a vertically polarized electric field ($\vec{E} = E_0\vec{a}_z$).

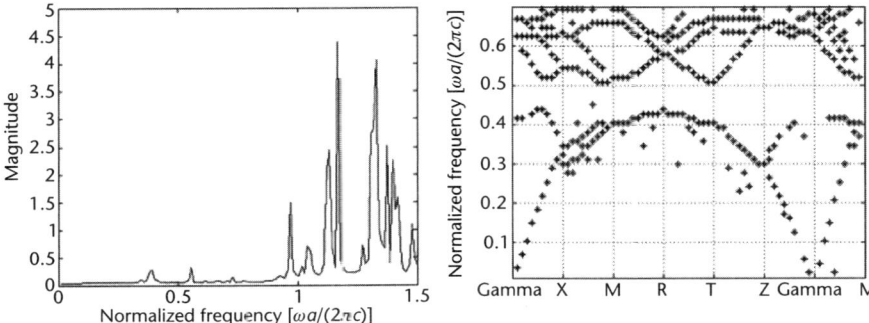

Figure 4.28 (a) Spectral distribution of the field observed inside the unit cell and (b) dispersion diagram of the designed woodpile.

The electric field distribution along the horizontal direction is shown in Figure 4.29. The attenuation of the electric field at 95 GHz is quite obvious, while the wave propagates with very little attenuation at frequencies outside the bandgap. The transmission characteristic of the woodpile has been simulated using FDTD and Ansoft HFSS; both results agree very well, as shown in Figure 4.30.

The designed woodpile structure has been fabricated using the freeforming technique as is shown in Figure 4.31. The fabricated woodpile structures were characterized by performing transmission measurements using a millimeter-wave transmitter and receiver. The millimeter-wave source used for the measurement is capable of sweeping from 75 GHz to 110 GHz. The fabricated woodpile structure was placed between two identical circular horn antennas and a frequency sweeping was performed to measure the transmission characteristic of the sample. In order to avoid diffractions, microwave absorbing materials were placed along the edges of the sample. The woodpile samples of 30×30 mm^2 were measured with 1 and 2 vertical periods, and the results are shown in Figure 4.32.

The simulated results are also plotted in the same figure for comparison. The simulated woodpile structure is assumed to be infinite. The actual physical parameters of the measured samples were $w = 0.41$ mm, $a = 1.67$ mm. The measured and simulated results show a good agreement with each other, except for the slight

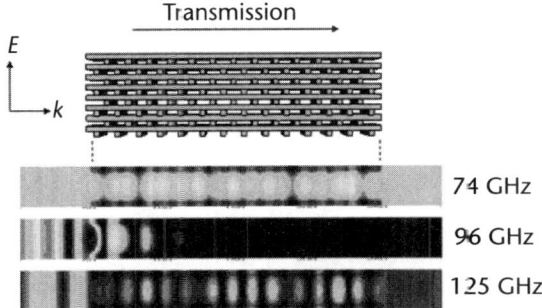

Figure 4.29 Electric field distribution inside the woodpile at different frequencies.

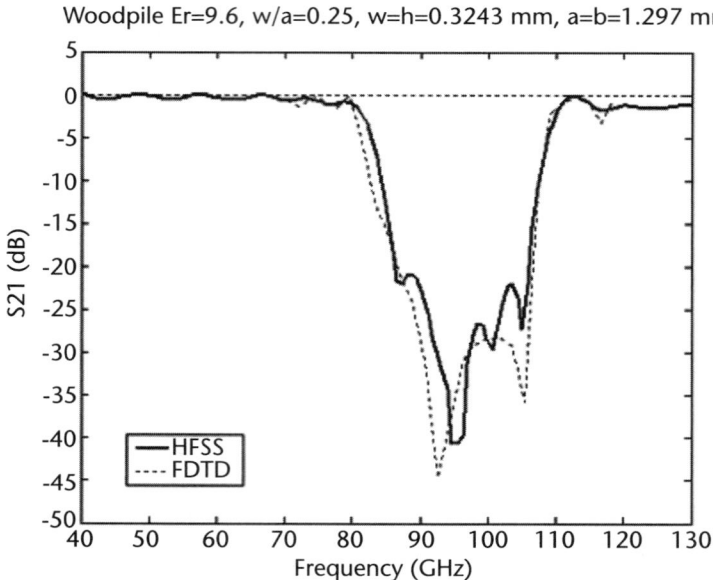

Figure 4.30 Transmission characteristic of woodpile structure: results comparison between FDTD and HFSS.

offset in the bandgap frequency and transmittance level. The frequency offset is caused mainly by the difference between the dielectric constant of the actual material and the simulated values, and a slight nonuniformity in the geometry (i.e., diameter and spacing) throughout the fabricated samples. The discrepancies in the transmittance level is attributed to the assumption (infinitely wide structure) made during the simulation, which shows a higher attenuation along the direction of the propagation.

In this section, we have analyzed a woodpile EBG structure for millimeter-wave applications fabricated by using a rapid prototyping technique based on extrusion freeforming. Numerical simulations have been performed for the design of this structure and the fabricated woodpile structures have been measured. Experimental results have shown a good agreement with the simulated results.

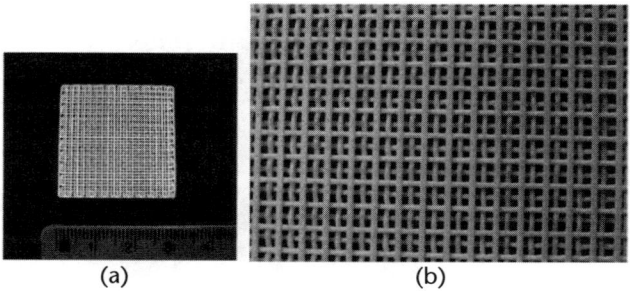

Figure 4.31 Photo of the fabricated woodpile: (a) fabricated sample and (b) details of filaments.

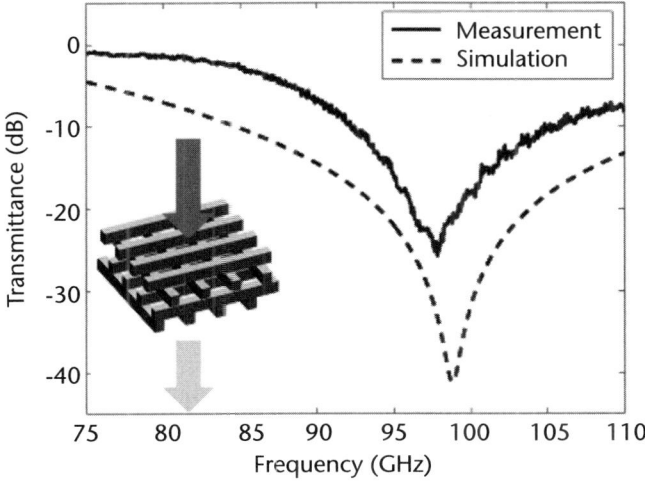

Figure 4.32 Simulated and measured transmittance of the woodpile structure.

4.4.3 A Millimeter-Wave EBG Antenna Based on a Woodpile Structure

In this section, we describe the design of an EBG resonator antenna at millimeter-wave frequencies (93–97 GHz). The overall geometry is shown in Figure 4.33.

In order to determine the separation distance (d) between the EBG and the antenna, we perform a parametric study. The separation distance (d) is normally chosen to be approximately $\lambda/2$, where λ is the free-space wavelength at the operating frequency. The size of the ground plane is 10 mm × 10 mm, and the rectangular aperture size is 1 mm × 3 mm with the longer side is aligned with x-axis.

Figure 4.34 shows the defect frequency and directivity as a function of the separation distance between the EBG and the antenna. The defect frequency, which is the operating frequency of the EBG antenna, decreases with an increasing d. However, the directivity shows the maximum value for a particular range of d, and it drops rapidly when d is too small.

It is also interesting to note that the actual separation distance in terms of the wavelength at the operating frequency increases an increasing d (see Figure 4.35).

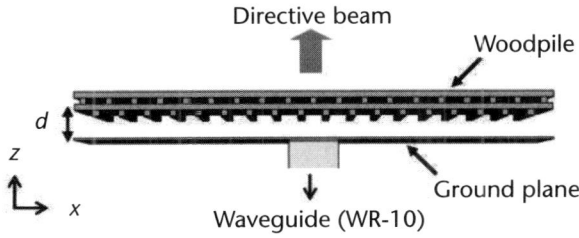

Figure 4.33 Geometry of the woodpile EBG resonator antenna.

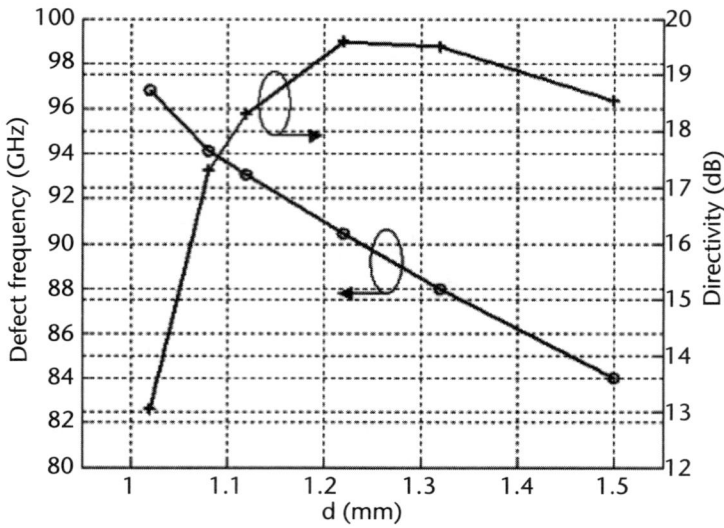

Figure 4.34 Antenna operating frequency (defect frequency) and directivity versus the separation distance between the EBG and the antenna.

This result implies that the wave is not exactly reflected at the surface of the EBG structure, but the location of the actual reflector is somewhere inside the EBG structure.

Also when the EBG is placed closely to the antenna, the energy penetrates deeper inside the EBG structure, which may reduce the directivity of the antenna.

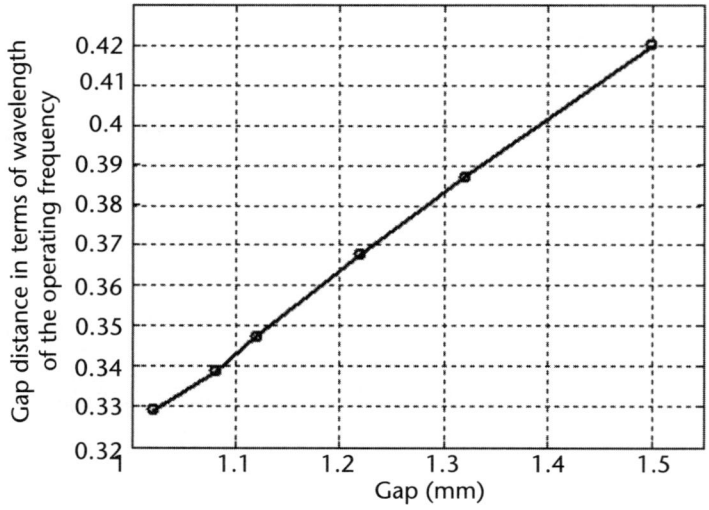

Figure 4.35 Separation distance between the EBG and the ground plane in terms of the operating wavelength.

4.4 Design and Modeling of Millimeter-Wave EBG Antennas

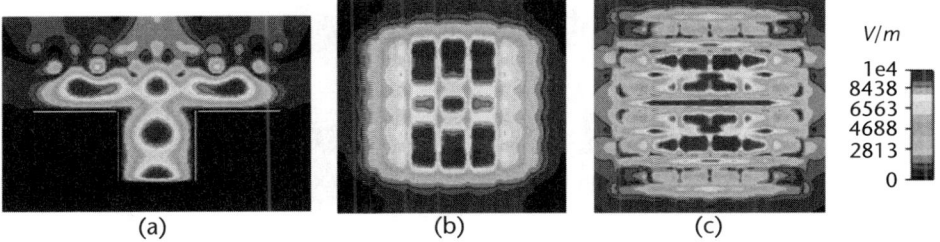

Figure 4.36 Simulated electric field distribution within the antenna structure at 93 GHz ($d = 1.32$ mm): (a) vertical cut, (b) horizontal cut for $z = 0.88$ mm (between the woodpile and ground plane), and (c) horizontal cut for $z = 2.48$ mm (top layer of the woodpile).

Figure 4.36 shows the simulated electric field distribution inside and outside the cavity created by the woodpile structure and the ground plane. It is shown that the entire area of the cavity is excited and the whole area of the woodpile surface is illuminated, which creates a high directive antenna pattern.

The simulated directivity pattern of the composite antenna is shown in Figure 4.37. The directivity achieved at 93 GHz is approximately 18 dBi. It is clearly seen that the EBG superstrate provides significant improvement in the directivity of the low directive antenna.

4.4.4 Experimental Results

The designed antenna has been fabricated by using the brass ground plane attached to a waveguide at the center and the woodpile EBG material fitted into the aluminium frame, as shown in Figure 4.38. It was found that the aluminium frame causes a distortion of the pattern, and it was later replaced with a nonmetallic frame. The separation distance between the ground plane and the frame was adjusted by using the screws at three points. The dimension of the woodpile structure used was 30 mm × 30 mm and the ground plane was 120 mm × 120 mm.

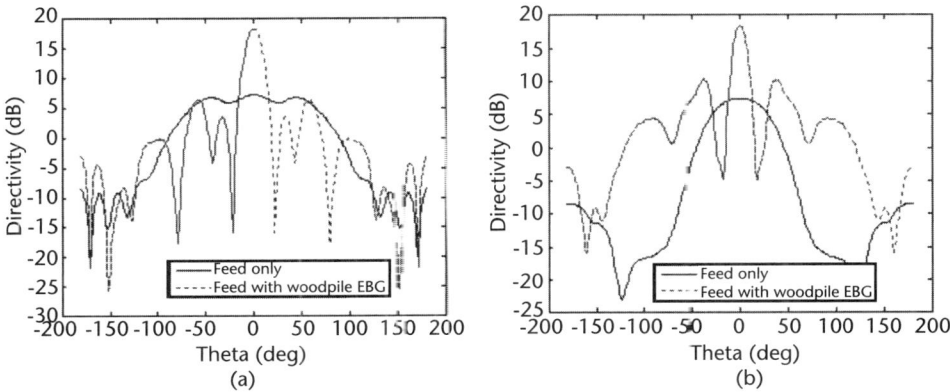

Figure 4.37 Simulated antenna pattern of the EBG resonator antenna at 93 GHz: (a) phi = 0 and (b) phi = 90.

Figure 4.38 Fabricated prototype antenna.

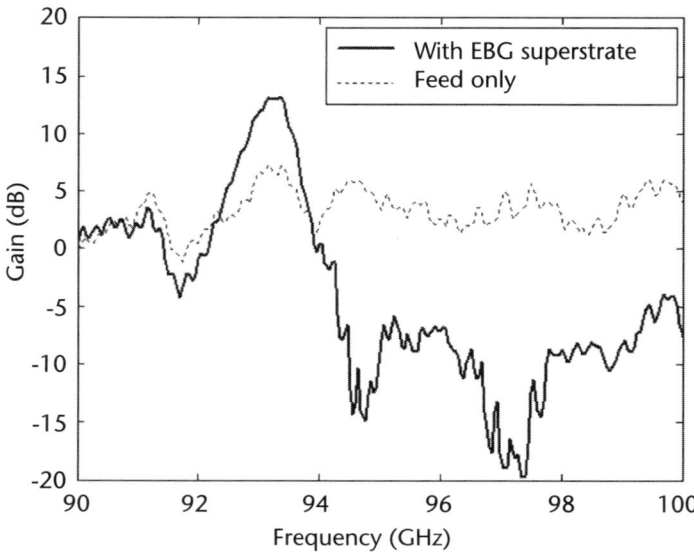

Figure 4.39 Received power level versus frequency.

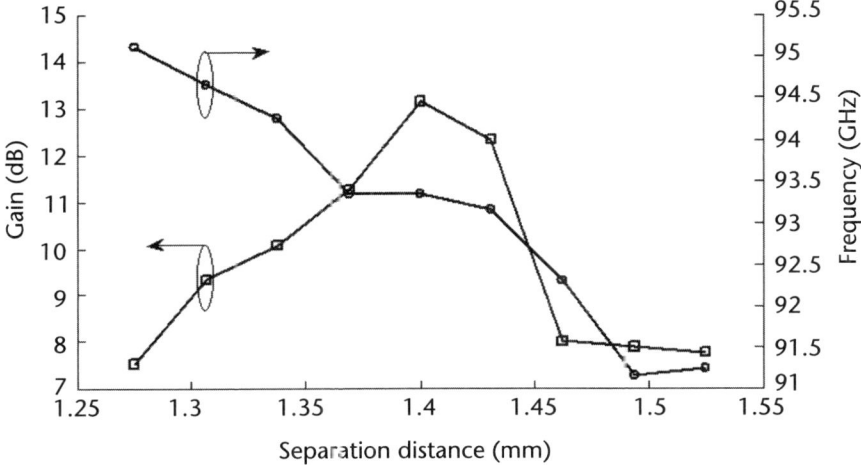

Figure 4.40 Received power level and maximum gain frequency versus separation distance.

The fabricated antenna has been mounted on a microwave test bench and the received power level has been measured as a function of the separation distance d over the millimeter-wave frequency band (75–110 GHz). Figure 4.39 shows the received power level with and without the woodpile EBG superstrate. The improvement of the broadside gain around 93 GHz is clearly seen from the figure. We have also investigated the antenna operating frequency (the frequency for which

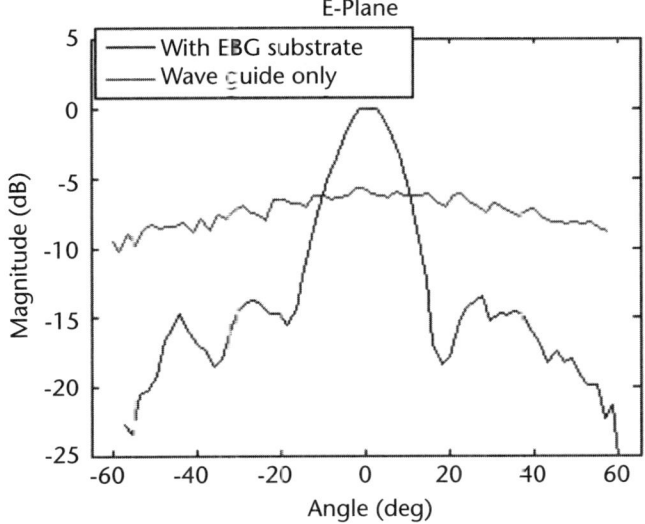

Figure 4.41 Measured E-plane pattern of woodpile EBG resonator antenna.

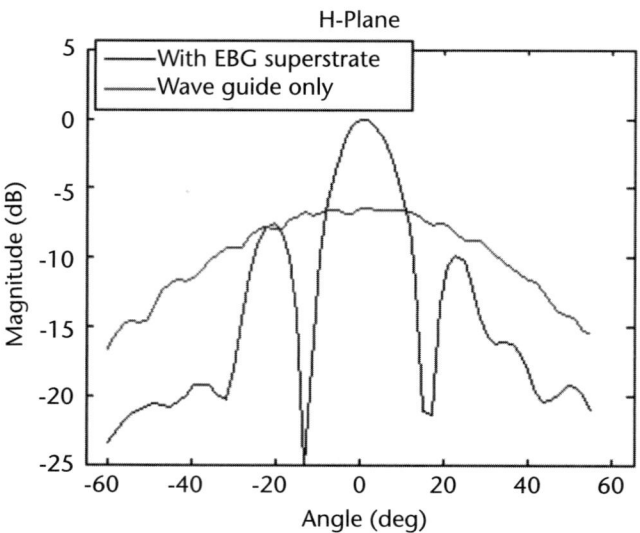

Figure 4.42 Measured *H*-plane pattern of woodpile EBG resonator antenna.

the highest gain is achieved) as well as the received power level as a function of the separation distance. The results are shown in Figure 4.40. We note that the experimental results confirm the simulated ones presented in the previous section. The operating frequency of the antenna decreases with an increase in the separation distance and the highest gain is achieved for a specific separation distance. The relative improvement of the gain over that of the waveguide with a ground plane was found to be approximately 6 dB.

The measured antenna patterns are shown in Figures 4.41 and 4.42. A clear improvement of the antenna directivity in both the *E*- and *H*-plane patterns is observed. The HPBW of the antenna and the received power level at the broadside are compared in Table 4.1.

The far-field pattern of the woodpile antenna has been simulated by using the FDTD method. The simulated and measured results are compared and a very good agreement has been observed between the simulated and the measured data, as may be seen from the results plotted in Figure 4.43.

Table 4.1 Comparison of HPBW and the Received Power Level at the Broadside With and Without the Woodpile EBG Superstrate

	E-plane		*H-plane*	
	HPBW (degrees)	*Received Power at Broadside (dB)*	*HPBW (degrees)*	*Received Power at Broadside (dB)*
EBG antenna	14	0.095	10	0.08
WG only	110	−5.672	68	−6.16

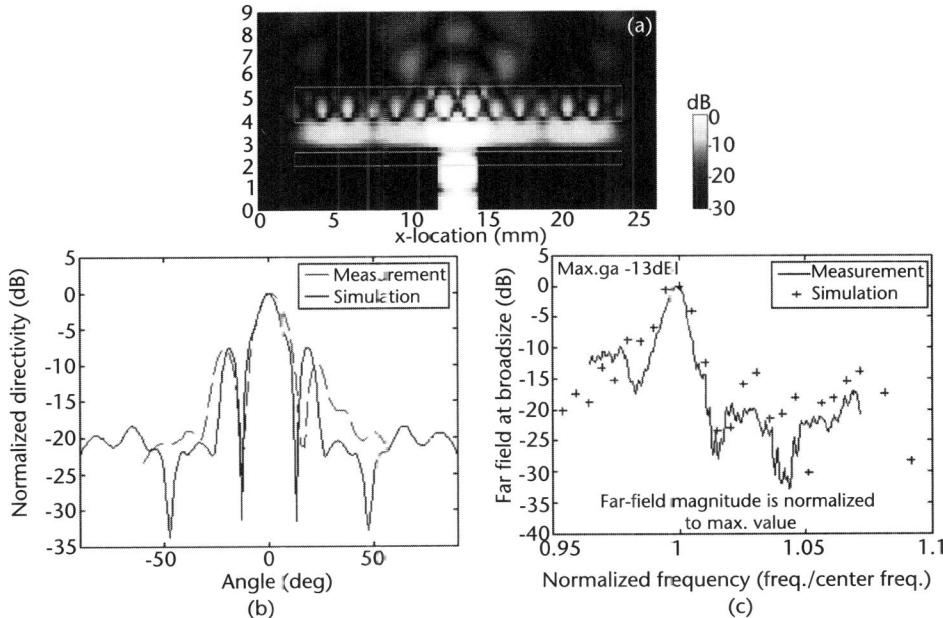

Figure 4.43 Comparison between the simulation and the measurement: (a) electric field distribution, (b) broadside gain versus frequency, and (c) comparison of normalized directivity pattern for the E-plane.

4.5 Conclusions

We have demonstrated some examples of EBG modeling based on Yee's and conformal algorithms. It has been shown that the "unit" cell approach can be applied to characterize the EBG, provided that the cell dimensions are small and the EBG structure is large in comparison to the wavelength of operation. The conformal FDTD provides a better numerical accuracy and efficiency over the conventional Yee's approach when modeling EBGs with curved elements. The design and modeling of a millimeter-wave EBG antenna based on woodpile structures have been presented. The designed antenna showed improved performances in terms of directivity and beamwidth. The composite antenna shows a 13-dBi gain and a 14- and 10-degree beamwidth in the E and H planes, respectively. The primary factor responsible for the operating frequency and the directivity is the separation distance between the EBG superstrate and the antenna. The results suggest that the separation distance must be chosen properly in order to achieve the maximum directivity at the intended operating frequency.

References

[1] R. Holland, "Finite-difference solution of Maxwell's equations in generalized nonorthogonal coordinates," *IEEE Trans. on Nuclear Science*, NS-30, 4589–4591, 1983.

[2] J.F. Lee, R. Palandech, and R. Mittra, "Modeling three-dimensional discontinuities in waveguides using nonorthogonal FDTD algorithm," *IEEE Transactions on Microwave Theory and Techniques,* Vol. 40, No. 2, 346-352, 1992.

[3] V. Douvalis, Y. Hao, and C.G. Parini, "Stable nonorthogonal FDTD method," *Electronics Letters,* Vol. 40, No. 14, 850-851, 2004.

[4] M. Sigalas, C.M. Soukoulis, E.N. Economou, C.T. Chan, and K.M. Ho, "Photonic band gaps and defects in two dimensions: Studies of the transmission coefficient," *Phys. Rev. B,* Vol. 48, Issue 19, 14121-14126, 1993.

[5] E. Yablonovitch, "Inhibited spontaneous emission in solid state physics and electronics," *Phys. Rev. Lett.,* Vol. 58, 2059, 1987.

[6] J.D. Joannopoulos, R.D. Meade, and J.N. Winn, *Photonic Crystals: Modeling the Flow of Light,* Princeton, NJ: Princeton University Press, 1995.

[7] Y.J. Lee, J. Yeo, K.D. Ko, R. Mittra, Y. Lee, and W. Park, "A novel design technique for control of defect frequencies of an electromagnetic bandgap superstrate for dual-band directivity enhancement," *Microwave and Optical Technology Letters,* Vol. 42, pp. 25-31, July 2004.

[8] A.R. Weily, L. Horvath, K.P. Essele, B. Sanders, and T. Bird, "A planar resonator antenna based on woodpile EBG material," *IEEE Transactions on Antennas and Propagation,* Vol. 53, pp. 216-223, January 2005.

[9] H. Yang, S. Yang, X. Chi, and J.R.G. Evans, "Fine ceramic lattices prepared by extrusion freeforming," *Journal of Biomedical Materials Research B,* Vol. 79B, October 2006, pp. 116-121.

[10] K.M. Ho, C.T. Chan, and C.M. Soukoulis, "Existence of photonic gap in periodic dielectric structures," *Phys. Rev. Lett.,* Vol. 65, 3152, 1990.

[11] K.M. Leung and Y.F. Liu, "Full vector wave calculation of photonic band structures in face-centered-cubic dielectric media," *Phys. Rev. Lett.,* Vol. 65, 2646, 1990.

[12] X. Wang, X.G. Zhang, Q. Yu, and B.N. Harmon, "Multiple scattering theory for electromagnetic waves," *Phys. Rev. Lett. B,* Vol. 47, 4161, 1993.

[13] J.B. Pendry and A. MacKinnon, "Calculation of photon dispersion relations," *Phys. Rev. Lett.,* Vol. 69, 2772, 1992.

[14] C.T. Chan and Q.L. Yu, "Order-N spectral method for electromagnetic waves," *Phys. Rev. Lett.,* Vol. 51, 16635, 1995.

[15] D.C. Dobson, J. Gopalakrishnan, and J.E. Pasciak, "An efficient method for band structure calculations in 3D photonic crystals," *J. Comp. Phys.,* 149, pp. 363-376, 1999.

CHAPTER 5
Left-Handed Metamaterials (LHMs) and Their Applications

5.1 Introduction

Metamaterials are often characterized in terms of their effective material parameters, such as electric permittivity and magnetic permeability. These constituent parameters can either be both negative, or only one of them may be negative, while the other is positive. The former is often referred to as LHM, DNG, or negative refractive index material (NRIM) [1–20]. The latter is called single negative material (SNG).

The concept of LHMs was first theorized by the Russian physicist Veselago in 1968 [1]. In this paper, Veselago speculated on the possible existence of LHMs and anticipated their unique electromagnetic properties such as the reversal of Snell's law, the Doppler effect, and the Vavilov Cherenkov effect. Veselago showed that the electric field, magnetic field, and wave vector of an electromagnetic wave in an LHM form an LH triad. As a result, LHMs support electromagnetic waves with group velocity and phase velocities that are antiparallel, known as backward waves. Consequently, while the energy still travels away from the source, so as to satisfy causality, wavefronts travel backward toward the source in an LHM, a phenomenon that is associated with negative refractive index of refraction.

5.2 Effective Medium Theory and Left-Handed Metamaterials

Every material is a composite in some sense, even if the individual ingredients consist of atoms and molecules. Typically, the motivation for introducing the concept of effective permittivity and permeability is to present a homogeneous view of the electromagnetic properties of a medium. For periodic structures, defined by a unit cell whose characteristic dimension is a, the following criterion must be satisfied in order for the structure to be viewed as a homogeneous medium:

$$a << \lambda = \frac{2\pi c_0}{\omega} \tag{5.1}$$

Should the above condition be violated, the possibility would exist that the internal structure of the medium would diffract as well as refract radiation, and thus invalidate the homogeneous medium assumption. It has been shown [3] that a structure consisting of very thin infinitely long metallic wires, arranged in a three-dimensional cubic lattice, could model the response of a diluted plasma, giving a negative ε_{eff} below a plasma frequency somewhere in the gigahertz range. Theoretical analysis of this structure has been confirmed by the experiment reported in [5]. Sievenpiper et al. have also investigated plasma-like effects in metallic structures [6, 7].

5.2.1 A Composite Medium of Metallic Wires and Split Ring Resonators

In a paper published in 1968 [1], Veselago predicted that electromagnetic plane waves in a medium having simultaneously negative permittivity and permeability would propagate in a direction opposite to that of the flow of energy. The direction of energy flow, given by $\vec{E} \times \vec{H}$, forms a right-handed system when the permittivity and permeability is both positive. Where the permittivity and permeability are negative, the direction of propagation is reversed with respect to the direction of energy flow, the vectors $\vec{E}, \vec{H}$, and $\vec{k}$ forming a left-handed system; thus, Veselago referred to such materials as "left-handed."

However, the research in this area was largely discontinued due to the absence of naturally occurring materials with negative μ. New discoveries of LH media was not made until recently, when a composite medium was demonstrated in which, both the effective ε and μ were purportedly shown to be simultaneously less than zero [2], over a finite frequency band. This physical medium was composed of distinct conducting elements, the size and spacing of which were on a scale much smaller than the wavelength in the frequency range of interest. Consequently, the composite medium was considered to be homogeneous at wavelengths of interest.

The composite medium (as shown in Figure 5.1) used in [2] made use of an array of metallic posts to create a frequency region with $\varepsilon_{eff} < 0$, interspersed with an array of split-ring resonators (SRR) for which $\mu_{eff} < 0$ was supposed in the frequency range of interest. The SRR and wire medium, both revisited by Pendry [3], have been extensively studied by a number of researchers.

The effective dielectric permittivity function of the thin wire medium and the effective magnetic permeability function of the SRR medium can be expressed as:

$$\varepsilon(\omega) = \varepsilon_0 \left(1 - \frac{\omega_{pe}^2}{\omega^2 - j\omega\gamma_e}\right) \quad (5.2)$$

$$\mu(\omega) = \mu_0 \left(1 - \frac{\omega_{pm}^2}{\omega^2 - j\omega\gamma_m}\right) \quad (5.3)$$

where ω_{pe} and ω_{pm} are electric and magnetic plasma frequencies and γ_e and γ_m are electric and magnetic collision frequencies, respectively.

If we assume that the wire and SRR arrays do not interact directly, the effective index of refraction of the composite medium can be expressed as $n(\omega) = \sqrt{\varepsilon_{eff}(\omega)\mu_{eff}(\omega)}$, with the material constants given by the expressions in (5.2) and (5.3). Thus, in a certain region both ε_{eff} and μ_{eff} can be simultaneously negative; the refractive index would then be real, and the fields would propagate in such a medium.

5.2 Effective Medium Theory and Left-Handed Metamaterials

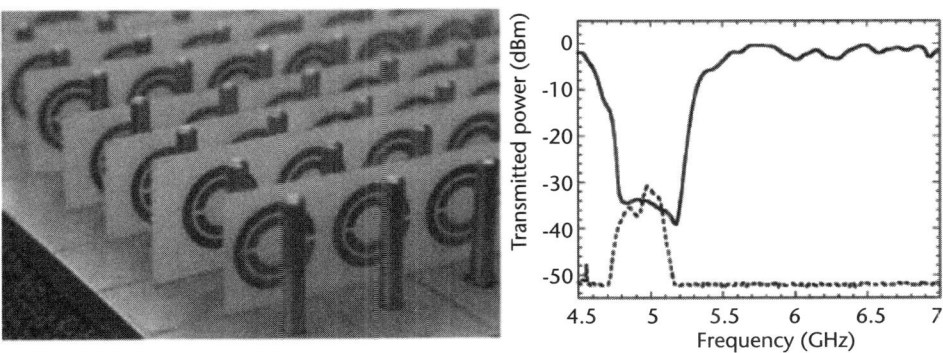

Figure 5.1 Left-handed medium and a transmission characteristic [2].

Figure 5.1 shows the results of transmission experiments on split rings alone (solid curve), and split rings with wires placed uniformly between (dashed curve). The square array of metal wires alone has a plasma frequency of 12 GHz; the region of negative permittivity this frequency attenuated the transmitted power to below the noise floor of the microwave detector (-52 dBm). When split rings are added to the wire array, a passband occurs, provided that the wave propagates along a direction such that its magnetic field is perpendicular to the plane of the rings.

5.2.2 Isotropic Three-Dimensional Left-Handed Metamaterials

A three-dimensional, nearly isotropic LHM design using the symmetric construction of unit cells that allows left-handed behavior for any direction of propagation and any polarization of the electromagnetic wave has been proposed by Koschny et al. [9], who have emphasized the importance of the symmetry issues in the design of isotropic metamaterials, as seen in Figure 5.2.

It has been shown that despite the square shape of the SRR, the scattering amplitude is independent of the orientation of the incident plane for both TE and TM modes. A reasonable isotropy has been observed. However, the geometry of the unit cell is very complicated for designing practical 3-D metamaterials.

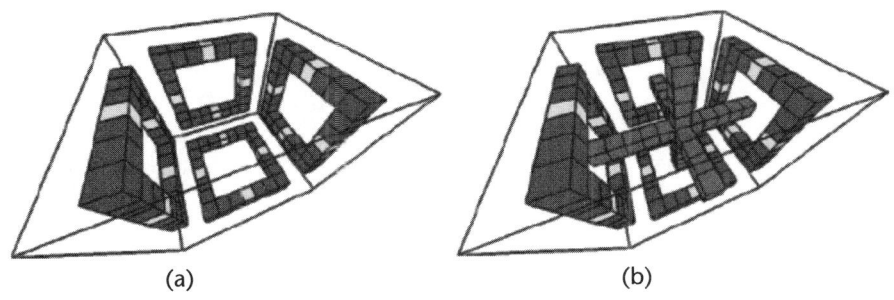

Figure 5.2 The design of a fully symmetric unit cell for a one-unit-cell thick slab of (a) an isotropic SRR and (b) an LHM [9].

5.2.3 Left-Handed Metamaterials Using Simple Short Wire Pairs

In the past few years, there has been ample demonstration of the existence of LHMs in the gigahertz frequency range, provided the propagation is strictly restricted to certain directions. Many groups have been able to fabricate such LHMs with an index of refraction $n = -1$ with losses of less than 1 dB/cm. However, to date, no materials with a negative n that is accompanied by a relatively small imaginary part have been found yet at the THz region. Currently, there is much interest in pushing the frequency range for LHM behavior into the infrared and optical regions of the spectrum. A recent theoretical work [10] has claimed that using pairs of finite length wires would not only allow replacing the SRRs as magnetic resonators but would also yield simultaneously, a negative ε and μ, and therefore, a negative n, without the need for additional continuous wires. However, the condition to obtain simultaneously negative μ and ε by pairs of finite metallic wires is very restrictive. Recent experiments [11] have not confirmed the existence of negative n in such short-wire pairs that have been investigated, refuting the claim [12] that one can realize negative n medium at tetrahertz frequencies.

The basic structure of the unit cell forming the wire-pair medium is shown in Figure 5.3. In it, the conventional SRR is replaced by a pair of short parallel wires while the continuous wire is preserved. The short wire pair consists of a pair of metal patches separated by a dielectric spacer of thickness t_s. In essence, the short wire pair is a "two-gap" SRR that has been flattened to yield the wire pair arrangement. For an electromagnetic wave incident with wave vector and field polarization as shown in Figure 5.3(a), the short wire pair will exhibit both inductive (along the wires) and capacitive (between the upper and lower adjacent ends of the short wires) behavior. It will thus have a magnetic resonance, which, in turn, would lead to a negative permeability.

Using the transmission and reflection results from a single layer, we can extract the effective refractive index for a periodic multilayer sample that utilizes the single-layer structure as a building block, as shown in Figure 5.4. The details of the

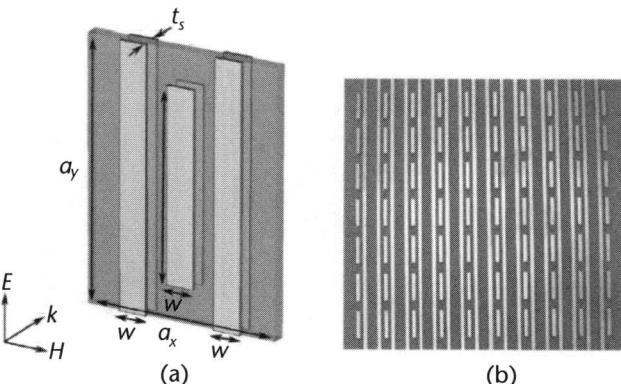

Figure 5.3 A wire pair structure: (a) schematic representation of one unit cell and (b) photograph of one side of a fabricated microwave-scale wire-pair sample.

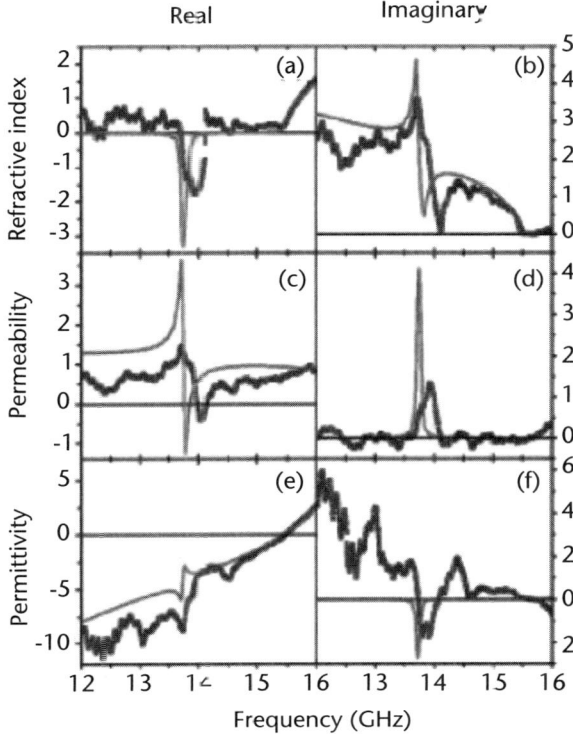

Figure 5.4 (a–f) Extracted electromagnetic properties of a periodic array of wire-pair unit cells, using the simulated and measured data [11].

numerical retrieval procedure have been described in [13–15]. We reiterate the fact that the retrieved parameters are only valid for a certain direction of propagation.

5.3 Applications of Left-Handed Metamaterials

5.3.1 Imaging by a Perfect LHM Lens

Conventional optical lenses have limitations. No lens can focus light onto an area smaller than a square wavelength. The reasons for the limitation in performance is that for larger values of transverse wave vectors, the evanescent waves decay exponentially with the axis of the lens and no phase correction will restore them to their original levels. They are effectively removed from the image, which is generally comprised of the propagating waves alone. The maximum resolution in the image can never be greater than

$$\Delta \approx \frac{2\pi}{k_{max}} = \frac{2\pi c}{\omega} = \lambda \tag{5.4}$$

However, a material with negative refractive index is in principle capable of focusing light even when in the form of a parallel-sided slab of material. Figure 5.5

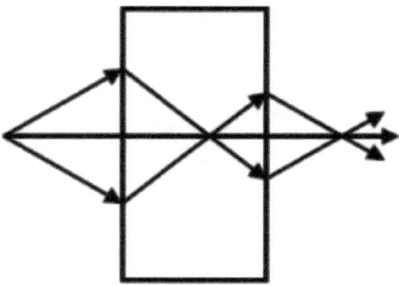

Figure 5.5 A negative refractive index lens.

shows such a lens with $n = -1$. A true negative refractive index lens would amplify evanescent waves, and both propagating and evanescent waves contribute to the resolution of the image (see Figure 5.6). Therefore, in principle, there is no physical obstacle to achieving a perfect reconstruction of the image beyond the practical limitations of apertures and the perfection of the lens surface. Details of imaging by a perfect LHM lens and its FDTD modeling can be found in Chapter 8.

5.3.2 Transmission Line Structures of Left-Handed Metamaterials

Several researchers have further studied the characteristics and applications of SRR-based LHMs. However, since the resonant structures such as SRRs are lossy, narrow-banded, and most importantly, anisotropic, they have not performed as expected, even at microwave frequency, let alone at higher frequencies. They have prompted several researchers to investigate a transmission line (TL) approach to realizing LHMs [16]. This approach has, in turn, led to nonresonant structures that have lower losses and wider bandwidths. In particular, metamaterials with right-handed (RH) and left-handed (LH) properties known as composite right/left hand

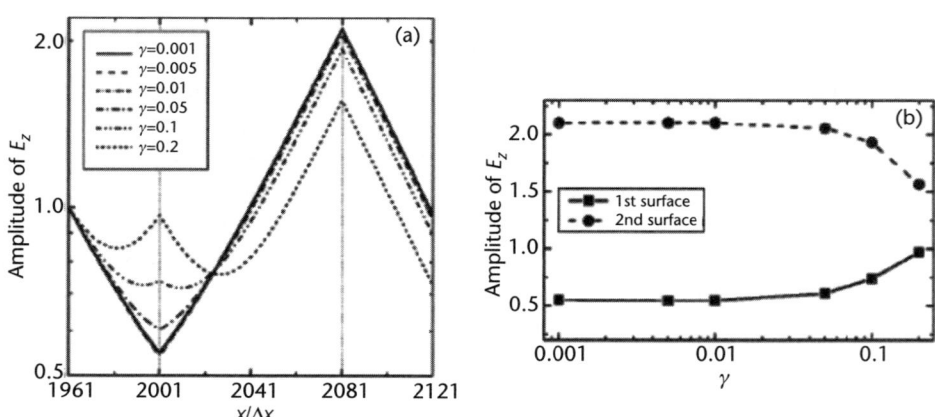

Figure 5.6 (a, b) Amplification of evanescent waves [21].

5.3 Applications of Left-Handed Metamaterials

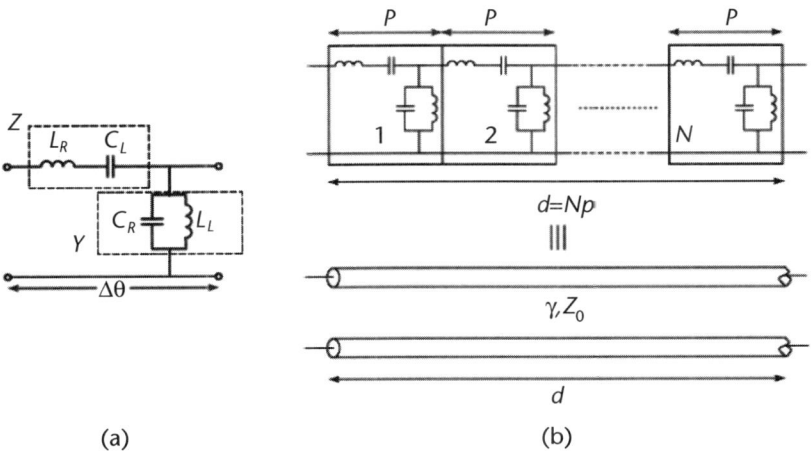

Figure 5.7 LC-based CRLH transmission line: (a) unit cell and (b) LC periodic network equivalent to a homogeneous CRLH TL of length d for $p = \Delta z \to 0$ [17].

(CRLH) metamaterials have led to the development of several novel microwave devices that we will discuss in this chapter. However, the realization of superlenses has still remained elusive.

The following configurations are most extensively studied:

1. CRLH TL using interdigital capacitors and shorted stub inductors, as shown in Figure 5.7;
2. Open 2-D CRLH mushroom structures.

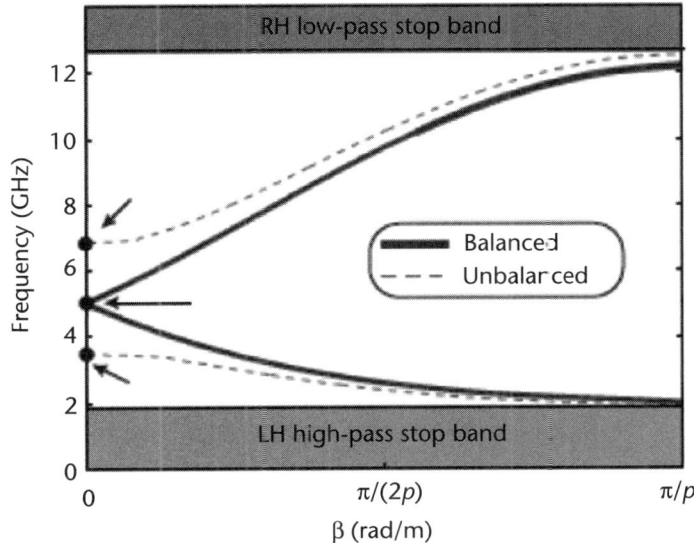

Figure 5.8 Dispersion diagram for the balanced and unbalanced LC-based CRLH TL. Balanced: $L_R=L_L=1$ nH, $C_R=C_L=1$ pF; unbalanced: $C_R=1$ pF, $C_L=2$ pF [18].

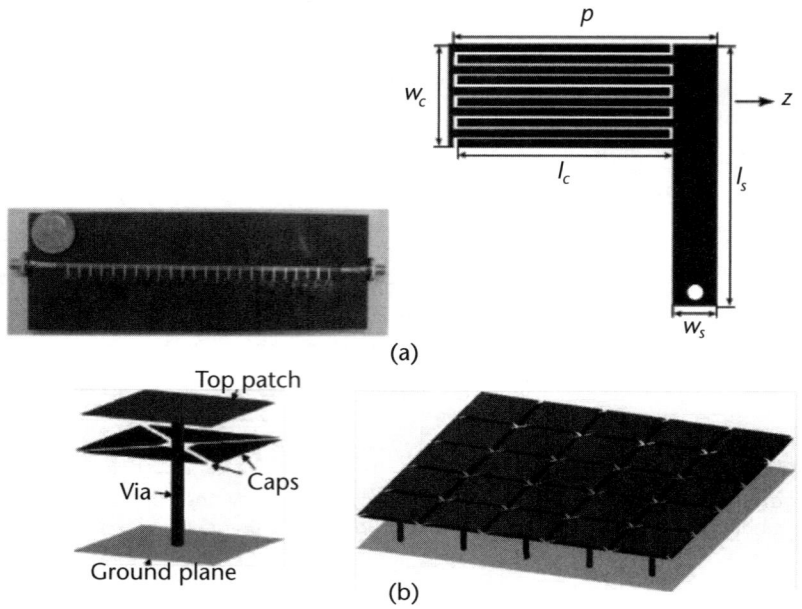

Figure 5.9 CRLH TL: (a) unit-cell 1-D CRLH TL and (b) open 2-D CRLH mushroom structure.

We can apply the PBCs related to the Bloch-Floquet theorem to the LC unit cell [17], as shown in Figure 5.7, to obtain the LC dispersion relation as follows:

$$\beta(\omega) = \frac{1}{p}\cos^{-1}\left(1 + \frac{ZY}{2}\right), \text{ where } Z(\omega) = j\left(\omega L_R - \frac{1}{\omega C_L}\right),$$
$$Y(\omega) = j\left(\omega C_R - \frac{1}{\omega L_L}\right) \tag{5.5}$$

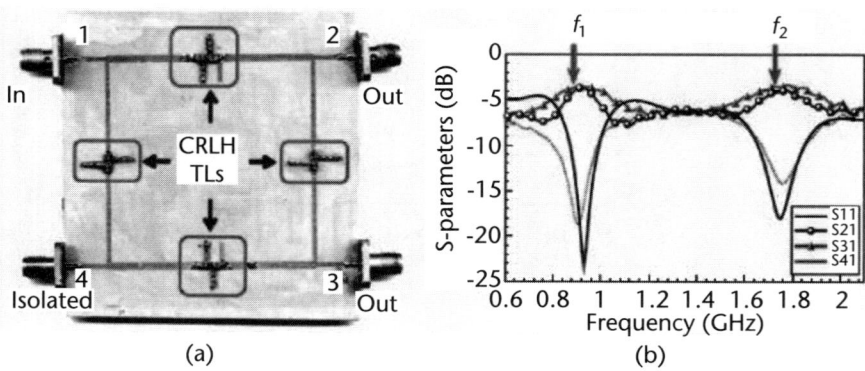

Figure 5.10 Dual band branch line coupler: (a) photograph and (b) measured S-parameter [18].

5.3 Applications of Left-Handed Metamaterials   131

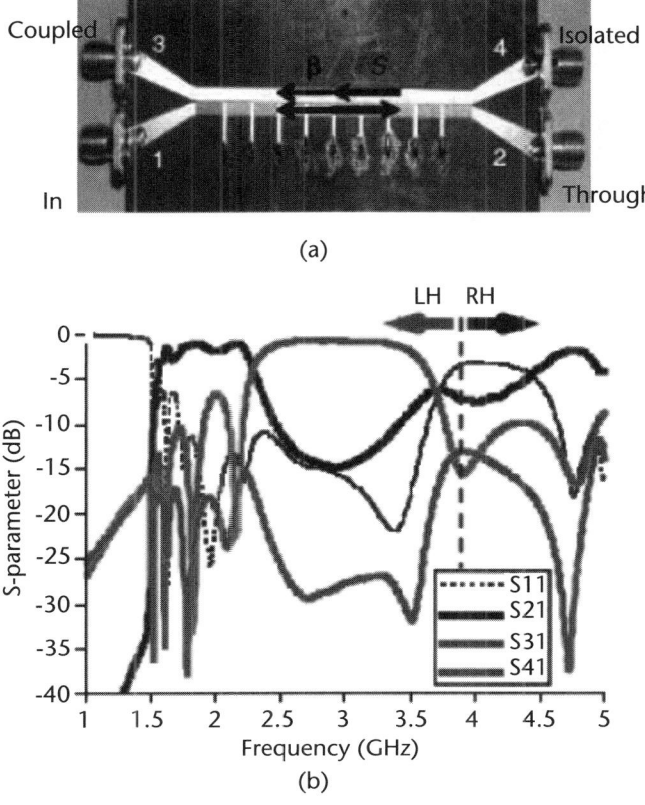

Figure 5.11 Asymmetric backward-wave directional coupler: (a) photograph and (b) measured S-parameter [17].

A computed dispersion diagram of an LC-based CRLHL-TL is shown in Figure 5.8. By selecting the LC values (balanced condition) appropriately, the bandgap between the LH and RH regions can be eliminated and a continuous change of phase velocity from negative to positive can be realized.

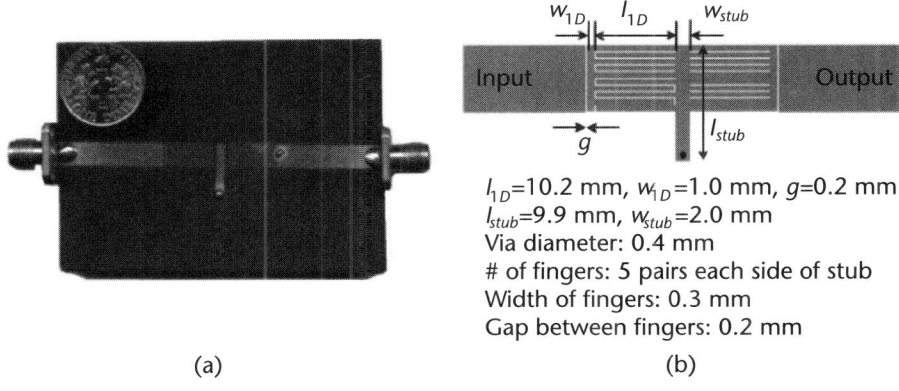

l_{1D}=10.2 mm, w_{1D}=1.0 mm, g=0.2 mm
l_{stub}=9.9 mm, w_{stub}=2.0 mm
Via diameter: 0.4 mm
# of fingers: 5 pairs each side of stub
Width of fingers: 0.3 mm
Gap between fingers: 0.2 mm

Figure 5.12 One-cell ZOR: (a) photograph and (b) layout with parameters [17].

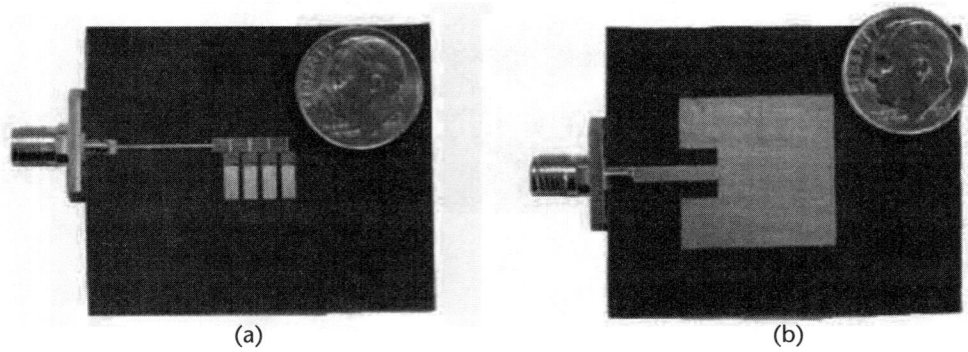

Figure 5.13 Antenna comparison: (a) ZORA and (b) conventional microstrip patch antenna [17].

Thy physical implementation of CRLH-TL can be accomplished by employing the surface-mount technology (SMT), chip components, or distributed components, to realize the LC network. Distributed components can be implemented via microstrip, stripline, coplanar waveguide, or another technology. 1-D and 2-D CRLH structures realized by using microstrip technology are shown in Figure 5.9.

5.3.2.1 Guided Wave Applications of CRLH-TL Metamaterials

The CRLH TL has led to several novel microwave applications and devices. Applications of TL-line–based LHM can be classified into three categories, namely guided, radiated, and refracted wave applications.

Conventional branch line couplers (BLCs) can only operate at their design frequency and odd harmonics. The conventional BLCs can be modified by replacing

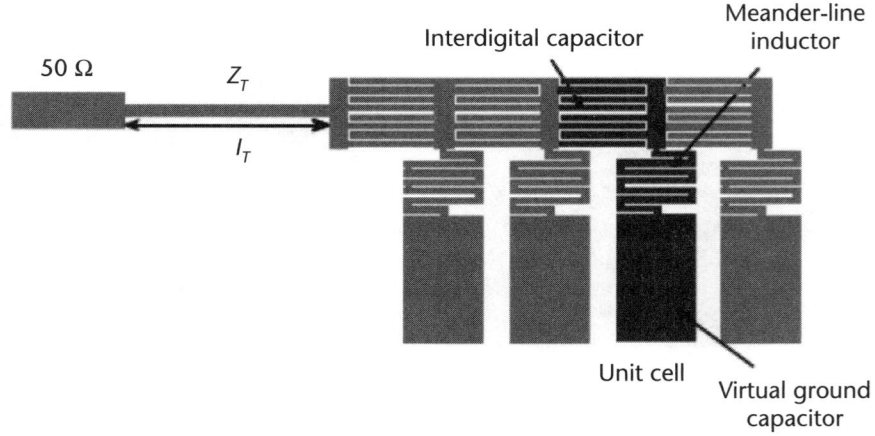

Figure 5.14 Four-cell ZORA [17].

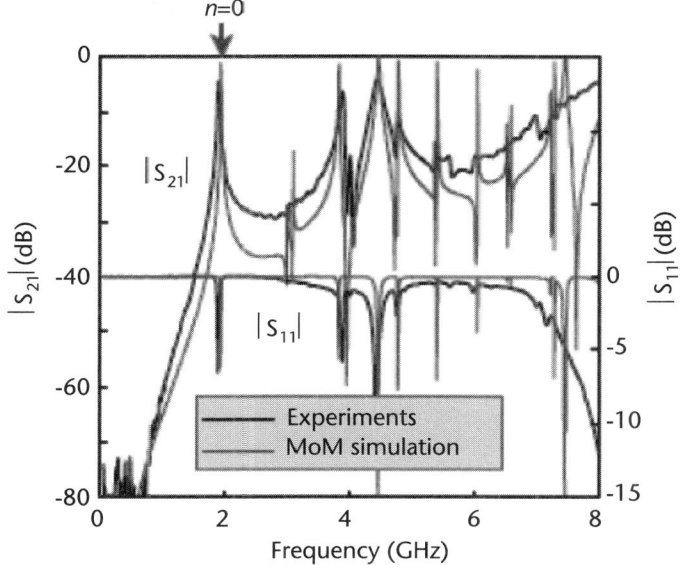

Figure 5.15 Measured and simulated S-parameters [17].

their RH TLs with CRLH TLs to yield a novel BLC with an arbitrary second operating frequency, as shown in Figure 5.10.

Asymmetric Backward-Wave Directional Coupler
Although conventional microstrip directional couplers are typically capable of operating over a broad bandwidth (> 25%), they typically have relatively low coupling levels, usually −10 dB or less. Noncoupled-line couplers, such as BLCs and ring couplers, offer tight coupling levels (−3 dB) but at the expense of a lower bandwidth (< 10%). The Lange coupler is able to achieve both a broad bandwidth and a tight coupling, though it requires the use of wire bonds that are cumbersome as well as expensive. The novel coupler of Figure 5.11 offers an alternative to these

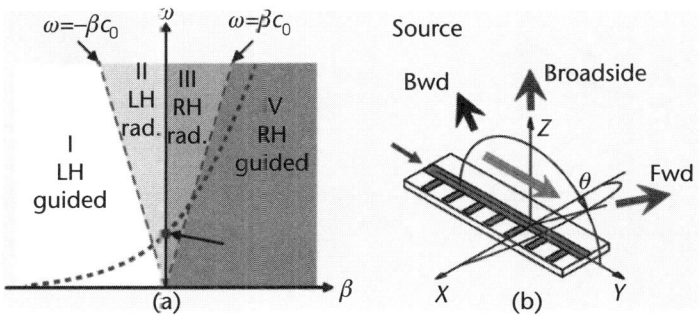

Figure 5.16 CRLH LW antenna: (a) typical dispersion diagram and (b) scanning operation [17].

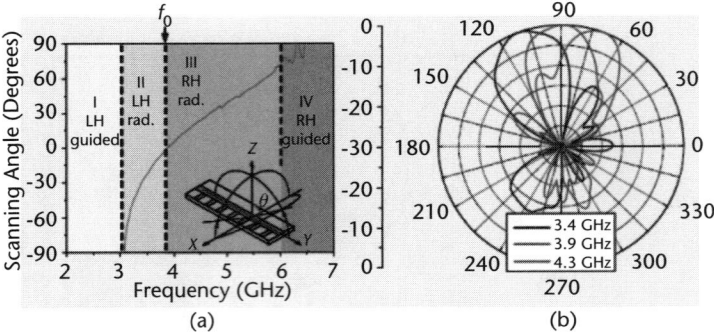

Figure 5.17 Measured results: (a) scanning angle versus frequency and (b) radiation patterns [17].

conventional couplers and is capable of achieving an arbitrary coupling level and a 50% bandwidth without the use of any wire bonds.

The coherent length (the length required for maximum coupling) of a conventional asymmetric coupler is given by $d_{max} = \frac{\pi}{|\beta_c \beta_\pi|}$, while that of the CRLH coupler is $d_{max} = \frac{\pi}{|\beta_{CRLH}| + \beta_{C\mu S}}$, where β_c and β_π are the c and π mode propagation constants.

5.3.2.2 A Zeroth-Order (ZOR) Resonator and Its Applications to Small Antennas

A unique feature of CRLH metamaterials is that of β, which we can use to achieve a wave number of zero at a nonzero frequency. This property can be used to create a novel zeroth-order resonator (ZOR), which is depicted in Figure 5.12. The resonance of ZOR is independent of the length of the structure but depends only on the reactive loadings. The unloaded quality factor Q_0 of the ZOR is $Q_0 = \frac{\sqrt{C_R L_L}}{G}$, where G is the shunt conductance of a lossy CRLH TL.

Since the resonance of the ZOR is independent of physical dimensions, the size of the antenna based on the ZOR can be smaller than a half wavelength.

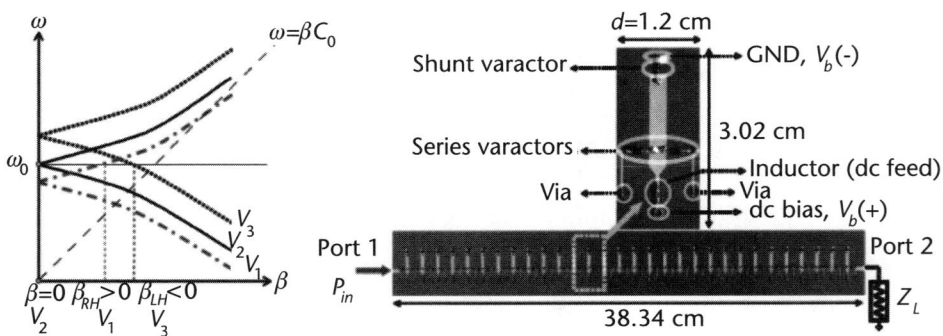

Figure 5.18 Principle of voltage-controlled CRLH unit cell and photograph of the actual antenna [17].

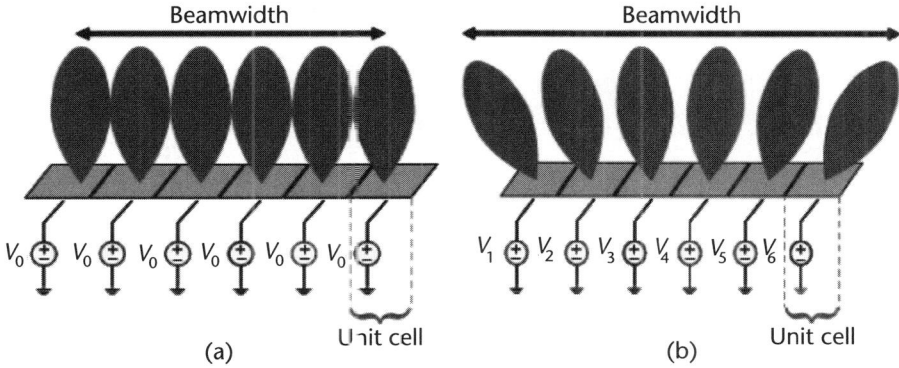

Figure 5.19 (a, b) Principles of beamwidth control [17].

Instead, the antenna's size is determined by the reactive loadings in its unit cells. Figure 5.13 shows the photographs of the two antennas with the same design frequency, 4.88 GHz. A detailed layout of the ZORA is shown in Figure 5.14. The LC components are implemented using interdigital capacitors and meander line inductors. Both simulated and measured S-parameters of ZORA are shown in Figure 5.15. It should be pointed out that the size reduction comes at a price, namely a reduction in the gain of the antenna.

5.3.2.3 A Backfire to Endfire Leaky-Wave (LW) Antenna

The balanced CRLH-TL can be used as an efficient, frequency-scanned leaky-wave (LW) antenna when optimally matched to the air impedance. A CRLH LW antenna has two distinct advantages over conventional LW antennas:

1. A CRLH LW antenna can operate at its fundamental mode, because this mode contains a radiation (or fast wave) region ($|\beta| < k_0$, where k_0 is the free-space propagation constant) in addition to a guided (or slow wave) region as shown in Figure 5.16. In contrast, since the fundamental mode

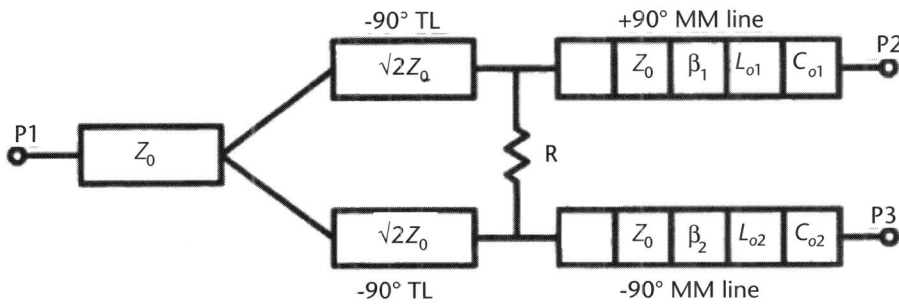

Figure 5.20 Block diagram of metamaterial balun [17].

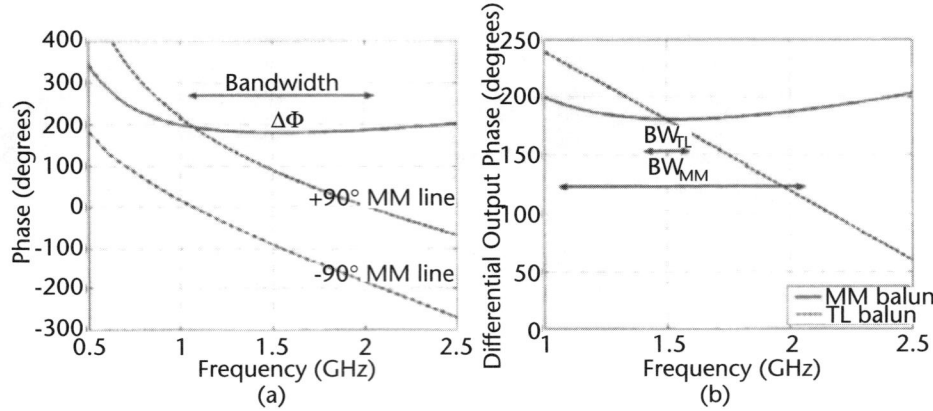

Figure 5.21 (a) Phase responses of S21 and (b) differential phase comparison [17].

of RH structures are always guided, they must be operated using higher-order modes in order to achieve radiation and, consequently, require a more complex and less-efficient feeding structure.

2. Unlike conventional LW antennas, the CRLH LW antenna is capable of continuous scanning from backward (backfire) to forward (endfire) angles. The LW antenna scanning angle is given by $\theta = \sin^{-1}\left(\frac{\beta_0 + 2n\pi/p}{k_0}\right)$, where β_0 is the propagation constant of the fundamental mode, and n is the index of spatial harmonics. Measured antenna radiation patterns for different scanning angles can be found in Figure 5.17.

The scanning angle of a LW antenna is frequency-dependent, which is not practical for most wireless systems. It has been reported that a frequency-independent

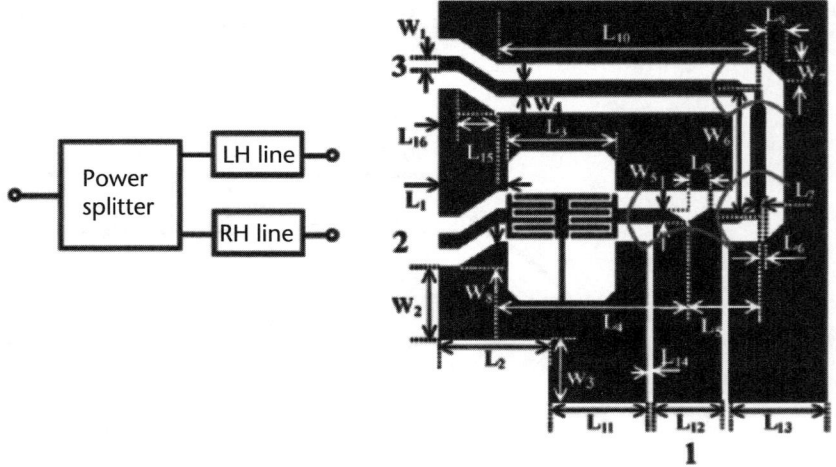

Figure 5.22 Block diagram of broadband arbitrary phase shifter and actual layout [17].

5.3 Applications of Left-Handed Metamaterials

LW antenna capable of continuous scanning and beamwidth control can be realized with a TL composed of voltage-controlled CRLH unit cells. By controlling the bias voltage of varactor diodes included in the CRLH unit cell, the capacitance of the cell can be changed, as shown in Figures 5.18 and 5.19. When using this approach, the propagation constant of the CRLH unit cell becomes a function of voltage. Since the voltage applied to each cell can be different, the voltage distribution on the antenna can be nonuniform, in general. Depending on the type of voltage distribution, the antenna can be used either as a scanning type or a beamwidth-controlled LW antenna.

5.3.2.4 A Broadband Balun and Phase Shifters Using CRLH-TL Metamaterials

Most balun designs are inherently narrowband, due to the frequency dependency of the components used in their construction. Broadband baluns usually require very long transmission lines or bulky ferrite cores and are not very compact. A broadband balun can be constructed by using a Wilkinson power divider connected to two ±90° phase-shifting lines that utilize CRLH-TL (see Figure 5.20). By adjusting the values of the loading elements, positive, negative, or zero insertion phase is realized. The fact that the frequency dependency of the positive and negative phase-shifting lines is very similar leads to the broadband characteristic of the balun, as shown in Figure 5.21.

A power splitter that exhibits a constant phase difference with arbitrary value over a wide bandwidth and provides advanced and retarded phase angles can be realized by using LH and RH transmission lines, as shown schematically in Figure 5.22. The RH transmission line is implemented by using conventional coplanar waveguide (CPW), and the LH transmission line is realized using the CRLH CPW comprising of interdigital capacitors (IDCs) and short-circuited stub inductors (SSIs). Typical delay line phase-shifters occupy large areas when operating at low frequencies; however this CRLH CPW is operated in the balanced condition by adjusting its equivalent inductances and capacitances, thus enabling the proposed

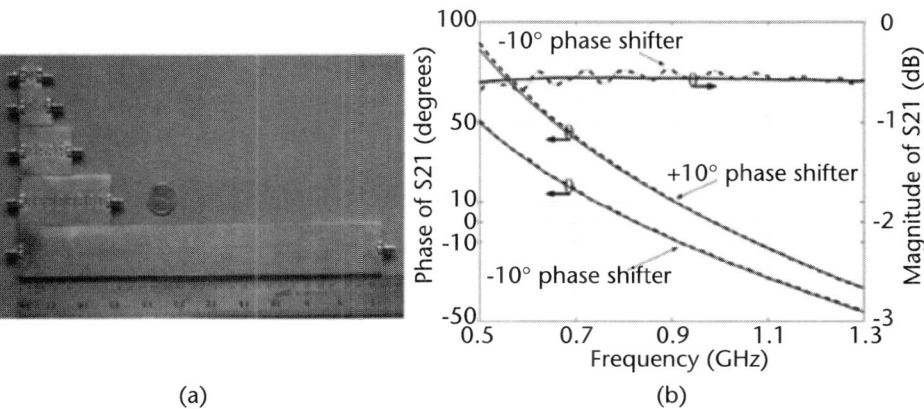

Figure 5.23 (a, b) Single-stage, two-stage, three-stage, and eight-stage 0° phase shifter compared to a conventional 360° TL at 0.9 GHz (right), phase and magnitude response (right) [22].

phase shifter to achieve the compact size and the arbitrary phase differences with broad bandwidth.

A compact one-dimensional phase shifter can be realized by using altering sections of LH transmission lines, as shown in Figure 5.23. The LH transmission line sections consist of lumped element capacitors and inductors. The amount of phase shift can be tailored to a given specification. Small variations in the LH transmission line element values can produce positive, negative, or zero phase shifts while maintaining the same short overall length. LHM phase shifters offer advantages over conventional delay lines: They are more compact in size and exhibit a linear phase response around the design frequency, enabling the realization of broadband components, as shown in Figure 5.24.

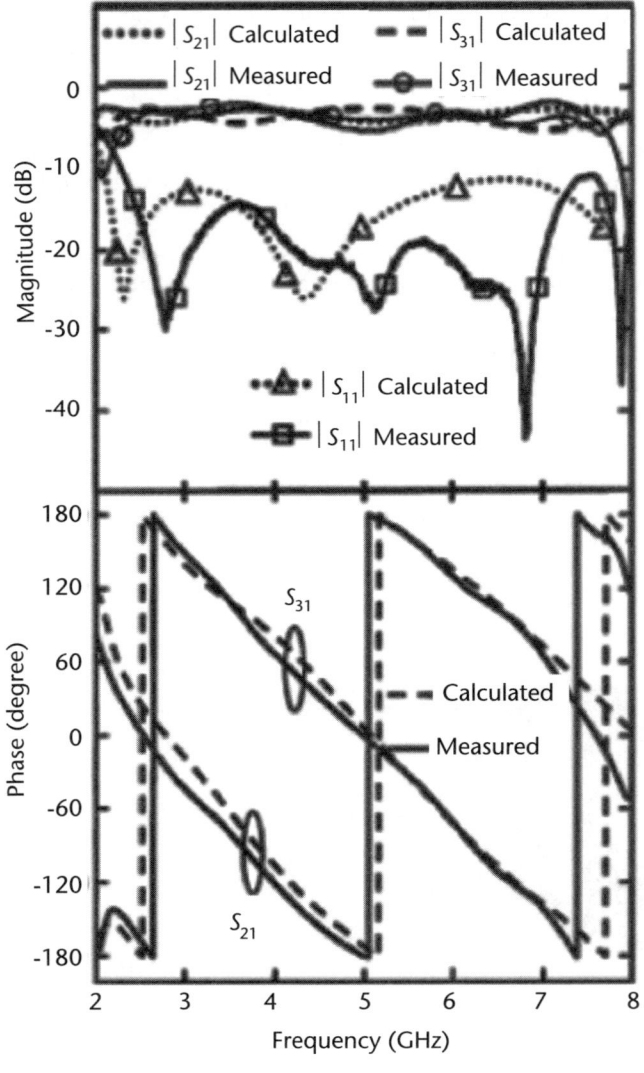

Figure 5.24 Magnitude and phase response of broadband LHM phase shifters [17].

5.3 Applications of Left-Handed Metamaterials

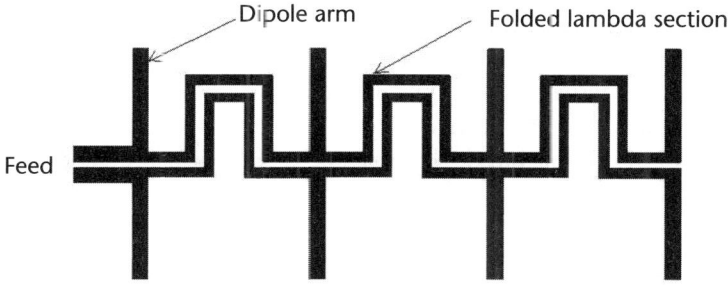

Figure 5.25 Two units of aperture-coupled microstrip array with H-slot.

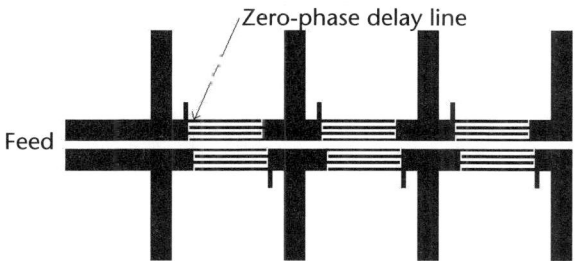

Figure 5.26 CRLH TL feeding line consisting of four LH-TL units.

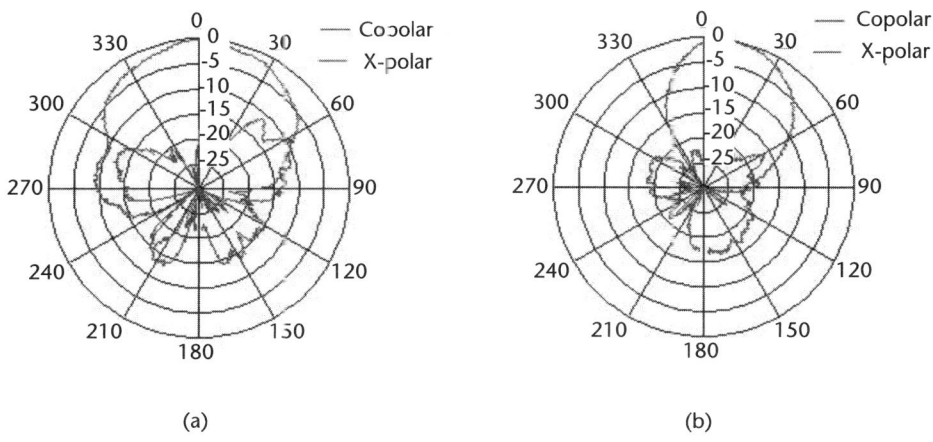

Figure 5.27 Comparison of radiation pattern between microstrip antenna arrays with (a) a traditional series-fed and (b) a CRLH TL–fed configuration at 94 GHz.

5.3.2.5 Applications of CRLH-TL in Antenna Array Design

Traditional feed configurations of microstrip array include parallel and series feeds. A parallel feed has the disadvantage that it requires a long transmission line between the radiators and input port. As a result, the insertion loss of the feed network becomes prohibitively large, and the overall efficiency of the array is reduced. On the other hand, the conventional series feed suffers from inherent beam shift with frequency due to the insertion phase shift of the patch in the series configuration. Based on the advance phase shift characteristics of LH transmission line, a CRLH-TL has been proposed, in which the insertion phase can be removed because of antiphase shift of LH and RH transmission lines. Figure 5.25 shows the

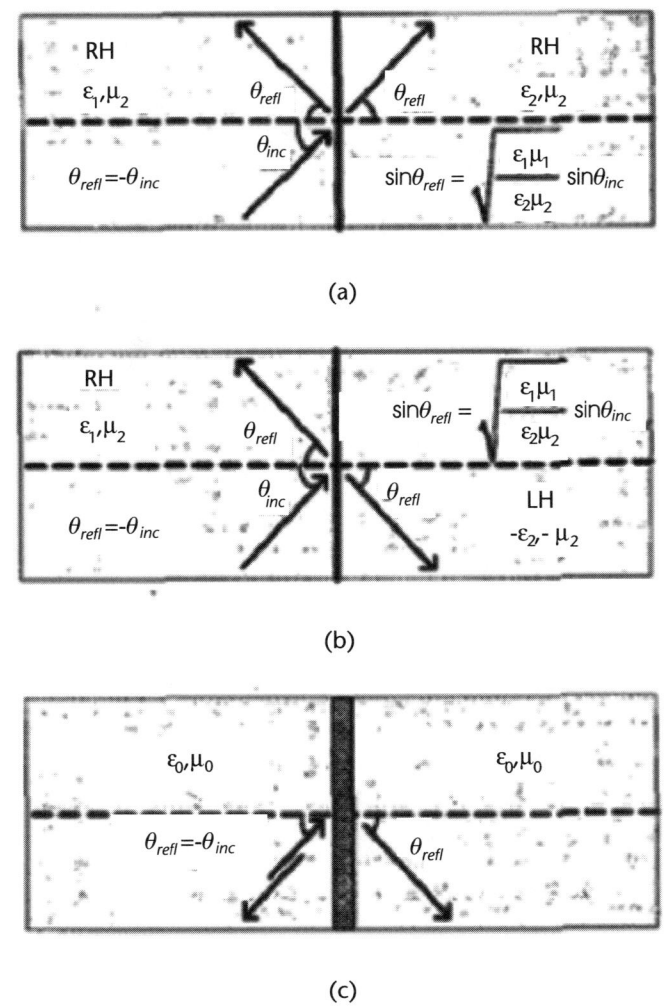

Figure 5.28 Reflection and refraction for (a) RH/RH, (b) RH/LH, and (c) proposed interface [22].

configuration of the millimeter-wave patch array using H-slots, and the CRLH-TL design is shown in Figure 5.26. A comparison of the radiation pattern with and without using the CRLH feed line shows improvement in the antenna gain. Reduced X-polar component and better beam shape is obtained with the array using CRLH Zero-phase delay line section (see Figure 5.27).

An artificial meta-interface has been implemented by using a phase-conjugating array. If a simple thin interface can produce effects similar to those of a complex interface between a conventional RH and a LH material, propagation will occur in real media and will thus circumvent the shortcomings of artificial structures. Figure 5.28 shows the possible reflection and refraction between RH/RH and RH/LH interfaces, and meta-interface. The normalized RCS of the meta-interface is shown in Figure 5.29. The actual structure is constructed by using two subcircuits, namely the antenna array (slot) and the mixer array as shown in Figure 5.30. Measured results have demonstrated the negative refraction and reflection in the farfield of the array and the displacement of refracted signal in the same direction as that of

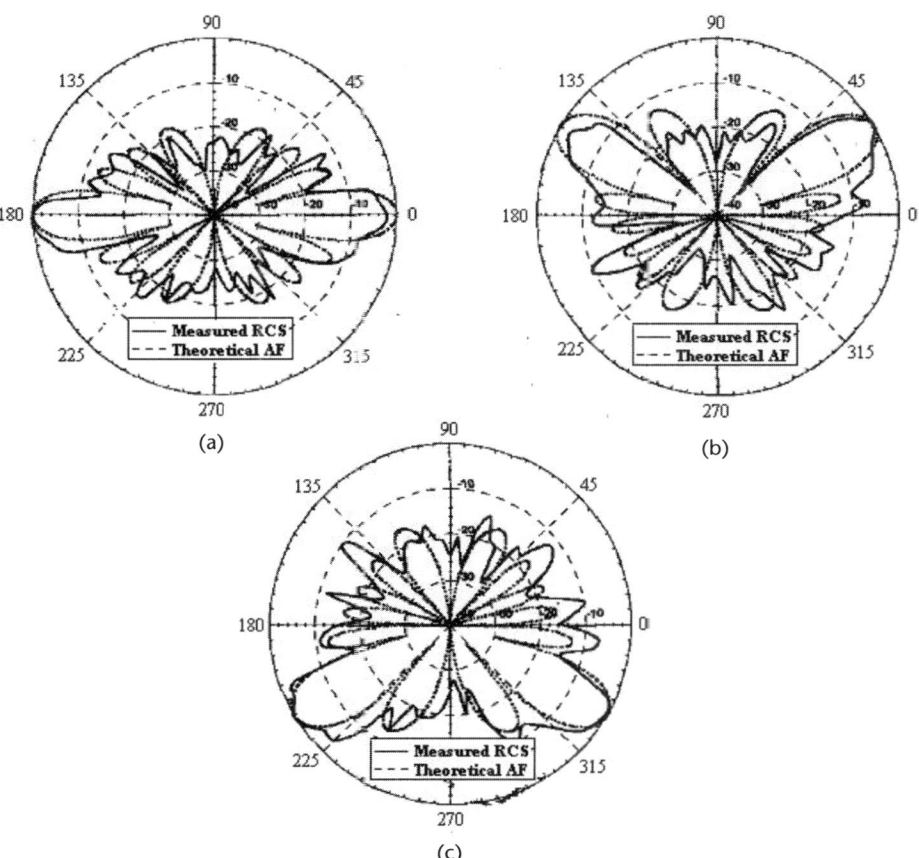

Figure 5.29 Measured normalized RCS of a meta-interface for a source located at (a) 0°, (b) 30°, and (c) −30° [22].

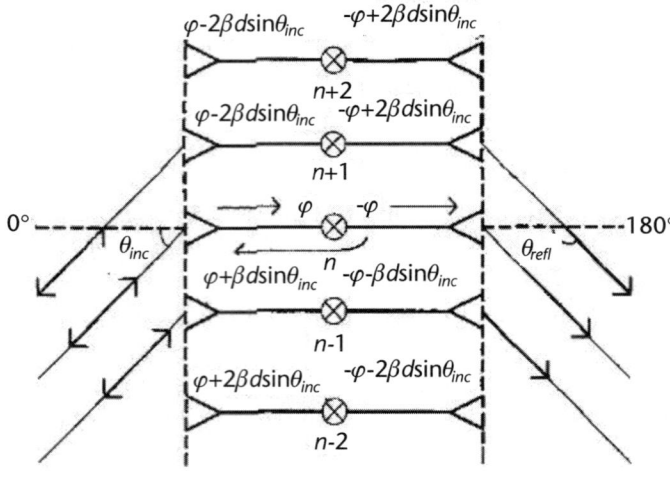

Figure 5.30 The principle underlying the implementation of a negative reflective/refractive interface and actual prototype [23].

the source parallel to the interface in the nearfield. These unusual effects may be exploited in quasi-optic beam-forming applications.

5.3.3 Directive Electromagnetic Scattering by an Infinite Conducting Cylinder Coated with LHMs

Electromagnetic scattering from metallic cylindrical structures coated with left-handed materials has been investigated by several groups [19, 20]. It has been observed that resonant peaks occur in the scattering width as a function of the frequency of the incident wave due to the surface paralitrons. The scattering characteristics of a conducting cylinder coated with a conventional dielectric material has been compared to those of a LHM-coated cylinder, assuming ideal LHM prop-

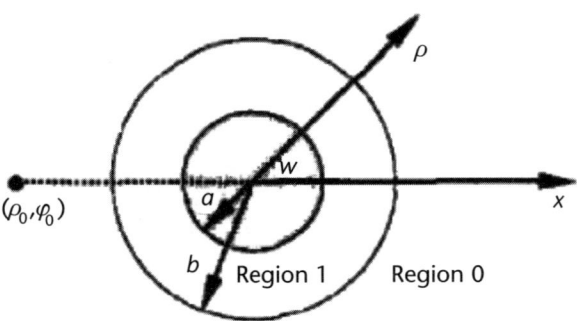

Figure 5.31 Geometry of the problem [19].

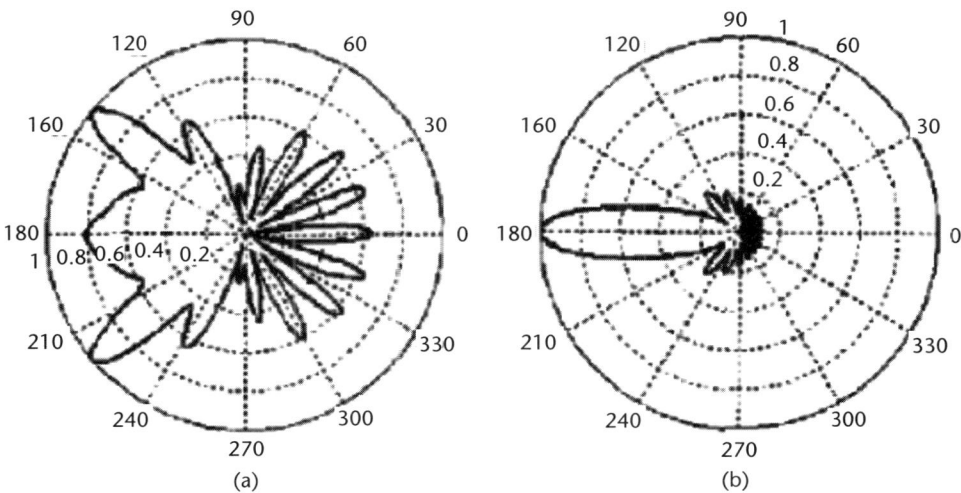

Figure 5.32 Far-field radiation pattern of a line source placed outside a conducting cylinder coated with (a) normal dielectric layer and (b) LHM [19].

erties. Improved directive scattering has been realized for the case of the idealized LHM-coated cylinder, and this has been attributed to the negative refraction of the LHM. The directivity depends on the size and the constitutive parameters of the LHM layer and the performance of the directive scattering, and it has been claimed that its performance could be further improved by moving the line source inside the coating dielectric layer.

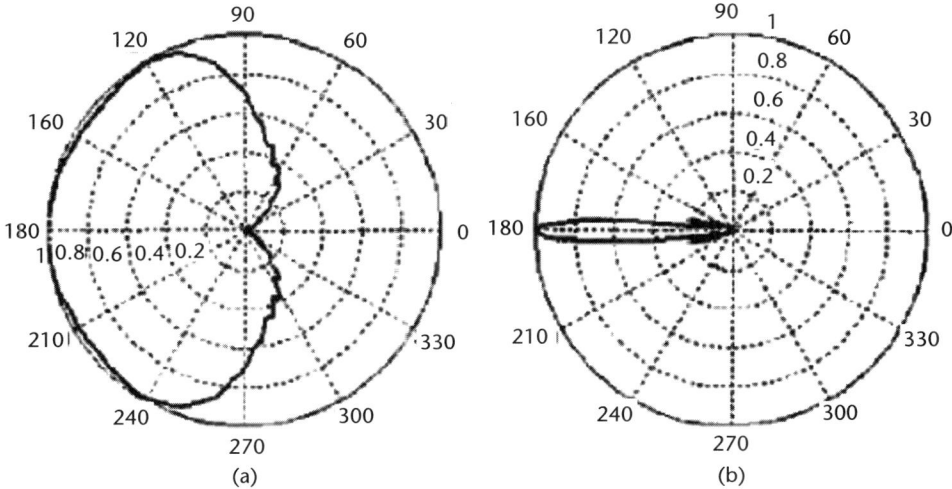

Figure 5.33 Far-field radiation pattern with the line source placed inside a conducting cylinder coated with (a) normal dielectric layer and (b) LHM [19].

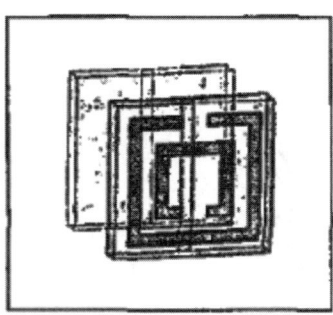

Figure 5.34 Unit cell of NIM.

Directive electromagnetic scattering by an LHM-coated conducting cylinder could be useful in potential applications of directive antennas (as shown in Figures 5.31–5.33), though this is again dependent upon the practical realization of such a coating.

5.3.4 Negative Index Materials (NIM) for Selective Angular Separation of Microwave by Polarization

It has been demonstrated that an anisotropic prism exhibits both positive and negative refractive indices and can split an incident beam into two components. The positive and negative indices are accessible by the choice of polarization of the electric field. The prism is constructed using the NIM unit-cell composed of metallic strip and split-ring resonators shown in Figure 5.34. For incident microwaves polarized so the electric field is parallel to the vertical metallic strip and the magnetic field is perpendicular to the boards, what appears to be negative refraction

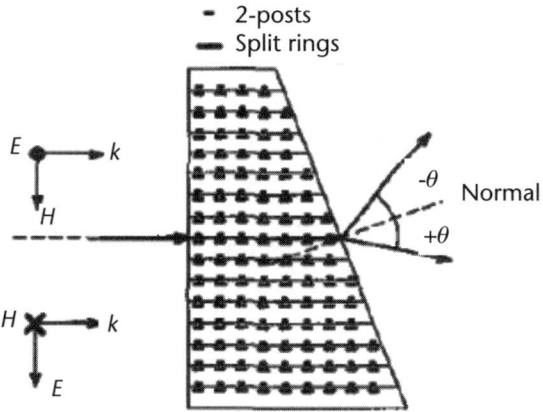

Figure 5.35 Top view of the prism measurement.

can be observed in a narrow frequency range. If the incident radiation is rotated by 90°, such a prism would have a positive index of refraction nearly unity at the same frequency for which the original polarization had a negative index. The use of intermediate polarizations can split the incident signals into two simultaneous output signals, one refracted negatively and one positively. The angular separation of the two output signals is determined by the prism angle and the negative index of refraction at the particular frequency.

It should be mentioned, however, that measurements of such prisms (see Figure 5.35) have shown that the level of the refracted signal is considerably lower, and this cannot be explained simply by taking the losses into account. Recently, an alternative explanation of this phenomenon has been advanced by Mittra et al. in terms of the Floquet harmonics of the periodic structure that makes up the prism, and the above explanation is consistent with the measured data for the prism.

References

[1] V. G. Veselago, "The electrodynamics of substances with simultaneously negative values of ε and μ," *Sov. Phys. Usp.* 10, 509 (1968).

[2] D. R. Smith, W. Padilla, D. C. Vier, S. C. Nemat-Nasser, and S. Schultz, "Composite medium with simultaneously negative permeability and permittivity," *Phys. Rev. Lett.*, Vol. 84, 4184-4187, 2000.

[3] J. B. Pendry, A. J. Holden, W. J. Stewart, and I. Youngs, "Extremely low frequency plasmons in metallic meso structures," *Phys. Rev. Lett.*, Vol. 76, pp. 4773-4776, 1996.

[4] D. R. Smith and N. Kroll, "Negative refractive index in left-handed materials," *Phys. Rev. Lett.*, Vol. 85, 2933, 2000.

[5] J. B. Pendry, A. J. Holden, D. J. Robins, and W. J. Stewart, "Low frequency plasmons in thin wire structures," *J. Phys. Condens. Matter*, vol. 10, pp. 4785-4809, 1998.

[6] D. F. Sievenpiper, M. E. Sickmiller, and E. Yablonivitch, "3D wire mesh photonic crystals" *Phys. Rev. Lett.*, Vol. 76, pp. 2480-2483, 1996.

[7] D. F. Sievenpiper, E. Yablonovitch, J. N. Winn, S. Fan, P. R. Vileneuve, and J.D. Joannopoulos, "3D metallo-dielectric photonic crystals with strong capacitive coupling between metallic islands," *Phys Rev. Lett.*, Vol. 80, pp. 2829-2832, 1998.

[8] J. B. Pendry, A. J. Holden, D. J. Robbins, and W. J. Stewart, "Magnetism from Conductors and Enhanced Nonlinear Phenomena," *IEEE Trans. Antennas and Propagat.*, Vol. 47, No. 11, pp. 2075-2084, November 1999.

[9] Th. Koschny, L. Zhang, and C M. Soukoulis, "Isotropic three dimensional left handed metamaterials," *Phys. Rev. Lett. B*, Vol. 71, 121103, 2003.

[10] V. A. Podlovsk, A. K. Sarychev, and V. M. Shalaev, "Plasmon modes and negative refraction in metal nanowire composite," *Opt. Express*, Vol. 11, 735-745, 2003.

[11] G. Dolling, C. Enkrich, M. Wegener, J. F. Zhou, C. M. Soukolis, and S. Linden, "Culture pairs and plate pairs as magnetic atoms for optical metamaterials," *Opt. Lett.*, 30, 3198 (2005).

[12] V. M. Shalaev, W. Cai, U. K. Chettiar, H. K. Yuan, A. K. Sarychev, V. P. Drachev, and A. V. Kildishev, "Negative index of refraction in optical metamaterials," *Opt. Lett.*, 30, 3356 (2005).

[13] D. R. Smith, S. Schultz, P. Markos, and C. M. Soukolis, "Determination of effective permittivity and permeability of metamaterials from reflection and transmission coefficient," *Phys. Rev. B*, Vol. 65, 195104 (2002).

[14] D. R. Smith, D. C. Vier, Th. Koschny, and C. M. Soukolis, "Electromagnetic parameter retrieval from inhomogeneous metamaterials," *Phys. Rev. E*, Vol. 71, 036617 (2005).

[15] Th. Koschny, P. Markos, E. N. Economou, D. R. Smith, D. C. Vier, and C. M Soukolis, "Impact of inherent periodic structure on effective medium description of left-handed and related metamaterials," *Phys. Rev. B*, 71, 245105 (2005).

[16] C. Caloz, H. Okabe, T. Iwai, and T. Itoh, "Transmission line approach of left-handed (LH) materials," *Proc. USNC/URSI National Radio Science Meeting*, San Antonio, TX, June 2002, Vol. 1, p. 39.

[17] C. Caloz and T. Itoh, *Electromagnetic Metamaterials: Transmission Line Theory and Microwave Applications*, New York: John Wiley & Sons, 2004.

[18] C. Caloz and T. Itoh, "A novel mixed conventional microstrip and composite right/left handed backward wave directional coupler with broadband and tight coupling characteristics," *IEEE Microwave Wireless Compon. Lett.*, vol. 14, pp. 31-33, January 2004.

[19] V. Kuzmak and A.A. Maradudin, "Scattering properties of a cylinder fabricated from a left-handed material," *Phy. Rev. B*, 66, 045116, 2002.

[20] R. Ruppin, "Surface polaritrons and extinction properties of a left-handed material cylinder," *J. Phys. Condens. Matter*, 16, pp. 5991-5998, 2004.

[21] X. Rao and C. K. Ong, "Amplification of evanescent waves in a lossy left-handed metamaterial slab," *Phys. Rev. B.*, 68, 113103, 2003.

[22] M. A. Antoniades and G. V. Eleftheriades, "Compact linear lead/lag metamaterial phase shifters for broadband applications," *IEEE Antennas and Propagation Letters*, Vol. 2, No. 1, 2003, pp. 103-106.

[23] C. A. Allen, K. Leong, and T. Itoh,"A negative reflective/refractive 'meta-interface' using bi-directional phase-conjugating array," *IEEE MTT-S International Microwave Symposium Digest*, 2003.

CHAPTER 6
Numerical Modeling of Left-Handed Material (LHM) Using a Dispersive FDTD Method

6.1 Introduction

Numerical techniques are designed to solve the relevant field equations in the computational domain, subject to the boundary constraints posed by the geometry. Without making a priori assumptions about which field interactions are most significant, numerical techniques analyze the entire geometry provided as input. They calculate the solution to a problem based on a full-wave analysis. The FDTD method [1–46], which is a powerful numerical modeling technique, has been widely used for modeling electromagnetic wave interaction with complex materials.

One of the most significant developments in the FDTD method is its capability of modeling frequency dispersive materials [1]. The existing frequency dispersive FDTD methods can be categorized as three types: the recursive convolution (RC) method, the auxiliary differential equation (ADE) method, and the Z-transform method.

In 1990 Luebbers et al. published the first frequency-dependent FDTD formulation for the modeling of Debye media [2] using a RC scheme by relating the electric flux density to the electric field through a convolution integral, and then discretizing the integral as a running sum. Soon the RC approach was extended for the study of wave propagation in a Drude material [3], Mth order dispersive media [4], an anisotropic magneto-active plasma [5], and ferrite material [6]. In 2004, the bi-isotropic/chiral media was modeled using the RC approach [7–9].

The ADE method was first used by Kashiwa and coworkers [10–12] in 1990 for Debye media, Lorentz media, and media obeying the Cole-Cole circular arc law, respectively. Joseph et al. [13] independently developed a similar ADE model for Debye media. Goorjian and Taflove [14] soon extended this model to include effects for nonlinear dispersive media. Independently, Gandhi et al. proposed the ADE method for treating M-th order dispersive media [15,16]. Later in 2004, the optical pulse propagation in 2-D Kerr and Raman nonlinear dispersive media was modeled using high-order FDTD and the ADE approach [17].

In 1992, Sullivan [18] proposed a dispersive FDTD formulation based on Z-transforms. Then the Z-transform approach was extended to treat nonlinear optical phenomena [19]. Recently, the chiral media was modeled using Z-transform method [20].

In [21], Feise et al. compared the ADE and Z-transform methods and applied the pseudo-spectral time-domain technique for the modeling of backward-wave metamaterials to avoid the numerical artifact due to the staggered grid in FDTD. Recently, Lee et al. used the piecewise linear RC (PLRC) method through an effective medium approach to model split-ring resonator LHM due to the similarity of its effective permittivity and permeability function to the Lorentz material model [22].

In addition to the frequency dispersion effect in all the above dispersive FDTD models, the inherent spatial dispersion effect also needs to be taken into account in order to accurately model the artificial media. Due to the similarity of the frequency and spatial dispersion effects, the ADE method can be directly applied in the dispersive FDTD model.

In this chapter, the ADE dispersive FDTD method will be reviewed in detail and applied to model LHMs for the demonstration of negative refraction, and the construction of flat-lens, zero-phase-delay wave transmission in layered LHM structures, conjugated LHM structures, and their applications in the design of transparent radomes.

6.2 The Effective Medium of Left-Handed Materials (LHMs)

It is worth mentioning here that frequency-independent negative material parameters are not physically realizable [23]. This can be verified by examining the relation between the energy density W, the electric field $\mathbf{E}$ and the magnetic field $\mathbf{H}$ [23]:

$$W = \frac{1}{2}\left(\varepsilon|\mathbf{E}|^2 + \mu|\mathbf{H}|^2\right) \tag{6.1}$$

which indicates that when the permittivity and permeability are negative, the total energy would have a negative value and the causality would be violated. If the medium is frequency-dependent, (6.1) would be replaced by [23]

$$W = \frac{1}{2}\left(\frac{\partial[\varepsilon(\omega)\omega]}{\partial\omega}|\mathbf{E}|^2 + \frac{\partial[\mu(\omega)\omega]}{\partial\omega}|\mathbf{H}|^2\right) \tag{6.2}$$

and the permittivity and permeability would be required to satisfy:

$$\frac{\partial[\varepsilon(\omega)\omega]}{\partial\omega} > 0, \quad \frac{\partial[\mu(\omega)\omega]}{\partial\omega} > 0 \tag{6.3}$$

It has been suggested in [24] that realistic LHMs can be characterized by using either a Lorentz or Drude dispersion model. It can be easily verified that both of these models do satisfy the criterion given in (6.3). For the simulations carried out in this chapter, we will use a Drude model for both the permittivity $\varepsilon(\omega)$ and permeability $\mu(\omega)$ with identical dispersion forms as follows:

$$\varepsilon(\omega) = \varepsilon_0\left(1 - \frac{\omega_{pe}^2}{\omega^2 - j\omega\gamma_e}\right) \tag{6.4}$$

$$\mu(\omega) = \mu_0\left(1 - \frac{\omega_{pm}^2}{\omega^2 - j\omega\gamma_m}\right) \tag{6.5}$$

6.2 The Effective Medium of Left-Handed Materials (LHMs)

Figure 6.1 The electric field intensity over the FDTD simulation space for the lossless LHM slab. The LHM slab is outlined and the source is located at the intersection of the horizontal and vertical lines [24].

where ω_{pe} and ω_{pm} are the electric and magnetic plasma frequencies and γ_e and γ_m are the corresponding collision frequencies, respectively.

There have been several attempts to verify the subwavelength imaging property of a LHM slab using the FDTD method. Ziolkowski and Heyman [24] have shown in Figure 6.1 that EM waves from a line source could be focused paraxially by a LHM slab, but no stable image could be formed. It was found that the image moved back and forth over time and sometimes vanished altogether.

A causal two-dimensional FDTD simulation is performed for the modeling of the scattering of a pulsed cylindrical wave by a matched, lossy Drude model LHM slab. These FDTD results conclusively demonstrate that the monochromatic electromagnetic power flow through the LHM slab is channeled into beams rather then being focused, as seen in Figure 6.2.

Mittra et al. conclude that the lens effect does not exist for any realistic dispersive, lossy LHM medium. No focal points either within the slab or in its exterior were found in any of the FDTD simulations. These simulations did, however, show a channeling or paraxial focusing of the wave energy due to the presence of a LHM slab, particularly when the index of refraction had large negative values.

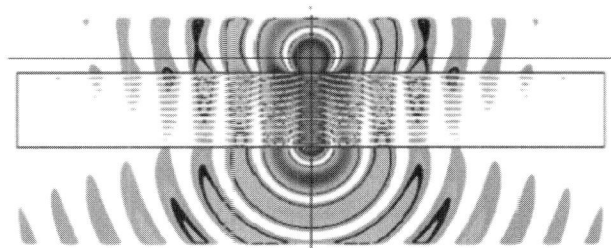

Figure 6.2 The electric field intensity over the FDTD simulation space for the LHM slab with a more negative refractive index. The LHM slab is outlined and the source is located at the intersection of the horizontal and vertical lines [24].

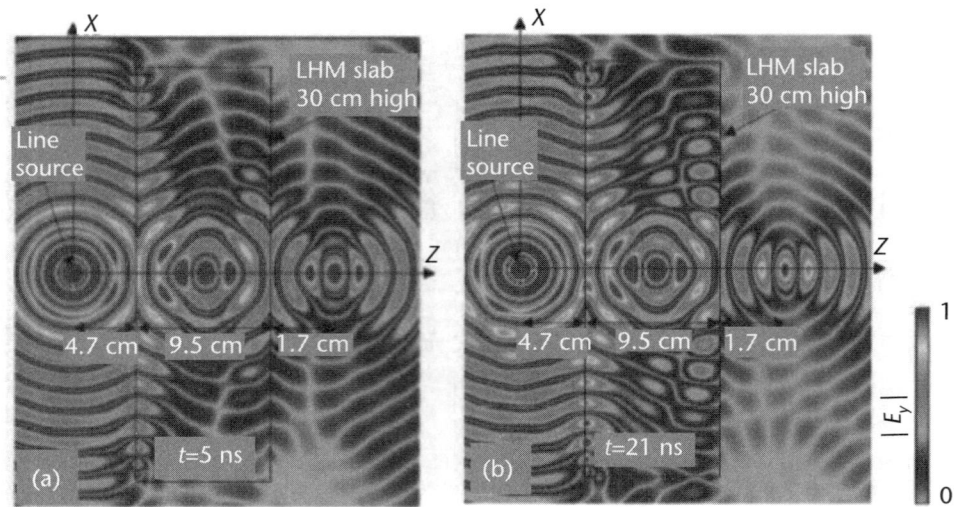

Figure 6.3 Snapshots of the electric field amplitude (absolute magnitude) produced by an infinite line source (y axis) excited by a cw impulse at 10 GHz. A LHM occupies the rectangular region denoted in the x-z plane and extends to infinite in the y-direction. At 10 GHz the index of the LHM is $-1.001 + i0.013$. (a) The snapshot at 5 ns after generation. (b) The same scene at 21 ns. Distinct focal areas are present at the center of the slab and equidistant to the right of the slab. Separate simulations for taller slabs indicate that edge diffraction plays a minimal role in the results [25].

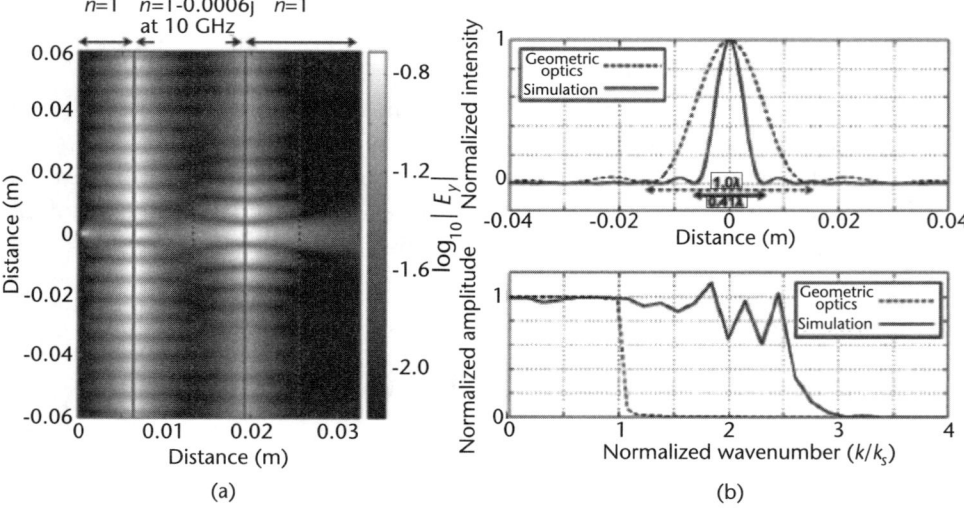

Figure 6.4 (a) Image of the sinusoidal steady state $|E_y|$ reached by the simulation. The solid lines show the material boundaries, and the dashed lines are the theoretical perfect focus planes. (b) Top panel: Optics-limited and simulated field intensity in the second focal plane, showing subwavelength focusing from the negative index slab. Bottom panel: Optics-limited and simulated transverse wave number spectrum in the same plane, showing that the amplitude of the evanescent transverse wave numbers ($k/k_{fs} > 1$) is restored by the negative index slab up to a limit of a $2.5 k_{fs}$ [26].

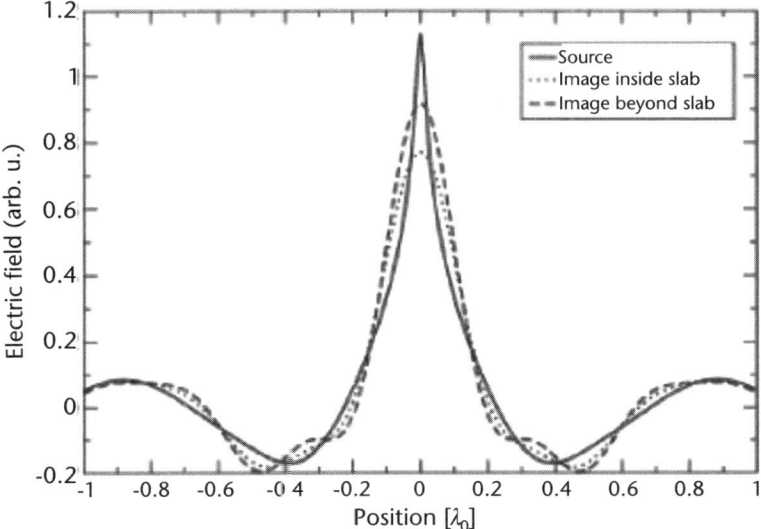

Figure 6.5 Snapshots of the electric field in the source plane (solid), the image plane inside the slab (dotted), and beyond the slab (dashed) with a single source.

Loschialpo et al. [25] performed numerical simulations using the FDTD method, by incorporating a causal Lorentzian form for the frequency-dependent material properties and observed a stable image of a line source formed by a LHM slab. They found that the image is of the order of the wavelength, showing no superlensing effect (Figure 6.3).

They demonstrated that a divergent line source spaced a distance H in front of a planar LHM slab and excited by either an impulse cw or a Gaussian frequency pulse is imaged at a distance H away, inside the LHM, and at H to the other side of the slab. The image size is λ consistent with limitations dictated by wave optics. They found no evidence of evanescent mode amplification.

However, a recent FDTD simulation by Cummer [26] reached a totally different result: subwavelength resolution of the image could be achieved by an LHM slab (Figure 6.4). He concluded that subwavelength focusing by negative refractive index slabs, as predicted by Pendry [27], is a real effect. Despite the limitations highlighted by this simulation and that have been demonstrated analytically elsewhere, subwavelength focusing should be observable in experiment. He also suggested that the difficulties of constructing a NIM with precisely $n = -1$ at the source frequency are similar to the difficulties of constructing a finite-difference approximation with precisely $n = -1$, and that simulations like this are thus a useful approximation of the degree of subwavelength focusing that may be observed in experiments.

In [28], Feise et al. used the pseudospectral time-domain method to study the unique features of imaging by a flat lens made of a LHM that possesses the property of negative refraction.

They confirmed the earlier finding that a left-handed flat lens can provide near-perfect imaging of a point source (see Figure 6.5) and a pair of point sources

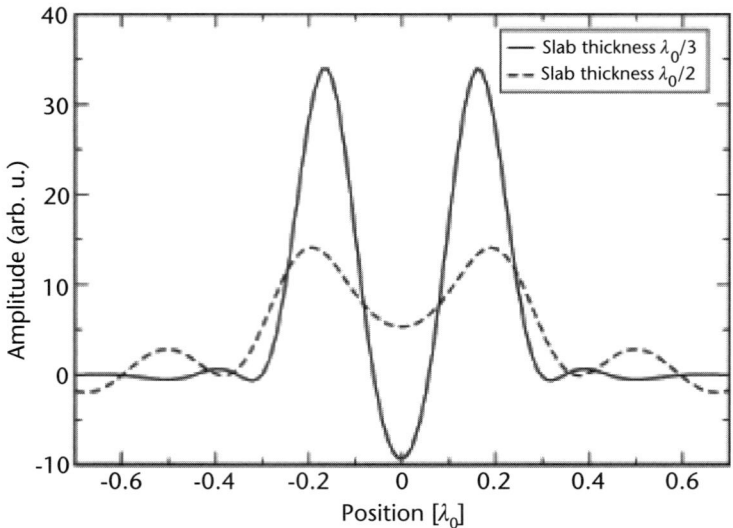

Figure 6.6 Snapshot of the y component of the Poynting vector in the image plane beyond the slab for a slab thickness of $\lambda_0/3$ (solid) and $\lambda_0/2$ (dashed).

with clear evidence of subwavelength resolution, as shown in Figure 6.6. They also illustrated the limitation of the resolution in the time-integrated image due to the presence of surface waves.

Rao et al. demonstrated in [29], by using the FDTD method, that subwavelength resolution can be achieved by LHM slabs with certain parameters (see

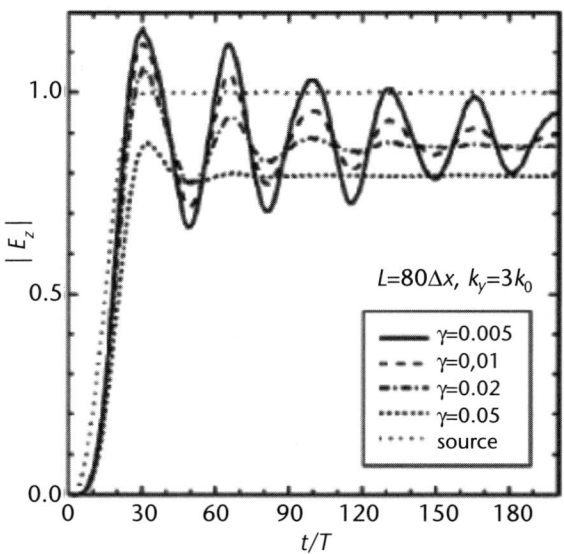

Figure 6.7 Time evolution of the E_z amplitude at the image plane for a LHM slab ($L = 80$ $\Delta x = 0.14\lambda_0$) with different values of $\gamma = \gamma_e = \gamma_m$, ranging from 0.005 to 0.5. The evanescent wave has $k_y = 3.0$. The time evolution of the source field is also present for [29].

6.2 The Effective Medium of Left-Handed Materials (LHMs)

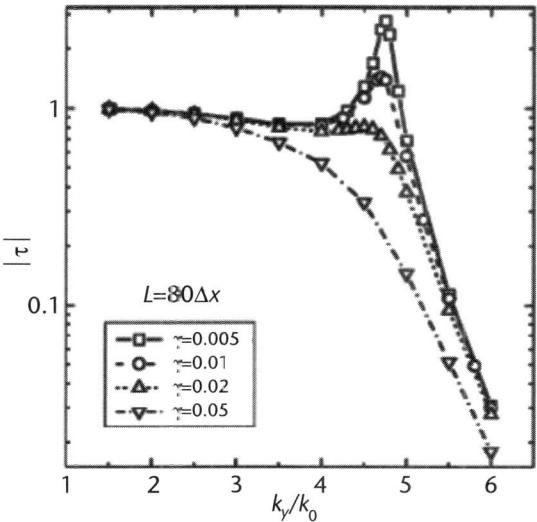

Figure 6.8 Transfer function of a LHM slab ($L = 80\Delta x = 0.14\lambda_0$) for different values of $\gamma = \gamma_e = \gamma_m$, ranging from 0.005 to 0.5, as a function of normalized transverse wave number k_y/k_0 [29].

Figure 6.7). They presented the dynamic feature of the imaging process and the dependence of physical parameters on the performance of the superlens.

They also showed that the achievable resolution is limited by the absorption and thickness of the LHM slabs, which introduces difficulties in practical applications of the superlens, as shown in Figures 6.8 and 6.9.

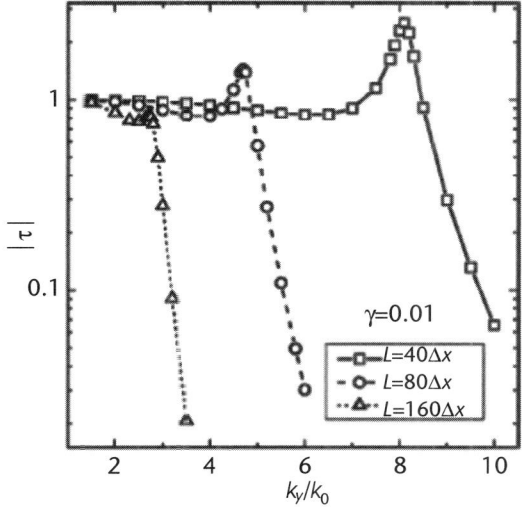

Figure 6.9 Transfer function for LHM slab ($\gamma = \gamma_e = \gamma_m = 0.01$ for all) of different thicknesses as a function of the normalized transverse wave number k_y/k_0 [29].

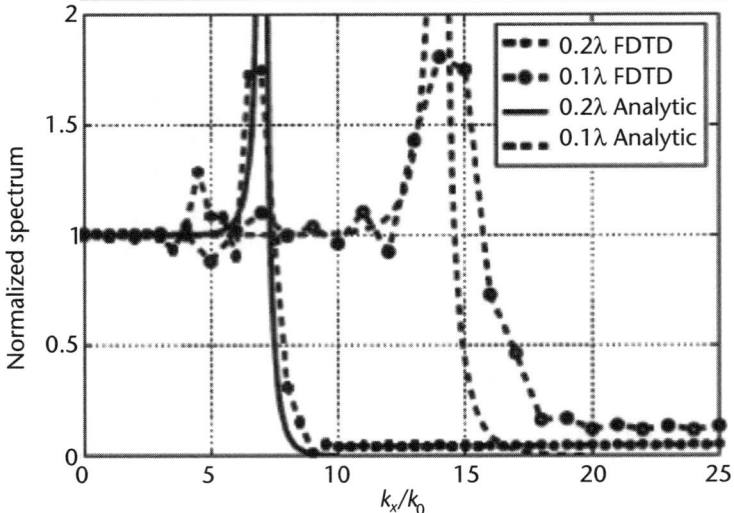

Figure 6.10 Comparison of E_y spectra at the image plane from the FDTD simulations and the analytical calculations for two slab configurations: one with a thickness of 0.2λ and the other with a thickness of 0.1λ. Both slabs are simulated with a same grid size of $\lambda/100$.

In [43], Chen et al. showed that because of the dispersive nature of the LHM medium, and the time discretization in FDTD modeling, an inherent mismatch in the constitutive parameters exists between the slab and its surrounding medium. This mismatch in the real part of the permittivity and permeability is found to have the same order of magnitude as the losses typically used in numerical simulations, as shown in Figures 6.10 and 6.11. Hence, when the LHM slab is lossless, this

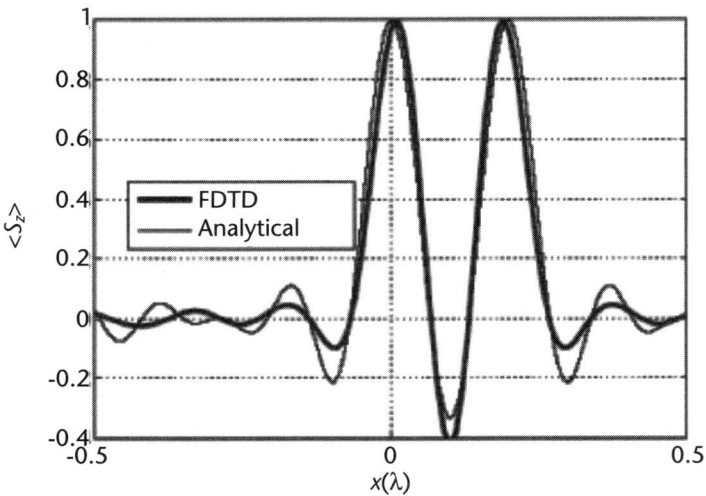

Figure 6.11 Comparison of time averaged Poynting power densities $<S_z>$ at the image plane from the FDTD simulation and the analytical calculation for the two line source imaging. The LHM slab is the same with Figure 6.10. The line sources are separated by 0.2λ [43].

6.2 The Effective Medium of Left-Handed Materials (LHMs)

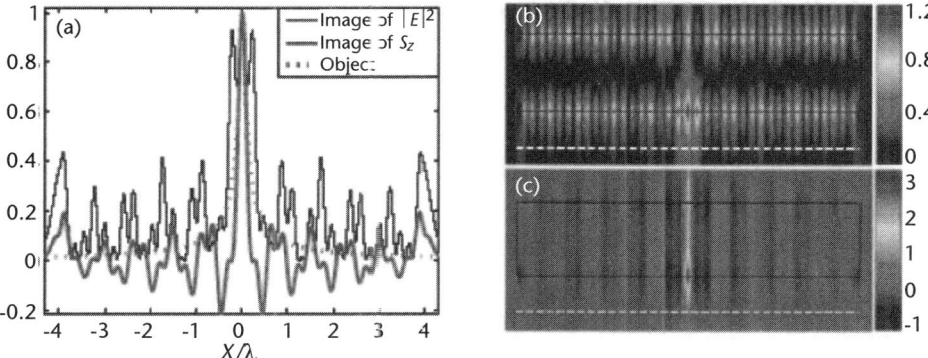

Figure 6.12 (a) Normalized distribution of field intensity (thin line) and the energy stream S_z (thick line) along the image plane for $L = 8\lambda$ and $d = 0.2\lambda$; here the normalized field intensity of the object is plotted (dotted line) for comparison. The corresponding 2-D distribution of (b) $|E|^2$ and (c) S_z. Here the LHM slab and the image plane are marked with a rectangle and a dashed line, respectively [44].

mismatch is shown to be the main factor contributing to the image resolution loss of the slab.

The characteristics of an imaging system formed by a LHM slab of finite length were studied in [44], and the influence of the finite length of the slab on the image quality was analyzed. Unusual phenomena such as surface bright spots and a negative energy stream at the image side are observed and explained as the cavity effects of surface plasmons excited by the evanescent components of the incident field, as shown in Figure 6.12.

For a thin LHM slab, the cavity effects are found to be rather sensitive to the variation in the length of the slab; the bright spots on the bottom surface of the

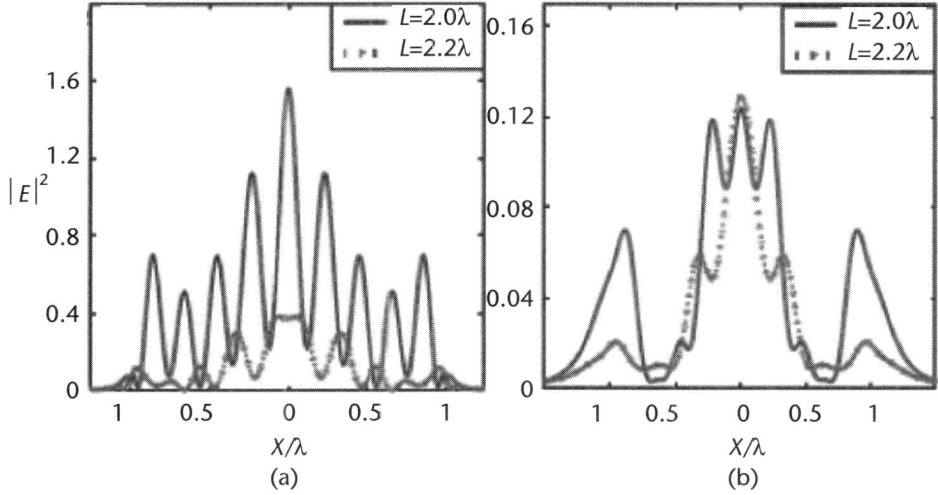

Figure 6.13 Field intensity profiles along the (a) bottom surface of the LHM slab, and (b) the image plane, when the incident field is generated by a point source. Here $d = 0.2\lambda$ [44].

slab may stretch to the image plane and degrade the image quality. It has been shown that both the length and the thickness of the LHM slab greatly influence the cavity effects of surface plasmons and, consequently, the image quality, as shown in Figure 6.13.

However, we shall demonstrate in late chapters that modeling LHMs based on the effective medium theory has its limitations: First, physical structures of isotropic, homogeneous and three-dimensional metamaterials are often not realizable, and second, modifications such as spatial averaging are frequently used at the material interfaces to improve the numerical accuracy due to the staggered grid in the FDTD domain.

6.3 Modeling of Left-Handed Metamaterials Using a Dispersive FDTD Method

Several techniques have been proposed to incorporate frequency dispersion into the FDTD methods. They can be roughly categorized into three types: the recursive convolution (RC) method [18, 30–32], the auxiliary differential equations (ADE) method [11, 13, 16, 33, 34], and the z-transform (ZT) method [18, 35, 36].

6.3.1 Two-Dimensional Dispersive FDTD with Auxiliary Differential Equations (ADEs)

In order to investigate electromagnetic wave interaction with the LH media, we start with the 2-D structures such as single or multilayer slabs and cylindrical shells. Consider a TE wave with field components E_z, H_x, and H_y; the 2-D FDTD formulation involves the solution of a set of equations as follows:

$$\frac{\partial D_z}{\partial t} = \frac{\partial H_y}{\partial t} - \frac{\partial H_x}{\partial y} \tag{6.6}$$

$$D_z = \varepsilon_0 \left(1 - \frac{\omega_{ep}^2}{\omega^2 - \omega_{e0}^2 - j\gamma_e \omega}\right) E_z \tag{6.7}$$

$$\frac{\partial B_x}{\partial t} = -\frac{\partial E_z}{\partial y} \tag{6.8}$$

$$B_x = \mu_0 \left(1 - \frac{\omega_{mp}^2}{\omega^2 - \omega_{m0}^2 - j\gamma_m \omega}\right) H_x \tag{6.9}$$

$$\frac{\partial B_y}{\partial t} = \frac{\partial E_z}{\partial x} \tag{6.10}$$

$$B_y = \mu_0 \left(1 - \frac{\omega_{mp}^2}{\omega^2 - \omega_{m0}^2 - j\gamma_m \omega}\right) H_y \tag{6.11}$$

6.3 Modeling of Left-Handed Metamaterials Using a Dispersive FDTD Method

Table 6.1 List of Symbols

Symbols	Representation
E_z	z component of electric field intensity (V/m)
H_x	x component of magnetic field intensity (A/m)
H_y	y component of magnetic field intensity (A/m)
D_z	z component of electric flux density (coulombs/square meter)
B_x	x component of magnetic flux density (webers/square meter)
B_y	y component of magnetic flux density (webers/square meter)
ω	angular frequency (rad/s)
ω_{ep}	electric plasma frequency (rad/s)
ω_{e0}	low frequency edge of electric forbidden band (rad/s)
γ_e	electric collision frequency (rad/s)
ω_{mp}	magnetic plasma frequency (rad/s)
ω_{m0}	low frequency edge of magnetic forbidden band (rad/s)
γ_m	magnetic collision frequency (rad/s)
ε_0	free space permittivity (8.854×10^{-12} F/m)
μ_0	free space permeability ($4\pi \times 10^{-7}$ H/m)

Since multiplication of $j\omega$ in the frequency domain is equivalent to time derivatives in the time domain, (6.7), (6.9), and (6.11) are equivalent to

$$\frac{\partial^2 D_z}{\partial t^2} + \gamma_e \frac{\partial D_z}{\partial t} + \omega_{e0}^2 D_z = \varepsilon_0 \frac{\partial^2 E_z}{\partial t^2} + \varepsilon_0 \gamma_e \frac{\partial E_z}{\partial t} + \varepsilon_0 \left(\omega_{e0}^2 + \omega_{ep}^2\right) E_z \quad (6.12)$$

$$\frac{\partial^2 B_x}{\partial t^2} + \gamma_m \frac{\partial B_x}{\partial t} + \omega_{m0}^2 B_x = \mu_0 \frac{\partial^2 H_x}{\partial t^2} + \mu_0 \gamma_m \frac{\partial H_x}{\partial t} + \mu_0 \left(\omega_{m0}^2 + \omega_{mp}^2\right) H_x \quad (6.13)$$

$$\frac{\partial^2 B_y}{\partial t^2} + \gamma_m \frac{\partial B_y}{\partial t} + \omega_{m0}^2 B_y = \mu_0 \frac{\partial^2 H_y}{\partial t^2} + \mu_0 \gamma_m \frac{\partial H_y}{\partial t} + \mu_0 \left(\omega_{m0}^2 + \omega_{mp}^2\right) H_y \quad (6.14)$$

Relevant symbols are defined in Table 6.1.

Applying the second-order FDTD discretization both in time and space to (6.6), (6.8), and (6.10), respectively, yield

$$D_z^{n+1} = (i+1/2, j+1/2) = D_z^n(i+1/2, j+1/2)$$
$$+ \Delta t \left[\frac{H_y^{n+1/2}(i+1, j+1/2) - H_y^{n+1/2}(i, j+1/2)}{\Delta x}\right]$$
$$- \Delta t \left[\frac{H_x^{n+1/2}(i+1/2, j+1) - H_x^{n+1/2}(i+1/2, j+1)}{\Delta y}\right] \quad (6.15)$$

$$B_x^{n+3/2} = (i+1/2, j+1) = B_x^{n+1/2}(i+1/2, j+1)$$
$$+ \Delta t \left[\frac{E_z^{n+1}(i+1/2, j+3/2) - E_z^{n+1}(i+1/2, j+1/2)}{\Delta y}\right] \quad (6.16)$$

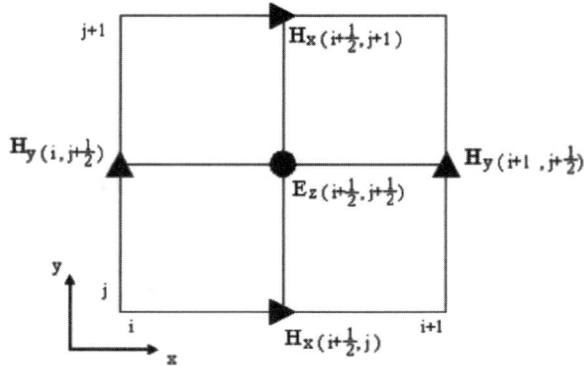

Figure 6.14 A Yee's cell used in modeling LHMs based on a dispersive FDTD algorithm.

$$B_y^{n+3/2} = (i+1, j+1/2) = B_y^{n+1/2}(i+1, j+1/2)$$
$$+ \Delta t \left[\frac{E_z^{n+1}(i+3/2, j+1/2) - E_z^{n+1}(i+1/2, j+1/2)}{\Delta x} \right] \quad (6.17)$$

As shown in Figure 6.14, the electric field E_z and electric flux density D_z are taken at the cell center with integer time steps, while the magnetic components $H_x(B_x)$ and $H_y(B_y)$ are taken at the edge of the cell with a half-step offset in time and space.

Applying a second-order accurate central-difference scheme centered at time-step n, arithmetic operators $\frac{\partial^2}{\partial t^2}$ and $\frac{\partial}{\partial t}$ in (6.15)–(6.17) can be represented as

$$\frac{\partial^2 F}{\partial t^2} = \frac{F^{n+1} - 2F^n + F^{n-1}}{(\Delta t)^2} \quad (6.18)$$

$$\frac{\partial F}{\partial t} = \frac{F^{n+1} - F^{n-1}}{2\Delta t} \quad (6.19)$$

For simplicity, F is introduced here to represent any one of the field components (D_z, E_z, B_x, H_x, B_y and H_y). Those fields F^n located at time point $t = n\Delta t$ are approximated by a semi-implicit scheme as

$$F = \frac{F^{n+1} + 2F^n + F^{n-1}}{4} \quad (6.20)$$

Applying (6.18)–(6.20) to (6.12), the explicit update equation for E_z is derived as follows:

$$E_z^{n+1}(i+1/2, j+1/2) = \sum_{m=0}^{M} b_m D_z^{n-m+1}(i+1/2, j+1/2)$$
$$- \sum_{m=1}^{M} a_m E_z^{n-m+1}(i+1/2, j+1/2) \quad (6.21)$$

6.3 Modeling of Left-Handed Metamaterials Using a Dispersive FDTD Method

where $M = 2$ and the coefficients a_m and b_m in (6.21) are

$$a_1 = \frac{2\varepsilon_0 \Delta t^2(\omega_{e0}^2 + \omega_{ep}^2) - 8\varepsilon_0}{A}; \quad a_2 = \frac{4\varepsilon_0 - 2\Delta t \varepsilon_0 \gamma_e + \varepsilon_0 \Delta t^2(\omega_{e0}^2 + \omega_{ep}^2)}{A} \tag{6.22}$$

$$b_0 = \frac{4 + 2\Delta t \gamma_e + \Delta t^2 \omega_{e0}^2}{A}; \quad b_1 = \frac{2\Delta t^2 \omega_{e0}^2 - 8}{A}; \quad b_2 = \frac{\Delta t^2 \omega_{e0}^2 + 4 - 2\Delta t \gamma_e}{A} \tag{6.23}$$

with

$$A = 4\varepsilon_0 + 2\Delta t \varepsilon_0 \gamma_e + \varepsilon_0 \Delta t^2(\omega_{e0}^2 + \omega_{ep}^2) \tag{6.24}$$

The same approach can be applied to (6.13) and (6.14) to yield H_x and H_y respectively.

$$H_x^{n+3/2}(i+1/2, j+1) = \sum_{m=0}^{M} d_m B_x^{n-m+3/2}(i+1/2, j+1)$$
$$- \sum_{m=1}^{M} c_m H_x^{n-m+3/2}(i+1/2, j+1) \tag{6.25}$$

$$H_y^{n+3/2}(i+1, j+1/2) = \sum_{m=0}^{M} d_m B_x^{n-m+3/2}(i+1, j+1/2)$$
$$- \sum_{m=1}^{M} c_m H_x^{n-m+3/2}(i+1, j+1/2) \tag{6.26}$$

where $M = 2$ in both (6.25) and (6.26), wherein the coefficients c_m and d_m are

$$c_1 = \frac{2\mu_0 \Delta t^2(\omega_{m0}^2 + \omega_{mp}^2) - 8\mu_0}{C}; \quad c_2 = \frac{4\mu_0 - 2\Delta t \mu_0 \gamma_m + \mu_0 \Delta t^2(\omega_{m0}^2 + \omega_{mp}^2)}{C} \tag{6.27}$$

$$d_0 = \frac{4 + 2\Delta t \gamma_m + \Delta t^2 \omega_{m0}^2}{C}; \quad d_1 = \frac{2\Delta t^2 \omega_{m0}^2 - 8}{C}; \quad c_2 = \frac{\Delta t^2 \omega_{m0}^2 + 4 - 2\Delta t \gamma_m}{C} \tag{6.28}$$

with

$$C = 4\mu_0 + 2\Delta t \mu_0 \gamma_m + \mu_0 \Delta t^2(\omega_{m0}^2 + \omega_{mp}^2) \tag{6.29}$$

We will apply the above FDTD iteration equations in the following sections to model two structures with alternating layered LHMs and dielectric slabs. Both structures demonstrate zero phase delay in wave transmission and with a proper selection of dimensions and electrical permittivity of slabs, material "transparency" can be achieved to waves propagating at any incidence angle.

6.3.2 Phase Compensation Through Layered LHM Structures

Although the thin LHM slab structure presented in [4] can produce a "perfect" near-field image, the "perfection" can easily be spoiled by the losses in the LHM that reduce the near-field resolution. In [41, 42], a multilayer stack consisting of thin alternating layers of conventional materials and LHMs is proposed to eliminate such dissipation. Currently, only equally spaced layered LHM structures have been investigated in the invisible light region. In this section, the effects of the spacing of layered LHM on evanescent wave amplification will be investigated in the microwave frequency region.

To verify the validity of our FDTD program, a stack with alternating positive and negative dielectric layers was designed to enhance evanescent wave transportation at microwave frequencies. Such a layered structure (see Figure 6.15) is considered with alternating "positive dielectrics" with $\varepsilon_{r+} = \mu_{r+} = 1$ and "negative dielectrics" with $\varepsilon_{r-}(\omega_0) = \mu_{r-}(\omega_0) = -1$ at the target frequency (10 GHz). For the negative dielectrics, $\omega_{pe} = \omega_{pm} = \sqrt{2}\omega_0$, $\gamma_e = \gamma_m = 0$ was used in the FDTD simulation. Currently, a fully stable algorithm of absorbing boundary conditions (ABCs) for LHM is not yet available, and hence only conventional ABCs are used in this chapter. The proposed multiple-layer structure is located in FDTD space with $\Delta x = \Delta y = \lambda/220$ and surrounded by an eight-cell-layer uniaxial perfectly

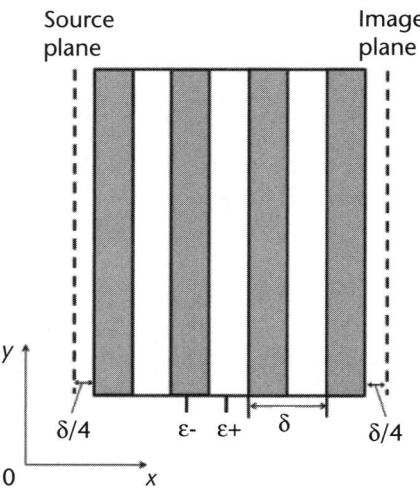

Figure 6.15 Schematic of the layered structure: the RHM (positive dielectric) and LHM (negative dielectric) layers have an equal thickness of $\delta/2$; the total length of the multiple layered slabs is $d = [(2N-1)\delta]/2$ where N is the number of LHM layers. The source plane is at a distance $\delta/4$ left to the structure while the image plane is at a distance $\delta/4$ right of the structure [46].

matched layer (UPML) [45] absorbing boundary. Here, a polynomial grading [9] is chosen as the UPML loss profile and the predicted reflection error from the boundary is of order 0.0001. Both the positive and negative dielectric layers are assumed to be of equal thickness ($\delta/2$), equivalent to $22\Delta x$. An evanescent source was used as the excitation and was located at a distance of $\delta/4$ ($11\Delta x$) from the multilayer slab. The number (N) of LHM layers is four, so the total thickness of the layered slabs is $d = [(2N-1)\delta]/2 = 154\Delta x$. As can be seen in Figure 6.16, unlike conventional evanescent waves that decay exponentially with distance away from the object, the evanescent waves will be enhanced in the LHM slabs due to the changed sign of the wave vector at the vacuum-LHM slab boundary [46].

6.3.3 Conjugate Dielectric and Metamaterial Slab as Radomes

Let a metamaterial slab of width d_2, permittivity $\varepsilon_2 = -|\kappa|\varepsilon_0$ and permeability $\mu_2 = -\mu_0$, be embedded between two semi-infinite dielectric media (ε_1, μ_0) and (ε_3, μ_0) as shown in Figure 6.17. Let an electromagnetic wave of angular frequency ω impinge on this slab at an angle θ_3 with the normal, in the TM mode. The incident, the reflected, and the transmitted waves are also shown in Figure 6.17. An $\exp(j\omega t)$ time variation is assumed throughout. Using the relationship between the tangential components of E and H fields on the slab interfaces, the reflection and transmission coefficients can be derived as follows:

$$R_2^s = \frac{r_{32}^s + r_{21}^s e^{j2\varphi_2}}{1 + r_{32}^s r_{21}^s e^{j2\varphi_2}} \tag{6.30}$$

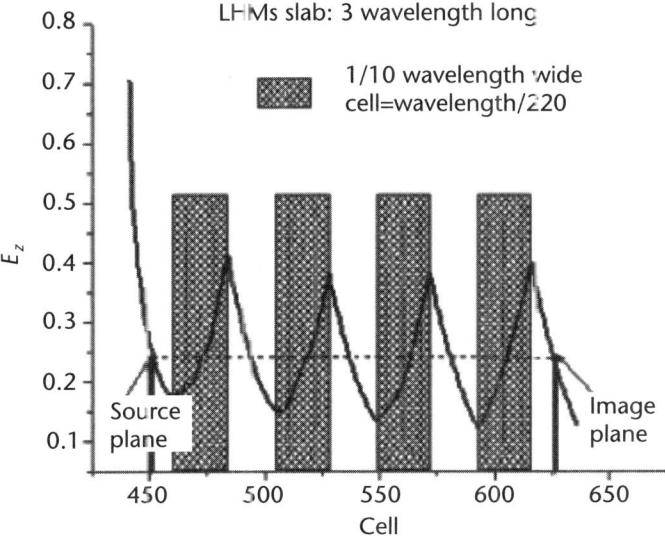

Figure 6.16 Near-field intensity through a multilayer LHM structure with refractive index $n = -1$ [46].

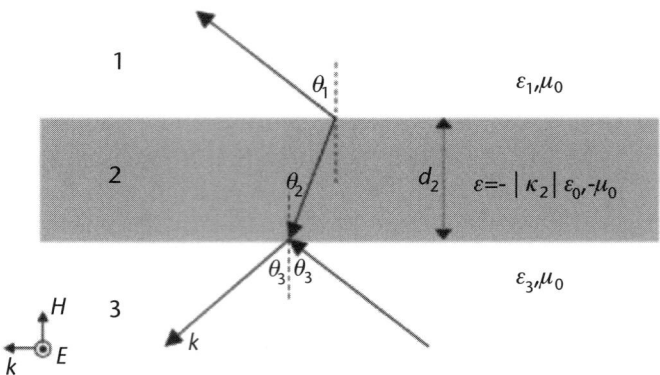

Figure 6.17 A metamaterial slab embedded between two semi-infinite dielectric media [47].

$$T_2^s = \frac{t_{32}^s + r_{21}^s e^{j2\varphi_2}}{1 + r_{32}^s r_{21}^s e^{j2\varphi_2}} \quad (6.31)$$

where $\varphi_i = k_{zi} d_i$, $k_{zi} = k_i \cos\theta_i$, $k_i = \omega\sqrt{\varepsilon_i \mu_i} = \omega n_i/c$. The variables φ_i, θ_i, k_{zi}, k_i, and n_i are positive real quantities while the interface reflection and transmission coefficients are given as follows:

$$r_{ij}^s = \frac{k_{zi} - k_{zj}}{k_{zi} + k_{zj}} \quad (6.32)$$

$$t_{ij}^s = \frac{2k_{zi}}{k_{zi} + k_{zj}} \quad (6.33)$$

The interface reflection and transmission coefficients are identical for dielectric and for metamaterials while the phase φ_2 is preceded by a + sign for a metamaterial slab and by a − sign for a dielectric slab. Adequate expressions can be found for a TE mode [37]. Similar expressions have been given in [38, 39].

Let us suppose that two slabs (one metamaterial and one dielectric) of widths d_3 and d_2, permittivities $\varepsilon_3 = -|\kappa_3|\varepsilon_0$ and $\varepsilon_2 = |\kappa_2|\varepsilon_0$, permeabilities $\mu_3 = -\mu_0$, are embedded between two semi-infinite dielectric media (ε_4, μ_0) and (ε_1, μ_0) as shown in Figure 6.18. Let an electromagnetic wave of angular frequency ω impinge on this pair of slabs at an angle θ_4 with the normal. The incident, the reflected, and the transmitted waves are also shown in Figure 6.18. The ratio of the reflection to the transmission coefficients for this pair of slabs is given as follows [37]:

$$\frac{R_3}{T_3} = \frac{r_{43} + r_{32}e^{j(2\varphi_3)} + r_{21}e^{j(2\varphi_3 - 2\varphi_2)} + r_{43}r_{32}r_{21}e^{j(-2\varphi_2)}}{t_{43}t_{32}t_{21}e^{j(\varphi_3 - \varphi_2)}} \quad (6.34)$$

Let us study a structure shown in Figure 6.17, consisting of a pair of metamaterial and dielectric slabs having the same width d, opposite permittivities $\varepsilon_3 = -|\kappa|\varepsilon_0$ and opposite permeabilities $\mu_3 = -\mu_0$ and $\mu_2 = \mu_0$. In this case, $\varphi_3 = \varphi_2$, $\theta_3 = \theta_2$, $r_{32}^s = 0$, $r_{43}^s = -r_{21}^2$ and $t_{32}^s = 1$. We will call it a pair of conjugate slabs. If $\varepsilon_1 = \varepsilon_4$, for any frequency and for any angle of incidence, $R_3^s = 0$,

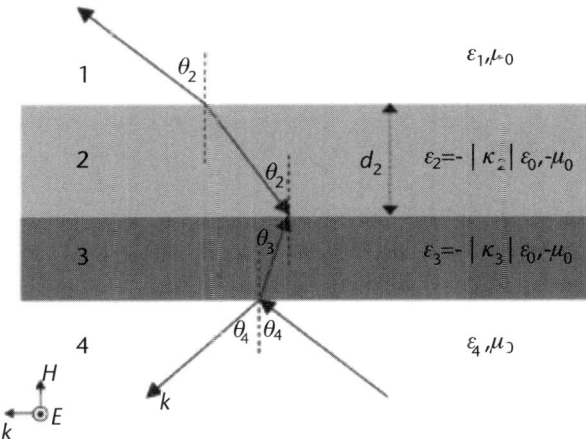

Figure 6.18 A metamaterial slab and a dielectric slab embedded between two semi-infinite dielectric media [47].

and the wave is wholly transmitted. If $\varepsilon_1 \neq \varepsilon_4$, for any frequency and for normal incidence, $\frac{R_3^s}{T_3^s} = \frac{r_{41}^s}{t_{41}^s}$ this case being akin to the case of the initial semi-infinite medium being adjacent to the final one. The same is true for TE polarization [37]. It is evident that any number of additional pairs of conjugate slabs, even if they are all different one from the other, would not change these results. The physical reason of this feature is that at the interface separating the two slabs of each pair, the reflection coefficient vanishes and the transmission coefficient is unity, while the phase of the wave after propagating through the first slab of each pair is completely canceled after propagating through the second slab of the pair, thus making the conjugate slabs a transparent structure. The advantage of these structures is that since the phase disappears, the reflection and transmission processes are independent of the phase, while they are independent of the angles of incidence as well, only if the initial and final semi-infinite media are identical. Therefore, these structures could be advantageously used as antenna radomes.

6.3.4 Numerical Results

For the simulation of the LHM slabs, in which the permittivity and permeability are negative, we use the FDTD method for dispersive material described in the previous section. We assume that the complex permittivity and permeability of LHM are described by the Drude dispersion relations.

$$\varepsilon(\omega) = \varepsilon_0 \left[1 - \frac{\omega_{pe}^2}{\omega(\omega - i\gamma_e)} \right] \quad (6.35)$$

$$\mu(\omega) = \mu_0 \left[1 - \frac{\omega_{pm}^2}{\omega(\omega - i\nu_m)} \right] \quad (6.36)$$

where ω_{pe} and ω_{pm} correspond to the electric and magnetic plasma frequencies, respectively. γ_e and γ_m are the electric and magnetic collision frequencies, which represent the losses in the medium. By choosing proper plasma frequencies for LHM, we set the constitutive parameters of LHM to be negative and hence obtain the corresponding negative refractive index.

In order to illustrate the ideas described in the previous section, a pair of conjugate slabs of identical width $d = 0.1\lambda$, permittivities $\varepsilon_r = -3, \varepsilon_0$ and $\varepsilon_r = 3\varepsilon_0$, permittivities $\mu_3 = -\mu_0$ and $\mu_2 = \mu_0$ has been studied first. Let a plane wave, simulated by an array of line sources, impinge on this structure. The cell size of the FDTD simulation was chosen to be $\Delta = \lambda/50$.

The four sides of the FDTD domain are terminated with a PML, and the code has been run until the steady state is reached. Figure 6.19 shows the wave propagation through the conjugate slab. Removing the pair of conjugate slabs and joining the region at the left of the structure to the region at its right, we obtain the same results with free space. This is more clearly illustrated in axis plot shown in Figure 6.20. It has been demonstrated that regardless of the slab thickness, the phase of the right side is always restored the same as the left side. The magnitudes of the electric field with and without the conjugate slabs show a good agreement except for minor errors caused by diffractions from the edges of the slab. Giving to d a value of one wavelength ensures that, at the same distances from the source, on either side of the structure, we obtain the same field values as for free space. This is illustrated in Figure 6.20. Therefore, if we wish the same wave to behave at the right of the structure, when it is present, as though it was absent (i.e., as though the propagation was in free space only), each slab should be n times ($n = 1, 2, 3, \ldots$) half a wavelength wide. The common reason for these various effects is that the transmission coefficient $T = 1$. Repeating the simulation with different conjugate slab characteristics leads to the same behavior.

Now we place the pair of conjugate slabs at an inclination of 45 degrees with respect to the wave normal of the incoming wave. The results are shown

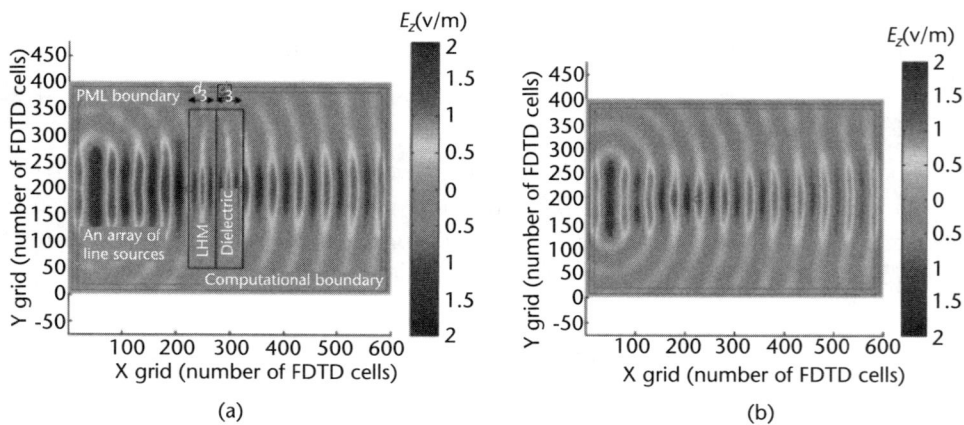

Figure 6.19 FDTD simulation of conjugate slab: (a) a pair of conjugate slabs if identical widths $d = 1.0\lambda$, permittivity $\varepsilon_3 = -3\varepsilon_0$ and $\varepsilon_2 = 3\varepsilon_0$, permeabilities $\mu_3 = -\mu_0$ and $\mu_2 = \mu_0$ and (b) removing the slabs from the simulation [47].

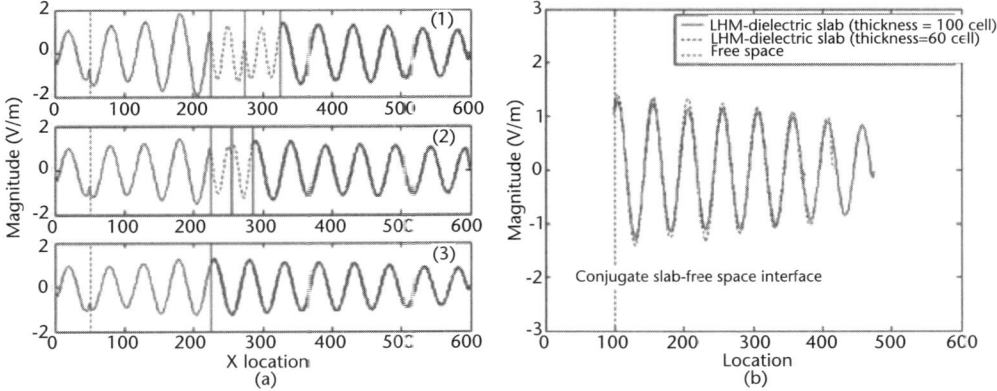

Figure 6.20 Comparison of axis plot of pairs of conjugate slabs of width $d = 1.0\lambda$, $d = 0.6\lambda$ each, immersed in vacuum and free space [47].

in Figure 6.21. In order to demonstrate the transmission and reflection characteristics of the conjugate slab, we compare the results for the simple dielectric, conjugate slab, and free-space cases with the same configuration for the incident wave. Figure 6.22 shows the electric field distribution in free space and the same

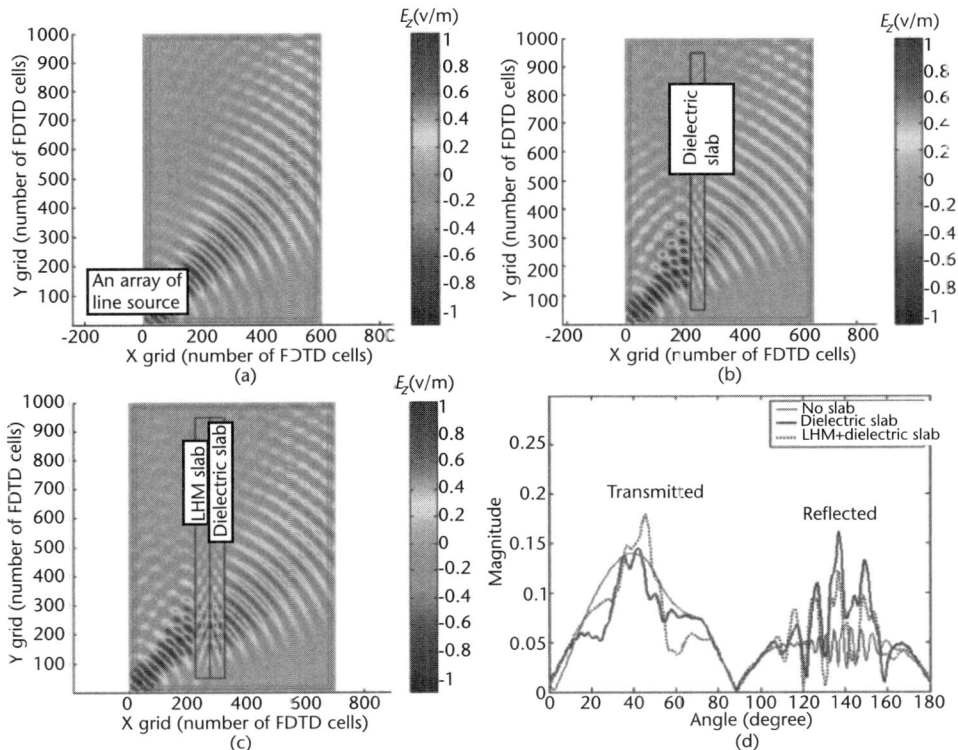

Figure 6.21 FDTD simulation of electric field incident (a) in free space, (b) on a dielectric slab, and (c) on conjugate slab and (d) comparison of far-field radiation patterns.

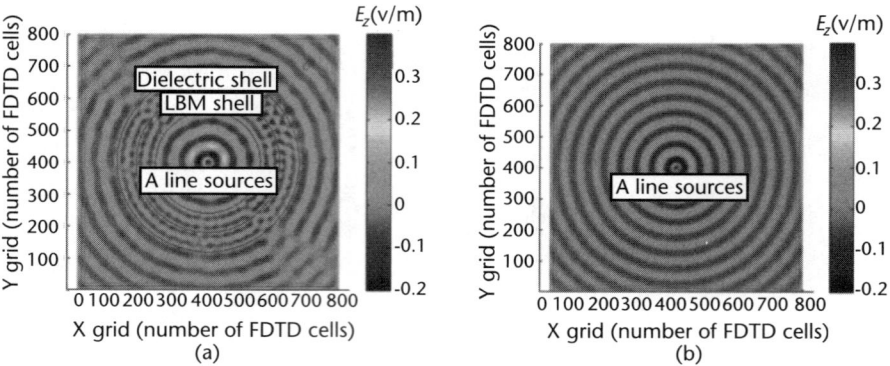

Figure 6.22 Electric field plot of line source (a) with two concentric conjugate shells and (b) in free space [47].

for the dielectric and conjugate slabs. Figure 6.22(d) shows the far-field pattern for the above cases. It shows a good transmission through the conjugate slabs at oblique incidence. As shown in the plot, the conjugate slab demonstrates a better transmission compared with the simple dielectric slab, reducing the reflection. We observe that the transmitted beamwidth for both dielectric and conjugate slabs are

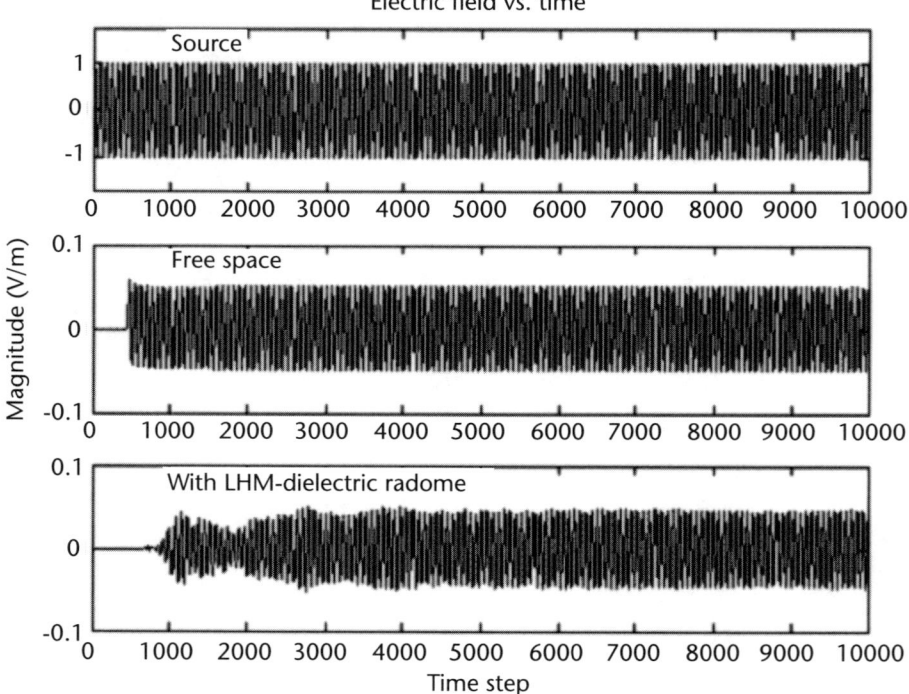

Figure 6.23 Comparison of the electric fields at the source and the observation point with and without the conjugate shell [47].

6.3 Modeling of Left-Handed Metamaterials Using a Dispersive FDTD Method 167

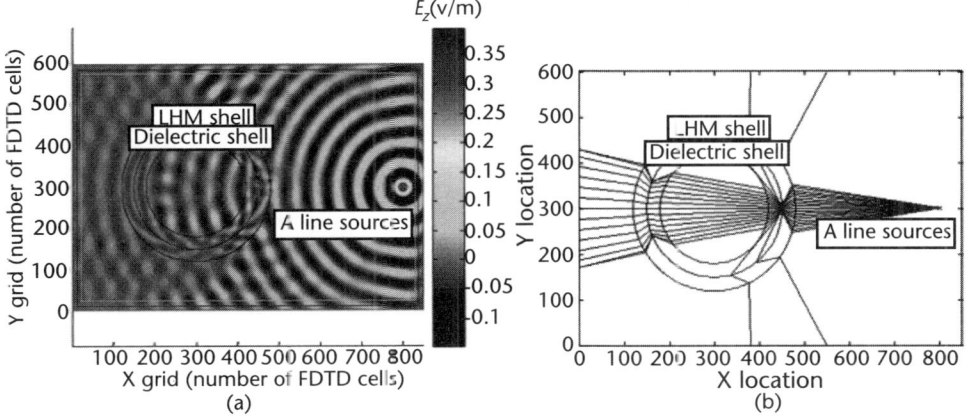

Figure 6.24 (a) Electric field plot for a line source at a distance outside the two concentric conjugate shells and (b) ray diagram [47].

narrower than one that without the slab. This is because a total reflection occurs for incident angles greater than the critical angle, which decreases the transmission in those directions.

We now turn to the cylindrical structure composed of two concentric conjugate cylinders of the same width instead of two straight conjugate slabs. We have placed a line source at the centre of the cylindrical shell and repeated the preceding simulations accordingly (Figure 6.22). The thickness of each slab is one wavelength. The electric field is recorded outside the cylindrical shell and compared with the results for free space. Figure 6.22(a) shows the electric field distribution for the cylindrical conjugate shell with a line source inside, and it is observed that the cylindrical wave front is well preserved. The transmission characteristic of the cylindrical conjugate shell shows slightly lower magnitude (Figure 6.23), and we believe that this is caused by the internal reflections due to the staircasing of FDTD cells.

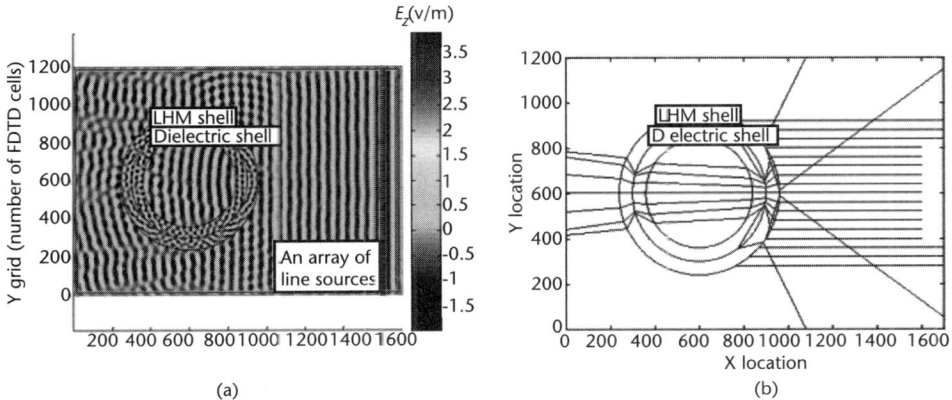

Figure 6.25 (a) Electric field plot for a plane wave impinging on two concentric conjugate shells and (b) ray diagram [47].

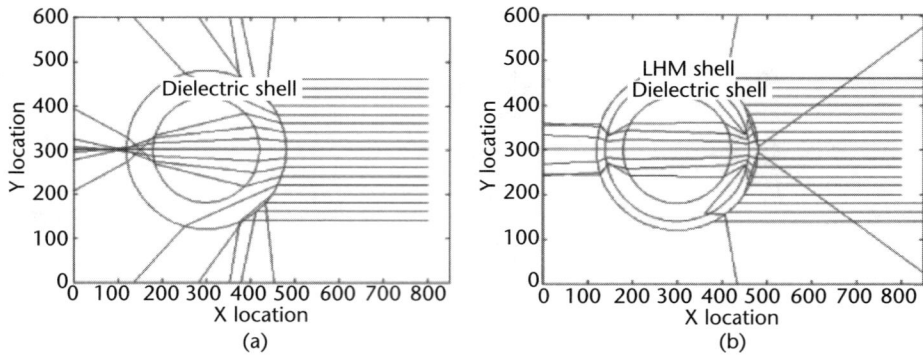

Figure 6.26 Ray diagram for a plane wave impinging (a) on a dielectric cylinder and (b) on two concentric conjugate cylinders with different thickness (outer shell=0.5λ, inner shell=0.7λ) [47].

Next, we investigate the scattering characteristics of the cylindrical conjugate shell. We fist examine the scattering due to the cylindrical wave front as shown in Figure 6.24(a). The thickness of the cylindrical shell is 0.6 wavelength and the negative material parameters $\varepsilon = -3\varepsilon_0$ and $\mu = -\mu_0$ are assigned for the outer shell and the conjugate parameters for the inner shell. We note that the strong reflection occurs in the edge region and the shape of the wave front inside the shell is cylindrical. A ray-tracing diagram is also shown in Figure 6.24(b).

We also perform simulations for plane wave incidence, and the results are plotted in Figure 6.25. Not like the cylindrical wave incidence, the wave front inside the shell no longer restores the plane wave, but slightly diverges. It should also be noted that the interesting internal reflections occur for regions off the center of the shell, which are different from the scattering from the conventional dielectric shell [Figure 6.26(a)]. It is also very interesting that the curvature of the wave front

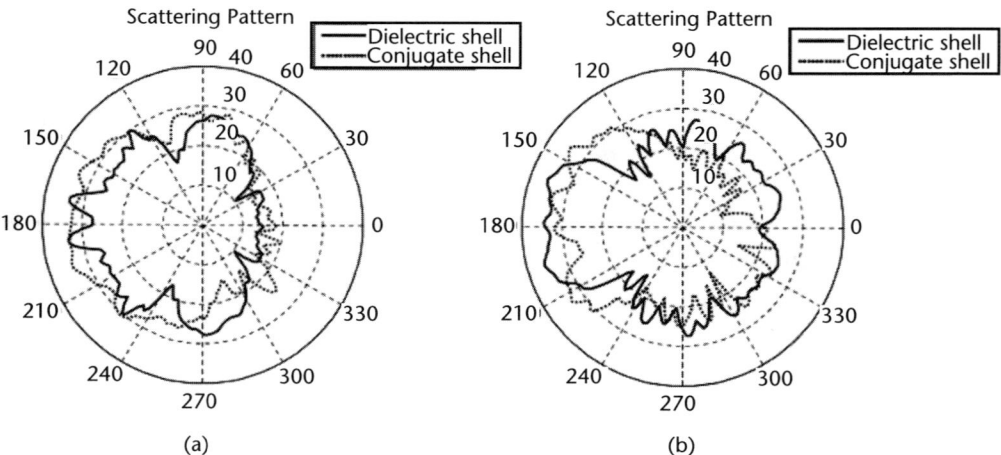

Figure 6.27 Comparison of scattering patterns of dielectric and two concentric conjugate cylinders with a width (a) $d = 0.6\lambda$ and (b) $d = 0.2\lambda$ each.

inside the shell can be controlled by using shells with different thicknesses. Figure 6.26(b) shows the ray diagram for the structure with different shell thickness (0.5 wavelength for outer shell and 0.7 wavelength for inner shell), and a fairly flat wave front has been achieved.

Finally, we compare the scattering pattern of the cylindrical conjugate shell with that of the conventional dielectric. Two different thicknesses (0.6 and 0.2 wavelengths) have been simulated, as shown in Figure 6.27. We observe that the thicker shell shows less front scattering and introduces higher scattering to the side. This is due to the internal reflections as shown in the above cases. Thinner shell shows higher front and back scattering, but lower magnitudes for the side. The overall results show that the cylindrical conjugate shell produces similar scattering sections compared with the dielectric shell of the same dimension.

6.4 Conclusions

We have studied the structures composed of two materials having opposite characteristics. The LHM coating on a conventional dielectric could increase the transmittance by reducing the reflections at the interface, which could be useful for antenna radome applications. In the conjugate slab, the LHM compensates the phase change due to the dielectric material, and makes the overall effect disappear. The cylindrical conjugate shell shows several interesting characteristics such as the multiple internal reflections and wave front changes. These behaviors could vary significantly depending on the diameter, thickness, and material parameters of each shell. The cylindrical dielectric-LHM conjugate structure shows comparable forward scattering sections. It is noted that the effect of staircasing is more adverse than when it only involves conventional materials. The dispersive FDTD method could be combined with a conformal scheme [40] in order to improve the accuracy of the results for curved structures.

The LHMs investigated in this section are idealized, as well as nonphysical, since no real materials with the characteristics we have attributed to the LHM slabs or cylinders is known to exist in the real world. Although extensive attempts have been made by researchers to synthesize the LHM artificially, success has been elusive to date, except perhaps under ideal conditions and in a narrow frequency range, as may be seen in Chapter 7.

It is recommended, therefore, that the real physical structure be analyzed decently to determine its characteristics, rather than predicting its performance as the basis of their equivalent medium, especially before drawing defective conclusions regarding the focusing characteristics of the metamaterial slabs, or their ability to magnify evanescent waves. Numerical investigations of some realistic structures may be found in Chapter 7.

References

[1] A. Taflove, *Computational Electrodynamics: The Finite-Difference Time-Domain Method*, Norwood, MA: Artech House, 1995.

[2] R. J. Luebbers, F. Hunsberger, K. S. Kunz, R. B. Standler, and M. Schneider, "A frequency-dependent finite-difference time-domain formulation for dispersive materials," *IEEE Trans. Electromagn. Compat.*, vol. 32, no. 3, pp. 222–227, 1990.

[3] R. J. Luebbers, F. Hunsberger, and K. S. Kunz, "A frequency-dependent finite-difference time-domain formulation for transient propagation in plasma," *IEEE Trans. Antennas Propagat*, vol. 39, no. 1, pp. 29–34, 1991.

[4] R. J. Luebbers and F. Hunsberger, "FDTD for Nth-order dispersive media," *IEEE Trans. Antennas Propagat*, vol. 40, no. 11, pp. 1297–1301, 1992.

[5] F. Hunsberger, R. J. Luebbers, and K. S. Kunz, "Finite-difference time-domain analysis of gyrotropic media. I: Magnetized plasma," *IEEE Trans. Antennas Propagat.*, vol. 40, no. 12, pp. 1489–1495, 1992.

[6] C. Melon, P. Leveque, T. Monediere, A. Reineix, and F. Jecko, "Frequency dependent finite-difference-time-domain formulation applied to ferrite material," *Microwave Opt. Technol. Lett.*, vol. 7, no. 12, pp. 577–579, 1994.

[7] Akyurtlu and D. H. Werner, "BI-FDTD: a novel finite-difference time-domain formulation for modeling wave propagation in bi-isotropic media," *IEEE Trans. Antennas Propagat.*, vol. 52, No. 2, pp. 416–425, 2004.

[8] A. Grande, I. Barba, A. Cabeceira, J. Represa, P. So, and W. Hoefer, "FDTD modeling of transient microwave signals in dispersive and lossy bi-isotropic media," *IEEE Trans. Microwave Theory Tech.*, vol. 52, no. 3, pp. 773–784, 2004.

[9] A. Akyurtlu and D. H. Werner, "A novel dispersive FDTD formulation for modelling transient propagation in chiral metamaterials," *IEEE Trans. Antennas Propagat.*, vol. 52, no. 9, pp. 2267–2276, 2004.

[10] T. Kashiwa, N. Yoshida, and I. Fukai, "A treatment by the finite-difference time-domain method of the dispersive characteristics associated with orientation polarization," *Trans. IEICE*, vol. E73, no. 8, pp. 1326–1328, 1990.

[11] T. Kashiwa and I. Fukai, "A treatment by the FD-TD method of the dispersive characteristics associated with electronic polarization," *Microwave Opt. Technol. Lett.*, vol. 3, no. 6, pp. 203–205, 1990.

[12] T. Kashiwa, Y. Ohtomo, and I. Fukai, "A finite-difference time-domain formulation for transient propagation in dispersive media associated with Cole-Cole's circular ARC law," *Microwave Opt. Technol. Lett.*, vol. 3, no. 12, pp. 416–419, 1990.

[13] R. M. Joseph, S. C. Hagness, and A. Taflove, "Direct time integration of Maxwell's equations in linear dispersive media with absorption for scattering and propagation of femtosecond electrogmagnetic pulses," *Optics Lett.*, vol. 16, no. 9, pp. 1412–1414, 1991.

[14] P. M. Goorjian and A. Taflove, "Direct time integration of Maxwell's equations in nonlinear dispersive media for propagation and scattering of femtosecond electromagnetic solitons," *Optics Lett.*, vol. 17, no. 3, pp. 180–182, 1992.

[15] P. Gandhi, B. Q. Gao, and J. Y. Chen, "A frequency-dependent finite-difference time-domain formulation for induced current calculations in human beings," *Bioelectromagnetics*, vol. 13, no. 6, pp. 543–556, 1992.

[16] P. Gandhi, B. Q. Gao, and J. Y. Chen, "A frequency-dependent finite-difference time-domain formulation for general dispersive media," *IEEE Trans. Microwave Theory Tech.*, vol. 41, no. 4, pp. 658–665, 1993.

[17] M. Fujii, M. Tahara, I. Sakagami, W. Freude, and P. Russer, "High-order FDTD and auxiliary differential equation formulation of optical pulse propagation in 2-D Kerr and Raman nonlinear dispersive media," *IEEE Journal of Quantum Electronics*, vol. 40, no. 2, pp. 175–182, 2004.

[18] D. M. Sullivan, "Frequency-dependent FDTD methods using Z transforms," *IEEE Trans. Antennas Propagat.*, vol. 40, no. 10, pp. 1223–1230, 1992.

[19] D. M. Sullivan, "Nonlinear FDTD formulations using Z transforms," *IEEE Trans. Microwave Theory Tech.*, vol. 43, no. 3, pp. 676–682, 1995.

[20] V. Demir, A. Z. Elsherbeni, and E. Arvas, "FDTD formulation for dispersive chiral media using the Z transform method," *IEEE Trans. Antennas Propagat.*, vol. 53, no. 10, pp. 3374–3384, 2005.

[21] M. W. Feise, J. B. Schneider and P. J. Bevelacqua, "Finite-difference and pseudospectral time-domain methods applied to backward-wave metamaterials," *IEEE Trans. Antennas Propagat.*, vol. 52, no. 11, pp. 2955–2962, 2004.

[22] J.-Y. Lee, J.-H. Lee. H.-S. Kim, N.-W. Kang and H.-K. Jung, "Effective medium approach of left-handed material using a dispersive FDTD method," *IEEE Trans. Magnetics*, vol. 41, no. 5, pp. 1484–1487, 2005.

[23] V. G. Veselago, "The electrodynamics of substance with simultaneously negative values of ε and μ," *Sov. Phys.*, 10, 509, 1968.

[24] R. W. Ziolkowski and E. Heyman, "Wave propagation in media having negative permittivity and permeability," *Phys. Rev. E*, 64, 056625, 2001

[25] P. F. Loschialpo, D. L. Smith, D. W. Forester, F. J. Rachford, and J. Schelleng, "Electromagnetic waves focused by a negative-index planar lens," *Phys. Rev. E*, 67, 025602, 2003.

[26] S. A. Cummer, "Simulated causal subwavelength focusing by a negative refractive index slab," *Appl. Phys. Lett.*, 82, 1503, 2003.

[27] J. B. Pendry, "Negative refraction makes a perfect lens,'" *Phys. Rev. Lett.*, 85, 3966, 2000.

[28] M. W. Feise and Y. S. Kivshar, "Sub-wavelength imaging with a left-handed material flat lens," *Phys. Lett. A*, 334, 326–323, 2005.

[29] X. S. Rao and C. K. Ong, "Subwavelength imaging by a left-handed material superlens," *Phys. Rev. E*, 68, 067601, 2003.

[30] R. J. Rubbers, F. Hunsberger, K. S. Kunz, R. B. Standler, and M. Schneider, "A frequency dependent finite-difference time-domain formulation for dispersive materials," *IEEE Transactions on Electromagnetic Compatibility*, vol. 32, pp. 222–227, August 1990.

[31] R. J. Rubbers, F. Hunsberger, and K. S. Kunz, "A frequency-dependent finite-difference time-domain formulation for transient propagation in plasma," *IEEE Transactions on Antennas and Propagation*, vol. 39, pp. 29–39, January 1991.

[32] R. J. Rubbers, F. Hunsberger, "FDTD for nth-order dispersive media," *IEEE Transactions on Antennas and Propagation*, vol. 140, pp. 1297–1301, November 1992.

[33] J. A. Pereda, L. A. Vielva, A. Vegas, and A. Prieto, "State-space approach to the FDTD formulation for dispersive media," *IEEE Transactions on Magnetics*, vol. 31, pp. 1602–1605, May 1995.

[34] J. L. Young, "Propagation in linear dispersive media: Finite difference time domain methodologies," *IEEE Transactions on Antennas and Propagation*, vol. 43, pp. 422–426, April 1995.

[35] D. M. Sullivan, "The use of z-transform theory for numerical simulation of dispersive and non-linear materials with the FDTD method," *Antennas and Propagation Society International Symposium*, AP-S, vol. 3, pp. 1450–1453, June 1994.

[36] D. M. Sullivan, "Z-transform theory and the FDTD method," *IEEE Transactions on Antennas and Propagation*, vol. 44, pp. 28–34, January 1996.

[37] H. Cory and C. Zach, "Wave propagation in metamaterial multilayered structures," *Microwave and Optical Technology Letters*, vol. 40, pp. 460–465, 2004.

[38] J. A. Kong, "Electromagnetic wave interaction with stratified magnetic isotropic media," *Prog. Electromag. Res. PIERS*, vol. 35, pp. 271–286, 2001.

[39] S. A. Ramakrishna, J. B. Pendry, D. Shurig, D. R. Smith, and S. Schultz, "The asymmetric lossy near perfect lens," *J. Modern Opt.*, vol. 49, pp. 1747–1762, 2002.

[40] Y. Hao and C. J. Railton, "Analyzing electromagnetic structures with curved boundaries on Cartesian FDTD meshes," *IEEE Transactions on Microwave Theory and Techniques*, vol. 46, pp. 92–98, 1998.

[41] S. A. Ramakrishna and J. B. Pendry, "Optical gain removes absorption and increases resolution in a near-field lens," *Physical Review B*, vol. 67, pp. 201101–201104, 2003.

[42] E. Shamonina, V. A. Kalinin, K. H. Ringhofer, and L. Solymar, "Imaging, compression and Poynting vector streamlines for negative permittivity materials," *Electronics Letters*, vol. 37, no. 20, pp. 1243–1244, September 2001.

[43] J. J. Chen, Tomasz M. Grzegorczyk, B.-I. Wu, and Jin Au Kong, "Limitation of FDTD in simulation of a perfect lens imaging system," *Opt. Express*, 13, 10840–10845, 2005.

[44] L. Chen, S. He, and L. Shen, "Finite-size effects of a left-handed material slab on the image quality," *Phys. Rev. Lett.*, 92, 107404, 2004.

[45] S. D. Gedney, "An Anisotropic Perfectly Matched Layer-Absorbing Medium for the Truncation of FDTD Lattices," *IEEE Trans. Antenna and Propagation*, vol. 44, no. 12, 1996.

[46] L. Lu, Y. Hao, and C. G. Parini, "Dispersive FDTD characterisation at no phase-delay radio transmission over layered left-handed metamaterials structure," *IEE Proceedings on Science, Measurement and Technology*, vol. 151, no. 6, pp. 403–406, 2004.

[47] H. Cory, X. J. Lee, Y. Hao, and C. G. Parini, "On the use of conjugate dielectric and metamaterial slabs as radomes," *IET Proceedings on Microwave, Antennas & Propagation*, 1(1), pp. 137–143, February 2007.

CHAPTER 7
FDTD Modeling and Figure-of-Merit (FOM) Analysis of Practical Metamaterials

7.1 Introduction

It is very common to make use of the effective medium representation to characterize the metamaterials. This representation is often based on the premise that the inclusions in the medium have sizes and periodicities that are sufficiently small in comparison to the wavelength of the impinging radiation. Under this condition, the fine structures of the charge and current distributions are considered to be indiscernible; therefore, the metamaterial structure is simply represented by a homogeneous one that can be represented by its effective material parameters, namely ε_{eff} and μ_{eff}, which are determined from the macroscopic EM response of the metamaterials structure, typically the reflection and transmission coefficients of a normally incident plane wave.

While the concept of effective material parameters of a slab of material is quite straightforward, and while it is relatively simple to use to characterize complex structures, there is a danger that the use of these parameters may oversimplify the EM properties of the metamaterials, so much so that the above parameters may fail to accurately predict the true performance of the antenna/metamaterials composite in real-world applications. For example, as mentioned earlier, the retrieval of these effective material parameters is usually based on the data obtained by illuminating an infinite, doubly periodic metamaterials slab of finite thickness with a normally incident plane wave, which is typically linearly polarized. As a consequence, these parameters often fail to predict the true response of the structure when the incident angle or the polarization of the impinging wave is varied, since the EM response at oblique incident angles, and/or for different polarizations, can be very different from that of the normal incidence case, because of the anisotropic nature of the inclusions. Since a practical antenna/metamaterials composite is typically excited by a localized source, whose spectrum comprises a set of plane waves, with different incident angles and polarizations, it is important to carry out a rigorous simulation of the original physical structure—instead of using one that is based on the effective medium representation—for an accurate analysis of the performance of the composite system.

In this chapter [1–29], we first analyze the EM response of an infinite, doubly periodic DNG slab comprised of arrays of a combination of split rings and wires, by using the PBC/FDTD technique. Next, we examine the concept of effective material parameters, by going through the retrieval process of these parameters using one of the most widely used techniques, namely, the inversion approach. We will also summarize the difficulties encountered in the retrieval process, and identify some issues and problem areas that may be encountered when using effective material parameters in real-world applications. We will simulate a finite, moderate-size DNG slab excited by a Gaussian beam, as well as by a small dipole, to search for the existence of two interesting phenomena, namely the negative refraction inside the slab and the superlensing effect, that has yet to be demonstrated by using rigorous numerical simulations of slabs containing real inclusions. Finally, we will present a critical study on the figure-of-merit (FoM) associated with loss and bandwidth of the metamaterials based on resonant particles such as the SRRs. The FoMs are calculated analytically and verified numerically for the metamaterials with various electrical sizes and volumetric densities for the constituent particles. High volumetric density and electrically large particles demonstrate superior FoMs for the construction of practical metamaterials. A loss tangent of 0.01 and fractional bandwidth of 1.2% can be achieved by using resonant particles with sizes close to 0.06 wavelength.

7.2 EM Response of the Infinite, Doubly Periodic DNG Slab with Plane Wave Illumination

7.2.1 Model Description of the Array Comprising of Split-Ring Resonators and Wires

The metamaterial slab considered in this investigation is an infinite, doubly periodic array comprised of a combination of SSRs and thin wires, beginning with one and going up to six layers in the longitudinal direction, that stand alone in free space. The configuration of this unit cell is similar to the one used in the first DNG structure realized by Smith et al., except that a somewhat simpler configuration of a "single" split ring is employed in the present study, as opposed to a double ring. It should be pointed out that the configuration of the unit cell and the separation distance between the layers are kept the same throughout this chapter, since we do not intend to perform a parametric study of the microscopic configuration of metamaterial structures.

The array has periodicities of 2.25 mm and 5 mm along the x- and y-directions, respectively, and a separation distance of 4 mm between layers for a multilayer configuration. A uniform FDTD cell-size Δ of 0.125 mm is employed throughout the computational domain, which requires a total of $18\Delta \times 40\Delta \times 84\Delta$ and $18\Delta \times 40\Delta \times 225\Delta$ to represent the one- and six-layer cases, respectively. The modeling has been carried out by using the GEMS code [26, 27], which is a fully 3-D FDTD code for analyzing arbitrary, complex, and multiscale geometries on parallel platform. Also, the periodic structures have been modeled with the PBC/FDTD code developed by Wu et al. [28, 29].

7.2.2 Scattering Parameters Measurements Obtained from the PBC/FDTD Code

The S-parameters for the slab of three different types of inclusions, specifically a combination of SRRs and wires, wires only, or SRRs only, are obtained by using the parallel PBC/FDTD technique, for both the one- and six-layer configurations, which are illuminated by a normally incident plane wave. Figures 7.1 and 7.2 plot the transmission magnitude and phase, while Figures 7.3 and 7.4 display the reflection magnitude and phase for all configurations, respectively.

As discussed in the previous section, one clear piece of evidence of the DNG behavior is the existence of a passband for the combined structure, over those frequency bands in which no transmission occurs for the individual structures, though this evidence alone is not conclusive. We can see from Figure 7.3(b) that the array of wires is nearly totally reflecting over the entire observed frequency band ranging from 10 to 20 GHz for both the one- and six-layer cases. This is expected, since the resonant frequency has been found to be around 30 GHz (not shown in the figure). For the SRR arrays, Figure 7.3(c) shows that the 3-dB stopbands are located between 14.8 to 20.0 GHz for the one-layer case, and between 13.8 to 19.9 GHz for the six-layer case, except for the two narrow gaps near 18.0 and 18.6 GHz for the six-layer case. For the combined structure, a 3-dB passband is observed, starting at 14.8 and going up to 20 GHz for the one-layer case, while two passbands are observed for the six-layer case, one ranging from 14.7 to 16.4 GHz and the other from 18.8 to 20 GHz. Since all of the passbands of the combined structure fall within the stopbands of the individual structures, it is possible to realize a DNG behavior of the combined structure in these passbands. However, as mentioned before, the transmission characteristic alone is not sufficient to conclude that the material is exhibiting a DNG behavior, since the interaction between the SRRs and wires may lead to a positive effective permittivity and permeability of the medium, which would then be the transmitting type in this frequency regime.

7.2.3 Phase Data Inside the DNG Slab

A more rigorous test to demonstrate the DNG nature of the slab is to directly measure the phase of a propagating wave inside the slab—which can be readily obtained from the FDTD simulations—and for the predicted negative phase velocity of a negative index medium.

In the FDTD simulation of a six-layer array, comprised of a combination of SRRs and wires, the time-domain fields are recorded along two lines along the z-direction (propagation direction) inside the slab, at different positions on the transverse plane. The DFT technique is then employed to extract the magnitude and phase at the desired frequency from the time-domain data. The normalized magnitude and phase of Ey (note: the incident field is Ey-polarized) on these two lines along the direction of propagation are plotted in Figures 7.5(a, b) at 15.2 and 19.6 GHz, respectively, where the circles are used to denote the boundaries between the adjacent layers. It is evident from the slope of the above phase plots that the phase propagates backward at 15.2 GHz, while it is forward-propagating at 19.6 GHz. Therefore, the DNG behavior is only observed at 15.2 GHz, within the first passband, but not at 19.6 GHz, which falls within the second passband.

Next, we compute the real part of the refractive index from the phase shift data for frequencies within the two passbands. The slope of the phase is computed as a function of the propagation distance by fitting a straight line to the phase

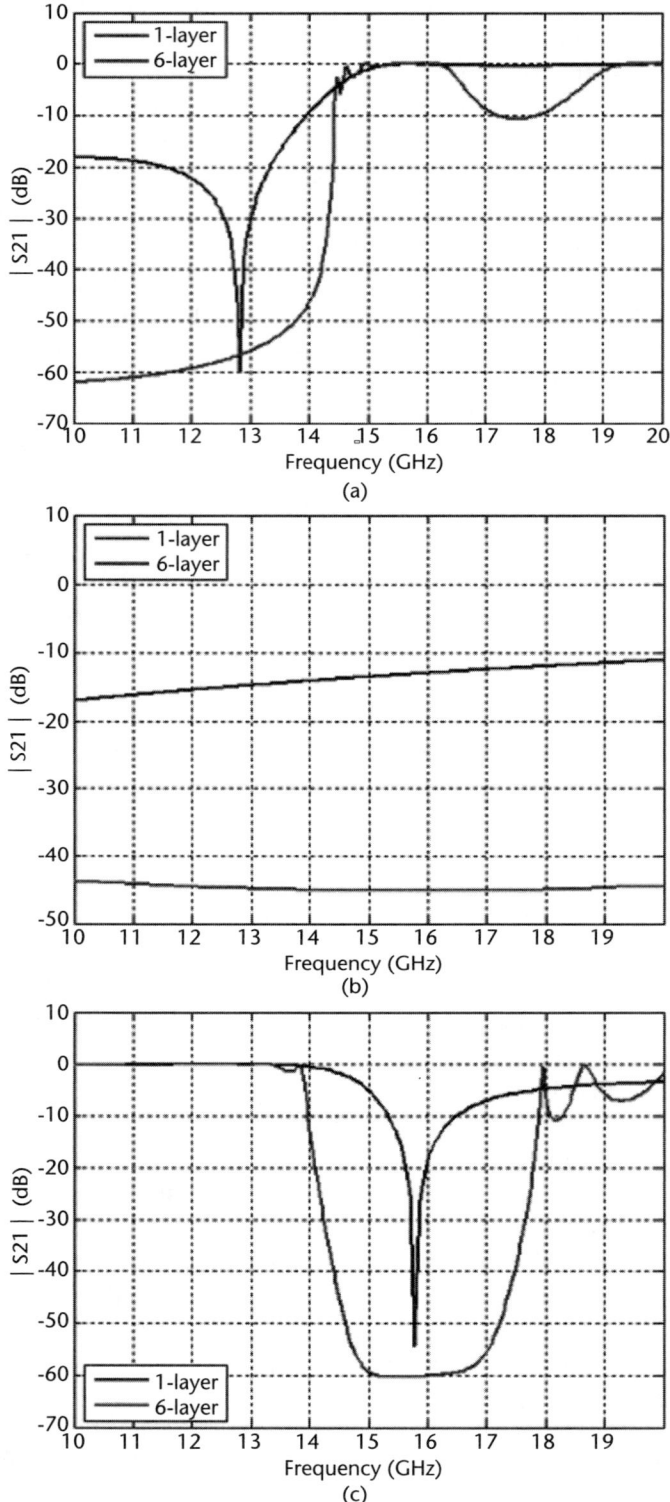

Figure 7.1 Magnitude of the transmission coefficient for the one-layer and six-layer slabs with different types of inclusions: (a) split ring and wires, (b) wires alone, and (c) split ring alone.

7.2 EM Response of the Infinite, Doubly Periodic DNG Slab with Plane Wave Illumination

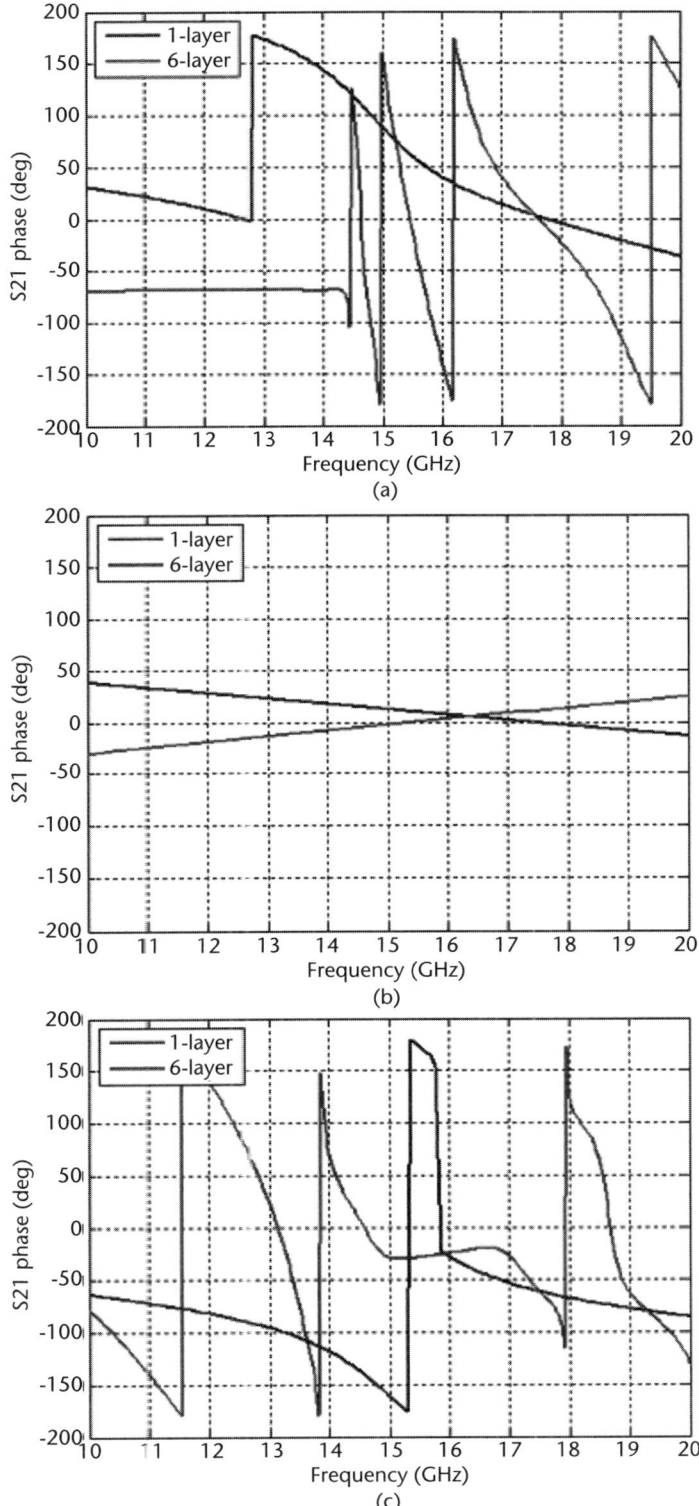

Figure 7.2 Phase of the transmission coefficient for the one-layer and six-layer slabs with different types of inclusions: (a) split ring and wires, (b) wires alone, and (c) split ring alone.

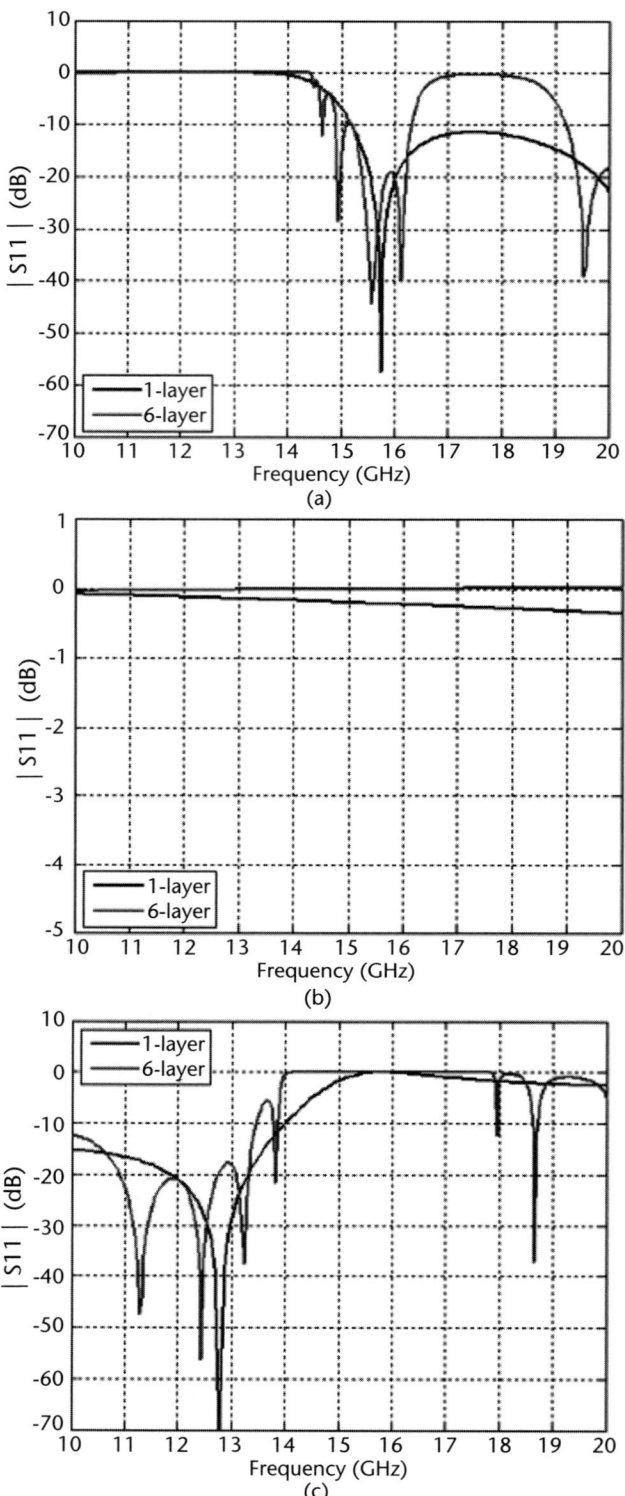

Figure 7.3 Magnitude of the reflection coefficient for the one-layer and six-layer slabs with different types of inclusions: (a) split ring and wires, (b) wires alone, and (c) split ring alone.

7.2 EM Response of the Infinite, Doubly Periodic DNG Slab with Plane Wave Illumination 179

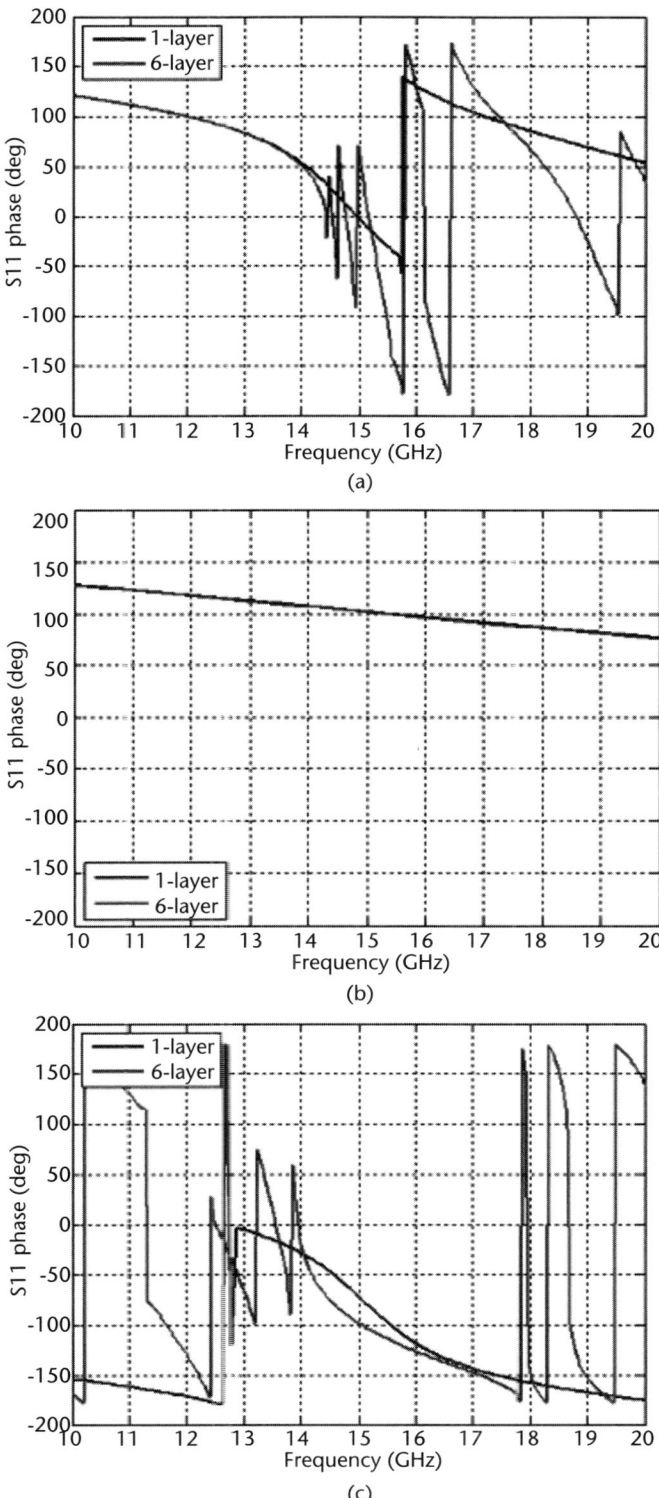

Figure 7.4 Phase of the reflection coefficient for the one-layer and six-layer slabs with different types of inclusions: (a) split ring and wires, (b) wires alone, and (c) split ring alone.

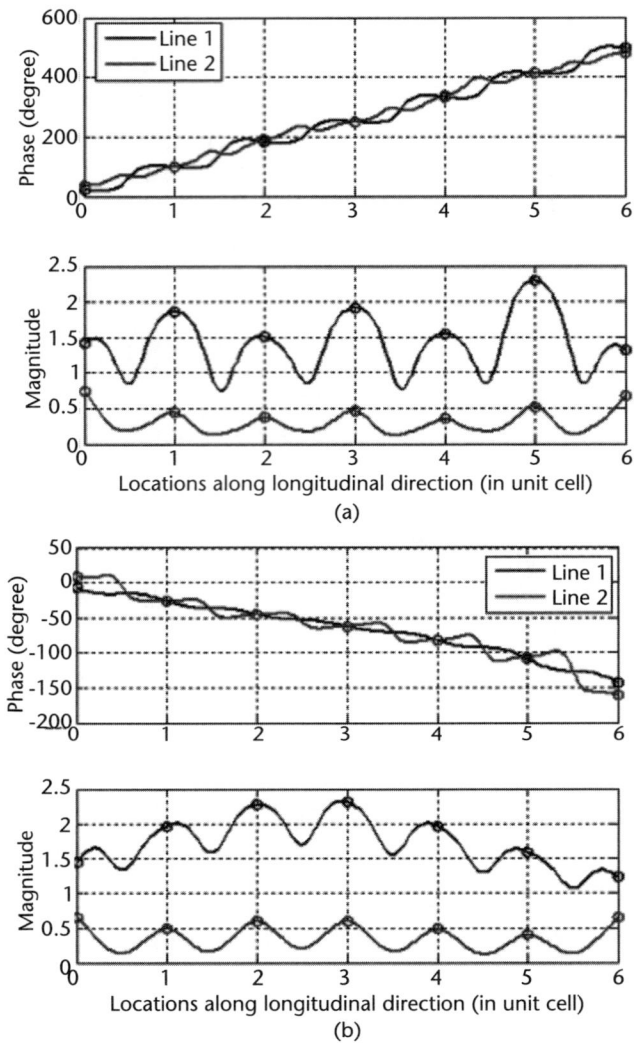

Figure 7.5 Magnitude and phase measured inside the six-layer DNG slab, along two different lines in the propagation direction at (a) 15.2 GHz and (b) 19.6 GHz. The circles represent the boundaries between the adjacent layers.

data at the boundaries between the adjacent layers (denoted by circles in Figure 7.5), and the results are plotted in Figure 7.6(a). Since we know that, if the reflection from the slab is negligible, and if we implicitly assume that the slab is homogeneous, then the phase change is equal to a product of n' (real part of refractive index), the free-space wave number (k_0), and the propagation distance (d), the computed values of n are plotted in Figure 7.6(b). It can be seen from this figure that, for the six-layer case, a negative refraction is observed in the first transmission band, while the refraction is positive in the second. In other words,

7.2 EM Response of the Infinite, Doubly Periodic DNG Slab with Plane Wave Illumination

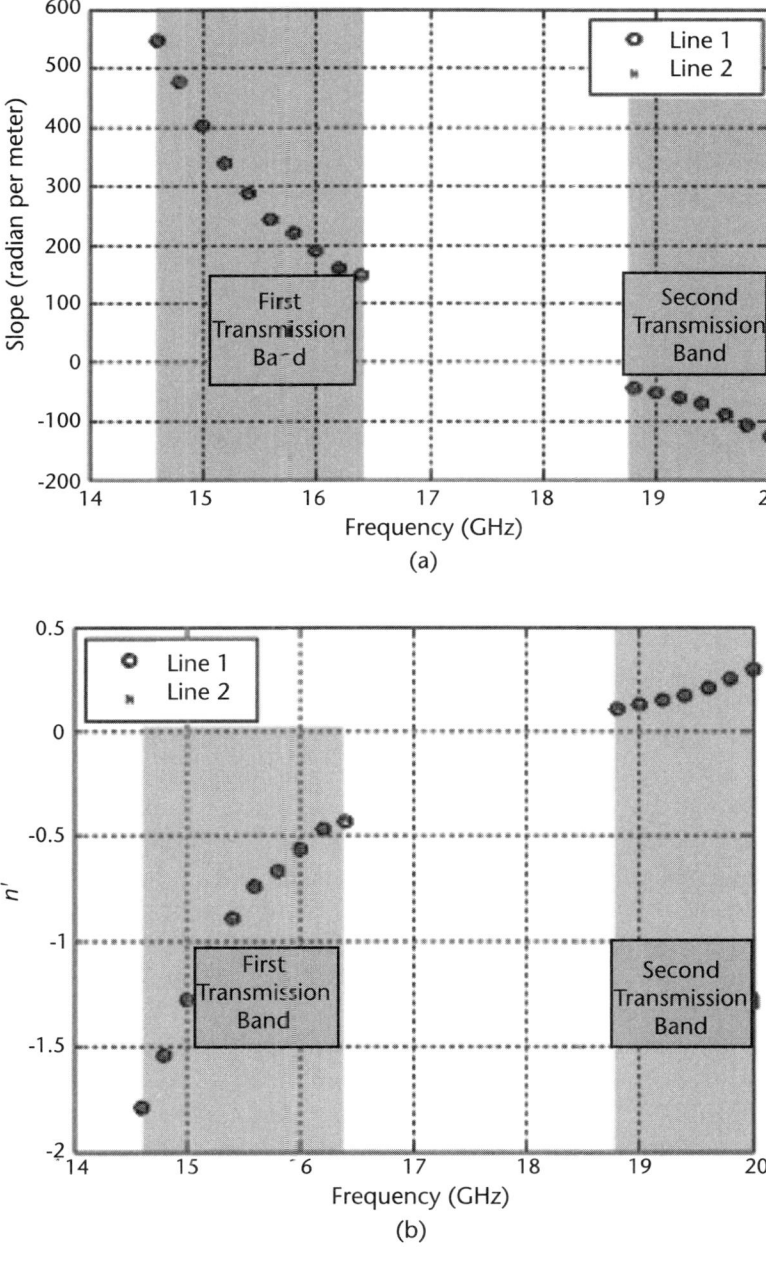

Figure 7.6 (a) The slope of the phase against propagation distance computed using the phase data obtained from FDTD simulations. (b) The real part of refractive index extracted from the slope in (a). The slope and the refractive index are only computed within the two 3-dB passbands (in gray) of the six-layer slab.

for the case of the six-layer slab, the phase data inside the slab confirms the DNG behavior in the first passband, which ranges from 14.7 to 16.4 GHz, while the permittivity and permeability are positive in the second passband that goes from 18.8 to 20 GHz.

7.3 Retrieval of Effective Material Constitutive Parameters Using the Inversion Approach

Availability of a robust method for characterizing an antenna system or its components is essential in engineering applications, since we always seek to achieve a design that meets the specifications for the applications we have in mind, in terms of the operating frequency, bandwidth, loss, and so on. The design process invariably relies upon characterization methods that must be both robust as well as reliable. Following the first experimental realization of DNG metamaterials, researchers began to search for ways to characterize their properties in a systematic way. A great deal of ongoing research has been directed towards determining the effective permittivity and permeability values of a composite that are both self-consistent and unambiguous. Some approaches reported in the literature to obtain the effective parameters from numerical data include inversion techniques [1–3]; averaging methods [4, 5] based on effective medium theory; and the use of results of phase velocity calculation [6] obtained from time-domain simulations. The inversion approach is one of the popular methodologies employed for this purpose. Unlike some of the other approaches, which make use of the knowledge of the field distribution inside the composite, the inversion method relies entirely upon the knowledge of the S-parameters of the slab illuminated by a normally incident wave. Note that the above parameters can be obtained either from experimental measurements, or from numerical simulations.

In the following sections, we will first discuss the basic premises upon which effective material parameters concepts are based. Next, we will review the inversion procedure for the retrieval of these parameters, and propose a modified one that helps us to choose the correct branch of the solution when certain approximations are found to be valid. Following this, we apply the modified approach to process the scattering parameters generated from the FDTD simulations, obtained by using the PBC/FDTD technique. The simulations are carried out for a number of different array settings, namely arrays of wires alone, SSR alone, and combinations thereof, beginning with one and going up to six layers. Finally, we will summarize the difficulties encountered in applying the inversion approach, and follow this up with a discussion on the use of effective material parameters for real-world applications.

7.3.1 Review of the Inversion Approach

The inversion approach [1–3] is the most widely used method for retrieving the effective material parameters. An important attribute of this approach is that it requires no additional assumption except, of course, that the effective medium concept be valid. However, this method does suffer from an inherent multiple-branch ambiguity problem arising from the multivalued nature of the logarithmic and square root functions appearing in the expressions for the refractive index n

and the impedance Z. To sort out this situation, the strategy is to eliminate the non-physical solutions by following a set of rules, and assuming that the only solution that survives is the correct one. The procedure for extracting the effective medium parameter from the S-parameters, as well as the selection rules for choosing the correct branches will be discussed in detail in the following paragraphs.

We assume that the electromagnetic fields have the $e^{j\omega t}$ time dependence. It should be noted that in most of the references on this topic, a time dependence of $e^{-j\omega t}$ is assumed though this is not explicitly stated. However, we adopt a time dependence of $e^{j\omega t}$ in this book, which is followed in the convention used in the FDTD as well as in most engineering disciplines. The scattering parameters of a homogeneous slab of thickness d in free space at normal incidence are given by:

$$S_{11} = \frac{\Gamma(1 - e^{-j2nk_0 d})}{1 - \Gamma^2 e^{-j2nk_0 d}} \tag{7.1}$$

$$S_{21} = \frac{(1 - \Gamma^2)e^{-jnk_0 d}}{1 - \Gamma^2 e^{-j2nk_0 d}} \tag{7.2}$$

where $\Gamma = \frac{Z-1}{Z+1}$ is the reflection coefficient at the boundary between the two media under study, as shown in Figure 7.7; Z ($Z = Z' + jZ''$ is assumed) is the wave impedance; n ($n = n' + jn''$ is assumed) is the refractive index of the slab; and k_0 is the free-space wave number. Once the reflection and transmission coefficients have been obtained, the refractive index as well as the wave impedance can be calculated by inverting (7.1) and (7.2) to get:

$$Z = \pm\sqrt{\frac{(1 + S_{11})^2 - S_{21}^2}{(1 - S_{11})^2 - S_{21}^2}} \tag{7.3}$$

$$Y = e^{-jnk_0 d} = X \pm j\sqrt{1 - X^2} \tag{7.4}$$

$$X = \frac{1}{S_{21}}(1 - S_{11}^2 + S_{21}^2) \tag{7.5}$$

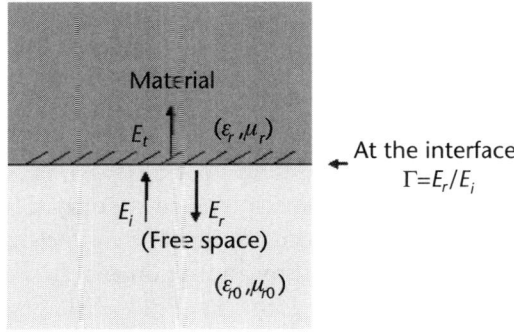

Figure 7.7 Wave reflection and transmission at normal incidence by a planar interface.

The refractive index n can be calculated from (7.4) and (7.5) by using the expression:

$$n = \frac{1}{k_0 d}\{[[\ln(e^{-jnk_0 d})]'' + 2m\pi] + j[\ln(e^{-jnk_0 d})]'\}$$

$$= \frac{1}{k_0 d}\{[[\ln(Y)]'' + 2m\pi] + j[\ln(Y)]'\} \quad (7.6)$$

where m is an integer related to the branch index of n' (real part) and is related to the thickness of the slab, expressed in terms of the wavelength. Once the refractive index and the wave impedance have been found, the permittivity and permeability can be obtained from:

$$\varepsilon_{eff} = n/Z \quad (7.7)$$

$$\mu_{eff} = nZ \quad (7.8)$$

The expressions appearing in (7.3), (7.4), and (7.6) contain complex and multi-valued functions. We must, therefore, impose additional constraints to obtain consistent and unambiguous results for the permittivity and permeability. The impedance is a square-root function, whose sign ambiguity can be resolved by imposing the condition Z' (real part) > 0, which must be satisfied by all passive materials. However, when the values of Z' are close to zero, even a slight perturbation in the values of the S-parameters can cause the signs of Z' to switch, and this in turn, causes the retrieval procedure to fail. To overcome this difficulty, it has been recommended that the expression that relates n and Z be employed to determine the sign of the impedance [3]. This relationship is given by:

$$e^{-jnk_0 d} = \frac{S_{21}}{1 - S_{11}\frac{Z-1}{Z+1}} \quad (7.9)$$

The sign of Z is chosen by using (7.9) such that the corresponding refractive index has a nonnegative imaginary part, or equivalently $\left|e^{-jnk_0 d}\right| \leq 1$. The impedance Z can be determined explicitly by following these two rules.

To satisfy the passivity condition, the imaginary part of the effective refractive index (i.e., n'') should either be zero or negative. Here we point out that the two roots in (7.4) result in two values of $\ln(e^{-jnk_0 d})$, and that only their signs are different. Thus, when n'' is not close to zero, only one root in (7.4) yields a nonpositive value for n'', and the other should be discarded. However, when n'' is close to zero, both the roots should be examined as possible choices when determining the real part of n. Nevertheless, n'' can be determined explicitly and unambiguously in both of these cases.

Up to this point, the impedance Z and the imaginary part of the refractive index n have been determined in an explicit manner. However, when n'' is close to zero, the real part of the refractive index still suffers from ambiguities introduced by the branches of the logarithmic function given in (7.6), and there exists the possibility of two acceptable roots in (7.4). The first ambiguity becomes even more problematic when the retrieval method is used for a thicker slab. This is because d, the thickness of the slab, appears in the denominator of the right-hand

7.3 Retrieval of Effective Material Constitutive Parameters Using the Inversion Approach

side of (7.6), and the term $\frac{2\pi m}{d}$ becomes small for a large d, making it difficult to choose the correct solution as all the branches now become very closely spaced. To circumvent this problem, it has been suggested that we use a slab whose thickness is relatively small and require that the $\varepsilon(f)$ and $\mu(f)$ be continuous functions of frequency [2]. An iterative approach has been suggested in [3] to implement this continuity condition. Assuming that we have obtained the value of the refractive index $n(f_0)$ at a frequency f_0, we still need to choose m, the branch index, from all the correct roots in (7.4) (note: more than one correct root exist only when n'' is close to zero) such that $n(f_1)$ for the chosen m is closest to $n(f_0)$, where f_1 is the next frequency adjacent to f_0. Additionally, the requirements $\mu'' \leq 0$ and $\varepsilon'' \leq 0$ must be imposed since the material is passive, and this helps to discard the nonphysical solutions of n'.

The initial solution of n' still needs to be determined at the starting frequency for this iterative approach to work. We can simplify this task by starting at a low frequency, as we will now explain. First, at low frequencies where $k_0 d$ is very small, the branch index m and the root in (7.4) should be chosen such that $[\ln(e^{-jnk_0^d})]'' + 2m\pi]$ is close to zero, since this term represents the total phase change across the slab at this frequency. Second, at frequencies well below the first resonance, we would not expect to see DNG-type behavior, which implies that the correct solution should be one for which $n' \geq 0$. The real part of n (i.e., n') can then be determined at the frequency of interest by imposing these two conditions.

As discussed above, the process for choosing the correct solution in the inversion approach is complex. In Ziolkowski's work [23], this difficult process was avoided by making the approximation of $e^{-jnk_0 d} \sim 1 - jnk_0 d$, which simplifies the expression for the refractive index to yield:

$$n = \frac{k}{k_0} = \frac{1}{jk_0 d} \frac{(1-V_1)(1+\Gamma)}{(1-\Gamma V_1)} \qquad (7.10)$$

where $V_1 = S_{11} + S_{21}$, and $\Gamma = \frac{Z-1}{Z+1}$ are the reflection coefficients at the two media interfaces, and Z is the wave impedance that can be computed by using the inversion approach in an unambiguous manner. However, we note that, according to [3], the above approximation is valid when

$$|n' k_0 d| \leq 1 \qquad (7.11)$$

while others have pointed out that the correct condition for this approximation to hold is:

$$|nk_0 d| \ll 1 \qquad (7.12)$$

For the sake of the following discussion, we will refer the solution in (7.10) as the small-phase-small-loss (SPSL) solution.

Even though (7.12) holds for a very limited range, we will employ the SPSL solution to assist the selection of the correct root in the inversion method, when all the solutions remaining after the elimination of the nonphysical ones satisfy (7.12).

We will now summarize the steps in our modified inversion procedure in the following:

1. Compute Z by using (7.3). The sign can be determined by using the following rules:
   - If Z' is not close to zero, choose the root with the positive real part.
   - If Z' is close to zero, choose the root, by using (7.9), such that $\left|e^{-jnk_0d}\right| \leq 1$.
2. Compute all possible solutions of n by using (7.4) to (7.6) for the two roots in (7.4) and a set of chosen branch indices $\{m\}$. Compute the effective ε and μ for all solutions of n and the Z extracted by using (7.7) and (7.8).
3. At this point, we have multiple sets of solutions, but only one of these sets is physical. Choose the correct solution by eliminating the nonphysical ones. These physical rules are:
   - At frequency well below the first resonance, choose the solution that yields $n'' <= 0$ and n' is closest to 0 from the positive side.
   - When n'' is not close to zero, discard the solutions computed from the root that yields a positive n''.
   - Examine ε'' and μ'' associated with the remaining solutions. Discard the solutions that give rise to positive ε'' or μ''.
   - Enforce the continuity of n'.
4. If more than one solution still satisfies the requirements set forth above in this stage, we check the values of $|nk_0d|$ for all of the remaining solutions. If $|nk_0d| \ll 1$ still holds for all of the remaining solutions, apply the SPSL criterion, given in (7.10), to choose the correct branch.

7.3.2 Retrieval of the Effective Material Parameters from the Numerical S-Parameters Obtained from FDTD Simulations of Metamaterials

The first step taken in the inversion approach is to define the location of the two effective boundaries of a metamaterial slab, which does not have a well-defined surface as does a homogeneous slab. Since we know that the impedance of a homogeneous slab does not depend on thickness of the slab, we can define the surfaces of the slab such that the results for the impedances extracted for the slabs show a consistent behavior as the number of layers is progressively increased. The S-parameters measurement uses these effective surfaces as the input and output ports, and the effective thickness (d) of slab is defined to be the separation distance between these surfaces.

We will now study the DNG array, described in Section 7.2, starting with one layer and going up to three, by using the PBC/FDTD code. Three pairs of measurement planes, positioned at one, two, and three FDTD cells, respectively, beyond the outermost edge of the metallic inclusions on each side of the slab, are tested for the optimal effective surfaces, as shown in Figure 7.7. For each pair of test planes (at the transmission and reflection sides, respectively), the wave impedance

7.3 Retrieval of Effective Material Constitutive Parameters Using the Inversion Approach

was computed by using the inversion approach, with the effective slab thickness defined as the separation distance between these planes. The optimal locations of the interfaces are found by minimizing the mismatch of the impedance for different number of layers. Figures 7.8(a–f) show the real and imaginary parts of the wave impedance for one- to three-layer cases, which are retrieved from the S-parameters computed for the three pairs of test planes described above, respectively. It can be seen that the mismatch of the wave impedance for different layers is the smallest when the test planes are located at three FDTD cells beyond the outermost edge of

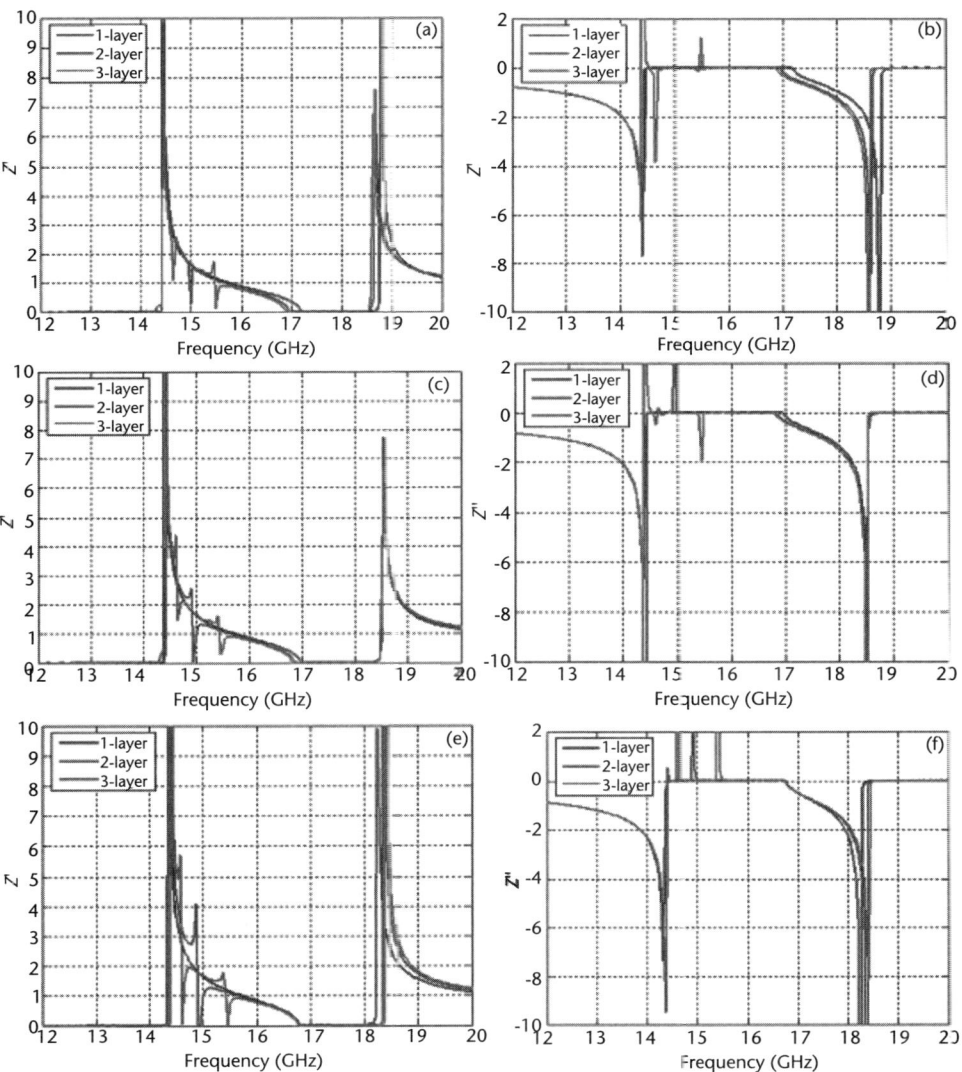

Figure 7.8 Comparison of the extracted impedance Z (left: real part, right: imaginary part) for slabs of one-, two-, and three-layers with imports and outports placed at different locations: (a) and (b) one FDTD cell; (c) and (d) two FDTD cells; and (e) and (f) three FDTD cells beyond the outermost edge of the metallic inclusions on each side of the slab.

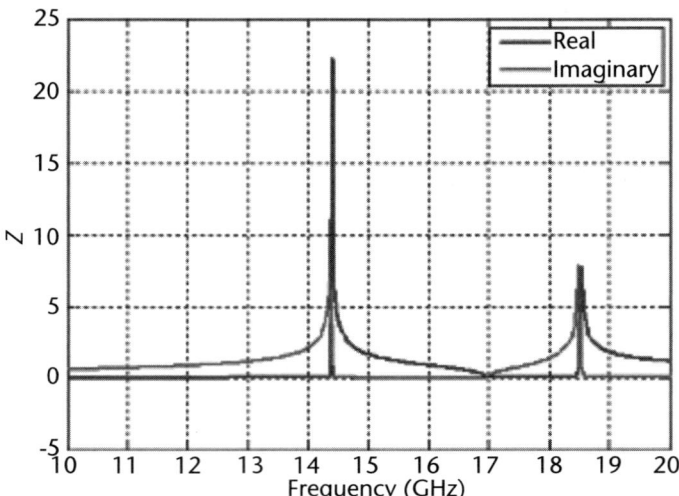

Figure 7.9 The wave impedance retrieved from the one-layer slab.

metallic inclusions [see Figures 7.8(c, d)]. For this reason, we define these planes to be the effective surfaces for the slab.

We now begin to demonstrate the retrieval process for the single-layer DNG array case. Figure 7.9 shows the wave impedance, which has been determined by using (7.3) and (7.9), without ambiguity. Figure 7.10 shows the solution of n'' derived from the two roots in (7.4), by using (7.6). As can be seen from the above

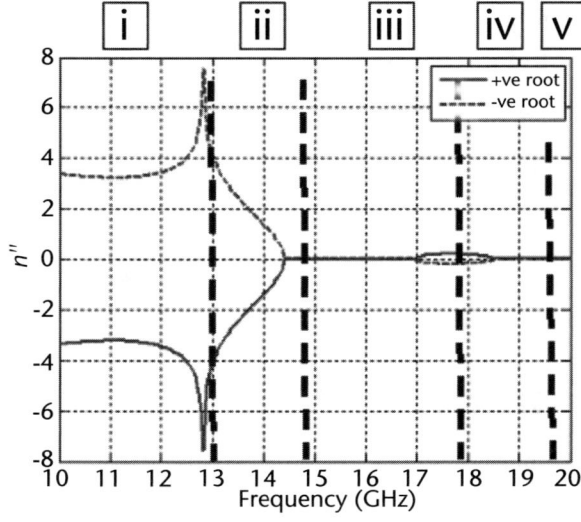

Figure 7.10 The solutions for n'' derived from the two roots in (7.4) for the one-layer slab.

figure, it is convenient to divide the span of the frequency spectrum, ranging from 10 to 20 GHz, into five regions, and then determine the correct solution in these regions separately. These regions are: (1) below 12.80 GHz; (2) 12.8–14.4 GHz; (3) 14.4–17.0 GHz; (4) 17.0–18.5 GHz; and (5) above 18.5 GHz. Since $n'' \leq 0$, the solutions with the negative roots in the frequency regions (1) and (2), as well as the solutions with positive roots in the frequency region (4) should be discarded.

Figure 7.11(a–c) shows the solutions for n', ε'', μ'', computed by using the positive root for the branch indices $m = 0, -1$, and $+1$, respectively. Figure 7.12(a–c) shows the corresponding solutions computed by using the negative root. In region (i), only three possible solutions of different branch indices with positive root survive. As can be seen from Figure 7.11(b, c), the solution with branch index $+1$ and -1 should be discarded, because either ε'' or μ'' are found to be positive. Thus, the only solution that survives is the one with positive root and $m = 0$, which is assumed to be the correct one. Similarly, in region (iv), only one solution—the negative root with $m = 0$—survives, when we examine the sign of ε'' or μ'' in Figure 7.12(b, c).

In regions (iii) and (v), all six solutions are still acceptable, insofar as the physical constraints are satisfied by n'', ε'', μ''. However, we can make use of the known solution in region (iii), together with the requirement on the continuity of n', to eliminate the four solutions that have a nonzero m. Then, only two solutions survive, both with $m = 0$, one from the positive root and the other from the negative one.

Next, we compute the value of $|nk_0d|$ for all possible solutions surviving in the frequency band (iii) through (v), and plot them in Figure 7.13(a). Let us use the condition $|nk_0d| < 0.2$ to define the frequencies at which the SPSL approximation is valid. Two narrow frequency bands, one ranging from 16.8 to 17.2 GHz and the other from 18.2 to 18.6 GHz, can be found to satisfy the above condition. Figure 7.13(b, c) plots the real and imaginary parts of n for all the acceptable solutions computed by using the inversion approach, along with the valid SPSL solution in these frequency bands. We can see that the two valid frequency bands happen to cover the upper end of region (iii) and the lower end of region (v), respectively; therefore, the branches can be correctly chosen in these regions by matching the above two solutions. In region (iii), the positive root with $m = 0$ is selected, whereas in region (v), the negative root with $m = 0$ is chosen. The SPSL solutions are also plotted in Figure 7.14 over the entire observed frequency band.

It should be pointed out that none of the solutions in region (ii) appear to be physical, since the condition that both ε'', $\mu'' \leq 0$ is violated by all solutions. It has been argued [22, 23] that such solutions are still physical, even when one of the ε'', μ'' is positive while the other is negative, as long as the electromagnetic losses, expressed in terms of the integral, $Q = \frac{1}{2\pi} \int d\omega \omega [-\varepsilon''|E|^2 - \mu''|H|^2]$, is still positive. Some have also suggested [24] that there are higher-order modes inside this resonance band, whose levels may be comparable to that of the first order one. The effective medium theory fails under these circumstances and, hence, the effective material constitutive parameters can no longer be defined. This issue is still controversial, and we will not dwell on it much longer. We just add the remark here that we still choose the positive root with $m = 0$ as our solution in this region,

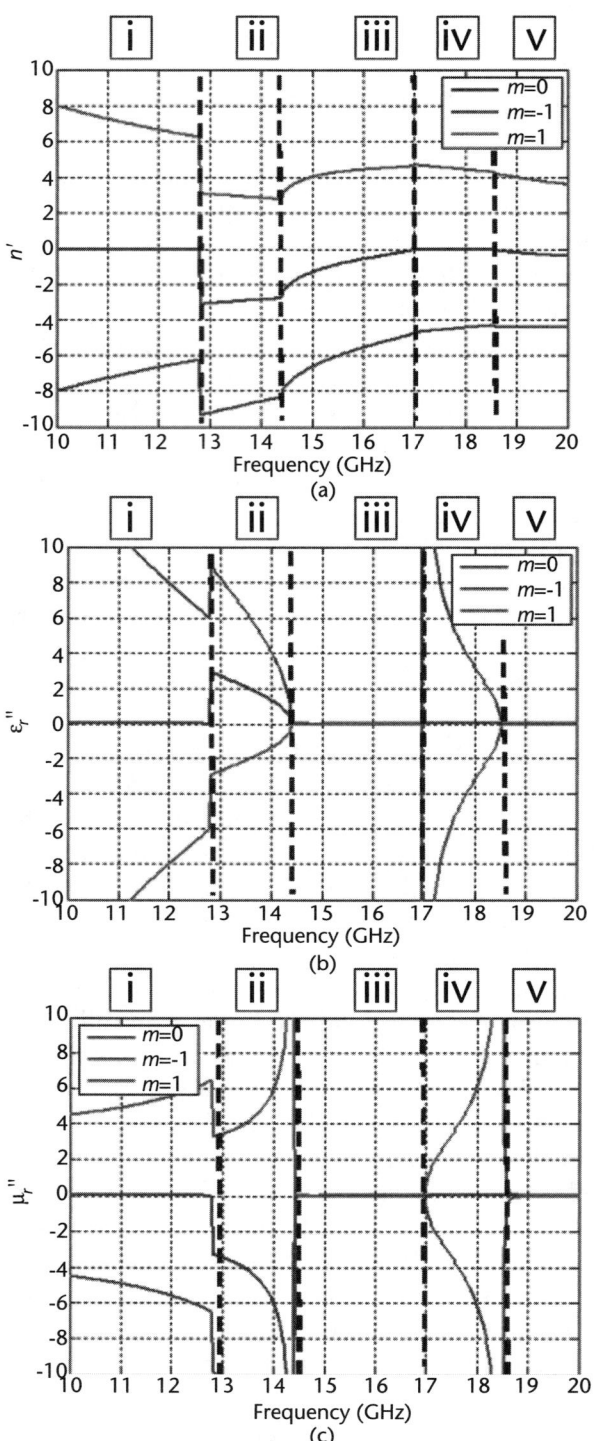

Figure 7.11 The solutions for (a) n', (b) ε'', and (c) μ''', computed by using the positive root in (7.4) with a branch index of $m = 0, -1$, and $+1$, respectively.

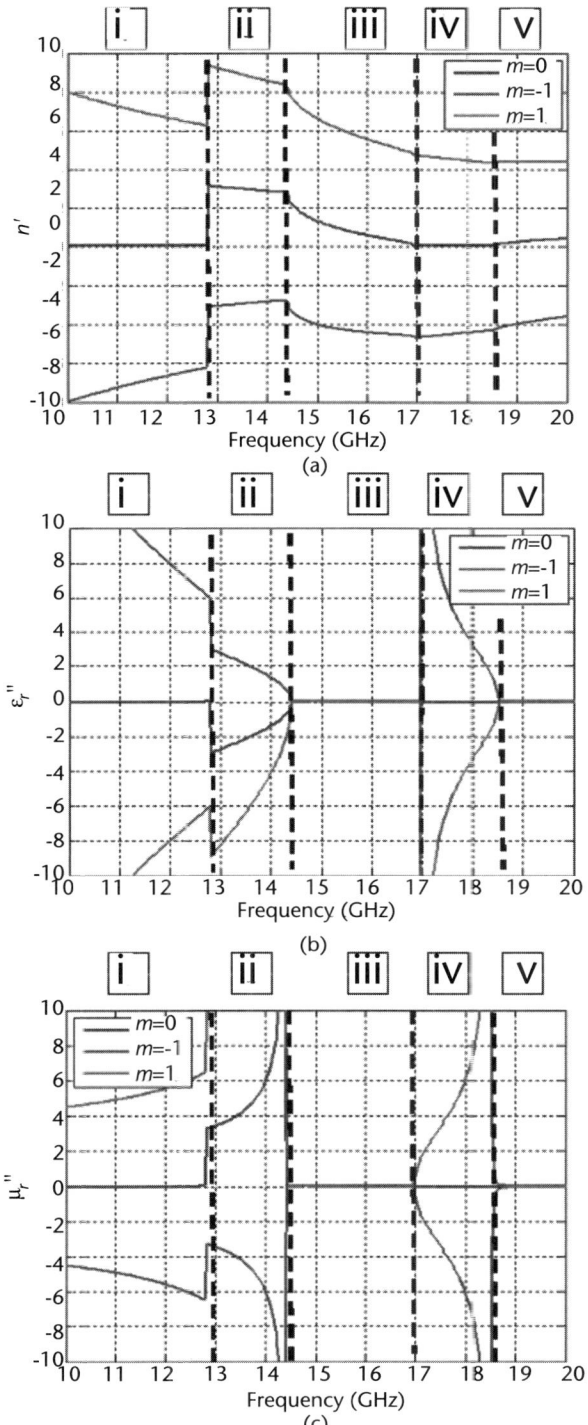

Figure 7.12 The solutions for (a) n', (b) ε_r'', and (c) μ_r'', computed by using the negative root in (7.4) with a branch index of $m = 0, -1$, and $+1$, respectively.

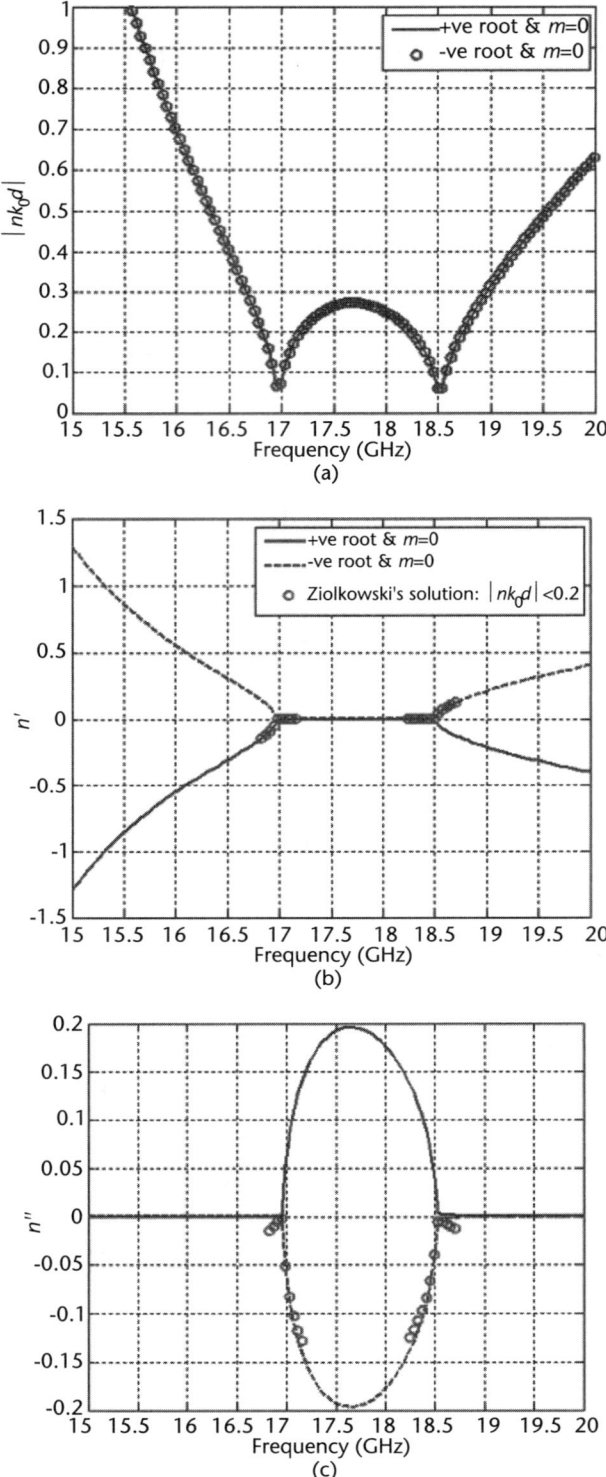

Figure 7.13 (a) The value of $|nk_0 d|$ for all remaining solutions obtained from the inversion approach; comparison of the (b) real and (c) imaginary parts of n for all remaining solutions obtained from the inversion approach (lines) and from the SPSL approximation (circles), in the frequency band (iii) through (v) as shown in Figure 7.10(b).

7.3 Retrieval of Effective Material Constitutive Parameters Using the Inversion Approach

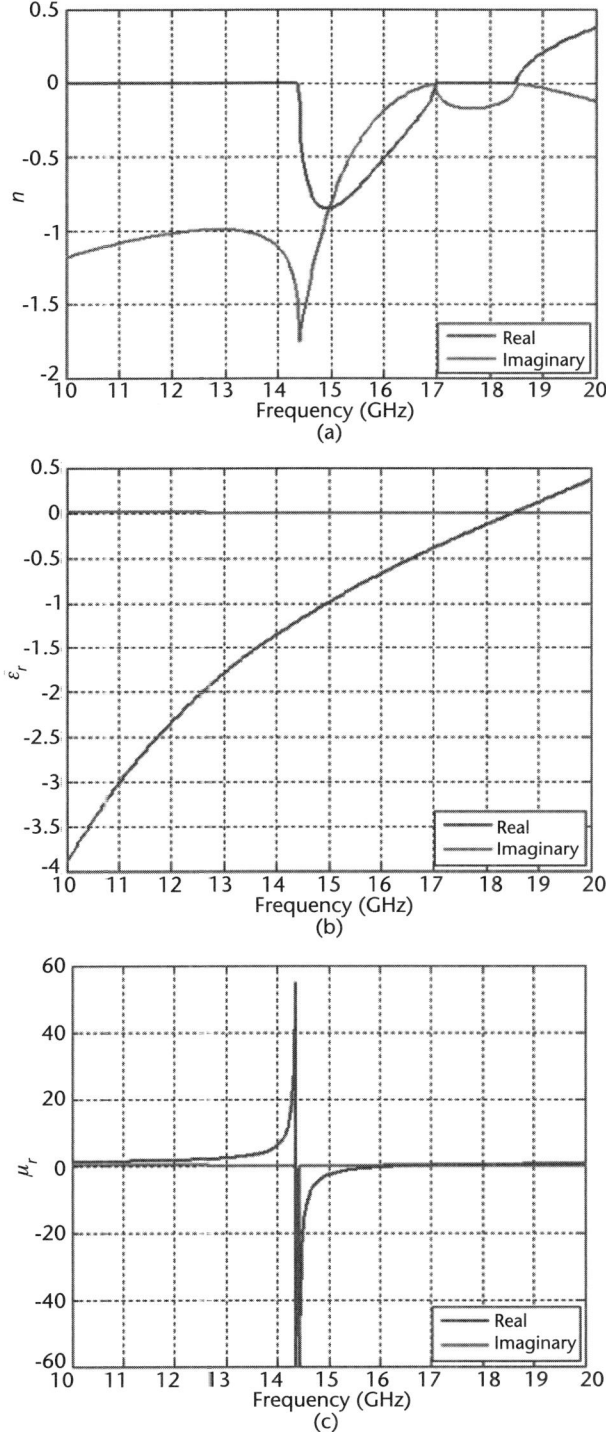

Figure 7.14 The SPSL solutions over the entire observation frequency band: effective (a) n; (b) ε; and (c) μ. Note that the SPSL condition is not satisfied over the entire band.

so as to render n' continuous at the interface between regions (ii) and (iii), while keeping $n'' \leq 0$.

The refractive index n, and the effective material parameters ε and μ of the medium all are extracted by using the inversion approach described above, and are plotted in Figure 7.15(a–c), with the gray area representing the nonphysical region, in which the criterion ε'', $\mu'' \leq 0$ is violated. The medium exhibits a DNG behavior in the frequency range of 14.4–17.0 GHz, which is consistent with the result presented earlier in Section 7.2.

We have also studied the S-parameters for the arrays of slabs comprising of either wires or SSR, by using the inversion approach for the single-layer case. Their wave impedances, refractive indices, and effective ε and μ parameters are plotted in Figures 7.16–7.19, respectively. Again, the frequency bands marked in grey represent the nonphysical regions, because the signs of ε'' and/or μ'' are positive in these regions. Not unexpectedly, the array of wires, or the SSRs, do not independently exhibit a DNG behavior, as shown Figure 7.17(b, c). A negative μ' region is found in the frequency band ranging from 15.8 to 17.0 GHz for the SSR array, while ε' is negative over the entire observed frequency band investigated, which ranges from 10.0 to 20.0 GHz for the array of wires. We also notice that the medium parameters of the SRR array exhibit a nonphysical behavior in the frequency band ranging from 14.4 to 15.8 GHz [see Figures 7.18(c) and 7.19(c)], but such regions are not present in the effective medium characteristics of the array of wires.

Next, we investigate the effective material parameters for a multilayer array comprised of a combination of SRRs and wires, by using the modified inversion approach discussed above. The effective surfaces are defined in the same way as before, and the effective thickness of the array is allowed to vary from two to six layers as follows: 7.75 mm, 11.75 mm, 13.75 mm, 17.75 mm, and 21.75 mm. Figures 7.20(a–c) through 7.23(a–c) show all the possible solutions of the refractive indices for the two-layer, four-layer, and six-layer cases. We see from Figure 7.20(a–c) that when we increase the number of layers, the entire frequency band of interest must be divided into an increasing number of subregions, since the spacing between the n' plots for different branch indices, that is proportional to $1/(k_0 d)$, becomes increasingly smaller. For the same reason, a larger range of branch index needs to be applied such that the set of all solutions can cover the correct solution. As a result, the selection process for the correct solution become very complex for structures with multiple layers as the number of layers is increased. Besides, as mentioned earlier, the higher-order Floquet harmonics are totally ignored when the slab thickness is increased by stacking multiple layers, and this can have serious consequences when we attempt to predict the performance of a thick metamaterial slab.

In view of the complexity of the selection process, we have attempted to apply the inversion approach only up to four-layer cases. Figures 7.24–7.27 plot the extracted wave impedance Z, effective refractive index n, ε, and μ, respectively, starting from a single layer and going up to four-layer cases in the frequency band ranging from 10 to 14.0 GHz. For the wave impedance, plotted in Figure 7.22(a, b), we observe more spikes in the frequency range between 14.4 to 17.0 GHz as the number of layers is increased. One might think that the occurrence of spikes is caused by the mutual coupling between the elements in different layers.

7.3 Retrieval of Effective Material Constitutive Parameters Using the Inversion Approach

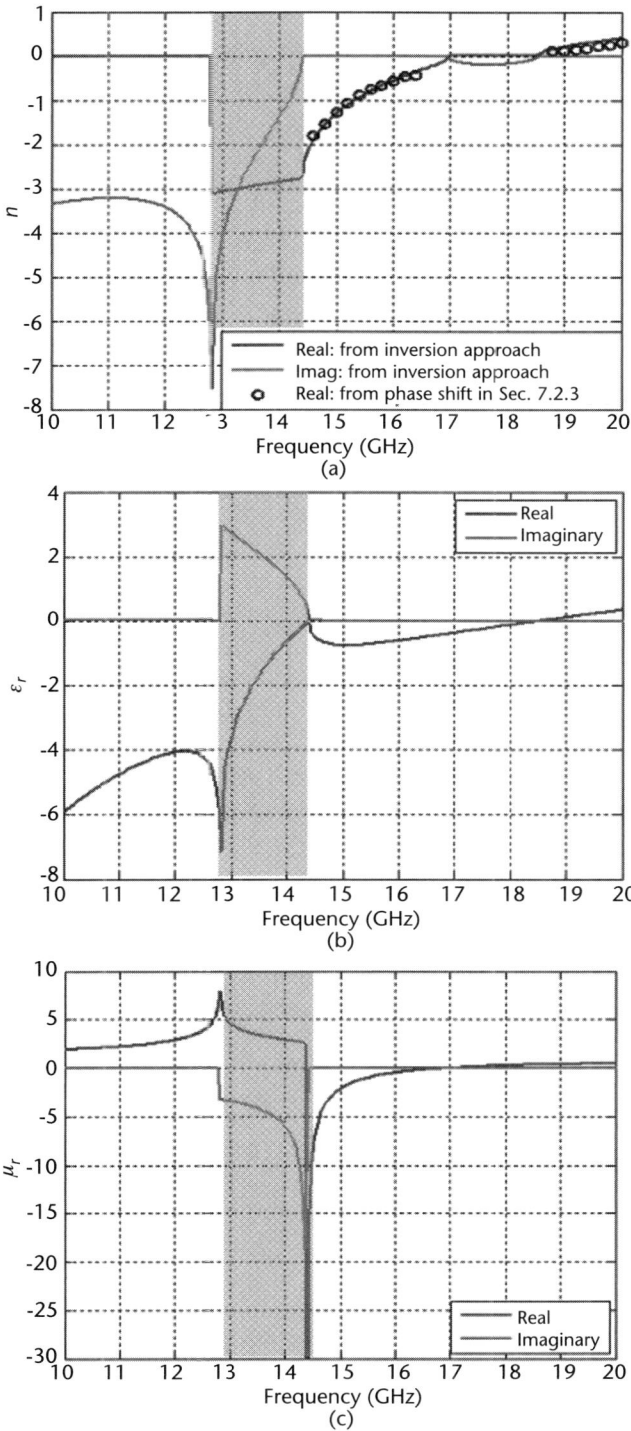

Figure 7.15 The extracted material parameters for the one-layer case: (a) n; (b) ε; and (c) μ. The gray area represents the nonphysical region where no solutions can be found to satisfy both ε'' and $\mu'' \leq 0$. The black circles in (a) represent the refractive index computed by using the phase data computed in Section 7.2.3.

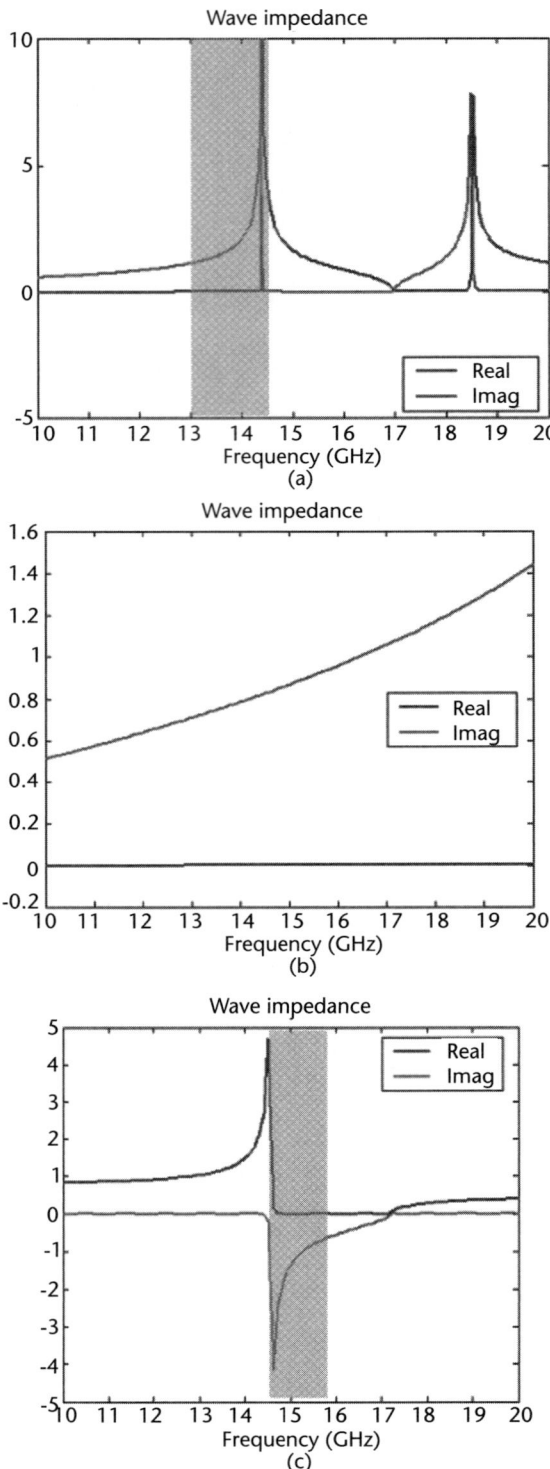

Figure 7.16 The wave impedance retrieved for the one-layer slab comprised of different types of inclusions: (a) SSRs and wires, (b) wires alone, and (c) SSRs alone. The gray area represents the nonphysical region where no solutions can be found to satisfy both ε'' and $\mu'' \leq 0$.

7.3 Retrieval of Effective Material Constitutive Parameters Using the Inversion Approach 197

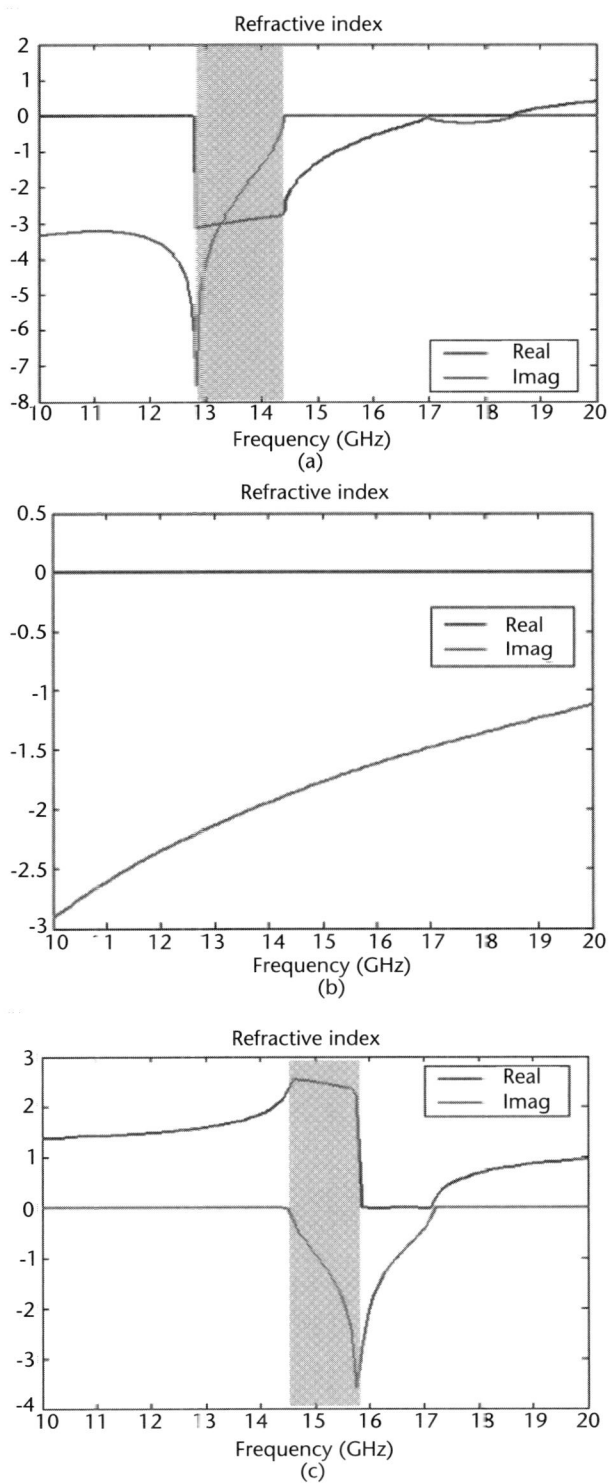

Figure 7.17 Refractive index retrieved for the one-layer slab, comprised of different types of inclusions: (a) SSRs and wires, (b) wires alone, and (c) SSRs alone. The gray area represents the nonphysical region where no solutions can be found to satisfy both ε'' and $\mu''\leq 0$.

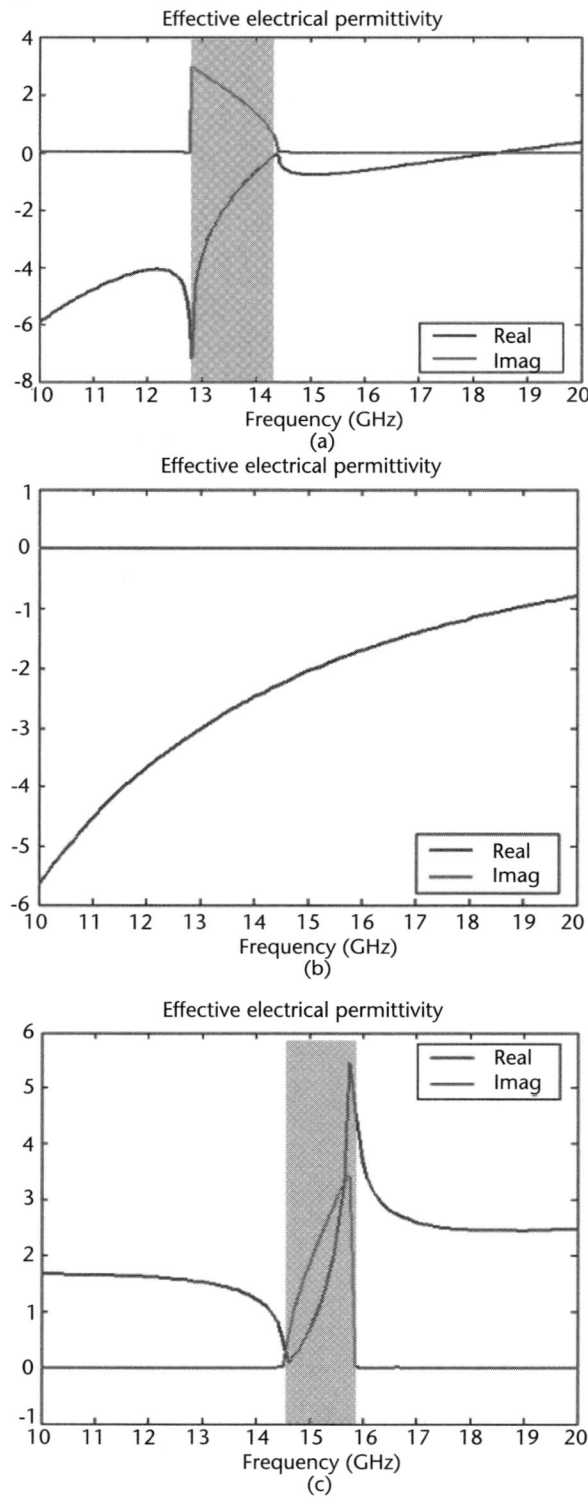

Figure 7.18 Effective electric permittivity retrieved for the one-layer slab, comprised of different types of inclusions: (a) SSRs and wires, (b) wires alone, and (c) SSRs alone. The gray area represents the nonphysical region where no solutions can be found to satisfy both ε'' and $\mu'' \leq 0$.

7.3 Retrieval of Effective Material Constitutive Parameters Using the Inversion Approach

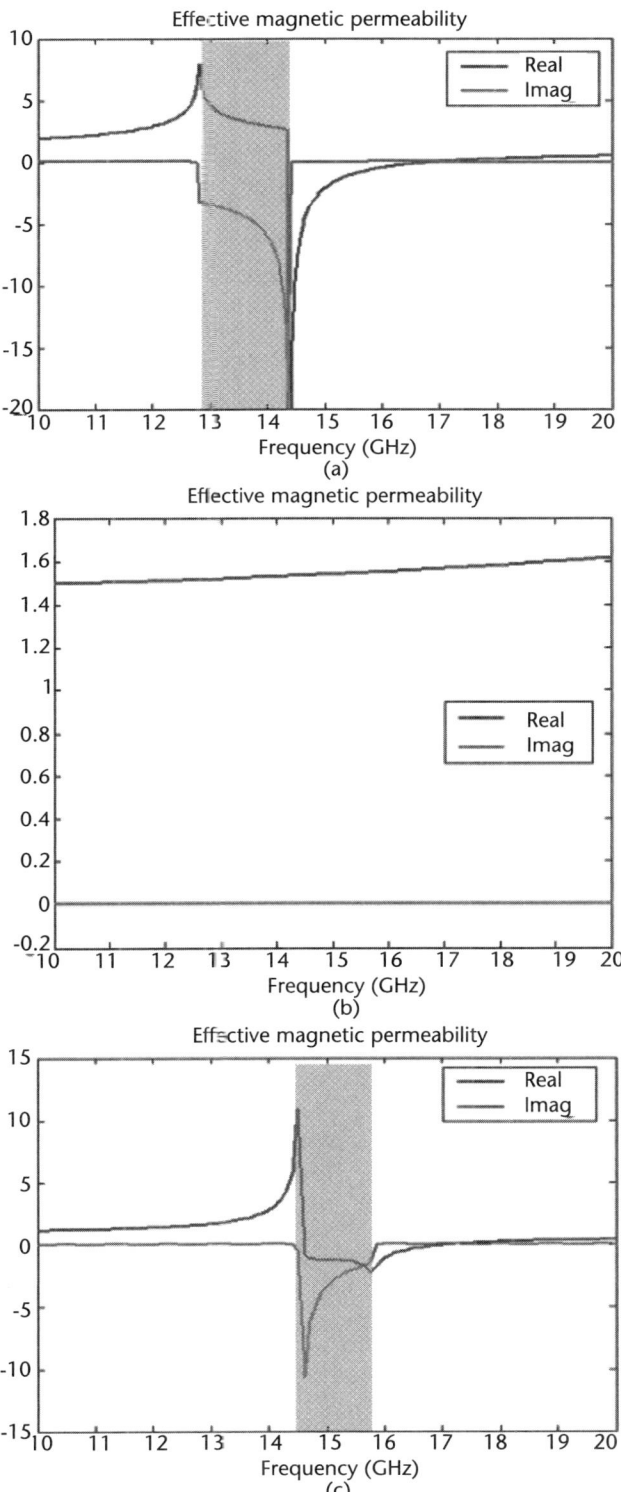

Figure 7.19 Effective magnetic permeability retrieved for the one-layer slab comprised of different types of inclusions: (a) SSRs and wires, (b) wires alone, and (c) SSRs alone. The gray area represents the nonphysical region where no solutions can be found to satisfy both ε'' and $\mu'' \leq 0$.

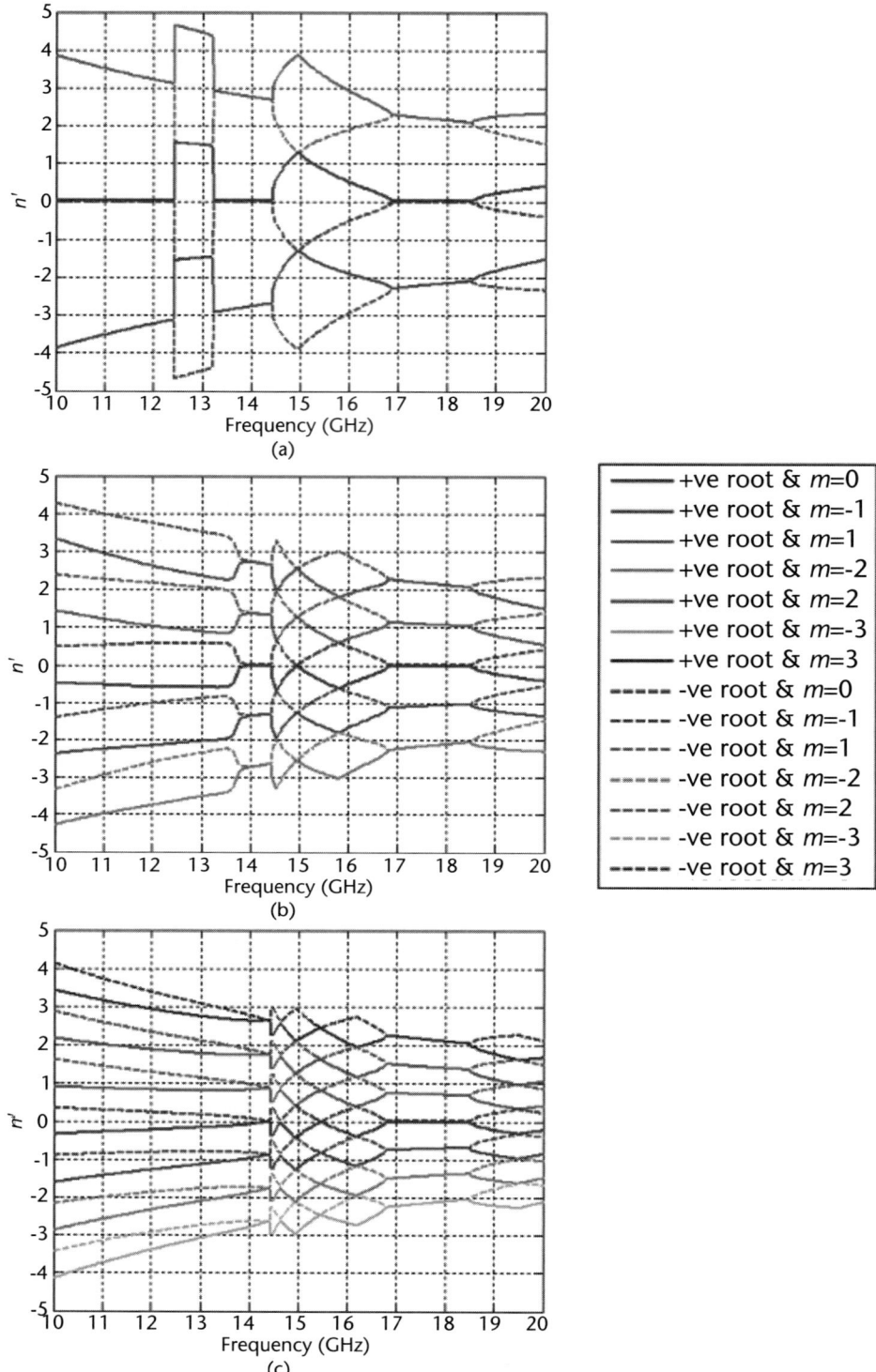

Figure 7.20 The set of all possible solutions of n' for the DNG slabs with multilayer inclusion: (a) two-layer, (b) three-layer, and (c) six-layer.

7.3 Retrieval of Effective Material Constitutive Parameters Using the Inversion Approach

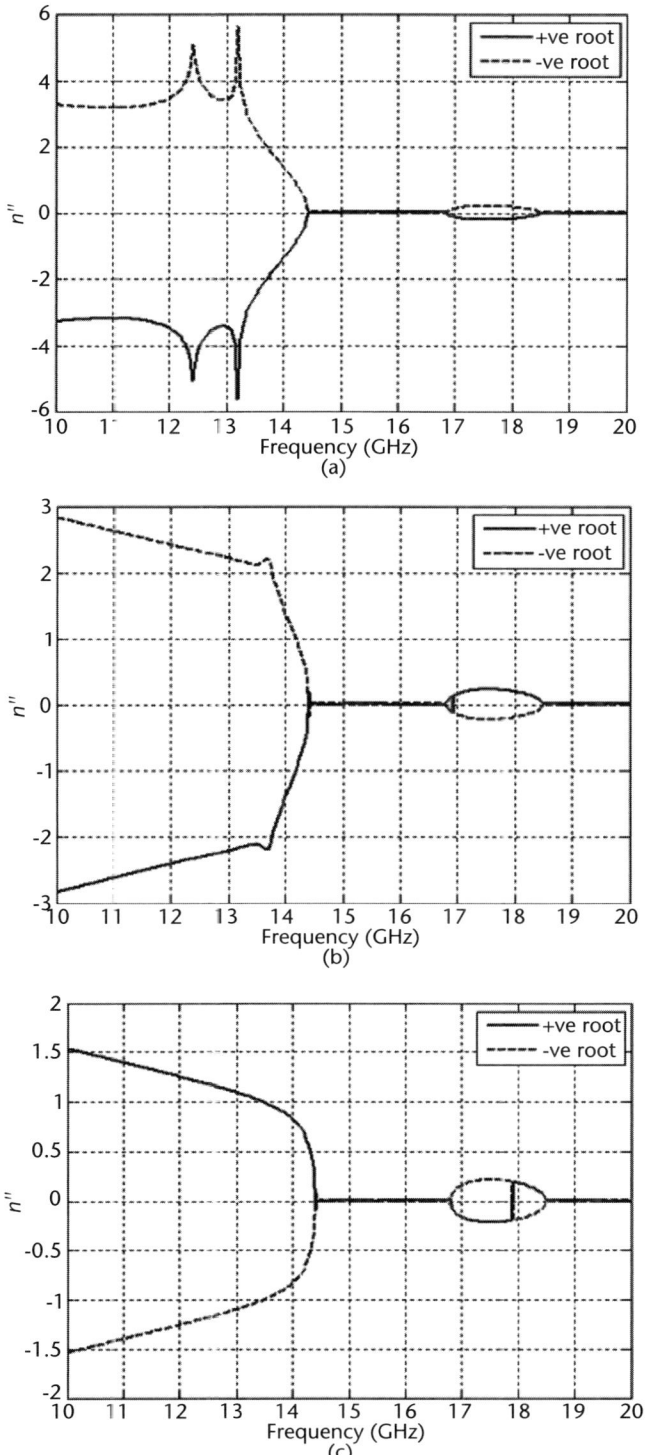

Figure 7.21 The set of all possible solutions of n'' for the DNG slabs with multilayer inclusion: (a) two-layer, (b) three-layer, and (c) six-layer.

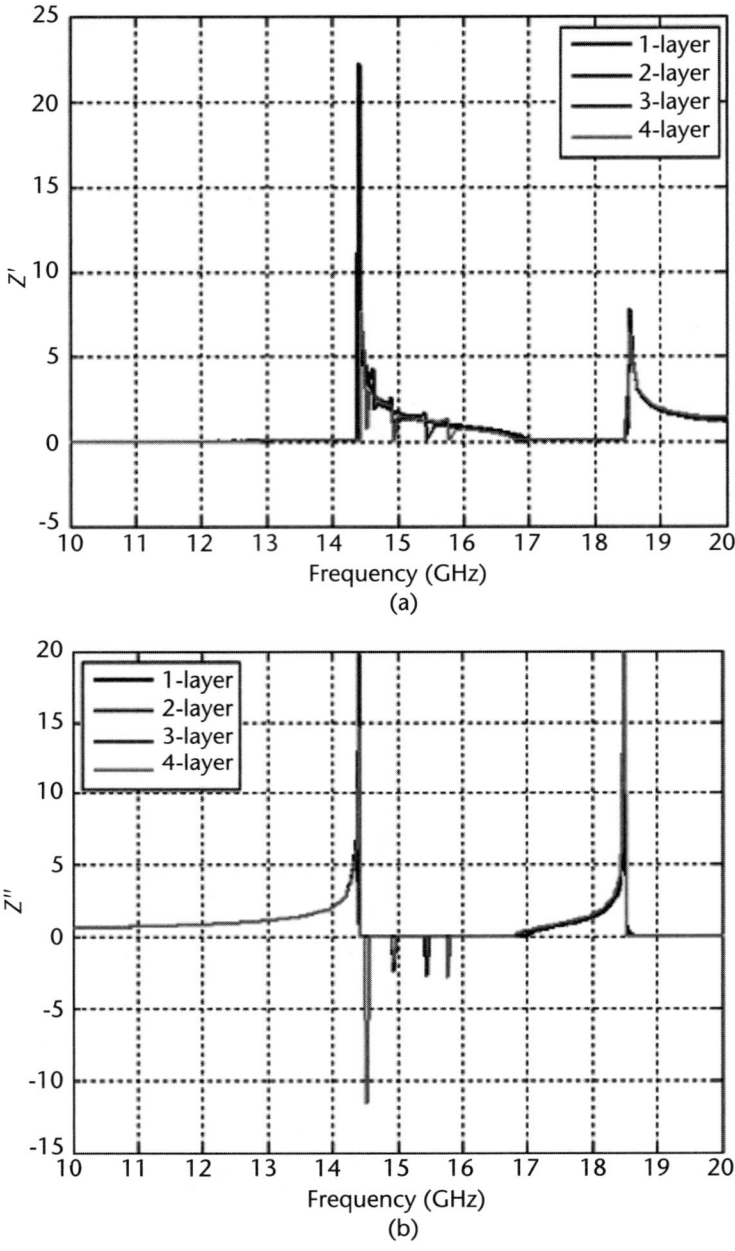

Figure 7.22 Comparison of the wave impedance extracted for the DNG slab, starting from one and going up to four layers: (a) real and (b) imaginary parts.

However, we also note that the locations of these spikes coincide with the dips of S_{11}; therefore, they might simply be caused by the numerical noise present in the computed S-parameters, since the wave impedance is inherently very sensitive to the variations in the S-parameters, whenever S_{11} is small. Even though the refractive index is more or less continuous in this frequency range, spikes in the

7.3 Retrieval of Effective Material Constitutive Parameters Using the Inversion Approach

wave impedance would, in turn, generate spikes in the effective ε and μ. No conclusion can be drawn at this point, as to whether the mutual coupling between the layers has a significant effect. Apart from the spikes, the effective material parameters match quite well as the number of layers is changed. For the refractive index n, the matching is good above 14.4 GHz for different number of layers [see

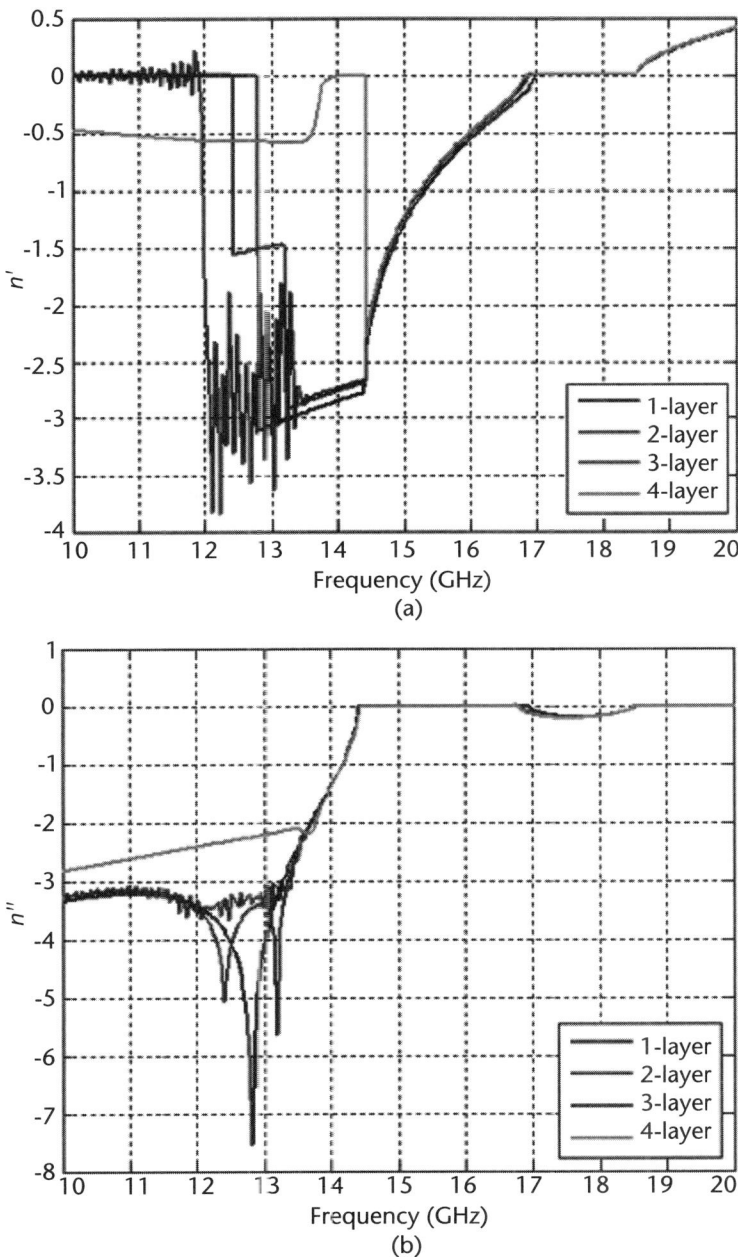

Figure 7.23 Comparison of the refractive index extracted for the DNG slab, starting from one and going up to four layers: (a) real and (b) imaginary parts.

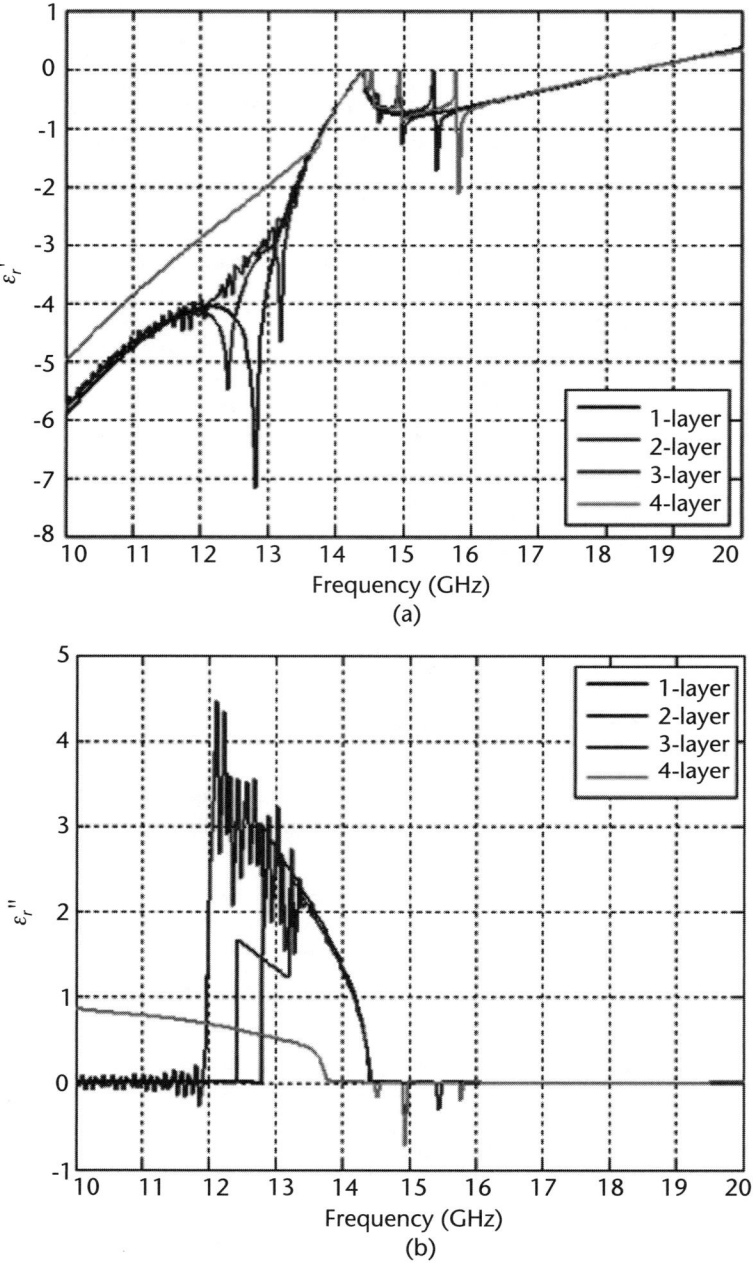

Figure 7.24 Comparison of the effective electric permittivity extracted for the DNG slab, starting from one and going up to four layers: (a) real and (b) imaginary parts.

Figure 7.23(a, b)]. It should be noticed that the DNG property, which is exhibited between 14.4 and 17.0 GHz, is consistent in all cases. However, some discrepancies of refractive index do occur between 12.8 to 14.4 GHz, which belongs to the nonphysical region.

7.3 Retrieval of Effective Material Constitutive Parameters Using the Inversion Approach

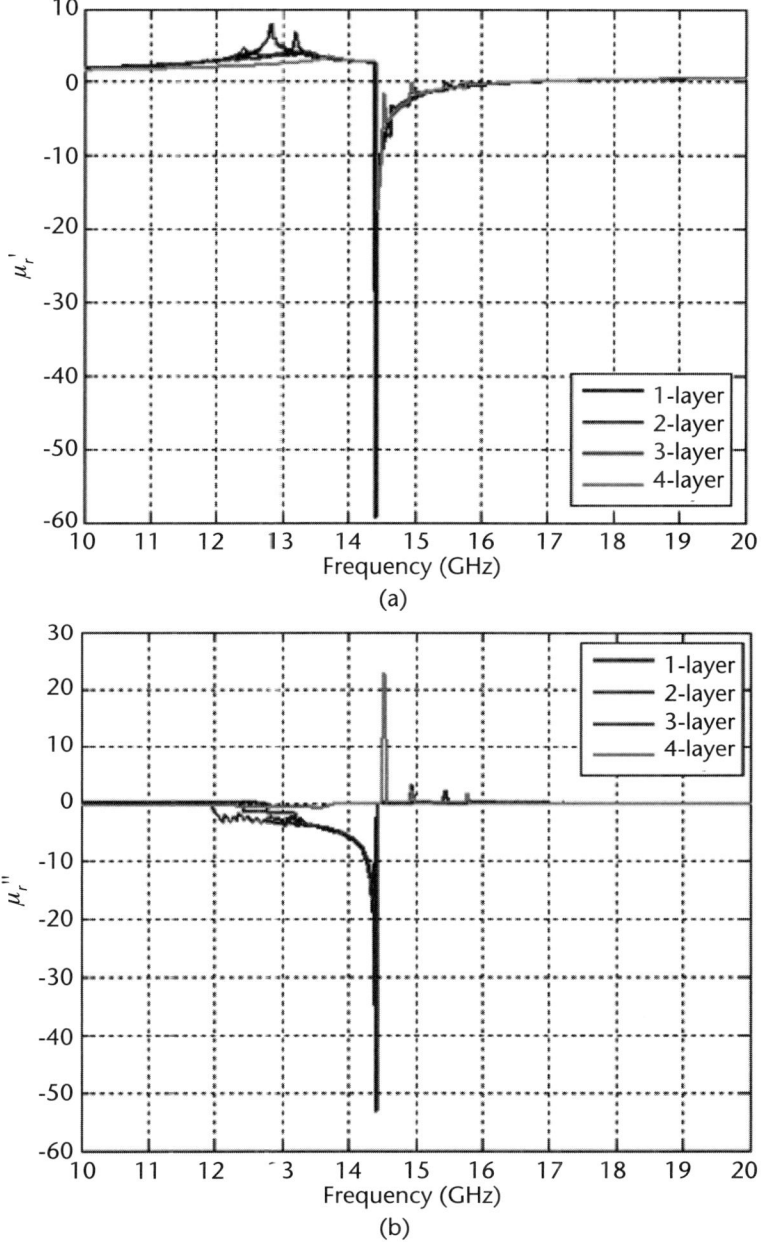

Figure 7.25 Comparison of the effective magnetic permeability extracted for the DNG slab, starting from one and going up to four layers: (a) real and (b) imaginary parts.

Finally, let us justify the condition $|n'k_0d| \leq 1$ proposed by Ziolkowski in [23], to approximate $e^{-jnk_0d} \sim 1 - jnk_0d$. We argue on the basis of the results presented below, that the condition should really be $|nk_0d| \ll 1$, by testing Ziolkowski's condition on the results for the single-layer DNG array. We first evaluate the values

$|nk_0d|$ and $|n'k_0d|$, where n is the effective refractive index extracted by using the modified inversion approach, as plotted in Figure 7.26(a). From these values, we locate the frequencies at which $|nk_0d|$ or $|n'k_0d|$ is small compared to 1, say less than 0.2. Next, we compute the SPSL solutions for the refractive indices, that are subsequently used to recover the S-parameters by using (7.1) and (7.2). Figure 7.26(b, c) plots the comparison of the recovered and original S-parameters for small values of $|nk_0d|$ and $|n'k_0d|$, respectively. It is clearly seen that the matching is excellent for small $|nk_0d|$, but not necessarily good for small $|n'k_0d|$. Therefore, we conclude that the correct condition for this solution to be valid should be $|nk_0d| \ll 1$. It should also be pointed out that the valid frequency range can be very limited, and even nonexistent sometimes, for the metamaterials slab we have investigated. Also, it is very important to guarantee that the condition $|nk_0d| \ll 1$ be satisfied when using this solution, in the alternative approach, for example, for the inversion method as stated before.

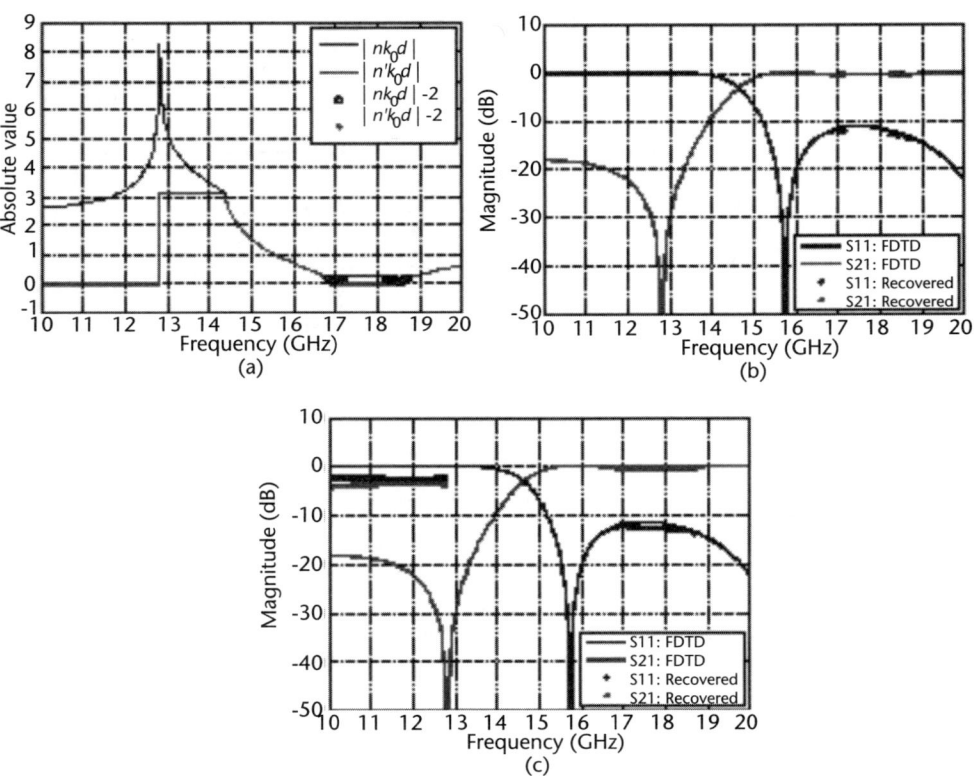

Figure 7.26 (a) The values of $|nk_0d|$ and $|n'k_0d|$ where n is computed by using the inversion approach; and the comparison of the S-parameters obtained by using FDTD (lines) and those recovered by using the SPSL solutions (dots) in the region where (b) $|nk_0d| < 0.2$ and (c) $|n'k_0d| < 0.2$.

7.3 Retrieval of Effective Material Constitutive Parameters Using the Inversion Approach

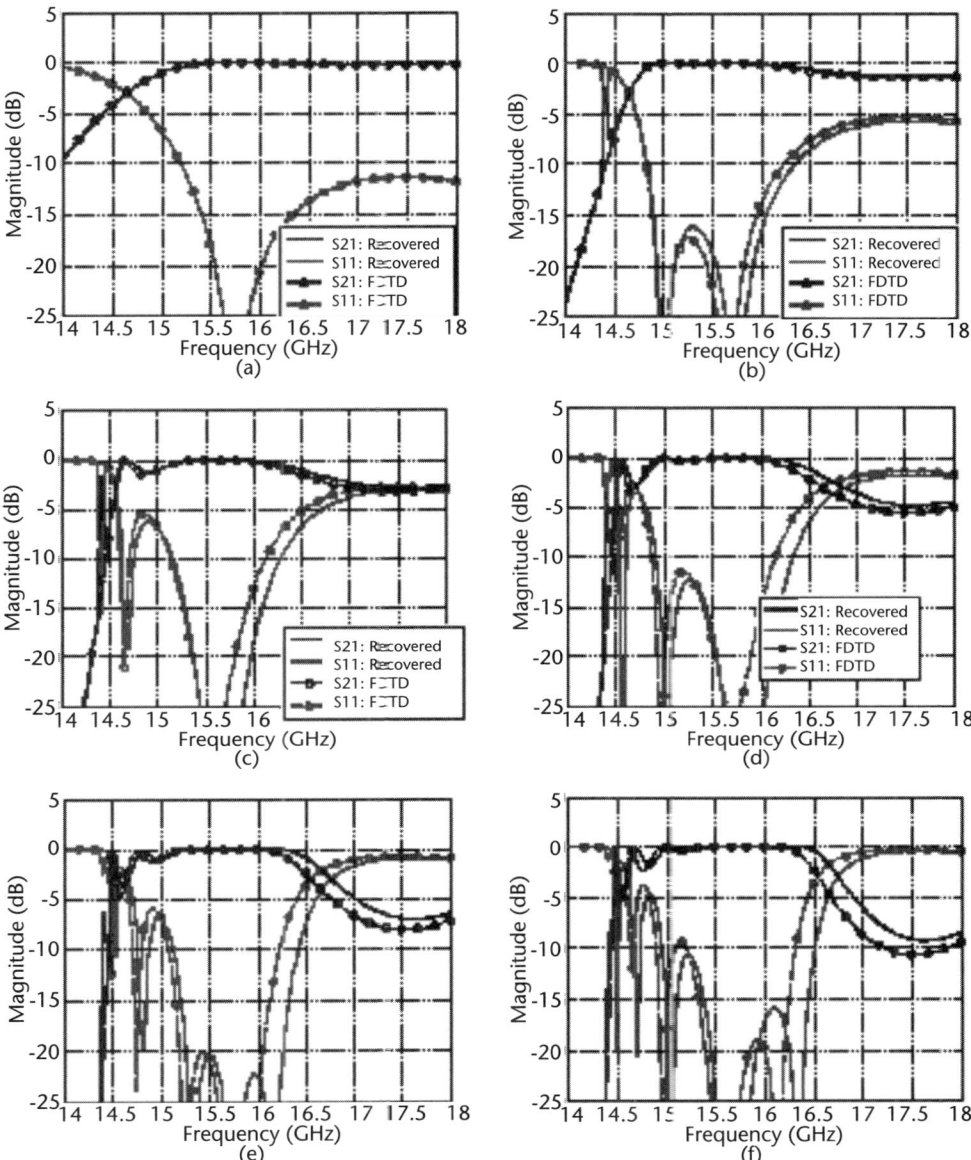

Figure 7.27 Comparison of the magnitude of the S-parameters, recovered from the retrieved n and Z of the one-layer slab (solid) and obtained from the direct FDTD simulations (circle dash) for one-layer up to six-layer slabs in (a) through (f).

7.3.3 Summary of the Difficulties Encountered Using the Inversion Approach for Effective Medium Characterization

In summary, the main difficulty encountered in the retrieval procedure arises during the process of selecting the correct branch from the multiple set of possible solutions. Although some guidelines are available for eliminating the nonphysical solutions, the unavoidable numerical artifacts present in the S-parameters, obtained either from experiments or from numerical simulations, often make it very difficult to apply these selection rules.

For example, enforcing the passivity condition on the extracted parameters, namely that n'', ε'', and $\mu'' \leq 0$, for all the admissible solutions, is not as easy as one would think. When values of n'', ε'', or μ'' are near zero, attempting to strictly enforce the condition that n'', ε'', and $\mu'' \leq 0$, simply leads to a rejection of all the solutions, because the parameters we desire may be so corrupted by the presence of numerical artifacts in the near-zero region, as to render them nonphysical. Consequently, from a practical point of view it may be better to work with a tolerance level and discard the solutions only when $n'' > \delta_1$, $\varepsilon'' > \delta_2$, or $\mu'' > \delta_3$, where δ_1, δ_2, and δ_3 are small positive numbers. We point out, however, that the choice of the values of these small parameters must be made with some care. For instance, if these parameters are chosen to be too large, an enforcement of the rules given above will not help eliminate the nonphysical solutions; on the other hand, if these parameters are too small, all solutions, including the correct ones, would be discarded. What is even worse, the optimal values for these parameters are problem-dependent, and they are strongly affected by the level of the noise present in the S-parameters, be they computed or measured.

Additionally, a frequency band can be found such that no solutions can satisfy the condition that either both ε'' and $\mu'' \leq 0$, or are smaller than some small positive numbers, in order to accommodate the numerical errors that are inevitably present in the S-parameters. This seemingly nonphysical region often appears adjacent to the DNG region, or near the frequency at which spikes and notches appear in the retrieved impedance, when S_{11} is small. Since none of the branches can be chosen to yield a physical solution in this region, the iterative process [3] that makes use of the continuity of the refractive index as a function of frequency fails as we move from one region to the next.

In addition, we have encountered some metamaterials structures for which multiple solutions still survive, even after applying all the selection rules alluded to above. In this situation, we cannot determine the correct solution by using the inversion approach we have described earlier.

The sensitivity analysis performed in [3] showed that the effective material parameters extracted for a given problem are very sensitive to the noise in the S-parameters, whenever either $|S_{11}|$ or $|S_{21}|$ is small. When $|S_{21}|$ is small, the extracted values of the refractive index turn out to be very sensitive to small perturbations of S_{21}. On the other hand, a small S_{11} generates spikes in the extracted impedance values, as may be clearly seen by referring to Figure 7.22 for the multilayer cases. As a result, the retrieved effective parameters turn out to be unreliable when either of the S-parameters is small.

7.4 EM Response of a Finite Artificial-DNG Slab with Localized Beam Illumination

In the following, we use the term "artificial-DNG" instead of simply "DNG" to describe the array we used in the simulations with localized beam or dipole illumination, in order to distinguish ours from the ideal DNG materials. In general, many physicists and engineers associate the usage of the latter term with a hypothetical material that has negative permittivity and permeability in nature. However, for most engineered structures, or metamaterials, such as the one studied in this work, their DNG nature is usually characterized by a set of effective material parameters, which are extracted from either the experimental data or simulation results for the normal incidence case. Therefore, these structures may not possess the DNG properties for oblique incidence angles and/or other polarizations, and the effective medium representation, derived for normal incidence, may not be valid in the general case.

7.4.1 Slab with Localized Beam Illumination

In this section, we study the EM response of a moderate-size, finite DNG array comprised of inclusions that are identical to those we studied previously, when the array is illuminated by a localized Gaussian beam incident either normally or from an oblique angle. We wish to investigate two interesting and unusual phenomena, namely the negative refraction and super-focusing effect of an artificial-DNG slab.

For the purpose of comparison, in terms of both the homogeneity of material and the sign of electric permittivity and magnetic permeability, we also study the EM response of a homogeneous dielectric slab ($\varepsilon_r = 4$), with the same dimensions as that of the artificial DNG array, when it is illuminated by the same Gaussian beam.

7.4.2 FDTD Model

The artificial DNG slab in our study is comprised of six layers, as shown in Figure 7.28, and consists of identical inclusions with the same periodicities as that studied previously. It contains a total of $38 \times 17 \times 6 = 3,876$ pairs of SSR and wire, and has the dimensions of 85.5 mm × 85 mm × 23.75 mm, corresponding to $4.3\lambda \times 4.3\lambda \times 1.2\lambda$ at the center frequency of interest (15 GHz). The slab, residing on the x-y plane, is illuminated by a Gaussian beam excitation located at one FDTD cell ($\Delta = 0.125$ mm) away from the bottom surface of the slab. The incident beam, whose beam maximum points toward the center of the slab, has a beam waist of 11 mm at the excitation plane, and covers a total of 34 pairs of inclusions (up to 10 and 5 inclusions along the x- and y-directions, respectively) in the first layer, within its spot size. Considering the anisotropic nature of the inclusions, we simulate the beam with three different settings of incident angles and polarizations, so as to trigger the DNG behavior of the slab at normal and oblique incidences. These three settings are described as follows:

1. Normal incidence: Wave is incident from $\theta = 180°$, with Ey polarization.
2. Oblique TM_z incidence: Wave is incident from ($\theta = 150°$, $\phi = 90°$) with Hx polarization.
3. Oblique TE_z incidence: Wave is incident from ($\theta = 150°$, $\phi = 0°$) with Ey polarization.

The computational domain has the dimensions of 85.5 mm × 85 mm × 72.75 mm, and it contains a total of $684\Delta \times 680\Delta \times 582\Delta$ with a uniform discretization

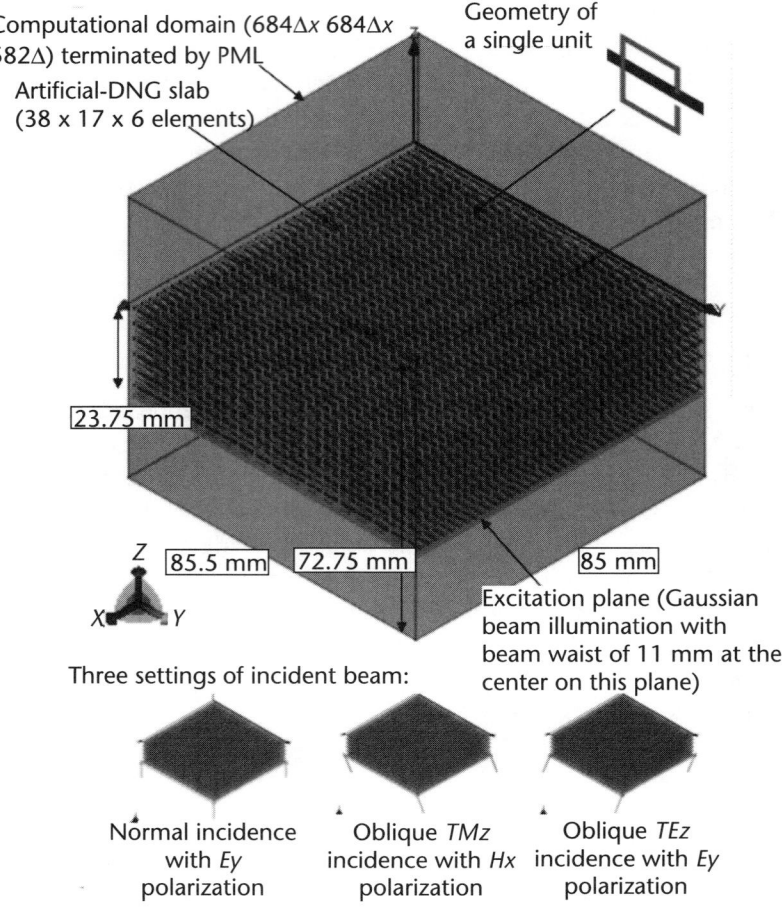

Figure 7.28 Geometry of a six-layer artificial DNG slab and three incident settings.

of 0.125 mm. At 15 GHz, the cell size is small, only $\lambda/160$, and this is necessary to accurately model the fine features of the inclusions. The computational domain is terminated by PMLs in all directions to absorb the outgoing waves. In this problem, the total number of FDTD cells, excluding the PMLs, is around 270 million, and the total number of unknown fields to be solved exceeds 1.6 billion. This, obviously, places a heavy burden on the computational resources that are not available in a single processor. Therefore, it is necessary to carry out the simulation by using a parallel FDTD code, running on multiple processors.

7.4.3 Total Transmission and Reflection Power Under Gaussian Beam Illumination

Before examining the detailed EM response of the artificial DNG array, it is necessary to first determine if the chosen spot size of the incident beam is suitable for the study of the macroscopic response of the metamaterial slab. For instance, if the spot size is too small, say smaller than or comparable to the periodicities of the inclusions, we can only observe the scattering characteristics contributed by a few inclusions or perhaps just by a single one. Various concepts commonly used to study the metamaterials, such as the effective medium theory, cannot be applied under these circumstances.

A convenient indicator of the macroscopic response of the array under certain illumination is the total transmitted or reflected power along different directions. The total transmitted and reflected powers along the specular angles can be computed by applying the near-to-far-transformation to the aperture fields on surfaces located at the transmission side of the slab and the scattered-field region behind the excitation plane, respectively. Similarly, the incident power along the propagation direction can be computed by using the aperture fields on the excitation plane when the array is removed. The ratios of the total transmission or reflection power and the incident power for the finite array case are then compared to the magnitude of transmission and reflection coefficients for the infinite array case, illuminated by a plane wave incident from the same angle. Figure 7.29 shows an excellent agreement between the transmission/reflection coefficients for the finite and infinite arrays for the normal incidence case. Figures 7.30(a, b) compare the results for the oblique incidence case of 30° off-normal, for TM_z and TE_z polarizations, respectively. As may be seen from the above figures, the match between these results is still good for an oblique incidence, though not as close as it was for the normal incidence case. It should also be noted that the scattering

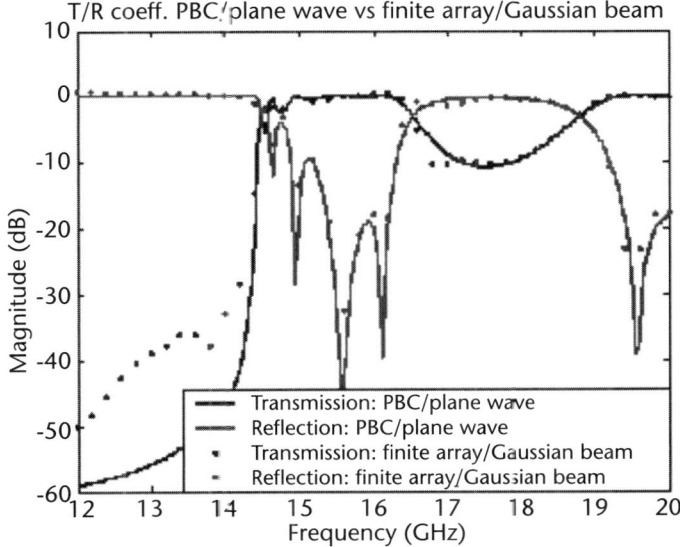

Figure 7.29 Comparison of the magnitude of transmission and coefficient coefficients for the infinite array under plane wave illumination (solid lines) and finite array under Gaussian beam illumination (dots) at normal incidence for Ey polarization.

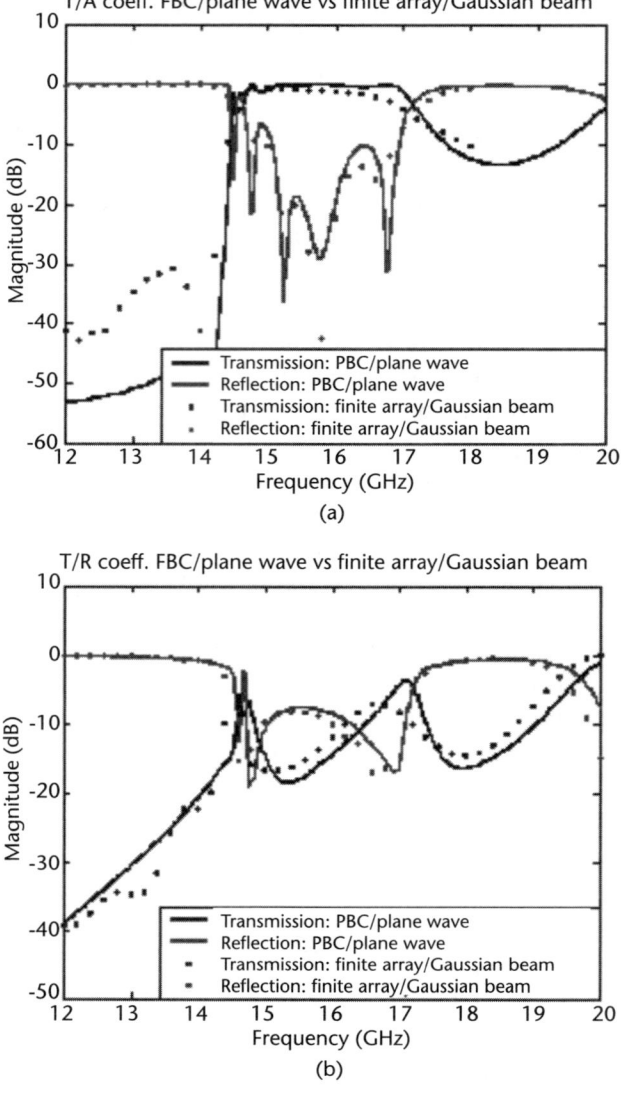

Figure 7.30 Comparison of the magnitude of transmission and coefficient coefficients for the infinite array under plane wave illumination (solid lines) and the finite array under Gaussian beam illumination (dots) at oblique incidence (30° off-normal), for (a) TM_z polarization and (b) TE_z polarization.

characteristics at oblique incidence with TE_z polarization is quite different from that of the normal incidence case; therefore, the DNG behavior might not be observed in the former case, for angles that are not close to normal.

Since a good agreement is observed between the magnitude of the transmission and reflection coefficients for the finite array case with a Gaussian beam illumination, and for the infinite array case with plane wave illumination for both the normal and oblique incidence cases, we conclude that the Gaussian beam with

7.4.4 EM Response of the Artificial-DNG Slab at Normal Incidence with Ey Polarization

We begin by examining the field distribution at 15.3 GHz, where the array is totally transmitting and has the refractive index closest to -1 ($n = -1.01 - 00004j$, $Z = 1.30 - 00004j$, $\varepsilon_{reff} = -0.776 - 0.0006j$ and $\mu_{reff} = -1.32 - 0.0001j$). The effective material parameters are computed by applying the inversion approach, described earlier in Section 7.3, to the S-parameters of the infinite, periodic array of single layer at normal incidence.

The magnitudes and phases of E_y at 15.3 GHz on the y-z (E-plane) and x-z planes (H-plane) are shown in Figures 7.31 and 7.32, respectively, for three different configurations: (a) and (b) with artificial-DNG slab; (c) and (d) for the artificial-DNG slab replaced by a dielectric slab of $\varepsilon_r = 4$ of the same size; and (e) with all structures removed (i.e., only for free space). The observation planes,

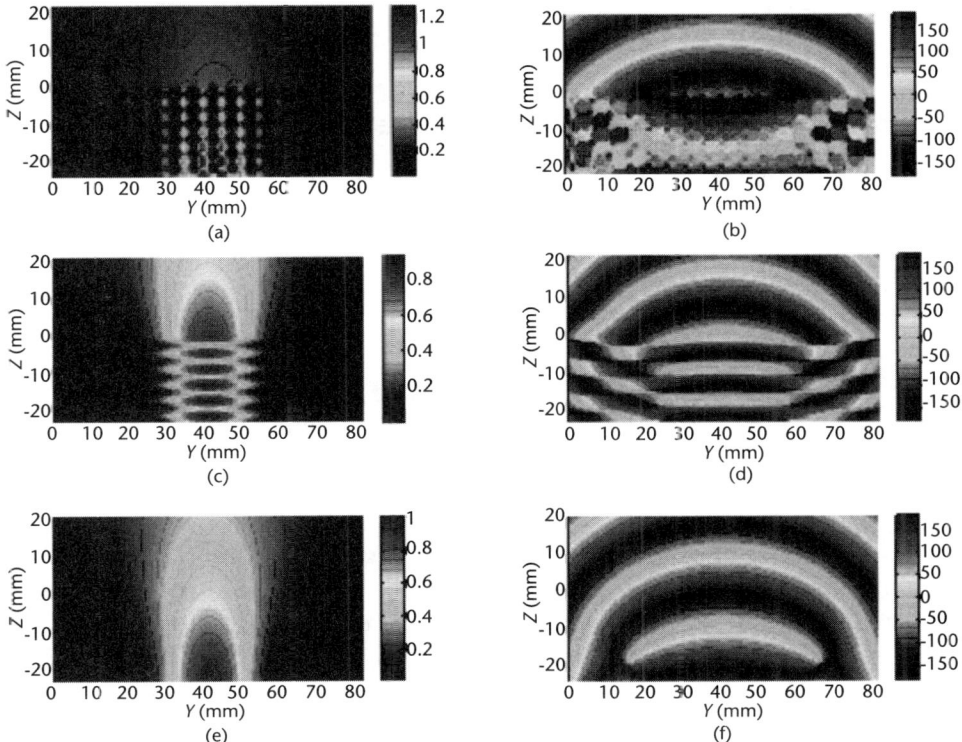

Figure 7.31 The magnitude (left) and phase (right) of Ey at 15.3 GHz on the YZ plane (E-plane), for three different configurations at normal incidence: (a) and (b) with DNG slab; (c) and (d) for the DNG slab replaced by a dielectric slab of $\varepsilon_r = 4$ of the same size; and (e) and (f) only for free space. The DNG or dielectric slab occupies the region between $z = -23.75$ and 0 mm; the excitation plane is on $z = 0$ mm; and the region above 0 mm is free space.

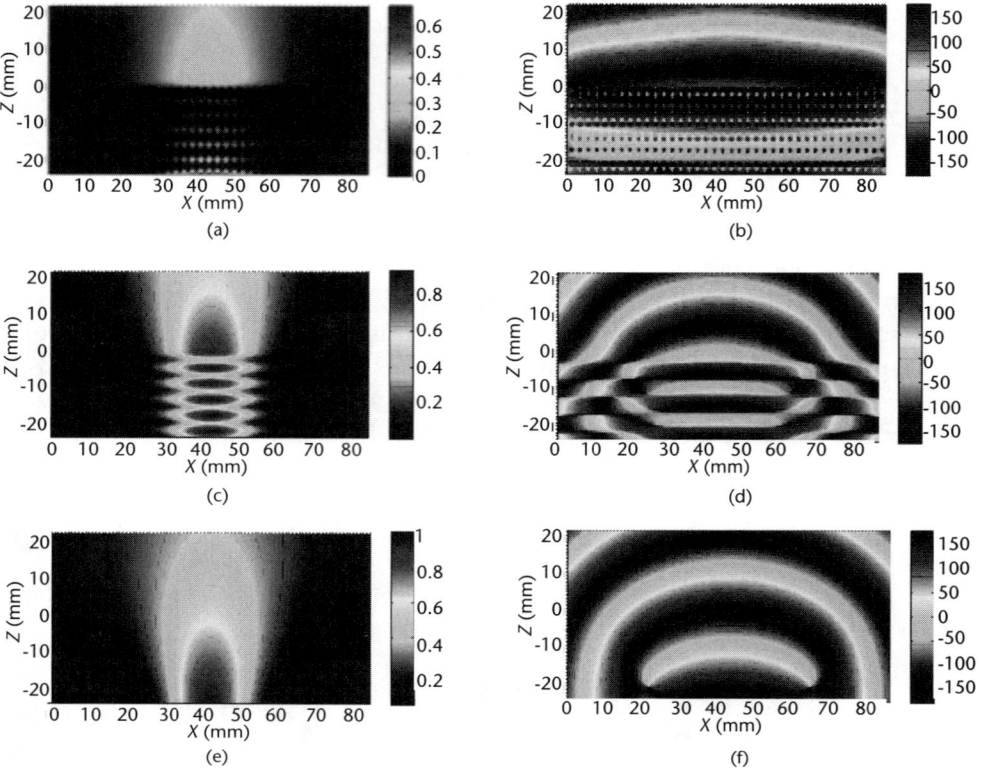

Figure 7.32 The magnitude (left) and phase (right) of Ey at 15.3 GHz on the XZ plane (H-plane), for three different configurations at normal incidence: (a) and (b) with DNG slab; (c) and (d) for the DNG slab replaced by a dielectric slab of $\varepsilon r = 4$ of the same size; and (e) and (f) only for free space. The DNG or dielectric slab occupies the region between $z = -23.75$ and 0 mm; the excitation plane is on $z = 0$ mm; and the region above 0 mm is free space.

with the axis of the Gaussian beam lying within, are located at the center of the computational domain, and their z-coordinates range from -24 mm to 22 mm, with the slab occupying the region between $z = -23.75$ mm to 0 mm. The region above $z = 0$ mm is free space, at the transmission side of the slab. From the plots for the magnitudes in both of these figures, it is evident that the magnitude decays smoothly in the free-space region. Inside the dielectric slab, we clearly see an interference pattern, formed by the standing waves that are caused by the reflection at the two dielectric/air interfaces. On the other hand, the distribution inside the artificial DNG array is dominated by strong fields arising from the discontinuities presented by the inclusions, and no clear interference patterns could be seen in these field plots.

However, the phase distribution clearly demonstrates the backward wave nature of the phase velocity, which is one of the unusual characteristics of the DNG materials as well as of backward leaky wave structures. A positive phase velocity is indicated by the decreasing trend of phase along the propagating direction (i.e., the positive z-direction) and vice versa for the negative phase velocity. From the

7.4 EM Response of a Finite Artificial-DNG Slab with Localized Beam Illumination 215

phase plots of the above figures, a positive phase velocity can be clearly observed in the free-space region. Inside the dielectric slab, the phase velocity is found to be smaller than that in the free space, which is indicated by the smaller spacing between the phase fronts, though it is still positive. Inside the artificial DNG slab, we can clearly see that the phase velocity is negative. Also, the shape of the phase front inside the artificial DNG slab curves inward towards the source, which is

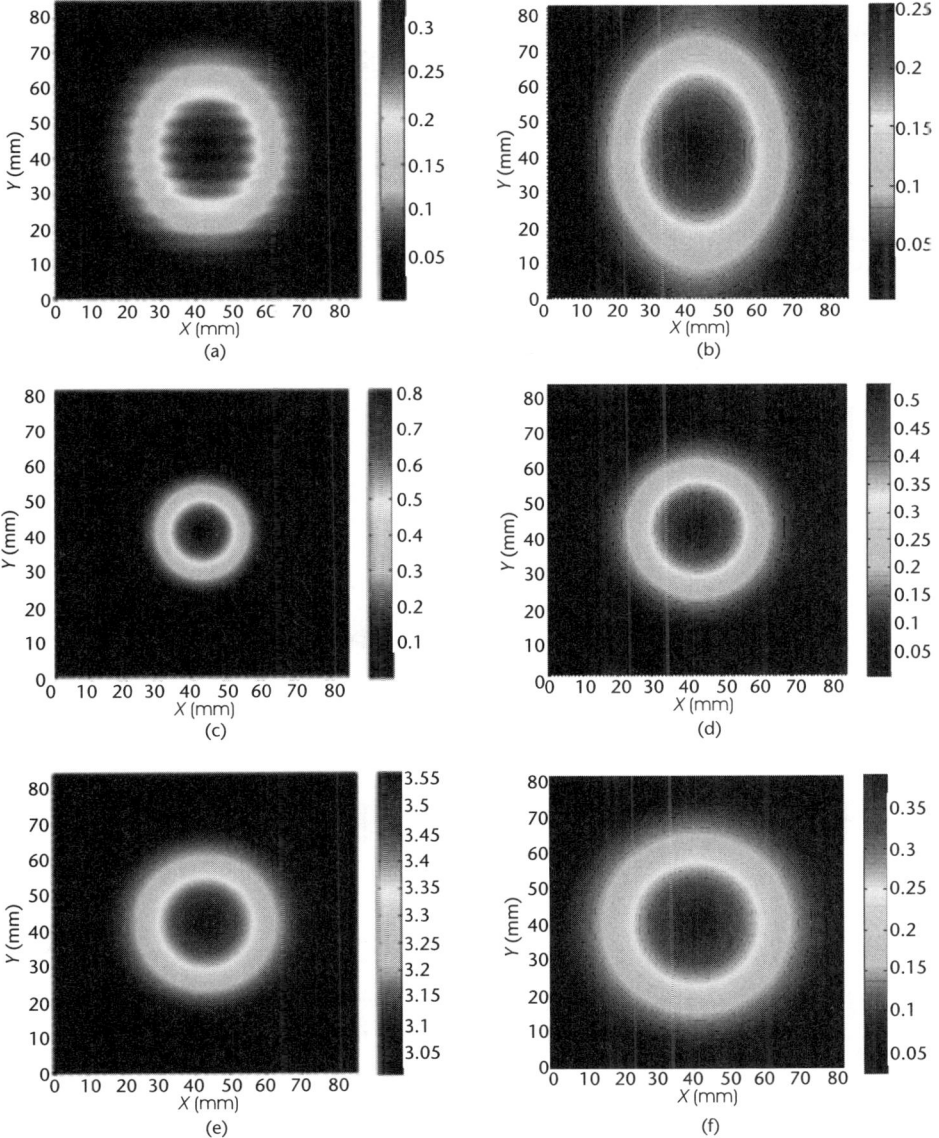

Figure 7.33 Magnitude of Ey at 15.3 GHz on the XY plane at 2 mm (left) and 19 mm (right) from the top surface of the slab, for three different configurations at normal incidence: (a) and (b) with DNG slab; (c) and (d) for the DNG slab replaced by a dielectric slab of $\varepsilon_r = 4$ of the same size; and (e) and (f) only for free space.

located at $z = -24$ mm [indicated by the planar wavefront near the bottom center of Figures 7.31(f) and 7.32(f) for incident field], while the shapes of the phase front inside the free space and the dielectric slab always curve away from the source.

The magnitudes of Ey at 15.3 GHz in the transverse plane at 2 mm and 19.5 mm away from the top surface of the slab are also plotted in Figures 7.33(a–f) for three different configurations. We see from Figure 7.33(a) that when the observation plane is only 0.1λ away from the slab, we can still observe the granularity of the field due to the inclusions inside the array, thought these variations appear to die out as we move the observation plane to about 1λ away [see Figure 7.33(b)]. For the other two configurations, no similar field variations can be observed because of the truly homogeneous nature of the materials themselves. It should be noted that the transverse field distribution for the artificial DNG array case is highly astigmatic, and the distribution is always found to suffer an elongation in the y-direction for this structure.

Next, we investigate the focusing effect by the artificial DNG slab under the illumination of Gaussian beam at normal incidence. It is known that when the refractive index of an isotropic, homogeneous slab is equal to -1 for a matched medium ($\varepsilon_r = \mu_r = -1$), an image can be formed at a distance h_2 from the slab, where $h_2 = d - h_1$, and h_1 is the distance of the source from the slab of thickness d. Its image properties would be quite different from those of the image formed by a conventional lens. For instance, it has been predicted that the phase at the image would be restored to its value at the source; and the image size can be smaller than a

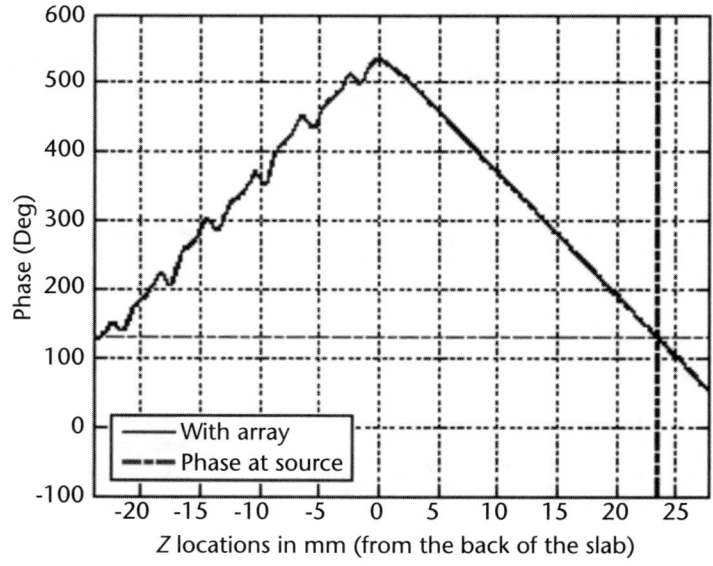

Figure 7.34 Phase of E_y at 15.3 GHz along the beam axis from $z = -24$ to 27.75 mm, with the DNG slab occupying the region between $z = -23.75$ and 0 mm. Notice that the measured phase restores to its value at the source at $z = 23.5$ mm, indicated by the vertical black dashed lines, which is the expected image location when the refractive index of the slab is equal to -1.

wavelength, because the evanescent waves are amplified by a real DNG slab to form the image. These two properties have encouraged many research groups throughout the world to attempt to realize this type of super lens that could possibly breach the diffraction limit associated with a conventional lens. Nevertheless, we point out that the slab we study is not a matched medium ($\varepsilon_{reff} = -0.776 - 0.0006j$ and $\mu_{reff} = -1.32 - 0.0001j$) at 15.3 GHz where the refractive index equals -1;

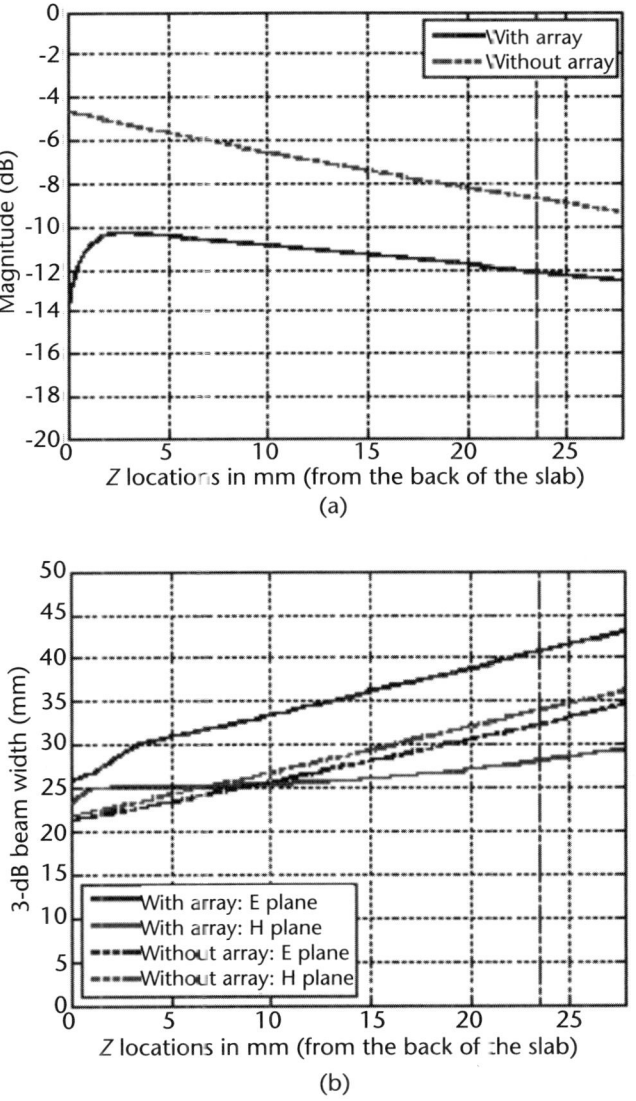

Figure 7.35 (a) Magnitude of Ey at the beam axis and (b) the 3-dB beam width on the E- and H-planes along the longitudinal direction at 15.3 GHz, inside the transmission region of the DNG slab. The corresponding magnitude and beam width with all structures removed are also plotted in dotted lines. The vertical dash line at z = 23.5 mm indicates the expected image location when the refractive index of the slab is equal to -1.

however, we still expect that the slab would form an image, though the amplitude may be weaker.

Figure 7.34 shows the phase of E_y at 15.3 GHz along the beam axis, starting from the source, which is at 0.25 mm ($z = -24$ mm) below the artificial DNG slab. As can be seen from the above figure, the phase propagates backward within the slab spanning the space between $z = -23.75$ and 0 mm, and then it becomes forward in free space, and restores to the same value as that of the source at $z = 23.5$ mm, which is the expected location of the image for this source location, if the slab had a refractive index of -1.

Next, we examine the change of the magnitude of E_y near the image position along both the longitudinal and transverse directions, by inspecting the magnitude of E_y along the beam axis [see Figure 7.35(a)], and the 3-dB beamwidths in the E- and H-planes [see Figure 7.35(b)], respectively, for the emerging beam. Neither of these plots indicates any localization of fields near the expected image position, at which all rays emanating from the source should pass. In addition, we observe a larger 3-dB beamwidth in the E-plane as compared to that on the H-plane, which

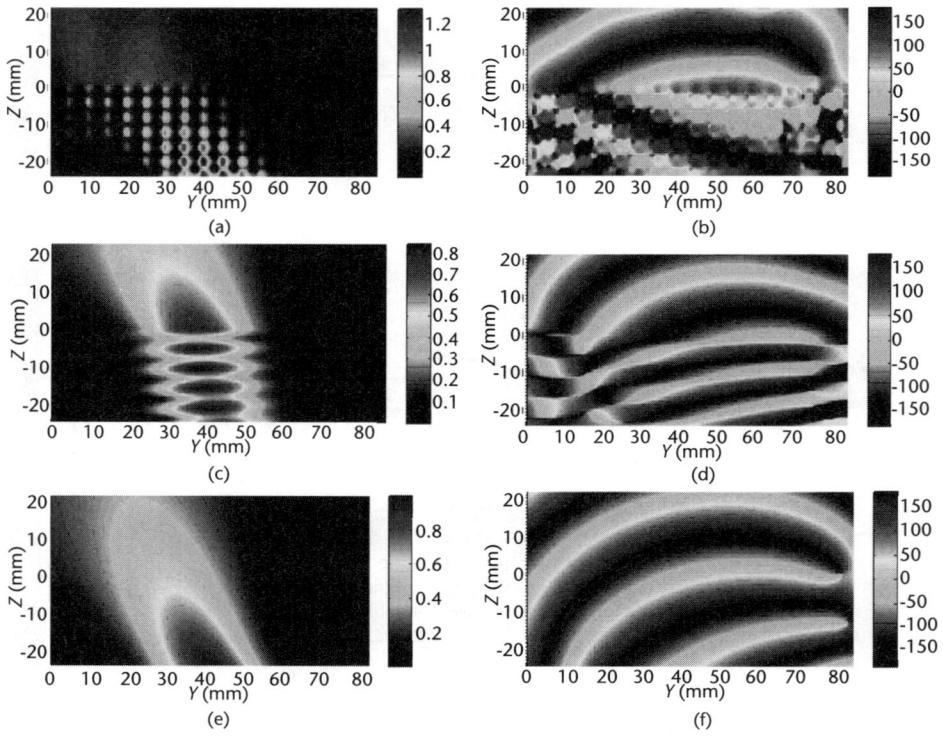

Figure 7.36 The magnitude (left) and phase (right) of E_y at 15.3 GHz on the YZ plane (E-plane), for three different configurations at oblique TM_z incidence (30° off-normal): (a) and (b) with DNG slab; (c) and (d) for the DNG slab replaced by a dielectric slab of $\varepsilon_r = 4$ of the same size; and (e) and (f) only for free space. The DNG or dielectric slab occupies the region between $z = -23.75$ and 0 mm, the excitation plane is on $z = 0$ mm, and the region above 0 mm is free space.

is consistent with the field distribution on the transverse plane, shown previously in Figure 7.33.

To summarize the results of our investigation on the focusing effect, we do observe a restoration of phase at the expected image position; however, we do not find any localization of field in its neighborhood. Therefore, we conclude that the focusing effect is not fully exhibited for this artificial DNG slab at the frequency where the effective refractive index has been found to be −1, under the illumination of the Gaussian beam at normal incidence.

7.4.5 EM Response of the Artificial-DNG Slab at Oblique TM_z Incidence Coming from ($\theta = 150°, \phi = 90°$) with Hx Polarization

Another unusual phenomenon associated with the negative refraction is the bending of rays towards the same side of the normal at the interface, when the rays enter a DNG slab from free space, or vice versa, at an oblique incidence, as shown in Figure 7.36. The Snell's law can still be applied by having negative values of refractive index, found inside a DNG medium. Since the transmission and reflection characteristics of the artificial DNG array at 30° off-normal incidence for the

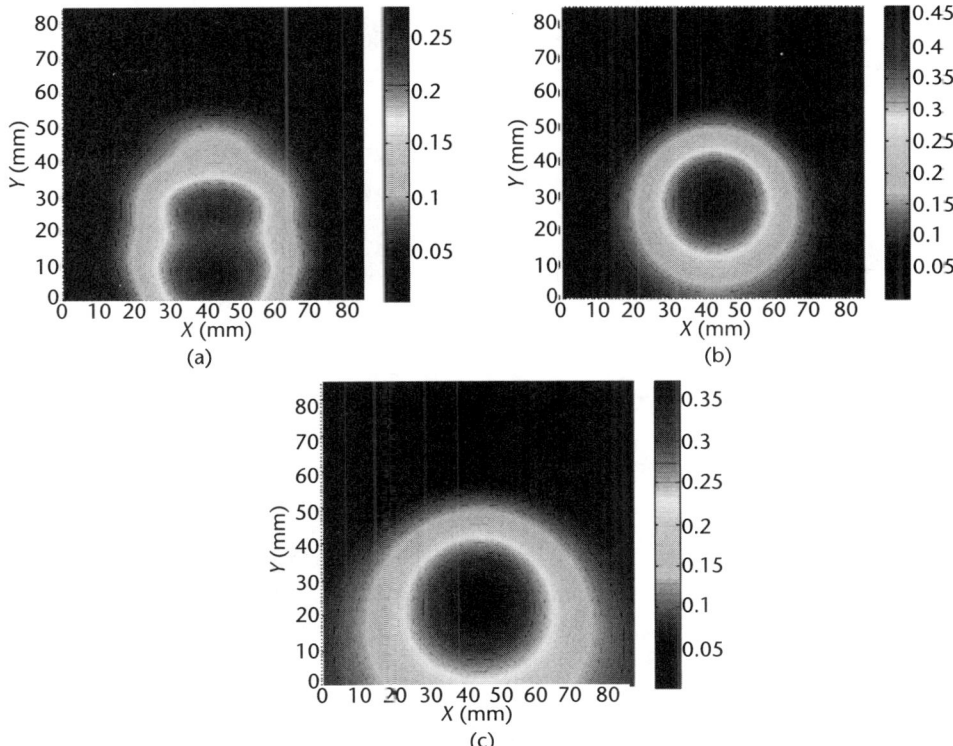

Figure 7.37 Magnitude of E_y at 15.3 GHz on the XY plane at 19 mm above the slab at oblique TM_z incidence (30° off-normal) for three different configurations: (a) with DNG slab; (b) for the DNG slab replaced by a dielectric slab of $\varepsilon_r = 4$ of the same size; and (c) only for free space.

TMz polarization are similar to those of the same array at normal incidence, the DNG behavior is expected to be retained for the oblique incidence case for TM_z polarization.

Figure 7.36(a–f) plot the magnitudes and phases of E_y at 15.3 GHz on the y-z plane (E-plane), which is also the plane of incidence, for three different configurations that are identical to those appearing in Figure 7.31. As shown in Figure 7.36(e, f), the incident beam, excited at $z = -24$ mm with its maximum at $y = 42.5$ mm (center of the computational domain), propagates along the positive z-direction, with decreasing y. A rough estimate of the beam direction can be made by tracking the beam maxima. For the dielectric slab, the beam first bends slightly towards the normal when it enters the slab, and then bends away from the

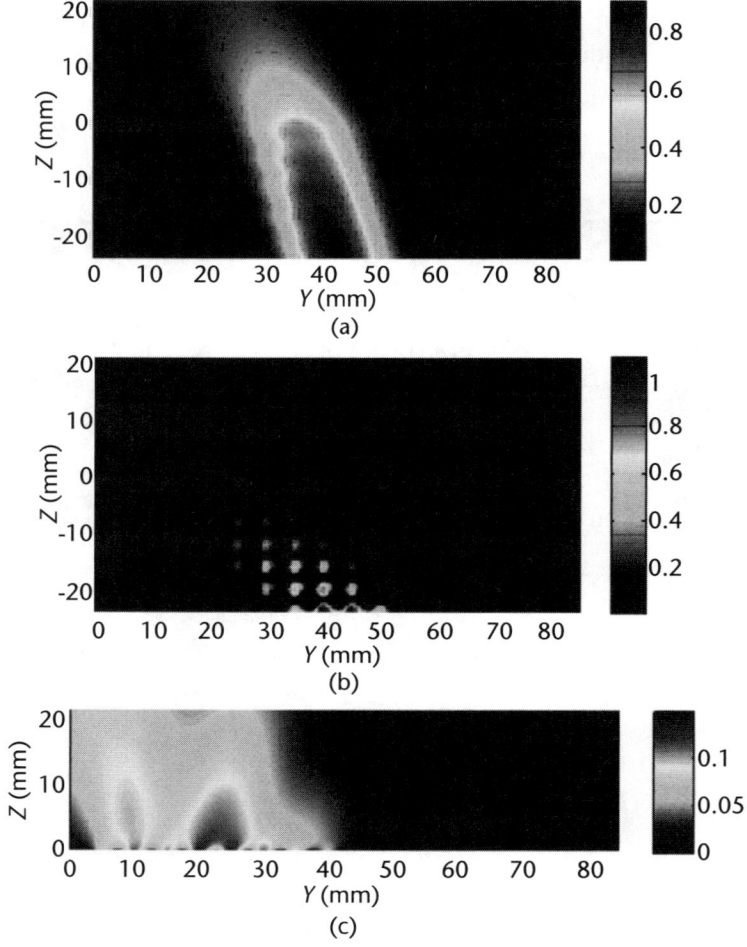

Figure 7.38 The normalized magnitude of S_z at 15.3 GHz on the YZ plane (E-plane) between $z = -24$ mm and 0 mm at oblique TM_z incidence (30° off-normal) for two different configurations: (a) dielectric slab and (b) DNG slab. (c) is the figure extracted from (b) over the free-space region above the DNG slab.

normal when it exits the slab, which is an indication of positive refraction of the medium. For the artificial DNG slab [see Figure 7.36(a)], it is unable to define a continuous beam path inside the slab because of the strong irregularity of fields in the inhomogeneously filled slab. Nonetheless we can still assert the fact that the beam does not bend towards the same side of the normal when it enters the slab, as we would expect when we have negative refraction, since the beam lies on the left half of the figure, all through the slab and, of course, when it exits the slab in the free-space region.

The magnitudes of E_y on the x-y plane at 19.5 mm above the slab are shown in Figure 7.37(a–c) for the same three configurations. From Figure 7.37(a), we observe the presence of more than one beam along the y-direction, which is different from those observed for the cases of homogeneous dielectric slab and with free space, or even for the artificial DNG slab illuminated by a normally incident beam. Since we do not observe any significant shift of the beam in the positive y-direction as compared to the incident beam in Figure 7.37(c), once again negative refraction is not seen to be present in this test example.

The distribution of the z-component of the Poynting vector, which is a more accurate representation of the energy flow, is also shown in Figure 7.38(a, b) for the dielectric and artificial DNG slabs, respectively. The bending of the beam at the entrance and exit the dielectric slab is seen much clearly because of the absence of an interference pattern; however, the path of the energy flow is still not defined precisely inside the artificial DNG slab. Figure 7.38(c) shows the z-component of the Poynting vector in the free-space region above the artificial DNG slab. We observe that there are multiple beams emerging from different positions at

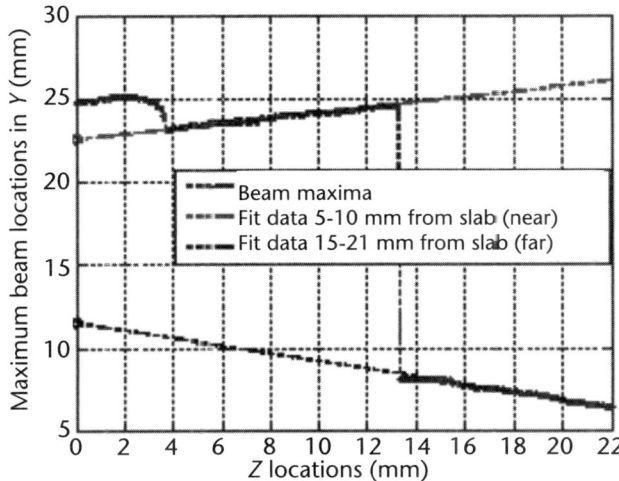

Figure 7.39 The transmitted beam maxima locations versus the longitudinal direction at 15.4 GHz for TM_z incidence of DNG slab. These data are separated into two groups according to the distances from the slab and utilized to track the exit location of the beam at the slab surface at $z = 0$ mm.

the surface of the slab and traveling in different directions. Nevertheless, all the emerging beams appear to exit from the left side of the slab, and negate, once again, the presence of negative refraction of the beam at the interface between the DNG medium and free space.

We further determine the exit locations of the transmitting beam at the artificial DNG slab by tracing the path at which the Poynting vector is at its maximum along the transverse direction, and then projecting the path backward to the top surface at the slab by using a linearly curve-fitting. Figure 7.39 plots the y-coordinates of the beam maximum along the longitudinal direction on the y-z plane at 15.3 GHz. Large jumps in the locations of the maxima occur near $z = 13$ mm, which can be explained by the existence of two emerging beams, as shown in Figure 7.38(c). Each beam decays at a different rate, causing one beam to dominate in the region near the slab and the other to be strong in the region further away from the slab. By employing the above steps on the fields in these two regions separately, we find that the two beams exit at $y = 11.5$ mm and 22.5 mm, respectively, which are both on the opposite side of the normal where the incident beam first enters the slab at $y = 42.5$ mm.

The exit locations of the beams for the artificial DNG slab, along with those for the dielectric slab as well as free space at $z = 0$ mm, are plotted in Figure 7.40 at frequencies within the first transmission band of the artificial DNG slab, ranging from 14.6 to 16.2 GHz. The refractive indices increase monotonically from -1.89 to -0.45 within this frequency band. We see that only a single beam is present for the dielectric slab as well as for the free-space case, and a single exit location can be determined by following the same steps as described earlier. From the above figure, we deduce that the refracted beam always bends towards the normal at the dielectric/free-space interface, which is expected for a positive-index medium. Except at 14.6 GHz, the refracted beam for the artificial DNG slab always bends

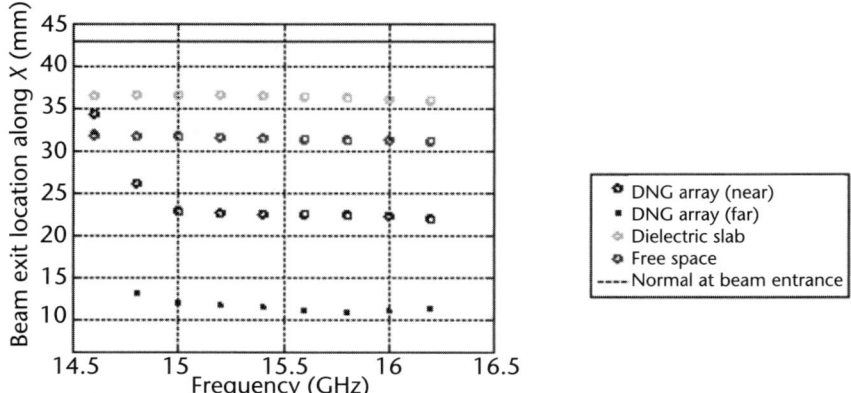

Figure 7.40 The exit location of the beam at the top surface of the slab on $z = 0$ at TM_z incidence, for three different configurations: the beam that dominates in the region near to and far away from the DNG slab; with dielectric slab; and only for free space. The incident beam enters the slab at $x = 42.75$ mm (black line).

away from the normal at the interface between the DNG medium and free space, which is contradictory to what is expected for a negative index medium.

In contrast to the magnitude distribution of the field, or the energy flow, the phase distribution inside the artificial DNG slab clearly exhibits negative refraction. As can be seen in Figure 7.36(b), the normal of the phase front bends towards the same side of the normal of the slab at both interfaces between the DNG medium and free space. In addition, we observe that the backward wave propagates within the artificial DNG slab, which was also true for the normal incidence case.

7.4.6 EM Response of the Artificial-DNG Slab at Oblique TE_z Incidence Coming from $\theta = 150°$, $\phi = 0°$ with E_y Polarization

Since the transmission and reflection characteristics for the incident angle of $\theta = 150°$ and $\phi = 0°$ and TE_z polarization are quite different from those for the normal incidence case, the DNG behavior may not be observed in this case.

We begin by examining the field distribution at 15.3 GHz, where the array is partially transmitting and partially reflecting for the present illumination. Figures 7.41(a–f) show the magnitudes and phases of E_y at 15.3 GHz on the x-z plane (H-plane), which is the plane of incidence, for the same configurations as studied

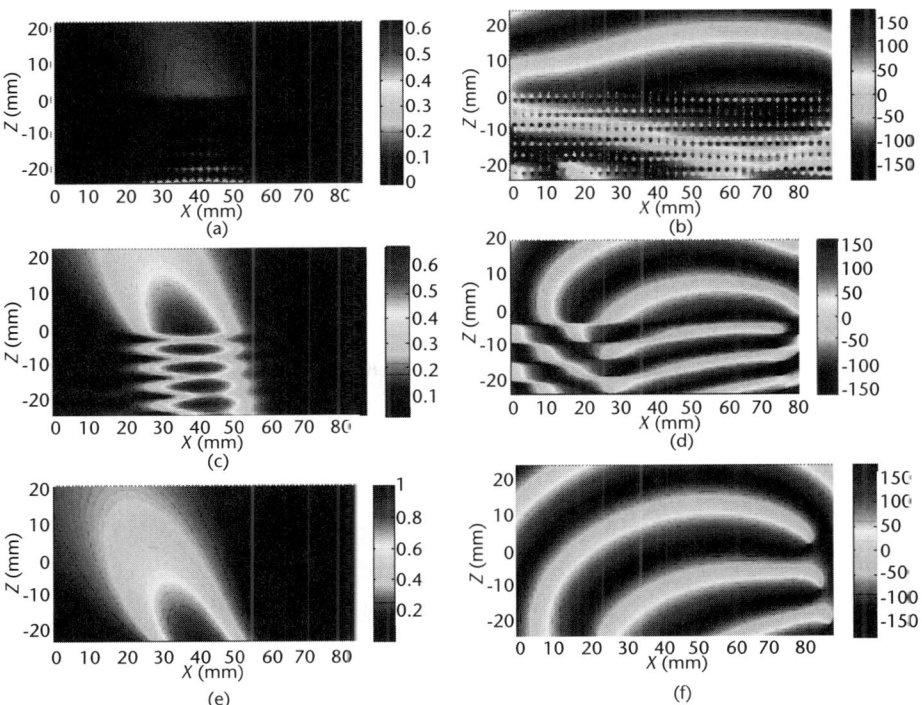

Figure 7.41 The magnitude (left) and phase (right) of E_y at 15.3 GHz on the x-z plane (H-plane) at oblique TE_z incidence (30° off-normal) for three different configurations: (a) and (b) with DNG slab; (c) and (d) for the DNG slab replaced by a dielectric slab of $\varepsilon_r = 4$ of the same size; and (e) and (f) only for free space. The DNG or dielectric slab occupies the region between $z = -23.75$ and 0 mm; the excitation plane is on $z = 0$ mm; and the region above 0 mm is free space.

previously. From the plots of Figure 7.41(a, b), we can see that the beam energy propagates in the direction close to the normal within the artificial DNG slab, while the phase propagates backwards within the slab, and the normal to the phase bends towards the same side of the normal of the slab at the interface between the DNG medium and free space. In contrast to the oblique incidence case when the polarization was H_x, only a single beam emerges from the artificial DNG slab. It is further confirmed by the E_y distribution on the transverse plane which is at 19.5 mm above the slab [see Figure 7.42(a)]. In addition, we can deduce from Figure 7.42(a, c) that the beam bends towards the normal when it enters the slab.

Figure 7.43 plots the exit locations of the transmitting beam at the artificial DNG slab, along with those for the dielectric slab, as well as free space, at $z = 0$ mm, for frequencies ranging from 14.6 to 16.2 GHz. From the results, we deduce that the refracted beam bends towards the normal at the DNG medium/free-space interface, indicating a positive refraction where the slab has a refractive index of greater than unity. Below 15.4 GHz, the beam travels close to the normal within

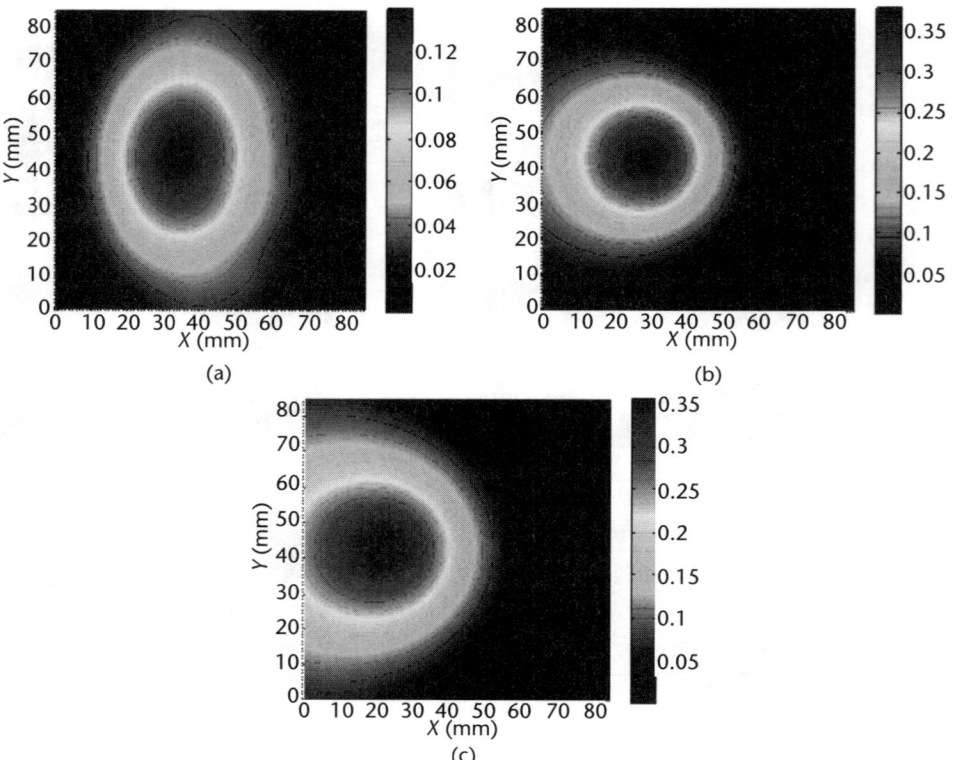

Figure 7.42 The exit location of the beam at the top surface of the slab on $z = 0$ for three different configurations: (a) the beam that dominates in the region near to and far away from the DNG slab; (b) with dielectric slab; and (c) only for free space. The incident beam enters the slab at $x = 42.75$ mm (black line).

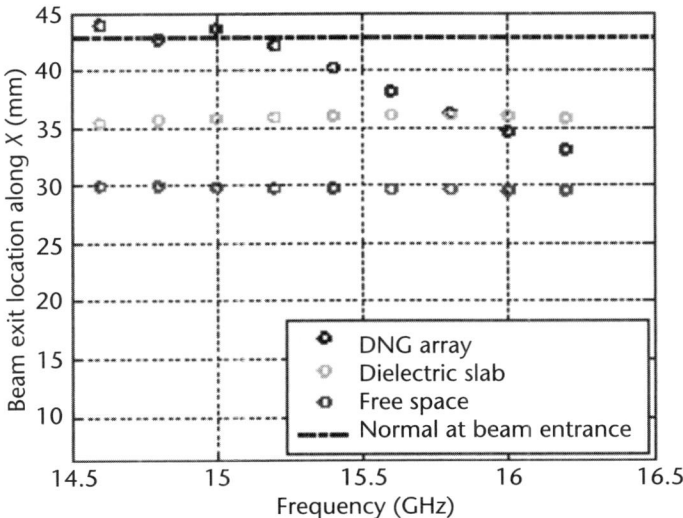

Figure 7.43 The exit location of the beam at the top surface of the slab on $z = 0$ at TEz incidence, for three different configurations: with DNG slab; with dielectric slab; and only for free space. The incident beam enters the slab at $y = 42.5$ mm (black line).

the artificial-DNG slab. So far, none of the above results has demonstrated the bending of beam towards the same side of the normal at the interface between the DNG medium and free space.

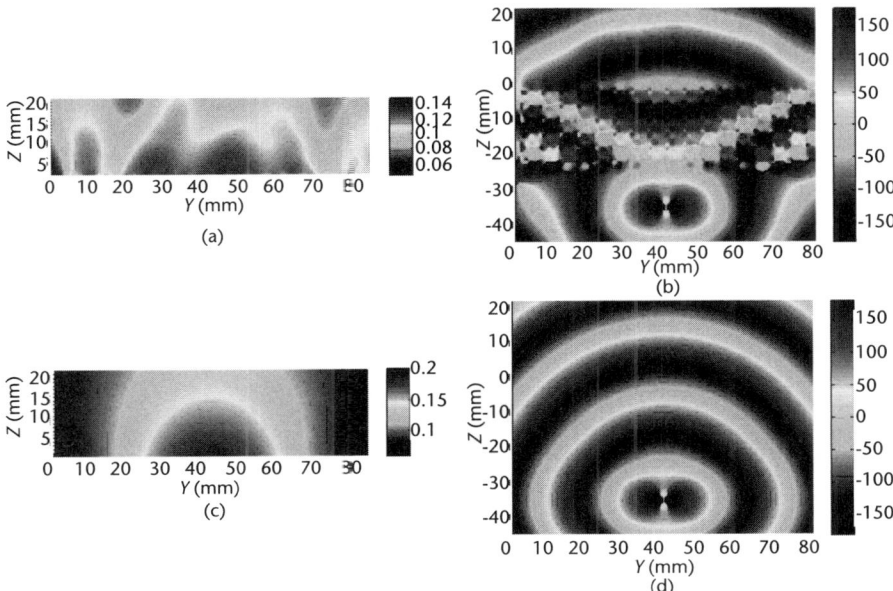

Figure 7.44 The magnitude (left) and phase (right) of E_y at 15.3 GHz on the YZ plane (E-plane), for two different configurations with dipole excitation at $z = -32.35$ mm: (a) and (b) with DNG slab; (c) and (d) only for free space. The DNG slab occupies the region between $z = -23.75$ and 0 mm, and the region above 0 mm is free space.

7.4.7 EM Response of a Finite Artificial-DNG Slab Excited by Small Dipole

In the previous sections, we have extensively studied the EM response of the artificial-DNG array illuminated by a Gaussian beam, with a spot size on the order of the wavelength. It is also worthwhile to study the focusing effect of the artificial DNG slab when a small source is used for excitation.

The artificial DNG array and all the FDTD settings remain the same as those employed previously in this section except that the Gaussian beam excitation is now replaced with a y-oriented dipole. The excitation is applied at the edge of the FDTD cell at the center of the x-y plane and located at 11.5 mm below the slab ($z = -32.25$ mm). At this source location, an image is expected to form at 12.25 mm above the slab with a refractive index of -1.

Figures 7.44(a, b) and 7.45(a, b) show the magnitudes and phases of E_y at 15.3 GHz on the y-z (E-plane) and x-z (H-plane) planes, respectively. The field distributions, with the artificial DNG array removed, are also shown in Figures 7.44(c, d) and 7.45(c, d) for comparison. Note that the magnitude is only shown for the transmission region 1.5 mm above the slab, in order to have a clear view of the transmitted field. From the above figures, we notice a highly astigmatic amplitude distribution in the transmission region of the artificial DNG slab. In the H-plane, the field distribution is much narrower than that when the artificial DNG slab is removed, while on the E-plane, the field distribution is much broader than that without the artificial DNG array. In addition, we note that there are multiple

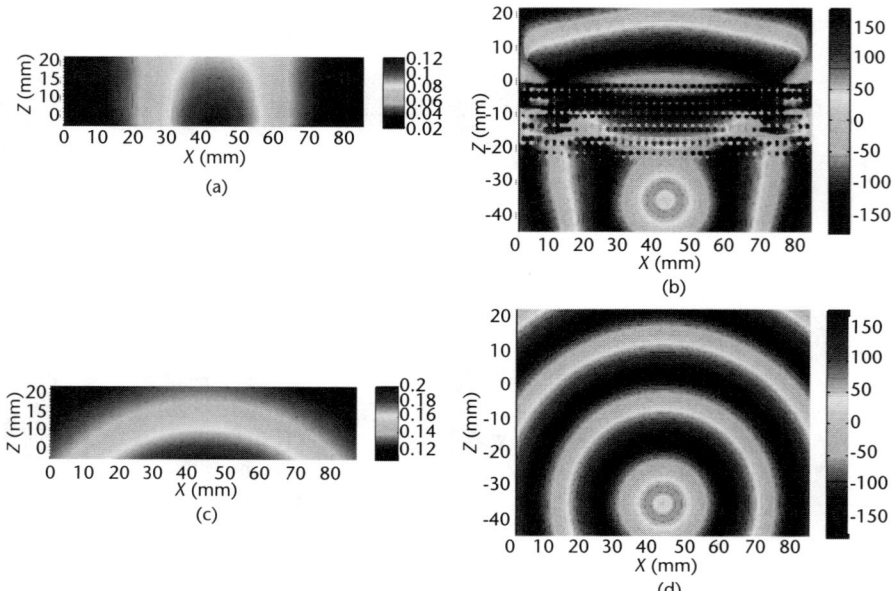

Figure 7.45 The magnitude (left) and phase (right) of E_y at 15.3 GHz on the XZ plane (H-plane), for two different configurations with dipole excitation at $z = -32.35$ mm: (a) and (b) with DNG slab; (c) and (d) only for free space. The DNG slab occupies the region between $z = -23.75$ and 0 mm, and the region above 0 mm is free space.

nodes on this plane. This broadening and narrowing of the field distribution along different directions are also observed when the array is illuminated by a Gaussian beam at normal and oblique incidence, while the existence of multiple nodes along the y-direction is only observed at oblique TM_z incidence of Gaussian beam. On the other hand, in common with the case of a Gaussian beam illumination, the phase distribution inside the artificial DNG slab clearly demonstrates the presence of negative refraction accompanied by a backward phase velocity.

Figure 7.46(a–c) shows the magnitude of E_y on the x-y plane at 12.25 mm above the slab at 14.8, 15.3, and 16.0 GHz, respectively. The corresponding refractive indices of the slab at these frequencies are -1.54, -1.01, and -0.554; therefore, an image is expected to form on the observation plane only at the second frequency. From the magnitude plots, we notice that multiple nodes, with a difference in the relative strength, are formed at nearly the same locations for all three frequencies, despite the large differences in the refractive indices. No clear image can be observed at 15.3 GHz. We also plot the magnitude of E_y at the centre axis, and the 3-dB beamwidth on the H-plane at 15.3 GHz in Figure 7.47(a, b), respectively. We find that none of these plots shows any localization of the field near the expected image location.

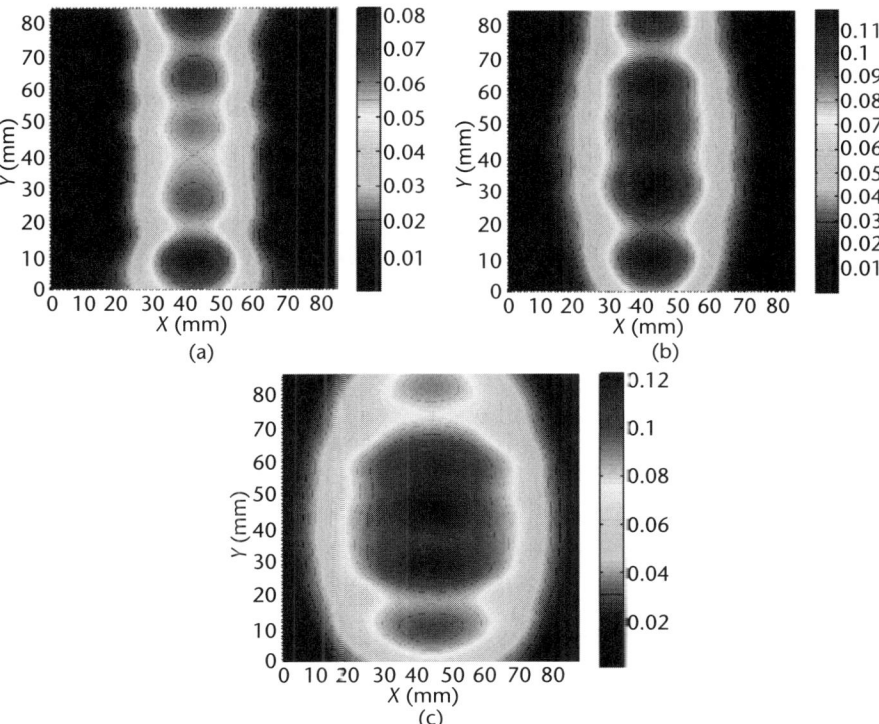

Figure 7.46 Magnitude of E_y on the XY plane at 12.25 mm above the DNG slab at frequencies of (a) 14.8 GHz, (b) 15.4 GHz, and (c) 16.0 GHz with dipole excitation at 11.5 mm below the bottom surface of the slab.

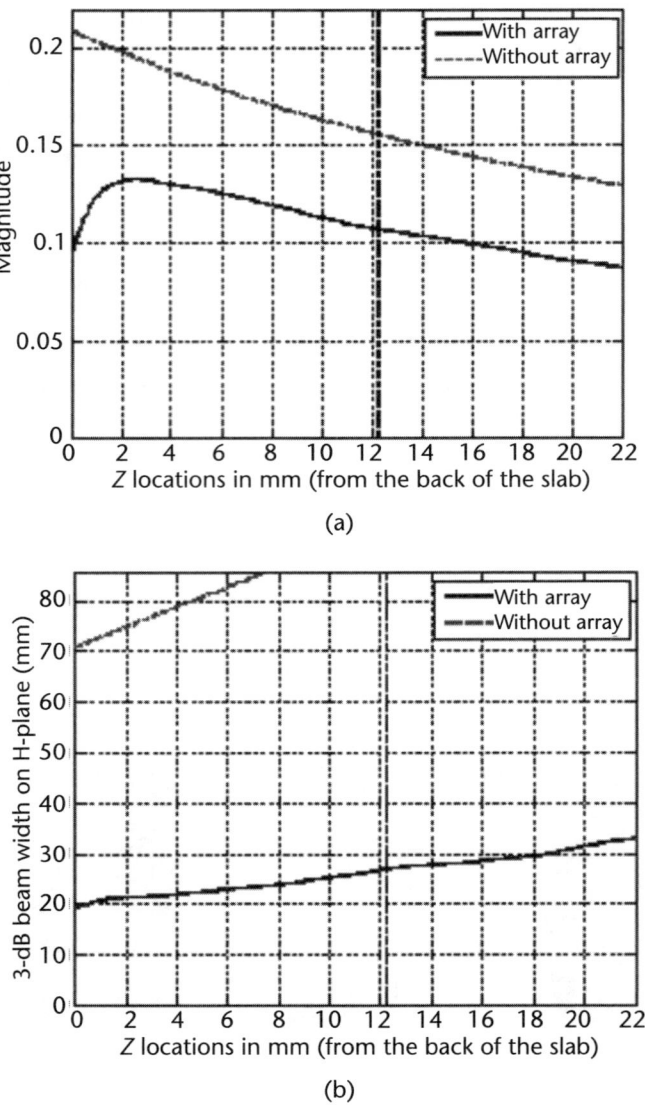

Figure 7.47 (a) Magnitude of E_y at the center axis and (b) the 3-dB beam width on the E- and H-planes along the longitudinal direction at 15.3 GHz, inside the transmission region of the DNG slab excited by small dipole. The corresponding magnitude and beamwidth with all structures removed are also plotted in dotted lines. The vertical dash line at $z = 12.25$ mm indicates the expected image location when the refractive index of the slab is equal to -1.

7.5 Figure-of-Merit (FOM) Analysis

Since the first experimental demonstration of LHM by Smith and Schultz et al. [21], the problem of the realization of SNG and DNG metamaterials using resonant particles such as SRRs and wire segments has been extensively studied by several groups [22, 23]. Due to the resonant behavior of electric or/and magnetic particles, the characteristics of the SNG or DNG metamaterials composed of those particles are highly frequency-dispersive. The nature of the negative property in such a medium stems from the singularity in the electromagnetic behavior of the constituent particle around the resonance where the real parts of the "effective" permittivity and permeability exhibit different signs at the frequencies below and above the resonance, and the loss term (imaginary part) shows the maximum value at the resonant frequency. Therefore, it is difficult to obtain large negative values of the permittivity and permeability from the SNG and DNG particles with a very small loss, but we need to compromise the value of negative property such that the other performance parameters are within a practical range of realization. These parameters include loss, bandwidth, and isotropy (angular dependency in 2-D or 3-D space). Until now, there have been no guidelines available for the design of constituent particles in order to construct metamaterials producing higher figure-of-merits. This work presents a comparison of the loss and bandwidth of magnetically negative (MNG) metamaterial consisting of split-ring type resonant particles in a two-dimensional lattice shown in Figure 7.48, with different volumetric densities and electrical sizes of the particle.

7.5.1 Loss and Bandwidth of Metamaterials with Different Electrical Sizes and Particle Densities

We start the investigation with the analytic model of SRRs. The effective magnetic permeability of the square array of the resonant particles is given by [7]

$$\mu_r = 1 - \frac{F}{1 + j\frac{2\rho}{\omega r \mu_0} - \frac{3}{\pi^2 \mu_0 \omega^2 C r^3}} \tag{7.13}$$

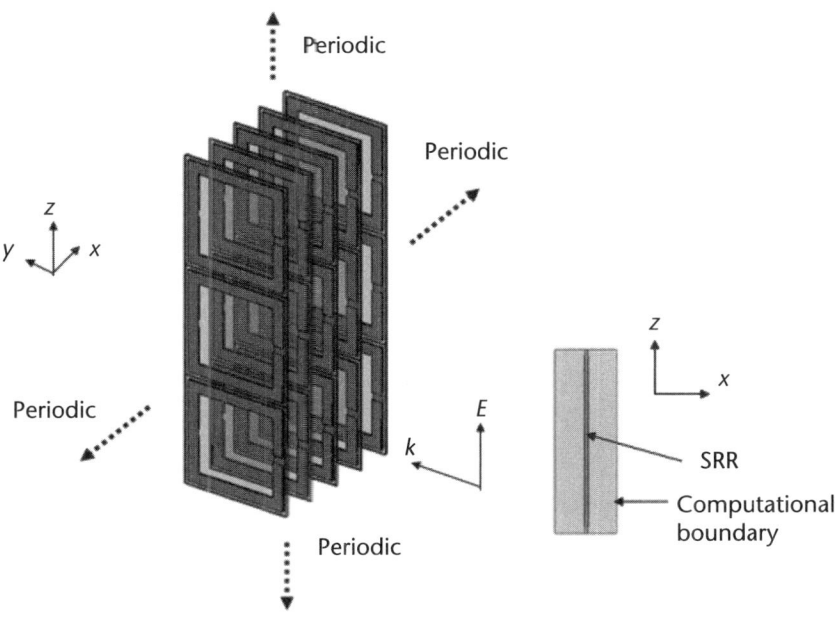

Figure 7.48 Geometry of the periodic split-ring resonator.

where F is the fractional volume of the cell occupied by the interior of the ring $F = \frac{\pi r^2}{a^2}$, ρ is the resistance of the conducting ring per unit area, and C is the capacitance per unit area between the two rings with separation distance (d), given by $C = \frac{\varepsilon_0}{d} = \frac{1}{dc_0^2 \mu_0}$.

It can be intuitively conceived that the volumetric density and the electrical size of the resonant particles, which comprise the metamaterial, directly affects the Q of the material as well as the loss behavior. Figure 7.49 shows the calculated effective

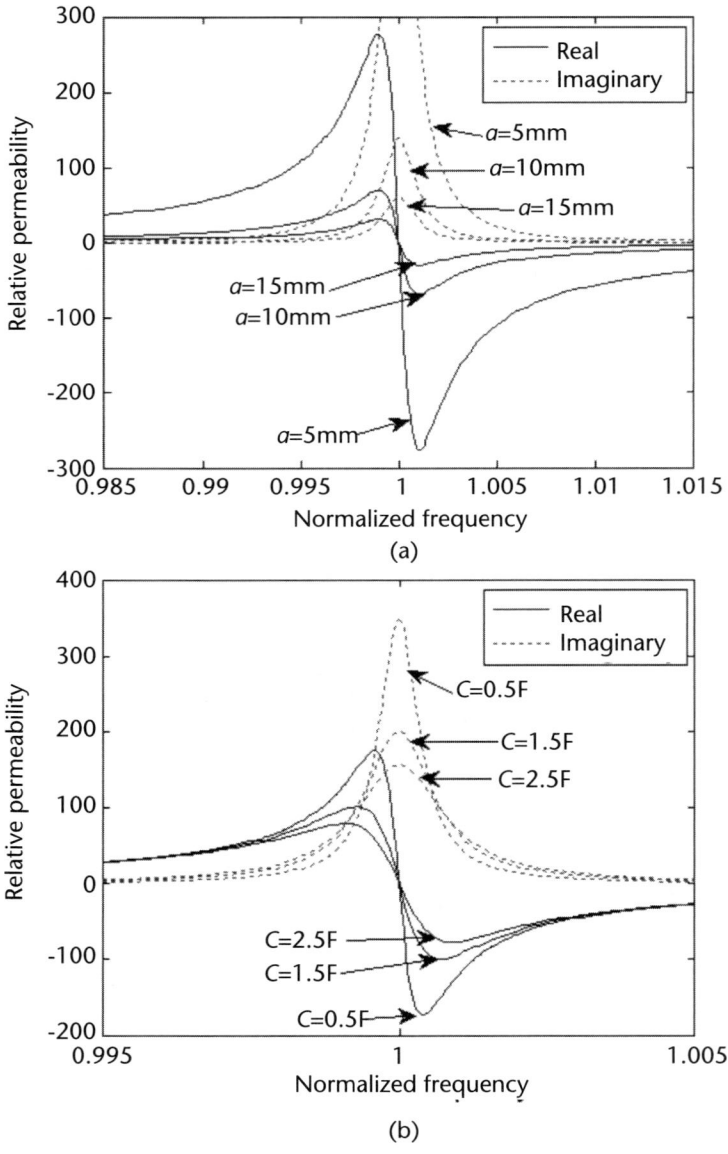

Figure 7.49 Relative permeability of SRR: (a) for different volumetric densities (a: spacing between SRRs) and (b) for different capacitance.

7.5 Figure-of-Merit (FOM) Analysis

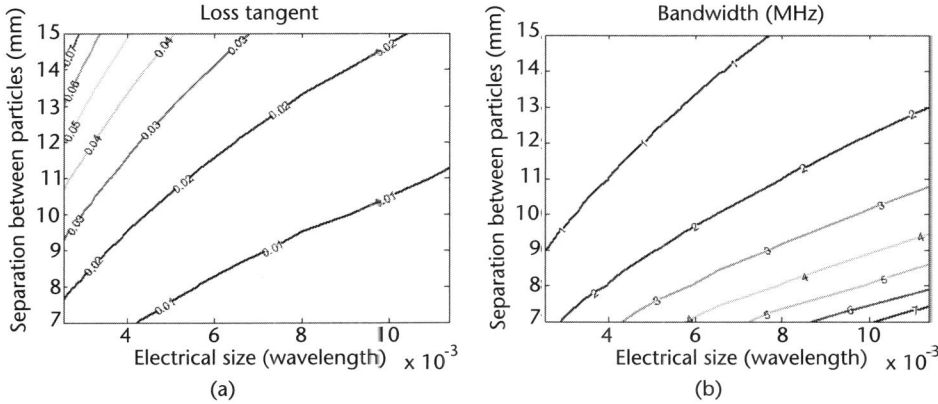

Figure 7.50 Loss and bandwidth of metamaterial consisting of resonant particle with different volumetric density and particle size for $\mu_r = -4$: (a) loss tangent and (b) bandwidth.

permeability of the SRR by (7.6) for different volumetric density and capacitance for the particle. The capacitance parameter in the SRR changes the resonant frequency of the particle, and it can be translated into different electrical sizes. We note that the high volumetric density results in greater magnetic properties, and smaller particles show increased Q. In order to carry out a realistic comparison, we consider the loss and bandwidth of the MNG material for a fixed negative real part, for example, $\mu_r = -4$, which are practically achievable. The bandwidth of the metamaterial in this work is defined by the change of frequency needed to make the unit variation of the material properties (i.e., $df/d\mu$). First, we start with calculating the theoretical permeability of the metamaterial composed of an SRR array using (7.13). After we find frequencies at which the material exhibits negative properties of the given value; then loss and bandwidth parameters of the material at those frequencies are calculated. Figure 7.50 shows the calculated loss

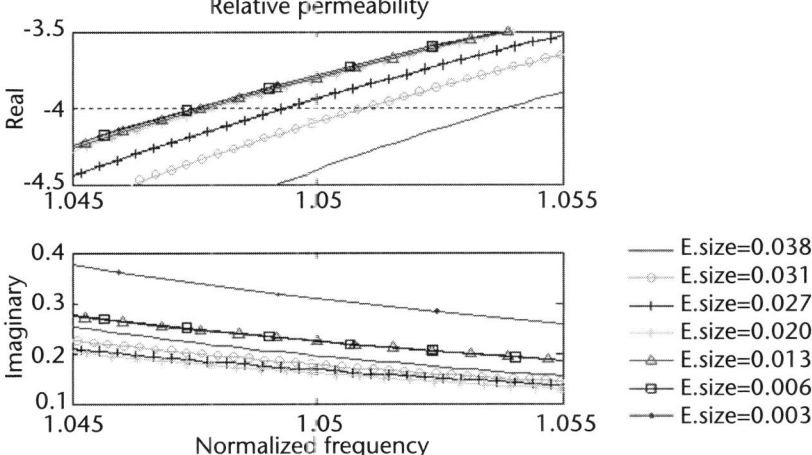

Figure 7.51 Effective permeability of SRR for different electrical sizes (full-wave simulation).

and bandwidth of the MNG material composed of SRRs. We observe that a higher density provides lower loss and larger bandwidth for a fixed electrical size of the particle. This is because the net magnetic property increases as the density of the particle increases; thus, the frequencies at which the property of the material exhibits the given value are further apart from the resonant frequency of the particle. In this case, the material achieves the intended negative permeability by using particles with smaller negative values of the permeability, which shows lower losses and larger bandwidths.

7.5.2 Figure-of-Merit Analysis by Numerical Experiments

7.5.2.1 Simulation Setup

In order to address the actual performance of the metamaterial, numerical experiments have been carried out by using Ansoft HFSS [25]. The simulated geometry is shown in Figure 7.48, where the SRRs are periodically arranged in a 2-D lattice. The array is assumed to be infinite, and the transmission and reflection characteristics of the unit cell have been simulated using the PBCs. The ring has finite conductivity (copper), and the dielectric involved between the rings is lossless. The physical parameters of the split ring are: strip width = 0.3 mm, thickness = 5 μm, size = 2.8 mm, and gap = 0.6 mm. In order to vary the resonant frequency of the SRR, a square dielectric substrate was employed in between the open rings with

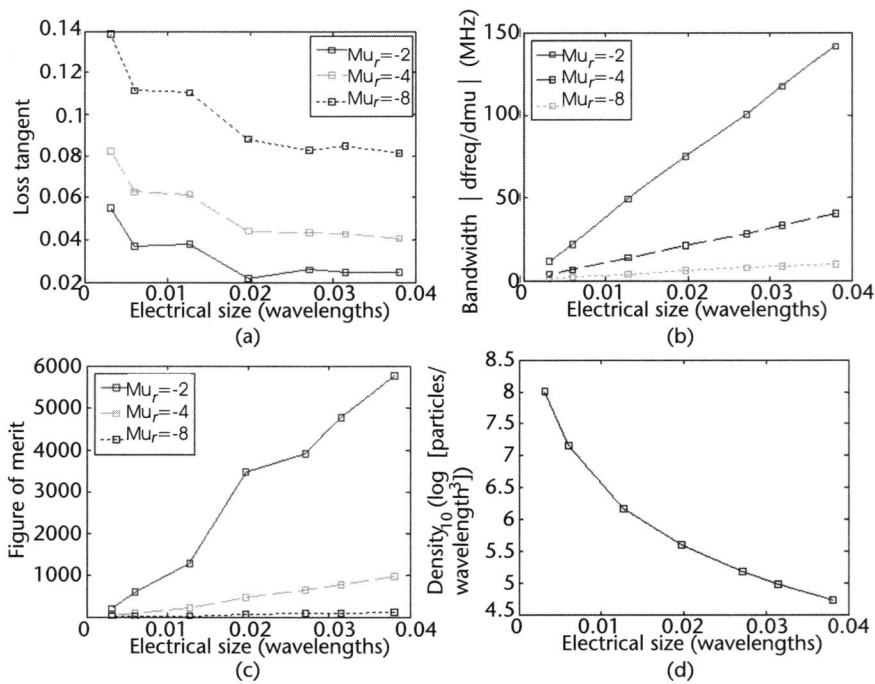

Figure 7.52 Characteristics of MNG material versus electrical size of the particle for different permeability value ($\mu_r = -2, -4, -8$) - particles have the same physical size, but resonant frequency varies: (a) loss tangent, (b) bandwidth, (c) figure of merit, and (d) volumetric density.

7.5 Figure-of-Merit (FOM) Analysis

a 50-μm thickness and a 3-mm length in side. The dielectric constant of the substrate material has been varied from unity to 50. The polarization of the electric field is parallel with the plane containing the ring, and the propagation vector is perpendicular to the axis of the ring. The effective permeability of the material has been extracted from the S-parameter using the standard technique [8].

7.5.2.2 Results and Discussion

In many cases, the artificial material is synthesized by using small resonant particles, and the resonant frequency of the particle can be controlled by the amount of loading involved in the particle—for example, the dielectric material or the number of turns in the solenoidal geometry. Figure 7.51 shows the numerically calculated effective permeability of the SRR array for such a case. The frequency

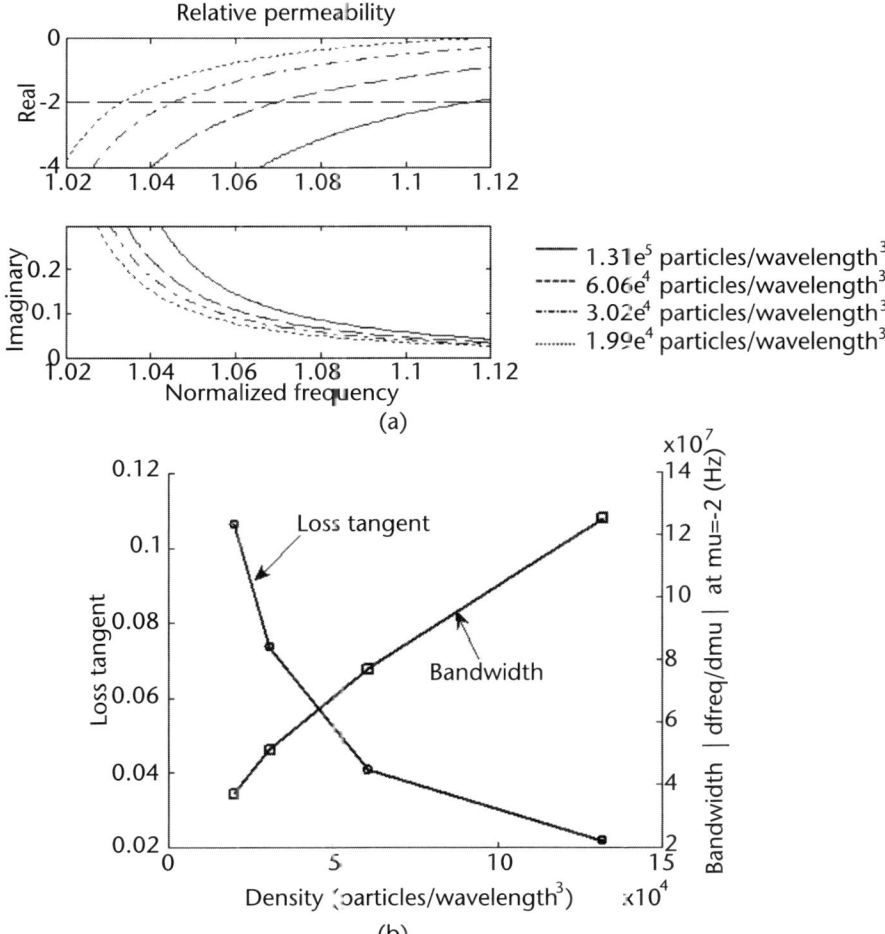

Figure 7.53 Characteristics of MNG material with different particle densities (particles have the same physical size): (a) effective permeability versus volumetric density of particle and (b) loss and bandwidth versus volumetric density of particle.

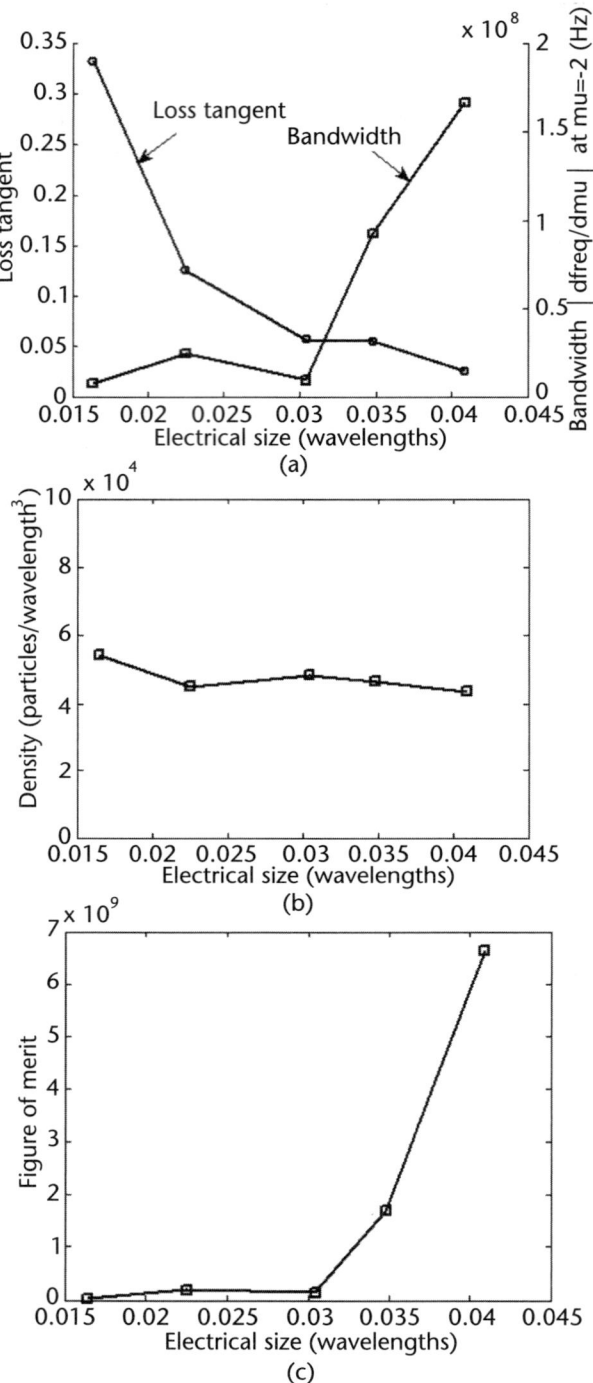

Figure 7.54 Loss and bandwidth (at $\mu_r = -2$) versus electrical size of the particle with similar volumetric density: (a) loss and bandwidth versus electrical size, (b) volumetric density versus electrical size, and (c) FOM versus electrical size.

has been normalized to the resonant frequency of the particle, and we observe characteristics that are similar to the ones for the analytical model discussed in Section 7.2. Next, we investigate the loss and bandwidth of the SRR array for the practical design values of the effective permeability (i.e., $\mu_r = -2, -4,$ and -8). For this experiment, the physical size and the spacing between the particles are fixed, but the resonant frequency is varied by increasing the capacitance between the split rings. The results are summarized in Figure 7.52. It is observed that both the loss and the bandwidth improve as the electrical size of the particle increases (less loading involved in the particle) and overall FOM (FOM=bandwidth/loss) increases for the arrays with fixed physical parameters. However, in this case, the electrical size of the particle is varied by changing the resonant frequency of the particle, causing the volumetric density of the particle in terms of the wavelength to decrease as the resonant frequency of the particle decreases, as shown in Figure 7.52(d). An interesting fact is that the bandwidth of the particle becomes less sensitive to the size of the particle when the particle is operating at the frequency band where the permeability value is relatively low [e.g., $(\mu_r = -2)$] or in other words, the operating frequency is further away from the resonance frequency of the particle. As indicated by the investigation in the previous section, the results of the numerical experiments shown in Figure 7.53 confirm that metamaterials with higher density of particles exhibit higher FOM. This implies that the FOM obtained for fixed physical parameters may not be as high as shown in Figure 7.52(c), but rather compromised because of the reduced volumetric density of the particle for larger electrical sizes. In order to take the density parameters for different electrical sizes into account, another set of experiments has been carried out, where the density of the particles within the metamaterial remains similar by adjusting the spacing between the particles, but only their electrical sizes are changed. The results are interesting in that the FOM still increases as the electrical size of the particle is increased, although this improvement becomes insignificant when the electrical size of the particle is very small. It is noted that the FOM remains rather unchanged when the electrical size of the particle is less than 0.03 wavelengths for the parameters chosen in this chapter (see Figure 7.54).

7.6 Conclusions

In this chapter, we have carried out an extensive study of the EM response of an artificial DNG slab, comprised of a combination of split rings and wires, by using the parallel FDTD code GEMS [26]. A preliminary analysis of the scattering characteristics of the infinite array, illuminated by a plane wave, has been performed by using the PBC/FDTD version of GEMS. The effective material parameters for the DNG slab have been extracted by using the modified inversion approach. We have also discussed the difficulties encountered in the retrieval process, identified some problem areas that may be encountered when using the effective material parameters in real-world applications, and pointed out the importance of performing a rigorous simulation of the physical structure, in which the inclusions inside the metamaterial structure are also modeled accurately, in order to evaluate the performance of such system in a reliable manner.

In addition, we have demonstrated the power of GEMS, which is a highly efficient, parallel FDTD code when rigorously simulating the EM response of the finite artificial DNG array when illuminated by a Gaussian beam, or a small dipole, where the total number of unknown fields to be solved typically exceeds 1.6 billion. Two unique phenomena of the DNG medium, namely the negative refraction and the focusing effect, have been investigated. At the frequency where the slab has a refractive index of -1, the phase clearly demonstrates a backward phase velocity inside the slab, and restores the value of the source at the expected image position. However, none of the results has demonstrated the bending of beam towards the same side of normal when the beam enters the artificial DNG slab comprised of SRRs and dipoles from free space, or any localization of fields near the image location. Also, the emerging beam has been found to be highly astigmatic in nature. In fact, multiple beams have been observed in the transmission region when the array is illuminated by a TM_z Gaussian beam at an oblique angle, or excited by a small dipole. None of these behaviors can be explained by using the effective medium concept, or deduced from the result of the simulation for the infinite, periodic array illuminated by the plane wave.

The main reason for the failure of our artificial DNG array to fully demonstrate the negative refraction and focusing effect is the anisotropic nature of the unit cell element we have chosen—the propagation in this type of medium is very dependent on the angle relative to the orientation of the SRR. However, according to the published literature, the SRR and wire combination is the distinct front-runner among the candidates for the metamaterial elements that exhibit a DNG behavior, as we have also conformed by evaluating its effective medium properties of the chosen element. We point out, once again, that the effective parameters are extracted by interrogating the slab.

We have verified that electrically small particles exhibit higher losses to produce the same negative permittivity or permeability values. Similarly, the bandwidth of the particle, defined by the inverse of the frequency derivative of the permittivity and permeability, decreases as the size of the particle is reduced. It has been observed that the bandwidth of the particle becomes less sensitive to the size of the particle when the particle is operating at the frequency band where the permeability value is relatively low. Finally, we have performed the FOM analysis for the square SRR based on the loss and bandwidth. The results indicated that larger particles have better FOMs in terms of loss and bandwidth. However, the benefit of using a larger particle becomes less significant when we require a higher value of permeability. For the electrically very small particles (less than 0.03 wavelength), the FOM becomes somewhat irrelevant to the electrical size.

References

[1] D. R. Smith, D. C. Vier, Th. Koschny, and C. M. Soukoulis, "Electromagnetic parameter retrieval from inhomogeneous metamaterials," *Phys. Rev. E*, 71, 036617, 2005.

[2] P. Markos and C. M. Soukoulis, "Transmission Properties and Effective Electromagnetic Parameters of Double Negative Metamaterials," *Opt. Express*, Vol. 11, No. 7, pp. 649–651, 2003.

[3]   X. Chen, T. M. Grzegorczyk, B. Wu, J. Pacheco, and J. A. Kong, "Robust Method to Retrieve the Constitutive Effective Parameters of Metamaterials," *Phys. Rev. E.*, Vol. 70, No. 1, 016608: 1–7, 2004.

[4]   J. B. Pendry, A. J. Holden, W. J. Stewart, and I. Youngs, "Extremely low frequency plasmons in metallic mesostructures," *Phys. Rev. Lett.*, Vol. 76, No. 25, pp. 4773–4776, 1996.

[5]   O. Archer, A. L. Adenot, and F. Duverger, "Fresnel Coefficients at an Interface with a Lamellar Composite Material," *Phys. Rev. B*, Vol. 62, No. 20, pp. 13748–13756, 2000.

[6]   C. D. Moss, T. M. Grzegorczyk, Y. Zhang, and J. A. Kong, "Numerical Studies of Left Handed Metamaterials," *Prog. Electromagn. Res.*, Vol. 35, pp. 315–334, 2002.

[7]   L. D. Landau, E. M. Lifshitz, and L. P. Pitaevskii, *Electrodynamics of Continuous Media*, 2nd ed., Oxford, England: Butterworth-Heinenann, 1984.

[8]   V. M. Shalaev, *Phys. Rep.*, 272, 61 1996.

[9]   D. J. Bergman, *Phys. Lett.*, C 43, 377, 1978.

[10] K. Bao, R. C. McPhedran, N. A. Nicorovici, C. G. Poulton, and L. C. Botten, *Physica B*, 279, 162, 2000.

[11] C. Grimes and D.M. Grimes, *Phys. Rev. B*, 43, 10780, 1991.

[12] A. Ishimaru, *Electromagnetic Wave Propagation, Radation, and Scattering*, Upper Saddle River, NJ: Prentice-Hall, 1991.

[13] J. B. Pendry, A. J. Holden, D. J. Robbins, and W. J. Stuart, *IEEE Trans. Microwave Theory Tech.*, 47, 2075, 1999.

[14] D. R. Smith, D. C. Vier, N. Kroll, and S. Schultz, *Appl. Phys. Lett.*, 77, 2246, 2000.

[15] D. R. Smith, S. Schultz, P. Markos, and C. M Soukoulis, *Phys. Rev. B*, 65, 195104, 2002.

[16] A. F. Starr, P. M. Rye, D. R. Smith, and S. Nemat-Nasser, *Phys. Rev. B*, 70, 115113, 2004.

[17] T. Koschny, P. Markos, D. R. Smith, and C. M. Soukoulis, *Phys. Rev. E*, 58, 065602, 2003.

[18] D. R. Smith, R. Dalichaouch, N. Kroll, S. Schultz, S. L. McCall, and P. M. Platzman, *J. Opt. Soc. Am. B*, 10, 314 (1993).

[19] D. M. Pozar, *Microwave Engineering*, 2nd ed., New York: John Wiley & Sons, 1998, p. 211.

[20] CST Microwave Studio Ver. 2006.04.

[21] D. R. Smith, W. J. Padilla, D. C. Vier, S. C. Nemat-Nasser, and S. Schultz, "Composite Medium with Simultaneously Negative Permeability and Permittivity," *Phys. Rev. Lett.*, 84, pp. 4184–4187, 2000.

[22] J. B. Pendry, A. J. Holden, D. J. Robbins, and W. J. Stewart, "Magnetism from Conductors and Enhanced Nonlinear Phenomena," *IEEE Transactions on Micrwave Theory and Techniques*, Vol. 47, No. 11, pp. 2075–2084, November 1999.

[23] R. W. Ziolkowski, "Design, Fabrication, and Testing of Double Negative Metamateirals," *IEEE Transactions on Antennas and Propagation*, Vol. 51, No. 7, July 2003.

[24] D. Seetharamdoo, R. Sauleau, K. Mahdjoubi and A. Tarot, "Effective Parameters of Negative Refractive Index Metamaterials: Interpretation and Validity," *J. Appl. Phys.*, Vol. 98, No. 6, 063505: 1–4, 2005.

[25] Ansoft HFSS 10.0.

[26] N. T. Huang, R. Mittra, R. Maaskant, W. Yu, "Investigation of the Vivaldi Array for Square Kilometer Array (SKA) Application Using the Parallelized FDTD Code GEMS on the LOFAR BlueGene/L Supercomputer," *2007 IEEE International Symposium on Antennas and Propagation*, Honolulu, HI, June 10–14, 2007.

[27] http://www.2comu.com/.

[28] W. Yu, R. Mittra, T. Su, Y. Liu, and X. Yang, *Parallel Finite-Difference Time-Domain Method*, Norwood, MA: Artech House, 2006.

[29] W. Yu and R. Mittra, *CFDTD: Conformal Finite-Difference Time-Domain Maxwell's Equation Solver—Software and User's Guide*, Norwood, MA: Artech House, 2004.

CHAPTER 8
Accurate FDTD Modeling of a Perfect Lens

8.1 Introduction

Recently a great deal of attention has been paid to the research of a new class of artificial electromagnetic material: LHMs. Introduced by Veselago in 1968 [1], LHMs possess simultaneously negative permittivity and permeability. LHMs are also often referred to as DNG material, backward-wave media, and NIMs. Because LHMs were not physically available in their early years, they did not attract much attention until in 1999 when Pendry et al. [2] demonstrated that materials with an array of SRRs behave as media with negative permeability over a limited frequency range. Furthermore, the combination of a 2-D array of SRRs with a 2-D array of metallic wires [3, 4] enables one to realize media that simultaneously display negative permittivity and permeability, and hence fall into the category of LHMs. In 2000, Smith et al. [5] demonstrated for the first time the experimental existence of LHMs, by using the structure shown in Figure 8.1 [6].

For an electromagnetic plane wave in LHMs, the electric field and magnetic field and the wave vector form a left-handed system of vectors, as illustrated in Figure 8.2(b).

The LHMs introduce unusual and interesting wave propagation properties, such as negative refraction, reversed Doppler effect and Cerenkov radiation. One of the most important applications of LHMs suggested by Pendry, is the "perfect lens" [7]. Unlike conventional curved lenses fabricated by using optically transparent materials, Pendry's planar perfect lenses (see Figure 8.3) have no optical axis. Furthermore, under ideal conditions, they reconstitute the near as well as the far fields of the source with subwavelength resolution. Apparently, this is accomplished by exciting the resonant surface waves (plasmons) at the interface between free space and the LHM slab.

Conventional imaging systems that utilize positive permittivity and permeability suffer from the so-called diffraction limit, and their minimum angular resolution (the space between two point sources) that an imaging system can resolve corresponds to approximately half of the incident wavelength of light that is used to produce the image. The reason for this is that the conventional systems are only capable of transmitting the propagating components emanating from the source [7], and the evanescent waves that carry subwavelength information about the object decay exponentially in such double positive media that they suffer significant decay before reaching the image plane. On the other hand, under ideal conditions, LHM lenses provide the unique properties of negative refraction for propagating waves,

239

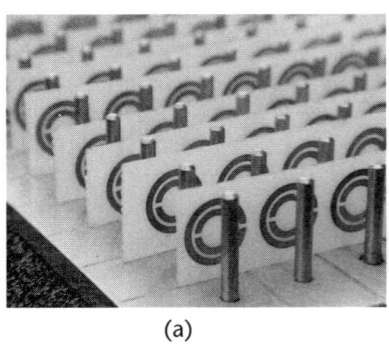 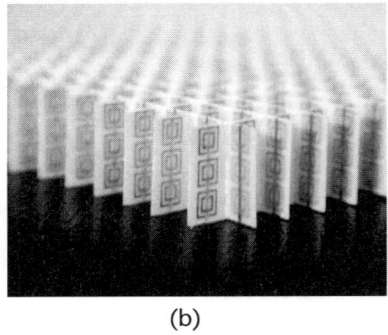

Figure 8.1 (a) First experimental demonstration of LHM [5] and (b) a 2-D isotropic LHM [6].

as shown in Figure 8.3(a), and amplify evanescent waves [see Figure 8.3(b)] [7]. This, in turn, enables them to overcome the diffraction limit and reconstruct the entire source field at the image plane.

There have been a number of attempts to model LHMs using the FDTD method [8–13]. The conventional FDTD has been generalized to apply to dispersive media in the literature, and the negative refraction effect that characterizes the boundary between free space and the LHM has been observed, and the planar superlens behavior has been successfully demonstrated under ideal conditions [8–10]. This implies that although the LHM has been modeled correctly only for propagating waves, the conventional implementation of the dispersive FDTD method may lead to inaccurate results when evanescent waves are involved. Since the evanescent waves typically suffer an exponential decay as functions of the distance, they are concentrated only in the close vicinity of sources, and the conventional FDTDs suffer little from a loss of inaccuracy when modeling a nondispersive media. However, in the case of LHM, the evanescent waves play a key role that has to be modeled accurately, because of reasons mentioned above [7]. This explains why early FDTD simulations failed to demonstrate the subwavelength imaging properties of LHM lenses [8, 9], namely, the amplification of evanescent waves, which normally decay in usual materials. Hence, the LHMs facilitate the transmission of subwavelength details of sources to significant distances.

The FDTD method has also been used to study the effect of losses on the transmission characteristics of LHM slabs [11] and how the material parameters and the thickness of the slab influence their imaging properties [13]. Besides the FDTD method, the pseudospectral time-domain (PSTD) method has been used to model

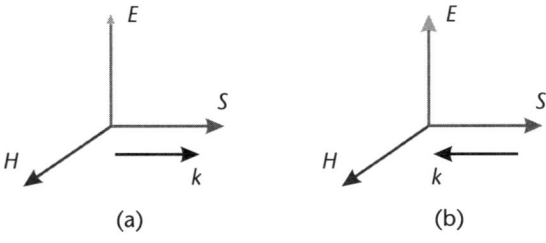

Figure 8.2 Illustration of electric, magnetic, Poynting, and wave vectors in conventional (a) RHMs and (b) LHMs where the wave vector and Poynting vector are in opposite directions.

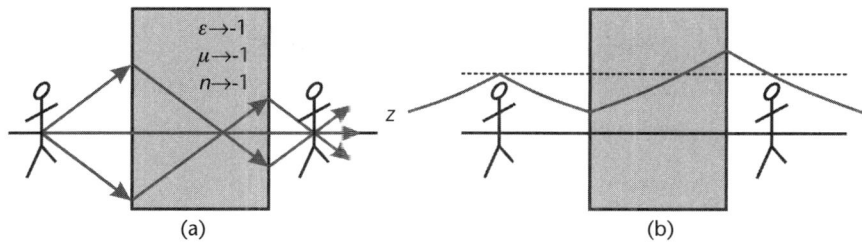

Figure 8.3 Illustration of Pendry's perfect lens formed by LHM with $\varepsilon = \mu = n = -1$ [7]: (a) negative refraction for the far field and (b) amplification of evanescent waves for the near field.

backward-wave metamaterials [14]. It is claimed in [14] that the FDTD method cannot be used to accurately model LHMs because of the numerical artifacts introduced by the staggered grid employed in the FDTD domain. However, as we will demonstrate later, with proper field averaging techniques [13, 15], the FDTD method can indeed be used to accurately characterize the behavior of both propagating and evanescent waves in LHM slabs. Furthermore, it has been reported in [16, 17] that with special treatments (i.e., averaging techniques) along material boundaries, accurate modeling of curved surfaces of conventional dielectrics as well as surface plasmon polaritons between metal-dielectric interfaces can be achieved without using a finely discretized FDTD grid.

In principle, an ideal and lossless LHM slab with infinite transverse dimensions can provide subwavelength resolution without limit. However, in realistic situations, the subwavelength resolution of LHM lenses is limited by losses [18], the thickness of the slab, and the mismatch of the slab with its surrounding medium [19]. It is important to understand these theoretical limitations and to compare the results of numerical simulations with analytical solutions for canonical problems, to validate the numerical algorithms, before applying them to more complex configurations for which analytical solutions are unavailable.

In this chapter, the infinite LHM slabs and their transmission characteristics have been modeled using the PBC and a material parameter averaging technique along the boundaries of the LHM slabs. In contrast to the FDTD modeling of conventional dielectric slabs, where the averaging is done by introducing a second-order correction that improves the accuracy of simulations, the averaging of permittivity is crucial to modeling the LHM slabs. Averaging the material parameters implemented in the proposed FDTD model is equivalent to that of averaging the current density, as originally introduced in [13], and the algorithm is detailed in this chapter. It is demonstrated that other aspects such as the material parameters and the switching time of the sources also have a considerable influence on the FDTD simulations.

8.2 Dispersive FDTD Modeling of LHMs with Spatial Averaging at the Boundaries

This section will model isotropic LHMs using the effective medium theory. Chapter 6 demonstrates that realistic LHMs can be characterized by using either a Lorentz or Drude dispersion model. Although there are various dispersive FDTD methods available for the modeling of LHMs, we have chosen to implement the auxiliary differential equations (ADEs) method, owing to its simplicity and efficiency. There are also different schemes involving ADEs in addition to the conventional FDTD updating equations. In this chapter, we introduce two general algorithms, namely the (E, J, H, M) [20] and (E, D, H, B) schemes [21], for this purpose.

8.2.1 The (E, D, H, B) Scheme

The (E, D, H, B) scheme is based on Faraday's and Ampere's laws:

$$\nabla \times \mathbf{E} = -\frac{\partial \mathbf{B}}{\partial t} \tag{8.1}$$

$$\nabla \times \mathbf{H} = \frac{\partial \mathbf{D}}{\partial t} \tag{8.2}$$

as well as the constitutive relationship $\mathbf{D} = \varepsilon \mathbf{E}$ and $\mathbf{B} = \mu \mathbf{H}$ where ε and μ are given by (6.4) and (6.5), respectively. Equations (8.1) and (8.2) can be discretized in the usual way [20] leading to the conventional FDTD updating equations:

$$\mathbf{B}^{n+1} = \mathbf{B}^n - \Delta t \cdot \widetilde{\nabla} \times \mathbf{E}^{n+\frac{1}{2}} \tag{8.3}$$

$$\mathbf{D}^{n+1} = \mathbf{D}^n + \Delta t \cdot \widetilde{\nabla} \times \mathbf{H}^{n+\frac{1}{2}} \tag{8.4}$$

where $\widetilde{\nabla}$ is the discrete curl operator, Δt is the FDTD time step, and n is the number of time steps at which the updating is performed.

In addition to the above updated equations, we need to enforce ADEs, which are discretized through the following steps. The constitutive relation between $\mathbf{D}$ and $\mathbf{E}$ reads

$$\left(\omega^2 - j\omega\gamma_e\right)\mathbf{D} = \varepsilon_0 \left(\omega^2 - j\omega\gamma_e - \omega_{pe}^2\right)\mathbf{E} \tag{8.5}$$

If we use the inverse Fourier transform, which leads to the following equivalence relations:

$$j\omega \to \frac{\partial}{\partial t}, \quad \omega^2 \to -\frac{\partial^2}{\partial t^2}, \tag{8.6}$$

We can write (8.5) in the time domain as

$$\left(\frac{\partial^2}{\partial t^2} + \frac{\partial}{\partial t}\gamma_e\right)\mathbf{D} = \varepsilon_0 \left(\frac{\partial^2}{\partial t^2} + \frac{\partial}{\partial t}\gamma_e + \omega_{pe}^2\right)\mathbf{E} \tag{8.7}$$

Next, we discretize the FDTD computational domain using an equally spaced 3-D grid with cell sizes of Δx, Δy, and Δz along the x-, y-, and z-directions, respectively. To discretize (8.7), we use the central finite difference operators in time to update δ_t and δ_t^2 and the central averaging operators μ_t and μ_t^2 in time as follows:

$$\frac{\partial^2}{\partial t^2} \to \frac{\delta_t^2}{(\Delta t)^2}, \quad \frac{\partial}{\partial t} \to \frac{\delta_t}{\Delta t}\mu_t, \quad \omega_{pe}^2 \to \omega_{pe}^2\mu_t^2$$

8.2 Dispersive FDTD Modeling of LHMs with Spatial Averaging at the Boundaries

where the operators δ_t, δ_t^2, μ_t, and μ_t^2 are defined in the same manner as in [22]:

$$\delta_t \mathbf{F}|_{m_x,m_y,m_z}^n \equiv \mathbf{F}|_{m_x,m_y,m_z}^{n+\frac{1}{2}} - \mathbf{F}|_{m_x,m_y,m_z}^{n-\frac{1}{2}}$$

$$\delta_t^2 \mathbf{F}|_{m_x,m_y,m_z}^n \equiv \mathbf{F}|_{m_x,m_y,m_z}^{n+1} - 2\mathbf{F}|_{m_x,m_y,m_z}^n + \mathbf{F}|_{m_x,m_y,m_z}^{n-1}$$

$$\mu_t \mathbf{F}|_{m_x,m_y,m_z}^n \equiv \frac{\mathbf{F}|_{m_x,m_y,m_z}^{n+\frac{1}{2}} + \mathbf{F}|_{m_x,m_y,m_z}^{n-\frac{1}{2}}}{2}$$

$$\mu_t^2 \mathbf{F}|_{m_x,m_y,m_z}^n \equiv \frac{\mathbf{F}|_{m_x,m_y,m_z}^{n+1} + 2\mathbf{F}|_{m_x,m_y,m_z}^n + \mathbf{F}|_{m_x,m_y,m_z}^{n-1}}{4} \quad (8.8)$$

Here the quantities in bold represent the field components and m_x, m_y, m_z are the indices corresponding to a certain discretization point in the FDTD domain. The use of the averaging operator μ_t enables a semi-implicit scheme and guarantees the FDTD simulations to be numerically stable and accurate [20, 22].

Discretizing (8.7), we get

$$\left[\frac{\delta_t^2}{(\Delta t)^2} + \frac{\delta_t}{\Delta t}\mu_t \gamma_e\right] \mathbf{D} = \varepsilon_0 \left[\frac{\delta_t^2}{(\Delta t)^2} + \frac{\delta_t}{\Delta t}\mu_t \gamma_e + \omega_{pe}^2 \mu_t^2\right] \mathbf{E} \quad (8.9)$$

Note that, in (8.9), the discretization of the term ω_{pe}^2 of (8.7) has been performed by using the central averaging operator μ_t^2 in order to guarantee the improved stability. The above operator μ_t is used for the term containing γ_e to preserve the second-order feature of this equation. Equation (8.9) can thus be written as

$$\frac{\mathbf{D}|_{m_x,m_y,m_z}^{n+1} - 2\mathbf{D}|_{m_x,m_y,m_z}^n + \mathbf{D}|_{m_x,m_y,m_z}^{n-1}}{(\Delta t)^2} + \gamma_e \frac{\mathbf{D}|_{m_x,m_y,m_z}^{n+1} - \mathbf{D}|_{m_x,m_y,m_z}^{n-1}}{2\Delta t}$$

$$= \varepsilon_0 \left[\frac{\mathbf{E}|_{m_x,m_y,m_z}^{n+1} - 2\mathbf{E}|_{m_x,m_y,m_z}^n + \mathbf{E}|_{m_x,m_y,m_z}^{n-1}}{(\Delta t)^2} + \gamma_e \frac{\mathbf{E}|_{m_x,m_y,m_z}^{n+1} - \mathbf{E}|_{m_x,m_y,m_z}^{n-1}}{2\Delta t}\right.$$

$$\left. + \omega_{pe}^2 \frac{\mathbf{E}|_{m_x,m_y,m_z}^{n+1} + 2\mathbf{E}|_{m_x,m_y,m_z}^n + \mathbf{E}|_{m_x,m_y,m_z}^{n-1}}{4}\right] \quad (8.10)$$

Using the above, the updating equation for $\mathbf{E}$ can be written in terms of $\mathbf{E}$ and $\mathbf{D}$ at previous time steps as follows:

$$\mathbf{E}^{n+1} = \left\{\left[\frac{1}{\varepsilon_0(\Delta t)^2} + \frac{\gamma_e}{2\varepsilon_0 \Delta t}\right]\mathbf{D}^{n+1} - \frac{2}{\varepsilon_0(\Delta t)^2}\mathbf{D}^n\right.$$

$$+ \left[\frac{2}{(\Delta t)^2} - \frac{\omega_{pe}^2}{2}\right]\mathbf{E}^n - \left[\frac{1}{(\Delta t)^2} - \frac{\gamma_e}{2\Delta t} + \frac{\omega_{pe}^2}{4}\right]\mathbf{E}^{n-1}$$

$$\left. + \left[\frac{1}{\varepsilon_0(\Delta t)^2} - \frac{\gamma_e}{2\varepsilon_0 \Delta t}\right]\mathbf{D}^{n-1}\right\} \bigg/ \left[\frac{1}{(\Delta t)^2} + \frac{\gamma_e}{2\Delta t} + \frac{\omega_{pe}^2}{4}\right] \quad (8.11)$$

The updating equation for **H** is in the same form as (8.11) by replacing **E**, **D**, ω_{pe}^2, and γ_e by **H**, **B**, ω_{pm}^2, and γ_m, respectively, such that

$$\mathbf{H}^{n+1} = \left\{ \left[\frac{1}{\mu_0 (\Delta t)^2} + \frac{\gamma_m}{2\mu_0 \Delta t} \right] \mathbf{B}^{n+1} - \frac{2}{\mu_0 (\Delta t)^2} \mathbf{B}^n \right.$$

$$+ \left[\frac{2}{(\Delta t)^2} - \frac{\omega_{pm}^2}{2} \right] \mathbf{H}^n - \left[\frac{1}{(\Delta t)^2} - \frac{\gamma_m}{2\Delta t} + \frac{\omega_{pm}^2}{4} \right] \mathbf{H}^{n-1}$$

$$+ \left. \left[\frac{1}{\mu_0 (\Delta t)^2} - \frac{\gamma_m}{2\mu_0 \Delta t} \right] \mathbf{B}^{n-1} \right\} \bigg/ \left[\frac{1}{(\Delta t)^2} + \frac{\gamma_m}{2\Delta t} + \frac{\omega_{pm}^2}{4} \right] \quad (8.12)$$

The set (8.3), (8.4), (8.11), and (8.12) form the relevant FDTD updating equations for the LHMs employing the (E, D, H, B) scheme. Not unexpectedly, if both the plasma and collision frequencies are equal to zero (i.e., $\omega_{pe} = \omega_{pm} = 0$ and $\gamma_e = \gamma_m = 0$), then the above updating equations reduce to the corresponding FDTD equations in free space.

8.2.2 The (E, J, H, M) Scheme

An alternative ADE FDTD scheme starts with different forms of Faraday's and Ampere's laws for LHMs:

$$\nabla \times \mathbf{E} = -\mu_0 \frac{\partial \mathbf{H}}{\partial t} - \mathbf{M} \quad (8.13)$$

$$\nabla \times \mathbf{H} = \varepsilon_0 \frac{\partial \mathbf{E}}{\partial t} + \mathbf{J} \quad (8.14)$$

where the electric and magnetic current densities, **J** and **M**, are defined as

$$\mathbf{J}(\omega) = j\omega \varepsilon_0 \frac{\omega_{pe}^2}{j\omega \gamma_e - \omega^2} \mathbf{E}(\omega), \quad (8.15)$$

$$\mathbf{M}(\omega) = j\omega \mu_0 \frac{\omega_{pm}^2}{j\omega \gamma_m - \omega^2} \mathbf{H}(\omega). \quad (8.16)$$

Following the same procedure as for the (E, D, H, B) scheme discussed in the previous section, (8.13)-(8.16) can be discretized as:

$$\mathbf{H}^{n+1} = \mathbf{H}^n - \frac{\Delta t}{\mu_0} \left[\widetilde{\nabla} \times \mathbf{E}^{n+\frac{1}{2}} + \mathbf{M}^{n+\frac{1}{2}} \right] \quad (8.17)$$

$$\mathbf{E}^{n+1} = \mathbf{E}^n + \frac{\Delta t}{\varepsilon_0} \left[\widetilde{\nabla} \times \mathbf{H}^{n+\frac{1}{2}} - \mathbf{J}^{n+\frac{1}{2}} \right] \quad (8.18)$$

8.2 Dispersive FDTD Modeling of LHMs with Spatial Averaging at the Boundaries

$$\mathbf{J}|_{m_x,m_y,m_z}^{n+1} = \frac{4}{\gamma_e \Delta t + 2} \mathbf{J}|_{m_x,m_y,m_z}^{n} + \frac{\gamma_e \Delta t - 2}{\gamma_e \Delta t + 2} \mathbf{J}|_{m_x,m_y,m_z}^{n-1}$$
$$+ \frac{\varepsilon_0 \omega_{pe}^2 \Delta t}{\gamma_e \Delta t + 2} \left(\mathbf{E}|_{m_x,m_y,m_z}^{n+1} - \mathbf{E}|_{m_x,m_y,m_z}^{n-1} \right) \quad (8.19)$$

$$\mathbf{M}|_{m_x,m_y,m_z}^{n+1} = \frac{4}{\gamma_m \Delta t + 2} \mathbf{M}|_{m_x,m_y,m_z}^{n} + \frac{\gamma_m \Delta t - 2}{\gamma_m \Delta t + 2} \mathbf{M}|_{m_x,m_y,m_z}^{n-1}$$
$$+ \frac{\mu_0 \omega_{pm}^2 \Delta t}{\gamma_m \Delta t + 2} \left(\mathbf{H}|_{m_x,m_y,m_z}^{n+1} - \mathbf{H}|_{m_x,m_y,m_z}^{n-1} \right) \quad (8.20)$$

Once again, (8.17)-(8.20) reduce to the free-space updating equations if both the plasma and the collision frequencies are set equal to zero (i.e., $\omega_{pe} = \omega_{pm} = 0$ and $\gamma_e = \gamma_m = 0$).

8.2.3 The Spatial Averaging Methods

In addition to the aforementioned ADE/FDTD schemes, a modification is frequently used at the material interfaces to improve the accuracy of FDTD simulations due to the staggered grid in the FDTD domain. It has been shown [23] that at material interfaces, the use of effective material parameters, namely permittivity and permeability, provides a second-order accuracy. The effective permittivity and permeability along the interface can be derived by analyzing the integral form of Ampere's and Faraday's laws [16, 23], such that

$$\frac{\partial}{\partial t} \int \mathbf{D} \cdot \mathbf{n} ds = \oint \mathbf{H} \cdot \mathbf{dl} \quad (8.21)$$

$$-\frac{\partial}{\partial t} \int \mathbf{B} \cdot \mathbf{n} ds = \oint \mathbf{E} \cdot \mathbf{dl} \quad (8.22)$$

For the case of 2-D TE polarization in the FDTD domain, the arrangement of electric and magnetic fields employed at a dielectric interface is illustrated in Figure 8.4. Using the above, the Ampere's law in (8.21) can be written as

$$\frac{\partial}{\partial t} \int_{(m_y-1)\Delta y}^{m_y \Delta y} D_x|_{m_x,y} dy = H_z|_{m_x,m_y} - H_z|_{m_x,m_y-1} \quad (8.23)$$

$$\frac{\partial}{\partial t} \int_{(m_x-1)\Delta x}^{m_x \Delta x} D_y|_{x,m_y} dx = H_z|_{m_x-1,m_y} - H_z|_{m_x,m_y} \quad (8.24)$$

Similarly, the Faraday's laws are written as

$$-\frac{\partial}{\partial t} \int_{(m_x-\frac{1}{2})\Delta x}^{(m_x+\frac{1}{2})\Delta x} \int_{(m_y-\frac{1}{2})\Delta y}^{(m_y+\frac{1}{2})\Delta y} B_z|_{x,y} dx dy$$
$$= \int_{(m_x-\frac{1}{2})\Delta x}^{(m_x+\frac{1}{2})\Delta x} \left(E_x|_{x,m_y-\frac{1}{2}} - E_x|_{x,m_y+\frac{1}{2}} \right) dx$$
$$+ \int_{(m_y-\frac{1}{2})\Delta y}^{(m_y+\frac{1}{2})\Delta y} \left(E_y|_{m_x+\frac{1}{2},y} - E_y|_{m_x-\frac{1}{2},y} \right) dy \quad (8.25)$$

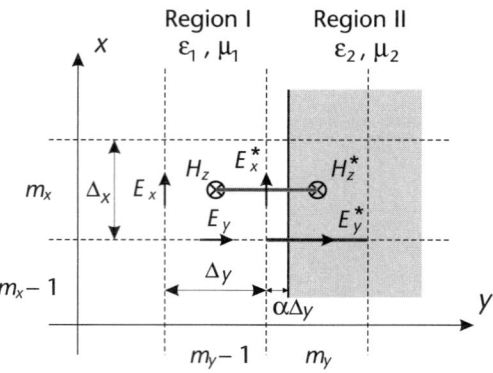

Figure 8.4 Arrangement of electric and magnetic fields near a dielectric interface for the case of 2-D TE polarization in the FDTD domain [23].

We can analyze (8.23) to calculate the effective permittivity of the electric field component E_x^*, which is tangential to the material interface. Since the material interface crosses the line of integration, we cannot assume that the electric field is constant along the line; though the magnetic field is not affected the tangential electric field component is continuous across the interface. We can rewrite the left-hand side of (8.23) as

$$\int_{(m_y-1)\Delta y}^{m_y \Delta y} D_x|_{m_x,y} dy = \int_{(m_y-1)\Delta y}^{(m_y-\frac{1}{2})\Delta y + \alpha \Delta y} D_x|_{m_x,y} dy + \int_{(m_y-\frac{1}{2})\Delta y + \alpha \Delta y}^{m_y \Delta y} D_x|_{m_x,y} dy$$

$$= \left[\left(\frac{1}{2} + \alpha\right) \varepsilon_1 + \left(\frac{1}{2} - \alpha\right) \varepsilon_2 \right] \Delta y E_x \big|_{m_x, m_y - \frac{1}{2}} \quad (8.26)$$

where α is defined as the distance from the interface to the location of the nearest tangential electric field component normalized by the grid size Δy (i.e., $0 \leq \alpha \leq 0.5$) (see Figure 8.4). Therefore, (8.26) leads us to the original Yee's algorithm, provided that we replace the permittivity of the cell (m_x, m_y) for the tangential field component with the effective permittivity as follows

$$\varepsilon_x^* = \left(\frac{1}{2} + \alpha\right) \varepsilon_1 + \left(\frac{1}{2} - \alpha\right) \varepsilon_2 \quad (8.27)$$

The effective permittivity for the electric field component E_y^*, which is perpendicular to the material interface, can be calculated in a similar way, and only the right side of Faraday's law (8.25) needs special treatment. We can apply the continuity of the electric flux density across the material interface; therefore the term containing E_y component on the right side of (8.25) can be evaluated by using

8.2 Dispersive FDTD Modeling of LHMs with Spatial Averaging at the Boundaries

$$\int_{(m_y-\frac{1}{2})\Delta y}^{(m_y+\frac{1}{2})\Delta y} \left(E_y \big|_{m_x+\frac{1}{2},y} - E_y \big|_{m_x-\frac{1}{2},y} \right) dy$$

$$= \int_{(m_y-\frac{1}{2})\Delta y}^{(m_y-\frac{1}{2}+\alpha)\Delta y} \left(E_y \big|_{m_x+\frac{1}{2},y} - E_y \big|_{m_x-\frac{1}{2},y} \right) dy$$

$$+ \int_{(m_y-\frac{1}{2}+\alpha)\Delta y}^{(m_y+\frac{1}{2})\Delta y} \frac{\varepsilon_1}{\varepsilon_2} \left(E_y \big|_{m_x+\frac{1}{2},y} - E_y \big|_{m_x-\frac{1}{2},y} \right) dy$$

$$= \left[\alpha + \frac{\varepsilon_1}{\varepsilon_2}(1-\alpha) \right] \Delta y \left(E_y \big|_{m_x+\frac{1}{2},y} - E_y \big|_{m_x-\frac{1}{2},y} \right) \quad (8.28)$$

Defining $E'_y \big|_{m_x \pm \frac{1}{2},y} = [\alpha + (1-\alpha)\varepsilon_1/\varepsilon_2] E_y \big|_{m_x \pm \frac{1}{2},y}$, (8.28) leads to the original Yee algorithm. The same is true if $E'_y \big|_{m_x - \frac{1}{2},y}$ is used in place of $E_y \big|_{m_x - \frac{1}{2},y}$, and the permittivity of the corresponding cell is replaced by the effective permittivity

$$\varepsilon_y^* = \left(\frac{\alpha}{\varepsilon_1} + \frac{1-\alpha}{\varepsilon_2} \right)^{-1} \quad (8.29)$$

For materials with magnetic properties such as LHMs, it is also necessary to find the effective permeability at the material interfaces. Similar to the procedure for the tangential electric field component E_x, the left side of (8.25) can be written as

$$-\frac{\partial}{\partial t} \int_{(m_x-\frac{1}{2})\Delta x}^{(m_x+\frac{1}{2})\Delta x} \int_{(m_y-\frac{1}{2})\Delta y}^{(m_y+\frac{1}{2})\Delta y} B_z \big|_{x,y} dx dy$$

$$= -\frac{\partial}{\partial t} \int_{(m_x-\frac{1}{2})\Delta x}^{(m_x+\frac{1}{2})\Delta x} \left(\int_{(m_y-\frac{1}{2})\Delta y}^{(m_y-\frac{1}{2}+\alpha)\Delta y} B_z \big|_{x,y} dy + \int_{(m_y-\frac{1}{2}+\alpha)\Delta y}^{(m_y+\frac{1}{2})\Delta y} B_z \big|_{x,y} dy \right) dx$$

$$= -\frac{\partial}{\partial t} \int_{(m_x-\frac{1}{2})\Delta x}^{(m_x+\frac{1}{2})\Delta x} [\alpha \mu_1 + (1-\alpha)\mu_2] \Delta y H_z \big|_{x,m_y} dx \quad (8.30)$$

which leads to the original Yee's algorithm if the permeability of the cell is replaced by the effective permeability

$$\mu_z^* = \alpha \mu_1 + (1-\alpha)\mu_2 \quad (8.31)$$

In the above derivations, it has been assumed that the material interface is located on the right side of its nearest tangential field component as shown in Figure 8.4. If instead, the interface lies on the left side of E_x^*, α is still defined as the distance between the interface and the location of E_x^* normalized to the cell size Δy, but the effective permittivity and permeability are changed to

$$\varepsilon_x^* = \left(\frac{1}{2} - \alpha \right) \varepsilon_1 + \left(\frac{1}{2} + \alpha \right) \varepsilon_2$$

$$\varepsilon_y^* = \left(\frac{1-\alpha}{\varepsilon_1} + \frac{\alpha}{\varepsilon_2} \right)^{-1}$$

$$\mu_z^* = (1-\alpha)\mu_1 + \alpha \mu_2 \quad (8.32)$$

and the location where the effective permittivity for E_y^* and effective permeability for H_z^* are applied, is shifted to $(m_x, m_y - 1)$.

Generally, the material interfaces can be located anywhere within the cell, and α is defined accordingly, as are the material parameters, effective permittivity, and permeability. However, in practice, a simpler way to implement the above parameters is to align the material interfaces the location of tangential field component E_x in the FDTD grid [i.e., let $\alpha = 0$, then $\varepsilon_y^* = \varepsilon_2$ in (8.29), $\mu_z^* = \mu_2$ in (8.31)] and calculate ε_x^* in (8.27) by a simple arithmetic mean of the permittivity of both the materials. In fact, for the case of $\alpha = 0$, the effective permittivity ε_x^* can also be calculated using either the harmonic or geometric means as follows [24–26]:

$$\varepsilon_x^* = \begin{cases} \dfrac{\varepsilon_1 + \varepsilon_2}{2}, & \text{(arithmetic mean)} \\ \dfrac{2\varepsilon_1 \varepsilon_2}{\varepsilon_1 + \varepsilon_2}, & \text{(harmonic mean)} \\ \sqrt{\varepsilon_1 \varepsilon_2}, & \text{(geometric mean)} \end{cases} \quad (8.33)$$

where it has been found that the arithmetic mean exhibits the best performance among above three schemes. Previous analyses of averaging techniques have only been performed for conventional dielectrics with positive permittivity and permeability. For materials with negative permittivity/permeability [e.g., $(\varepsilon = -1)/(\mu = -1)$], one of the simplest ways to implement the averaging scheme is to use the arithmetic mean, since the harmonic and geometric mean cannot be clearly defined.

Since the averaging is only applied when dealing with the field components tangential to the material interfaces [assuming $\alpha = 0$ in (8.27) and (8.29)]. Hence depending upon the configuration of the domain of the FDTD simulation, for instance, 2-D TE, 2-D TM, or 3-D, the averaging needs to be performed in different ways. In this study, we only consider a 2-D simulation domain as shown in Figure 8.4. For the interfaces between the LHM slab and free space along the x-direction, the averaged permittivity for the tangential electric field component E_x is given by

$$<\varepsilon_x> = \frac{\varepsilon_0 + \varepsilon_x}{2} = \varepsilon_0 \left[1 - \frac{\omega_{pe}^2}{2(\omega^2 - j\omega\gamma_e)} \right] \quad (8.34)$$

which is equivalent to replacing the plasma frequency ω_{pe} by $\omega'_{pe} = \omega_{pe}/\sqrt{2}$ in (6.4). Therefore along the boundaries, the updating equation for E_x reads

$$E_x^{n+1} = \left\{ \left[\frac{1}{\varepsilon_0 (\Delta t)^2} + \frac{\gamma_e}{2\varepsilon_0 \Delta t} \right] D_x^{n+1} - \frac{2}{\varepsilon_0 (\Delta t)^2} D_x^n \right.$$

$$+ \left[\frac{2}{(\Delta t)^2} - \frac{\omega_{pe}^2}{4} \right] E_x^n - \left[\frac{1}{(\Delta t)^2} - \frac{\gamma_e}{2\Delta t} + \frac{\omega_{pe}^2}{8} \right] E_x^{n-1}$$

$$\left. + \left[\frac{1}{\varepsilon_0 (\Delta t)^2} - \frac{\gamma_e}{2\varepsilon_0 \Delta t} \right] D_x^{n-1} \right\} \bigg/ \left[\frac{1}{(\Delta t)^2} + \frac{\gamma_e}{2\Delta t} + \frac{\omega_{pe}^2}{8} \right] \quad (8.35)$$

The updating equation in (8.35) is used at the locations marked by the arrows in Figure 8.5.

The averaging of permittivity can be implemented for the (E, D, H, B) scheme by the procedure mentioned earlier. However, for the (E, J, H, M) scheme, it is proposed in [13] that an averaging of the tangential current density along the boundaries of the LHM slab be used. Since the free-space current density $J_0 = 0$, the averaged current density can be calculated from:

$$<J_x> = \frac{J_0 + J_x}{2} = \frac{J_x}{2} \tag{8.36}$$

Then the updating equation for E_x along the boundaries of LHM slab becomes an expanded version of (8.18) as follows:

$$E_x|^{n+1}_{m_x,m_y} = \left[E_x|^n_{m_x,m_y} + \frac{\Delta t}{\varepsilon_0 \Delta y} \left(H_z|^{n+1}_{m_x,m_y} - H_z|^{n+1}_{m_x,m_y-1} \right) \right.$$

$$- \frac{\Delta t(\gamma_e \Delta t + 6)}{4\varepsilon_0(\gamma_e \Delta t + 2)} J_x|^n_{m_x,m_y} - \frac{\Delta t(\gamma_e \Delta t - 2)}{4\varepsilon_0(\gamma_e \Delta t + 2)} J_x|^{n-1}_{m_x,m_y}$$

$$\left. + \frac{\omega_{pe}^2 (\Delta t)^2}{4\gamma_e \Delta t + 8} E_x|^{n-1}_{m_x,m_y} \right] \bigg/ \left[1 + \frac{\omega_{pe}^2 (\Delta t)^2}{4\gamma_e \Delta t + 8} \right] \tag{8.37}$$

Theoretically, the above two averaging methods can be used interchangeably, since the relations

$$\mathbf{D} = \varepsilon \mathbf{E} = \varepsilon_0 \mathbf{E} + \frac{1}{j\omega} \mathbf{J}, \quad \mathbf{B} = \mu \mathbf{H} = \mu_0 \mathbf{H} + \frac{1}{j\omega} \mathbf{M} \tag{8.38}$$

are linear; therefore, averaging of the current density is identical to the averaging of the permeability. In this chapter, the (E, D, H, B) scheme is used in all simulations because of its simplicity in implementation. To demonstrate the advantage of the averaging technique, we compare the results from the simulations with and without averaged permittivity along the material boundaries. When we do not carry out an

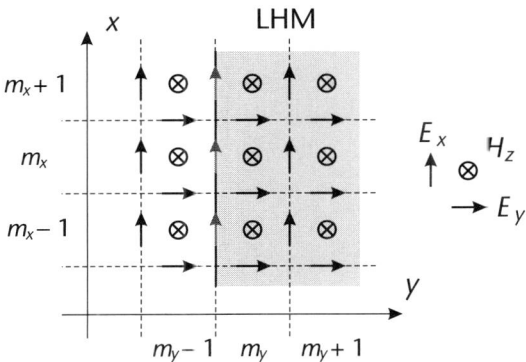

Figure 8.5 Layout of FDTD grid illustrating the arrangement of material boundaries along the x-direction. The FDTD unit cell is shown on the right side.

averaging of the material properties, we update the tangential electric fields using the update equations in free space.

The above averaging procedure applies only to the field components tangential to the material interfaces and for the case of TE polarization considered in the simulations. If we need to apply the averaging scheme to materials with planar boundaries for the TM and 3-D cases, or even for structures with curved surfaces, we can follow the procedures introduced in [16, 17].

8.3 Numerical Implementation

For the sake of simplicity, the plasma frequencies are assumed to be $\omega_{pe} = \omega_{pm} = \omega_p$ and the operating frequency is chosen as $\omega = \omega_p/\sqrt{2}$. This ensures, in turn, that the LHM slab is almost matched to the free space at this frequency in simulations. We introduce a loss (i.e., $\gamma_e = \gamma_m = \gamma = 0.0005\,\omega$) that results in a relative permittivity and permeability $\varepsilon_r = \mu_r = -1 - 0.001j$ and helps to speed up the convergence of the simulations. It is worth mentioning that there is a small amount of mismatch between the numerical (in the FDTD domain) and analytical permittivities (6.4), which is caused by the temporal discretization in FDTD [13]. However, this mismatch, although slight, still causes an amplification of the transmission coefficient though only for the lossless LHM slabs, or when the losses are very small. In the simulations below, we use sufficient losses so as to damp out amplifications of the transmission coefficient (the effect of FDTD cell size on this mismatch will be analyzed in the next section).

As shown in Figure 8.6, an infinite LHM slab is modeled by applying the Bloch type PBCs. For a periodic structure, the field satisfies the PBCs, also called the Bloch condition, given by

$$\mathbf{E}(x+L) = \mathbf{E}(x)e^{jk_xL}, \quad \mathbf{H}(x+L) = \mathbf{H}(x)e^{jk_xL} \qquad (8.39)$$

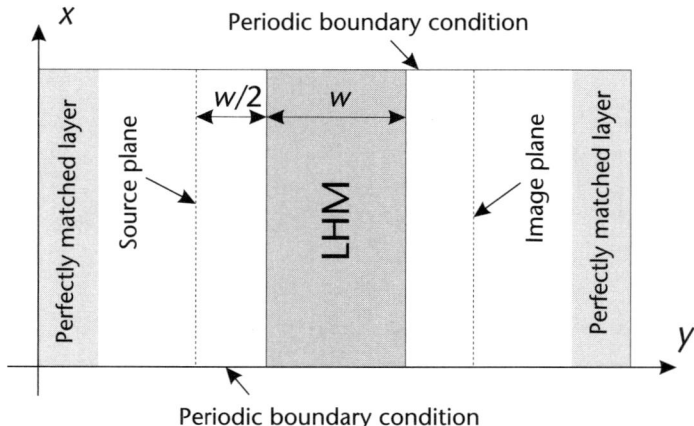

Figure 8.6 Schematic diagram of 2-D FDTD simulation domain for calculation of numerical transmission coefficient.

where k_x is the wave number in the x-direction and L is the lattice period along the direction of periodicity (see Figure 8.6). When updating the fields at the boundaries of the computational domain by using the FDTD method, it is possible to relate the required fields from outside the computational domain by using the known field value inside the domain by invoking (8.39). Since infinite structures have no periodicity, we truncate them arbitrarily, and save computation time by using only four FDTD cells in the x-direction ($L = 4\Delta x$). We employ the Berenger's original PML given in [27] in the y-direction, to absorb propagating waves (with $k_x < k_0$), and a modified PML [28] is used to calculate the transmission coefficient for evanescent waves ($k_x > k_0$). We use a soft, plane-wave, sinusoidal current-sheet source, which allows the scattered waves to pass through, and we insert phase delays corresponding to different wave numbers. The excitation is given by:

$$H_z(i, j_s) = H_z(i, j_s) + s(t)e^{-jk_x i \Delta x} \tag{8.40}$$

where j_s is the location of the source along the y-direction, $s(t)$ is a sinusoidal wave function; $i \in [1, I]$ is the index of cell location and I is the total number of cells in the x-direction ($I = 4$ in this case). Note that we can excite either pure propagating waves ($k_x < k_0$) or pure evanescent waves ($k_x > k_0$) by changing the values of the wave number k_x. The sinusoidal wave function $s(t)$ has to be ramped up slowly and smoothly to its maximum amplitude to minimize the excitation of other frequency components. The sinusoidal function used in the FDTD simulations is given by

$$s(t) = \begin{cases} g_{on}(t)e^{j\omega t} & \text{for } 0 \leq t < mT_0 \\ e^{j\omega t} & \text{for } t \geq mT_0 \end{cases} \tag{8.41}$$

where the switching function g_{on} is the turn-on part of the m-n-m pulse (a sinusoidal signal that has a smooth windowed turn-on for m cycles, a constant amplitude for n cycles, and then a smooth windowed turn-off for m cycles [8]). The switching function is given by

$$g_{on}(t) = 10.0 x_{on}^3 - 15.0 x_{on}^4 + 6.0 x_{on}^5 \tag{8.42}$$

where $x_{on} = 1.0 - (mT_0 - t)/mT_0$ and T_0 is the period of the sinusoidal wave function. An example case with $m = 10$ is shown in Figure 8.7.

The spatial discretization in the FDTD simulation is chosen to be $\Delta x = \Delta y = \lambda/100$, where λ is free-space wavelength at the operating frequency. According to the stability criterion [20], the discretized time step is $\Delta t = \Delta x/\sqrt{2}c$ where c is the velocity of light in free space. As illustrated in Figure 8.6, the source plane is located at a distance of $w/2$ from the front interface of the LHM slab where w is its thickness. Therefore, the first image plane is located at the center of the LHM slab, whereas the second image plane is formed at the same distance of $w/2$ beyond the slab. The spatial transmission coefficient is calculated as a ratio of the field intensity at the second image plane to the source plane for different transverse wave numbers k_x after the simulations have reached steady state.

Figure 8.8 shows the transmission coefficient for an infinite planar LHM slab whose thickness is $w = 0.2\lambda$, which has been simulated by using the FDTD method, with and without using an averaging of the permittivity along the boundaries and its comparison with a rigorous analytical solution. It is clearly

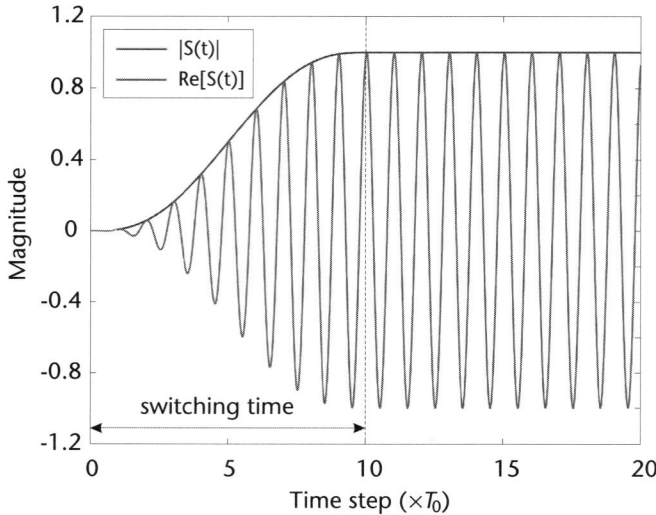

Figure 8.7 A time domain excitation function with a smooth switching time of $10T_0$ where T_0 is the period of the sinusoidal wave function.

evident that numerical results obtained by using the conventional dispersive FDTD, without an averaging of the permittivity, are correct only for $k_x < 2k_0$. This range of k_x not only covers the propagating waves ($k_x < k_0$) and but also a small part of weakly decaying evanescent waves ($k_0 < k_x < 2k_0$). Note, however,

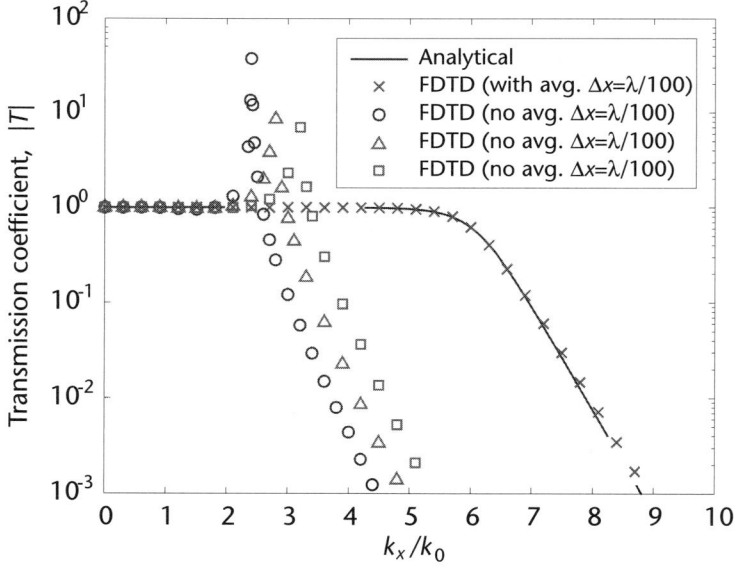

Figure 8.8 Comparison of transmission coefficient as a function of transverse wave vector k_x for infinite planar LHM slabs with $\varepsilon_r = \mu_r = -1 - 0.001j$ and thickness $w = 0.2\lambda$, calculated from exact analytical solution and dispersive FDTD method with and without averaging of permittivity along the boundaries of LHM slabs.

that for evanescent waves with $k_x > 2k_0$, the numerical results are significantly different from the analytical ones and, that the former shows a resonant behavior with a strong peak at $k_x = 2.4k_0$. This effect can be explained as a numerical artifact (which is caused by the incorrect modeling of material boundaries and only exists in inaccurate numerical simulations) at the back interface of the LHM slab. However, a similar behavior can also be observed when the slabs are metallic [7] or for unmatched LHMs [19], but in this particular case it is a purely a numerical artifact. This incorrect behavior of the numerical solutions continuous to prevail even when the FDTD cell size is reduced to $\lambda/200$ and $\lambda/400$, but the resonance shifts to $k_x = 2.8k_0$ and $k_x = 3.2k_0$ (as may be seen from Figure 8.8 where these two results are indicated by triangles and squares), respectively.

The above comparison provides evidence that the boundaries between the LHM and the free space have not been modeled accurately. If at the boundaries the mean value of the permittivity of LHM and the free space is used for updating the tangential component of the electric field (which is equivalent to the spatial averaging suggested in [13]) then the spurious resonant behavior no longer exists and the modeling results becomes very accurate. The transmission coefficient calculated by using the proposed spatial averaging at the boundaries is depicted in Figure 8.8 by using crosses. It is evident that the results agree well with estimated analytical values with a very good accuracy for the entire range of spatial spectra of the waves. The calculation has been carried out by using $\Delta x = \Delta y = \lambda/100$, and it remains accurate even for larger cell sizes (e.g., $\Delta x = \Delta y = \lambda/80$). The above simple test leads us to conclude that the conventional dispersive FDTD method fails to describe the propagation of high-order evanescent waves along the boundaries of LHM slabs if we do not apply the interface condition correctly at the boundaries of the LHM slab. Consequently, a number of results previously obtained by using the conventional dispersive FDTD approaches have to be examined. This is particularly relevant when we model the phenomenon of subwavelength imaging by LHM slabs [7] that involves the contribution of evanescent waves. Note that the simulations that are concerned only with propagating waves, as for instance, when attempting to demonstrate negative refraction for an obliquely incident plane wave, are not affected by this difficulty encountered during the process of dealing with the evanescent waves.

The results shown in Figure 8.8 may help to explain why some of results obtained previously for the LHM slab problem were incorrect. For instance, we can argue that the amplification in the transmission coefficient is solely due to the mismatch of the real part of permittivity/permeability across the interface [19]. This is confirmed by the plot in Figure 8.9, which is in good agreement with the theoretical result given in [19]. The above figure also shows that the mismatch causes the cutoff of transmission to move to a lower value, in other words, regardless of the amount of losses, if the LHM slab is matched to free space [i.e., $(\text{Re}[\varepsilon] = -\varepsilon_0)$], there is no amplification in the transmission coefficient [18]. Furthermore, increasing the amount of losses lowers the cutoff of the transmission coefficient [18], which is also observed in the FDTD simulations with averaged permittivity, as shown in Figure 8.10. Therefore it is conjectured that the inaccuracy in the numerical transmission coefficient, derived by using the FDTD method in [11, 12] is attributed to an incorrect modeling of the boundaries of LHM slabs in FDTD.

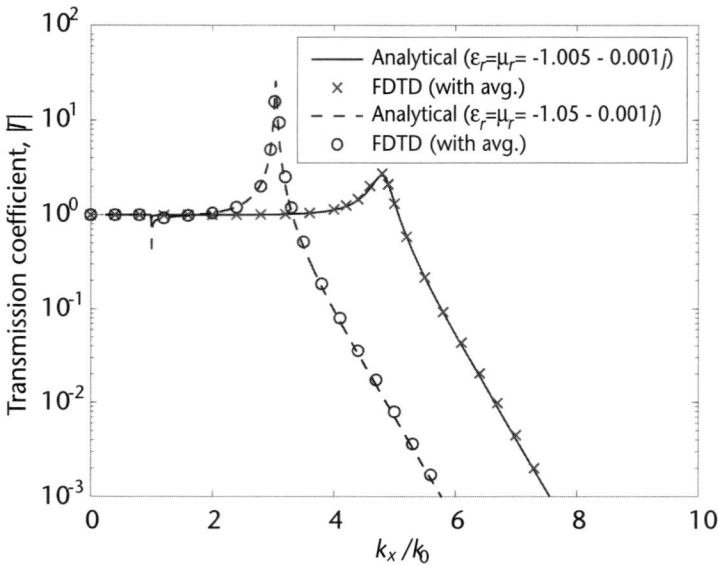

Figure 8.9 Comparison of transmission coefficient as a function of transverse wave vector k_x for infinite planar LHM slabs with thickness $w = 0.2\lambda$ and different amount of mismatch of the real part of material parameters ($\varepsilon_r = \mu_r = -1.005 - 0.001j$ and $\varepsilon_r = \mu_r = -1.05 - 0.001j$) calculated from exact analytical solution and dispersive FDTD method with averaging of permittivity along the boundaries of LHM slabs.

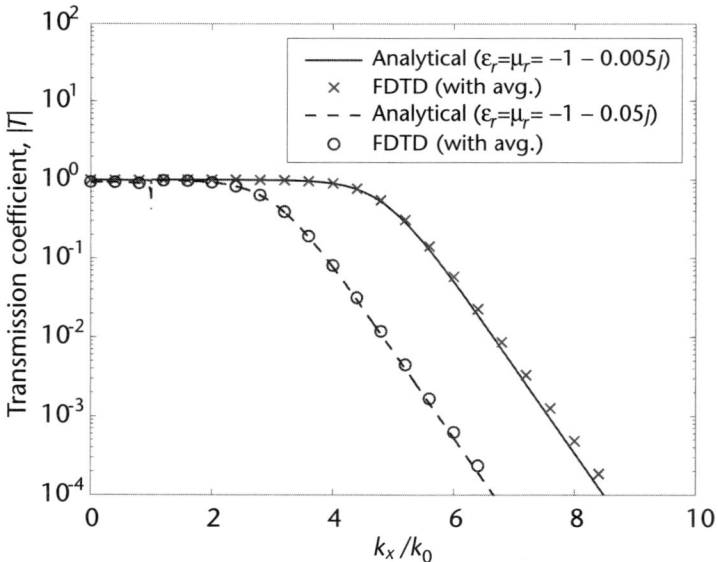

Figure 8.10 Comparison of transmission coefficient as a function of transverse wave vector k_x for infinite planar LHM slabs with thickness $w = 0.2\lambda$ and different amount of losses ($\varepsilon_r = \mu_r = -1 - 0.005j$ and $\varepsilon_r = \mu_r = -1 - 0.05j$) calculated from exact analytical solution and dispersive FDTD method with averaging of permittivity along the boundaries of LHM slabs.

To calculate the transmission coefficient, we use PBCs in the x-direction to simulate an infinite structures and average the permittivity along both the boundaries in the y-direction. Also we show in a later part of this chapter that we use an averaging of the permittivity for the corresponding tangential field components along the boundaries in both the x- and y-directions, when modeling finite-sized structures.

Besides the averaging technique mentioned above and used along the boundaries of LHM slabs, there are other numerical aspects of the FDTD simulations, identified below, that influence the correctness of the results pertaining to the LHM slabs. In the rest of this chapter, unless otherwise specified, the field averaging scheme developed below is always applied at the material interfaces for the accurate FDTD modeling of LHM slabs with other media.

8.4 Effects of Material Parameters on the Accuracy of Numerical Simulation

Typically, when modeling conventional dielectrics, it is assumed that the results are sufficiently accurate and that the effect of numerical material parameters is negligible, provided that we use an FDTD cell size that is less than $\Delta x = \lambda/10$. However, when we are dealing with evanescent waves, the true value of permittivity and permeability and their counterparts that are strongly affected by the spatial resolution of the FDTD grid, have a significant impact on the accuracy of the simulation results. The above effect was originally identified in [13] for lossless LHMs using the (E, J, H, M) scheme. We can follow a similar procedure to obtain the numerical permittivity/permeability for lossy LHMs, and below we derive for the (E, D, H, B) scheme. Note that since we have assumed that $\varepsilon_r = \mu_r$, we limit the derivation of the numerical material parameter to the permittivity alone.

If we revert to updating (8.11), substitute the expressions,

$$\mathbf{E}^{n\pm 1} = \mathbf{E}^n e^{\pm j\omega\Delta t}, \quad \mathbf{D}^{n\pm 1} = \mathbf{D}^n e^{\pm j\omega\Delta t} \tag{8.43}$$

and utilize the trigonometric relations,

$$e^{j\omega\Delta t} + e^{-j\omega\Delta t} = 2\cos(\omega\Delta t), \quad e^{j\omega\Delta t} - e^{-j\omega\Delta t} = 2j\sin(\omega\Delta t) \tag{8.44}$$

We can rewrite (8.11) as

$$\mathbf{E}^n \left[\frac{\omega_p^2}{2} + \frac{2}{(\Delta t)^2} \right] \cos(\omega\Delta t) + \mathbf{E}^n \left[\frac{\omega_p^2}{2} - \frac{2}{(\Delta t)^2} \right] + \mathbf{E}^n \frac{\gamma}{\Delta t} j\sin(\omega\Delta t)$$

$$= \mathbf{D}^n \frac{2}{\varepsilon_0 (\Delta t)^2} \cos(\omega\Delta t) - \mathbf{D}^n \frac{2}{\varepsilon_0 (\Delta t)^2} + \mathbf{D}^n \frac{\gamma}{\varepsilon_0 \Delta t} j\sin(\omega\Delta t) \tag{8.45}$$

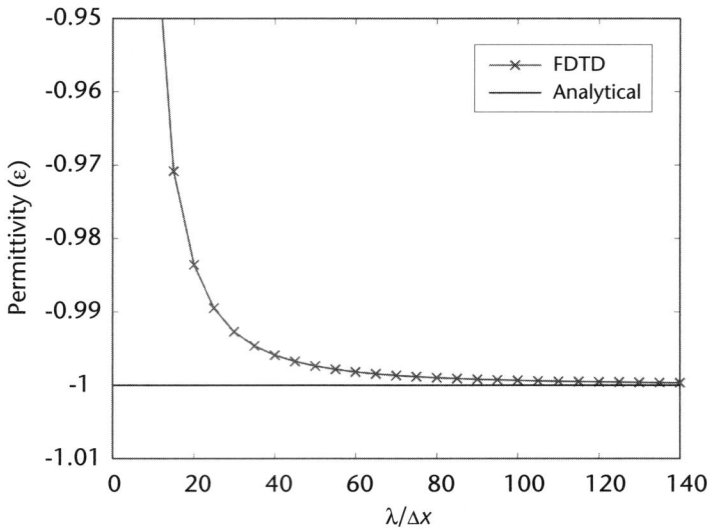

Figure 8.11 Comparison of the real part of analytical permittivity (6.4) and numerical permittivity (8.46) for different FDTD spatial resolutions. The parameters (i.e., ω_p, γ_p, and ω) are chosen so that the real part of analytical permittivity is equal to -1.

Using straightforward manipulations, we can derive the numerical permittivity ($\tilde{\varepsilon} = \mathbf{D}^n/\mathbf{E}^n$) from the above equation to get

$$\tilde{\varepsilon} = \varepsilon_0 \left[1 - \frac{\omega_p^2 (\Delta t)^2 \cos^2 \frac{\omega \Delta t}{2}}{2 \sin \frac{\omega \Delta t}{2} \left(2 \sin \frac{\omega \Delta t}{2} - j\gamma \Delta t \cos \frac{\omega \Delta t}{2} \right)} \right] \quad (8.46)$$

If we let the collision frequency $\gamma = 0$, then (8.46) reduces to the numerical permittivity for lossless LHMs given in [13]. If $\Delta t \to 0$, then (8.46) reduces to the permittivity for the analytical study given in (6.4).

To study how different spatial FDTD spatial resolutions affect the numerical permittivity, we first choose the parameters in (6.4) that yield a permittivity equal to -1 and then compute the numerical permittivities from (8.46). The comparison between the two results is shown in Figure 8.11. It can be evident that while the analytical permittivity is always equal to -1, the same is not true for its numerical counterpart. For instance, when the FDTD spatial resolutions of $\Delta x > \lambda/60$, the numerical permittivity shows significant deviations from the true value, and this, in turn, leads to an amplification of the transmission coefficient as we show in the following section. This leads us to conclude that the spatial resolution typically employed in the conventional FDTD simulations of frequency-independent dielectrics, namely $\Delta x < \lambda/20$, is inadequate when analyzing the LHMs because a coarse discretization fails to deal with evanescent waves accurately. The plot in Figure 8.11 also suggests that the spatial resolution of less than $\Delta x < \lambda/80$ is necessary for the accurate modeling of the LHMs.

Previously an FDTD cell size of $\Delta x = \lambda/100$ was used in the simulations. By substituting the corresponding time step $\Delta t = \Delta x/\sqrt{2}c$, and the operating frequency, we obtain the numerical value for the relative permittivity from (8.46) to

8.4 Effects of Material Parameters on the Accuracy of Numerical Simulation

get $\tilde{\varepsilon}_r = -0.9993 - 0.0010j$. Although there is a very slight mismatch between the real part of the numerical relative permittivity and its true value of -1, the losses in the LHMs damp out the spurious results of the fields introduced by the mismatch and the simulation results are shown be very accurate. However, if we increase the FDTD cell size to $\Delta x = \lambda/40$, we observe a much more severe effect on the numerical permittivity and considerable discrepancy between the FDTD simulation result and the exact solution. The mismatch between the material parameters serves to introduce an amplification in the transmission coefficient in the simulation as shown in Figure 8.12. We can again estimate this mismatch by using (8.46) which yields the numerical relative permittivity of $\tilde{\varepsilon}_r = -0.9959 - 0.0010j$, and the corresponding transmission coefficient is plotted in Figure 8.12 and compared with the analytical result. Good agreement between the two is seen for high-wave-vector k_x regime, which corresponds to the evanescent wave region.

Another advantage of estimating the numerical permittivity while modeling LHMs is that the effects of mismatch in the FDTD simulations can be corrected. After simple derivations using (8.46), the corrected plasma and collision frequencies can be obtained as follows:

$$\tilde{\omega}_p^2 = \frac{2\sin\frac{\omega\Delta t}{2}\left[-2(\varepsilon_r' - 1)\sin\frac{\omega\Delta t}{2} - \varepsilon_r''\gamma\Delta t \cos\frac{\omega\Delta t}{2}\right]}{(\Delta t)^2 \cos^2\frac{\omega\Delta t}{2}} \tag{8.47}$$

$$\tilde{\gamma} = \frac{2\varepsilon_r''\sin\frac{\omega\Delta t}{2}}{(\varepsilon_r' - 1)\Delta t \cos\frac{\omega\Delta t}{2}} \tag{8.48}$$

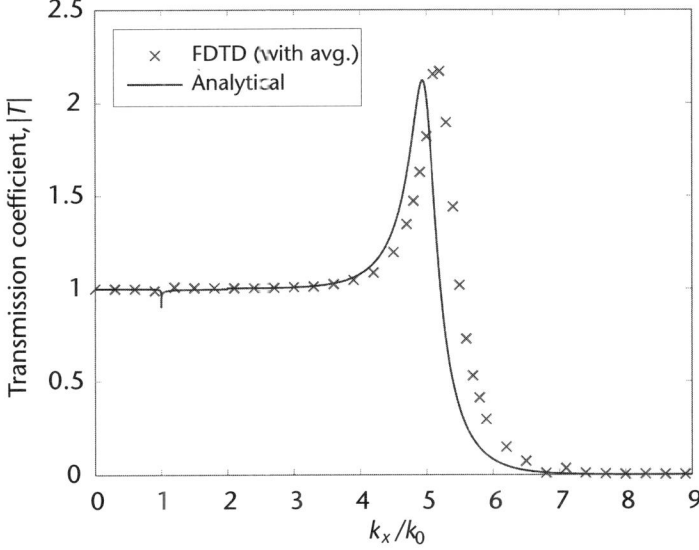

Figure 8.12 Transmission coefficient of infinite planar LHM slabs using proposed FDTD method with averaged permittivity and without the correction of material parameters. The amplification of transmission coefficient is caused by the mismatch introduced by time discretization in FDTD ($\tilde{\varepsilon}_r = -0.9959 - 0.0010j$) with $\Delta x = \lambda/40$. The same permittivity is used to obtain an analytical solution for comparison.

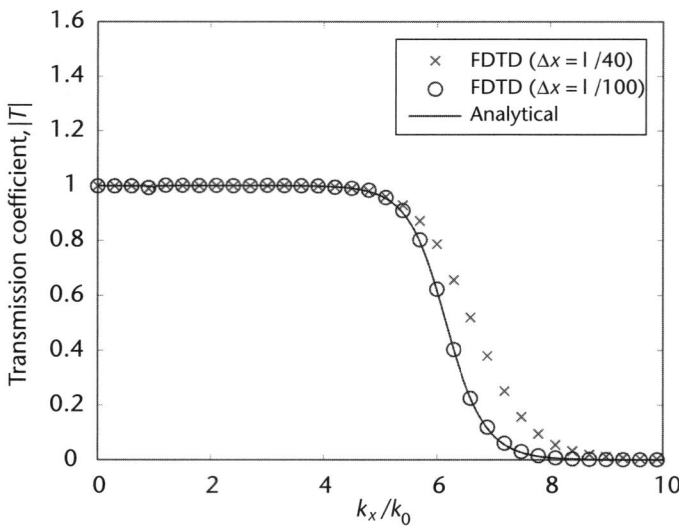

Figure 8.13 Transmission coefficient of infinite planar LHM slabs using proposed FDTD method with averaged permittivity and with the correction of material parameters for different FDTD spatial resolutions: $\Delta x = \lambda/40$ and $\Delta x = \lambda/100$. The numerical permittivity in FDTD is $\tilde{\varepsilon}_r = -1 - 0.001j$. The same permittivity is used to obtain analytical solution for comparison.

where ε_r' and ε_r'' are the real and imaginary parts of the relative permittivity ε_r, respectively. For $\varepsilon_r = -1 - 0.001j$, we substitute $\varepsilon_r' = -1$ and $\varepsilon_r'' = -0.001$ into (8.47) and (8.48) to get $\tilde{\omega}_p = 1.4157\omega$ and $\tilde{\gamma} = 5.0051 \times 10^{-4}\omega$. The FDTD simulation results based on these corrected material parameters and its comparison with the analytical solution are shown in Figure 8.13. It can be seen that the mismatch is no longer noticeable in the FDTD simulations, and consequently, there is no amplification in the transmission coefficient. We reiterate that the discrepancy between the exact and numerical solutions in the high-wave-vector region is caused by an inadequate level of discretization in the FDTD where a spatial resolution of $\Delta x = \lambda/40$ was chosen. However, if we use a finer spatial resolution of $\Delta x = \lambda/100$, the FDTD simulation results show good agreement with the analytical solution. Therefore, it is recommended that an FDTD cell size smaller than $\Delta x = \lambda/80$ be used for modeling LHMs especially when evanescent waves must be accounted accurately.

8.5 Effects of Switching Time

When the FDTD is used for single frequency simulations, the time domain source should be turned on and ramped up smoothly to its maximum value to avoid the excitation of other frequency components [8]. The switching time has an even more significant effect, when modeling the LHMs, on single frequency results. It is well known that the switching time influences the oscillation of images in the simulation of a perfect lens, and often 30 time periods are used to switch on the source [10, 11]. Recently, it was reported in [29] that using a switching time of at least 100 is needed to obtain a stabilized image for lossless LHMs, and

8.5 Effects of Switching Time

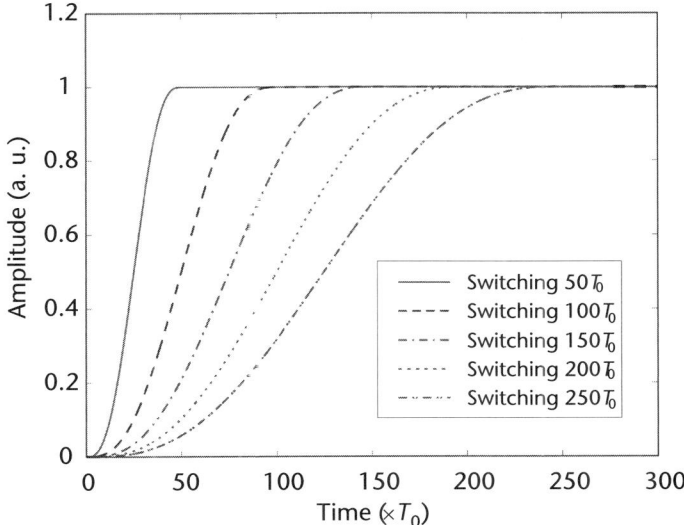

Figure 8.14 The amplitude of source for different switching times in FDTD simulation of infinite LHM slabs for a fixed wave number $k_x = 3k_0$. T_0 is the period of sinusoidal wave function at the operating frequency.

we conjecture that the reason why no stable images were obtained was that the switching time did not include a sufficient number of cycles.

On the basis of the FDTD simulations performed in this study, we have also found that the switching time has a significant influence on the nature of oscillation of the field intensity at the image plane and, hence, affects the convergence of the

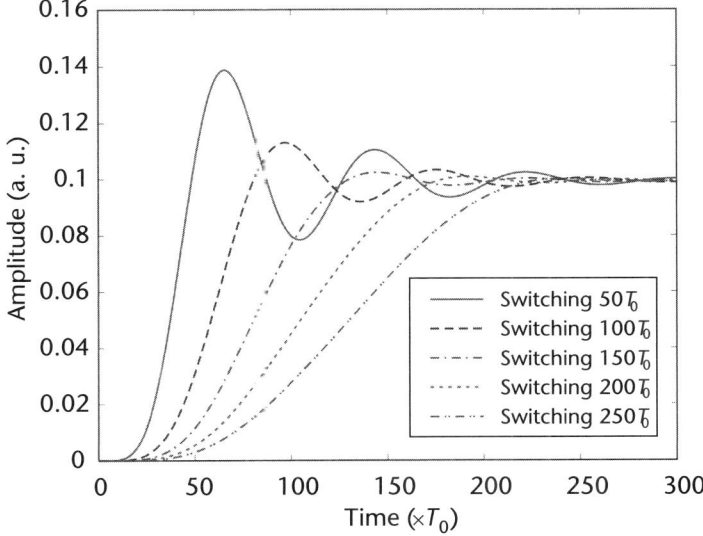

Figure 8.15 The influence of different switching times on convergence time in FDTD simulation of infinite LHM slabs for a fixed wave number $k_x = 3k_0$. T_0 is the period of sinusoidal wave function at the operating frequency. The field amplitude is taken at the second image plane of LHM slab.

time domain simulations. We choose the switching times in the FDTD simulation to be $50T_0$, $100T_0$, $150T_0$, $200T_0$, and $250T_0$ where T_0 is the period of the sinusoidal wave function. The FDTD cell size is $\Delta x = \lambda/100$ and corrected material parameters derived from (8.47) and (8.48) are used. In order to ensure faster convergence, larger amounts of losses and $\tilde{\varepsilon}_r = -1 - 0.01j$ are used in simulations. It should be noted that because high-wave-vector components travel very slowly in LHM slabs, it takes a very long time for evanescent waves to grow fully and to reach the steady-state. For this reason, the simulations should be continued into the late times until the field values have fully converged. For $\tilde{\varepsilon}_r = -1 - 0.01j$, we use a convergence criterion of 0.001% to terminate the simulations. The temporal signatures of the sources for different switching times for a fixed wave number $k_x = 3k_0$ are displayed in Figure 8.14 and the corresponding amplitudes of the field at the second image plane of the LHM slab are presented in Figure 8.15. It is evident from Figure 8.15 that for a fixed wave number of $k_x = 3k_0$, the oscillation in the field intensity can be significantly suppressed by using protracted switching time. A large amount of losses ($\tilde{\varepsilon}_r = -1 - 0.01j$) has been used to demonstrate the effect of switching time. The switching time has even more significant impact on the convergence time in FDTD simulations if we reduce the losses to lower levels.

It is understandable that when the oscillation can be neglected, the convergence time increases with the switching time. For the demonstration of the impact of the switching time on the convergence time, FDTD simulations with various switching times are performed and the collected data is plotted in Figure 8.16. It can be seen that there exists an optimum switching time when the minimum convergence time can be achieved for the case of $k_x = 3k_0$. However, for different wave vectors and material parameters, the behavior of oscillation varies considerably and in certain cases the oscillation may last for a very long time. For simulations pertaining to subwavelength imaging by a line source, it is necessary to switch the source sufficiently slow to ensure the convergence of the time-domain simulation, since the source is comprised of a wide spectrum of wave vectors.

8.6 Effects of Transverse Dimensions on Image Quality

As we mentioned earlier, a number of results reported in previous works need to be evaluated carefully and checked for validity. Recently, a rather puzzling result related to the quality of imaging provided by LHM slabs has been reported in the literature [30]. It is claimed that the performance of a finite-sized LHM structure is significantly affected by its transverse dimensions. We reexamine this statement by using an FDTD code that utilizes spatial averaging at the boundaries of the LHM slabs, and has been proven to be accurate. The finite-sized slabs of LHM excited by soft magnetic current line sources are modeled for three different transverse dimensions, $L = \lambda$, 2λ, and 4λ, as illustrated in Figure 8.17. The parameters of the LHM slab ($\varepsilon = \mu = -1 - 0.001j$, $w = 0.2\lambda$), and the distance between the source and front interface equals $w/2$ for all these simulations. The computational domain is truncated by PMLs located at a distance of $\lambda/2$ away from the LHM slab, and in both source and image planes. The FDTD cell size $\Delta x = \Delta y = \lambda/100$, and time step $\Delta t = \Delta x/\sqrt{2}c$ are the same as those used for the previous plane-wave

8.6 Effects of Transverse Dimensions on Image Quality

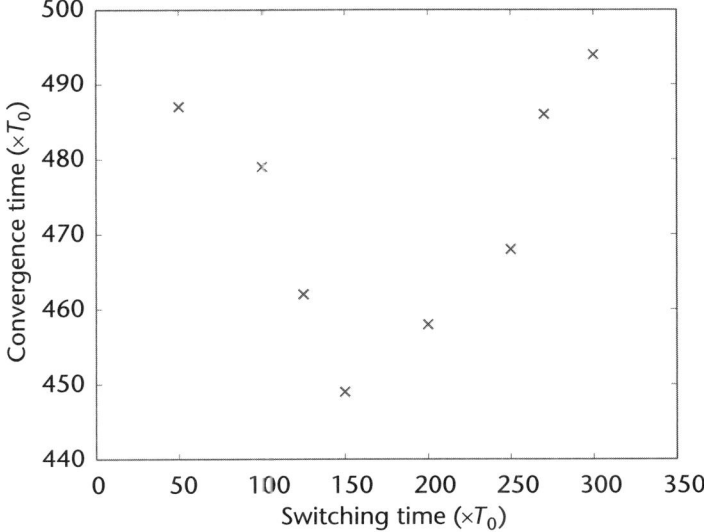

Figure 8.16 The dependence of convergence time on switching time in FDTD simulations of infinite LHM slabs for a fixed wave number $k_x = 3k_0$. A criterion of 0.001% is used to detect iteration errors and terminate simulations.

simulations. The source is switched slowly and smoothly to avoid the contributions from undesired frequency components, as in [8], and the time stepping is continued until the steady state is reached. The intensity distributions obtained in the image plane are plotted in Figure 8.18 for all three cases of different transverse dimensions mentioned above. It is evident from this figure that the image quality is almost unaffected by the transverse size of the LHM slab. The fields in the image plane are seen to replicate the source distribution, which is also plotted in the same figure. The subwavelength resolution is seen to be good and the finiteness of LHM slab

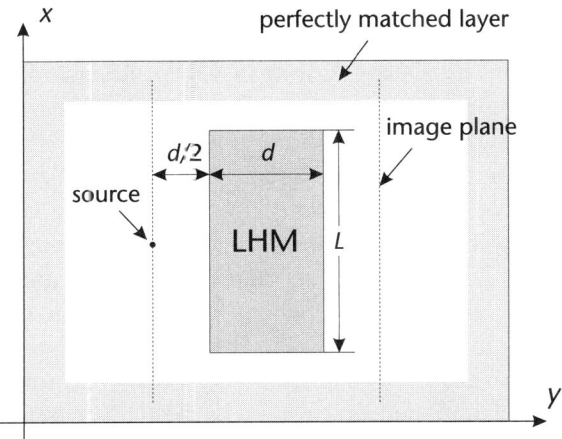

Figure 8.17 Schematic diagram of 2-D FDTD simulation domain for modeling the imaging of magnetic line sources by finite-sized LHM slabs.

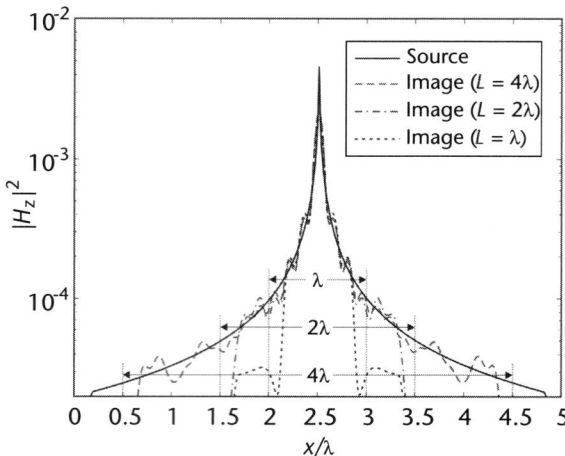

Figure 8.18 Comparison of field intensities of magnetic field $|H_z|^2$ at image planes of planar LHM lenses ($\varepsilon_r = \mu_r = -1 - 0.001j$, $w = 0.2\lambda$) with various transverse dimensions: $L = \lambda$, 2λ, and 4λ.

does not distort the image, which is contrary to the conclusions reported in [30]. The slight disagreement between the image and source distribution is attributable to losses in the LHM. The resonant effects and image distortions reported in [30] may have been caused by spurious excitations of "numerical surface plasmon" at the interfaces between the LHM slab and free space, since these artifacts also arise when the FDTD simulations are used without spatial averaging of material properties at the boundaries introducing undesirable inaccuracies in the process. Incidentally, the fact that the imaging performance of finite-sized LHM slabs is unaffected by their transverse dimensions has been confirmed independently by full-wave electromagnetic simulation using the Ansoft HFSS package.

8.7 Modeling of Subwavelength Imaging

To further validate the assertion that the subwavelength imaging property of the LHM lenses is insensitive to their transverse dimensions, FDTD simulations with the spatial averaging at the boundaries have been performed for an LHM slab whose transverse size is $L = \lambda$, and that is excited by two magnetic line sources $\lambda/8$ between each other. The rest of the parameters for this study are the same as those used in the previous investigation of the single-source case. The distance between the two sources exceeds the resolution of the LHM lens, which is estimated to be better than $\lambda/12$ based on the fact that the transfer function plotted in Figure 8.8 is close to unity for $k_x/k < 6$; hence, for the present case, we expect two well-resolved maxima in the image plane. The distribution of magnetic field intensity in a subdomain near the LHM slab (the actual FDTD domain is larger) in the steady state is presented in Figure 8.19(a). The field distribution in the image plane is shown in Figure 8.19(c) together with the source. Two maxima at the distance of $\lambda/8$ are clearly visible in the image plane. This confirms the subwavelength imaging capability of the LHM lens whose width is only one wavelength. This

8.7 Modeling of Subwavelength Imaging

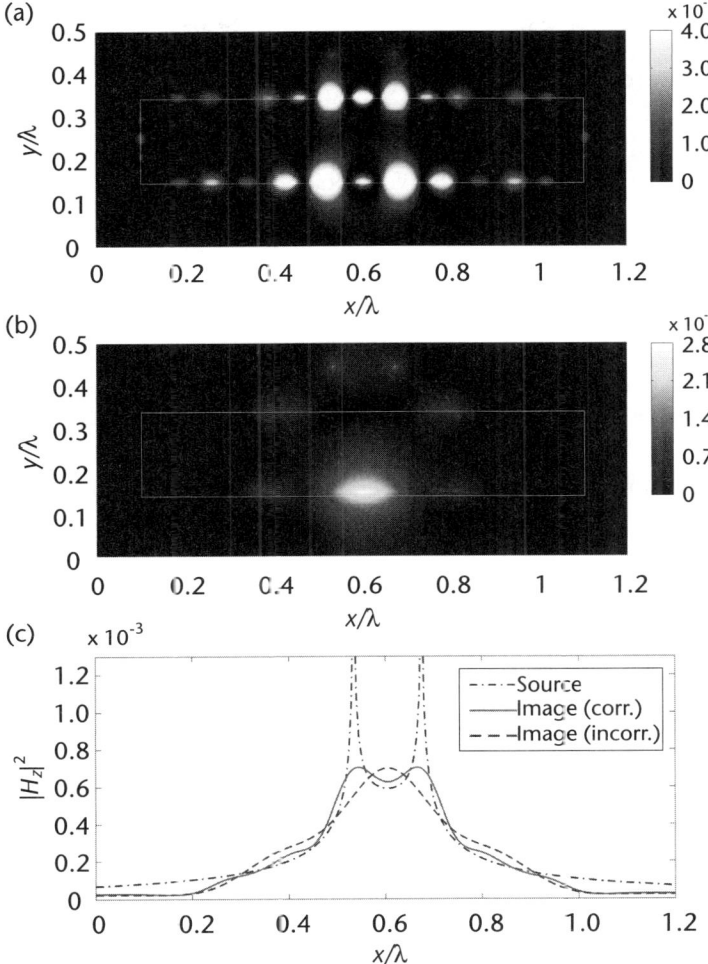

Figure 8.19 The magnetic field intensity distributions around a LHM slab with transverse size $L = \lambda$ excited by two magnetic line sources placed at $\lambda/8$ distance between each other obtained using the FDTD method (a) with and (b) without spatial averaging; and (c) comparison of magnetic field intensities at the source and image planes of the LHM slab.

leads us to conclude that the truncation of the transverse dimensions of an LHM slab does not appear to have an overt effect on its functionality as a subwavelength imaging device. However, when the same system is incorrectly modeled via the FDTD method without using boundary corrections, the field distribution around the slab can be significantly different, as shown in Figure 8.19(b). For instance, the field distribution along the slab interface is found to be smooth, which once again confirms that higher-order evanescent waves are not correctly modeled in the simulation without boundary corrections. As a consequence, the subwavelength details of the source are not resolved in the image plane, which only shows a single and relatively wide maximum [see Figure 8.19(c)], as opposed to two closely-spaced maxima.

8.8 Conclusions

In this chapter, numerical simulations of LHMs have been performed using the dispersive FDTD method. Two ADE methods for simulation, namely the (E, D, H, B) and the (E, J, H, M) schemes, which yield identical results, have been introduced. It is demonstrated that the conventional FDTD method for the modeling of LHM introduces inaccuracies when dealing with higher-order evanescent waves and is unable to simulate the subwavelength imaging phenomena correctly. The simulations introduce spurious artifacts of "numerical surface plasmons," which appear at the interfaces between the LHM slab and the free space. To solve this problem, and to ensure the accuracy of the FDTD modeling, a spatial averaging scheme at the boundaries of LHM slabs has been applied to a test problem, which is analytically tractable. The results of the simulation have been found to have good accuracy for all angles of incidence even when higher-order evanescent waves are involved.

The numerical permittivity in FDTD has been derived to handle the simulation where a mismatch between the numerical and analytical permittivities is introduced by the discretization in FDTD, and a way to correct such a mismatch has been described. The oscillatory behavior of field intensity has been studied as a function of the switching time, and it is shown that there exists an optimum switching time that serves to reduce the convergence time in the FDTD simulations to a minimum.

Finite-sized LHM slabs excited by a single line source have been investigated by using the proposed technique for various transverse dimensions of the slab, and their subwavelength imaging capability has been confirmed. It is shown that the finite transverse dimensions of the structure does not introduce distortion in the image and is thus contrary to the results reported in [30]. It is suspected that the resonant effects due to the finite transverse dimensions of LHM slabs reported in [30] are caused by the incorrect modeling of the material boundaries and are numerical artifacts. These effects are not seen to be present in real structures, and there appear to be almost no restrictions on the functionality of the LHM subwavelength lenses that are tied to their transverse dimensions. In fact, it is demonstrated that the FDTD simulations with spatial averaging at material boundaries can successfully image two line sources separated by a subwavelength distance between each other, even when the transverse dimension is only one wavelength.

References

[1] V. G. Veselago, "The electrodynamics of substances with simultaneously negative values of ε and μ," *Sov. Phys. Usp.*, vol. 10, pp. 509–514, 1968.

[2] J. B. Pendry, A. J. Holden, W. J. Stewart, and I. Youngs, "Extremely low frequency plasmons in metallic mesostructures," *IEEE Trans. Microw. Theory and Tech.*, vol. 47, pp. 2075–2084, 1999.

[3] J. B. Pendry, A. J. Holden, and D. J. Robbins, "Magnetism from conductors and enhanced nonlinear phenomena," *Phys. Rev. Lett.*, vol. 76, p. 4773, 1996.

[4] A. L. Pokrovsky and A. L. Efros, "Electrodynamic of metallic photonic crystals and the problem of left-handed materials," *Phys. Rev. Lett.*, vol. 89, p. 093901, 2002.

[5] D. R. Smith, W. J. Padilla, D. C. Vier, S. C. Nemat-Nasser, and S. Schultz, "Composite media with simultaneously negative permeability and permittivity," *Phys. Rev. Lett.*, vol. 84, pp. 4184–4187, 2000.

[6] R. A. Shelby, D. R. Smith, S. C. Nemat-Nasser, and S. Schultz, "Microwave transmission through a two-dimensional, isotropic, left-handed metamaterial," *Appl. Phys. Lett.*, vol. 78, p. 489, 2001.

[7] J. B. Pendry, "Negative refraction makes a perfect lens," *Phys. Rev. Lett.*, vol. 85, no. 28, pp. 3966–3969, 2000.

[8] R. W. Ziolkowski and E. Heyman, "Wave propagation in media having negative permittivity and permeability," *Phys. Rev. E*, vol. 64, p. 056625, 2001.

[9] P. F. Loschialpo, D. L. Smith, D. W. Forester, F. J. Rachford, and J. Schelleng, "Electromagnetic waves focused by a negative-index planar lens," *Phys. Rev. E*, vol. 67, p. 025602, 2003.

[10] S. A. Cummer, "Simulated causal subwavelength focusing by a negative refractive index slab," *Appl. Phys. Lett.*, vol. 82, p. 1503, 2003.

[11] X. S. Rao and C. K. Ong, "Subwavelength imaging by a left-handed material superlens," *Phys. Rev. E*, vol. 68, p. 067601, 2003.

[12] M. W. Feise and Y. S. Kivshar, "Sub-wavelength imaging with a left-handed material flat lens," *Phys. Lett. A*, vol. 334, pp. 323–326, 2005.

[13] J. J. Chen, T. M. Grzegorczyk, B.-I. Wu, and J. A. Kong, "Limitation of FDTD in simulation of a perfect lens imaging system," *Opt. Express*, vol. 13, pp. 10840–10845, 2005.

[14] M. W. Feise, J. B. Schneider, and P. J. Bevelacqua, "Finite-difference and pseudospectral time-domain methods applied to backward-wave metamaterials," *IEEE Trans. Antennas Propagat.*, vol. 52, no. 11, pp. 2955–2962, 2004.

[15] Y. Zhao, P. Belov, and Y. Hao, "Accurate modelling of the optical properties of left-handed media using a finite-difference time-domain method," *Phys. Rev. E*, vol. 75, p. 037602, 2006.

[16] A. Mohammadi, H. Nadgaran, and M. Agio, "Contour-path effective permittivities for the two-dimensional finite-difference time-domain method," *Opt. Express*, vol. 13, pp. 10367–10381, 2005.

[17] A. Mohammadi and M. Agio, "Dispersive contour-path finite-difference time-domain algorithm for modelling surface plasmon polaritons at flat interfaces," *Opt. Express*, vol. 14, pp. 11330–11338, 2006.

[18] V. A. Podolskiy and E. E. Narimanov, "Near-sighted superlens," *Opt. Lett.*, vol. 30, pp. 75–77, 2005.

[19] D. R. Smith, D. Schurig, M. Rosenbluth, S. Schultz, S. A. Ramakrishna, and J. B. Pendry, "Limitations on subdiffraction imaging with a negative refractive index slab," *Appl. Phys. Lett.*, vol. 82, pp. 1506–1508, 2003.

[20] A. Taflove and S. C. Hagness, *Computational Electrodynamics: The Finite-Difference Time-Domain Method*, 2nd ed., Norwood, MA: Artech House, 2000.

[21] O. P. Gandhi, B. Q. Gao, and J. Y. Chen, "A frequency-dependent finite-difference time-domain formulation for induced current calculations in human beings," *Bioelectromagnetics*, vol. 13, no. 6, pp. 543–556, 1992.

[22] F. B. Hildebrand, *Introduction to Numerical Analysis*, New York: McGraw-Hill, 1956.

[23] K.-P. Hwang and A. C. Cangellaris, "Effective permittivities for second-order accurate FDTD equations at dielectric interfaces," *IEEE Microw. Wireless Components Lett.*, vol. 11, pp. 158–160, 2000.

[24] A. Taflove, *Advances in Computational Electrodynamics: The Finite-Difference Time-Domain Method*, Norwood, MA: Artech House, 1998.

[25] P. Wesseling, *An Introduction to Multigrid Methods*, New York: John Wiley & Sons, 1992.

[26] E. Luo and H. O. Kreiss, "Pseudospectral vs. finite difference methods for initial value problems with discontinuous coefficients," *SIAM J. Sci. Comput.*, vol. 20, pp. 148–163, 1999.

[27] J.-P. Berenger, "A perfectly matched layer for the absorption of electromagnetic waves," *J. Computat. Phys.*, vol. 114, no. 2, pp. 185–200, 1994.

[28] J. Fang and Z. Wu, "Generalized perfectly matched layer for the absorption of propagating and evanescent waves in lossless and lossy media," *IEEE Trans. Microw. Theory Tech.*, vol. 44, No. 12, pp. 2216–2222, 1996.

[29] X. Huang and L. Zhou, "Modulating image oscillations in focusing by a metamaterial lens: time-dependent Green's function approach," *Phys. Rev. B*, vol. 74, p. 045123, 2006.

[30] L. Chen, S. He, and L. Shen, "Finite-size effects of a left-handed material slab on the image quality," *Phys. Rev. Lett.*, vol. 92, p. 107404, 2004.

CHAPTER 9
Spatially Dispersive FDTD Modeling of Wire Medium

9.1 Introduction

In the previous chapter, we have investigated, through numerical simulation whether or not the finite transverse dimensions of LHM slabs influence the quality of their subwavelength imaging. However, in practice, the fabrication of LH media remains problematic, since it requires the creation of negative permeability, which does not exist in nature. Furthermore, currently available designs of the LH media are very lossy at both microwave and optical frequencies, which restricts and even prevents their use in subwavelength imaging applications. There is an alternative approach to subwavelength imaging in the sense of mapping the source distribution in one plane onto another (the imaging plane). This approach involves neither the use of LH media nor does it capitalize on negative refraction or amplification of evanescent waves, which has been referred to as *canalization* in [1–42]. It is based on transporting both the propagating and evanescent spectra of a source by transforming them into propagating waves inside a slab of specially designed materials. Then, these propagating modes are capable of transporting subwavelength images from one interface of the slab to the other. The source must be placed very close to the front interface of the slab in order to minimize the degradation of its spectrum that occurs when the fields propagate in the free space. It is also necessary to minimize the reflection from the slab via an appropriate choice of its thickness. This is done by tuning the slab thickness for Fabry-Perot resonance to reduce reflections from its interface for a wide range of angles of incidence and minimize the interfering interactions between the source and the slab that can distort the image. The material operating in the canalization regime should have a flat isofrequency contour, implying that it should support waves traveling in a certain direction with fixed phase velocity for any transverse wave vectors. The materials that fulfill this requirement are available at both microwave [1, 2, 6] and optical [4] frequency ranges. One such artificial material, suggested in [6] is the wire medium comprised of a regular array of parallel metallic wires [5] (see Figure 9.1 for its geometry).

Originally, it was thought that the transmission devices formed by wire media would function only at microwave frequencies, where most of the metals behave as ideal conductors. However, recently, the use of similar structures for subwavelength imaging at terahertz and infrared frequency bands, where metals display a plasma-type behavior [6] have been recently proposed.

In this chapter, the performance of subwavelength imaging by the transmission devices formed by a wire medium at microwave frequencies is investigated.

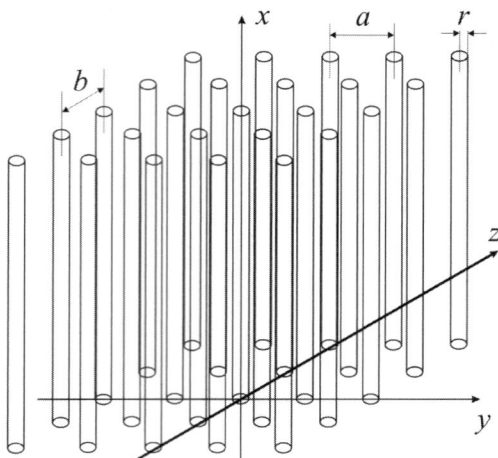

Figure 9.1 The wire medium: a rectangular lattice of parallel ideally conducting thin wires.

The finite-sized planar slabs of the wire media excited by subwavelength sources are modeled by using a new spatially dispersive FDTD scheme. The simulation of the wire medium can be performed either by modeling its physical structure, namely, an array of parallel conducting wires [6] or via the use of the effective medium approach, provided that the dimensions of the inclusions is sufficiently small compared with the wavelength of operation. To model the physical structures, we can either use an extremely fine FDTD grid in the conventional Yee algorithm (which requires an excessive amount of computer resources); spectral domain methods that utilize modified telegraph equations; use the subcell method [7–12]; the method based on the contour-path integral formulations of FDTD [11,13,14]; the method by discretizing a second-order wave equation for the current [15]; or the method based on the weighted residual (WR) interpretation of the FDTD algorithm [16–18]. Among all of these available techniques, we explore its use below for the imaging problem at hand.

Numerical simulations of the structure consisting of an array of 21 by 21 aluminium wires excited by a source in the form of a "P" letter were performed using the CST Microwave Studio utilizing the finite integral technique [6]. The dimensions of the wire medium simulated in [6] were chosen as $a = b = 10$ mm, $r = 1$ mm and the thickness of the wire medium is $d = 150$ mm (half-wavelength at the operating frequency, $f = 1.0$ GHz). If such a structure is simulated in FDTD by modeling its physical details, excessive computational resources are required. On the other hand, the effective medium approach is seen as a simple and efficient alternative when exploring the abundant applications of the wire medium in antenna and microwave engineering.

One of the applications of the wire medium as a transmission device is to increase the capacity of digital versatile discs (DVDs), where closer located bright spots can be distinguished compared to the conventional case. Using the wire medium also avoids the touching and destroying of data surface when performing the scanning. Other potential applications of the wire medium include the construction of directive antennas [1] and antenna substrates/superstrates [2].

9.2 Spatial Dispersion in the Wire Medium

As mentioned earlier, the wire medium is an artificial material formed by using parallel, perfectly conducting wires, arranged in a rectangular lattice (see Figure 9.1). This medium has been well known as an artificial dielectric with plasma-like properties at microwave frequencies [20, 21, 41], but only recently it has been demonstrated that the wire medium has a strong spatial dispersion [5], and hence, it is a nonlocal material and in the spectral domain, it cannot be described by using only a frequency-dependent permittivity tensor. The permittivity tensor of the wire medium depends on the wave vector as well. In this study, we employ the following expression given in [5] to represent the permittivity tensor of the medium, as follows:

$$\bar{\bar{\varepsilon}}_r(\omega,\mathbf{q}) = \varepsilon_r(\omega,q_x)\hat{x}\hat{x} + \hat{y}\hat{y} + \hat{z}\hat{z}, \quad \varepsilon_r(\omega,q_x) = 1 - \frac{k_p^2}{k^2 - q_x^2} \quad (9.1)$$

where the x-axis is oriented along the wires, q_x is the x-component of the wave vector $\mathbf{q}$; $k = \omega/c$ is the wave number of free space; c is the speed of light; and k_p is the wave number corresponding to the plasma frequency of the wire medium. The wave number k_p depends upon the lattice periods a and b, and on the radius of the wires r [22]:

$$k_p^2 = \frac{2\pi/(ab)}{\log\frac{\sqrt{ab}}{2\pi r} + F(a/b)}, \quad F(\xi) = -\frac{1}{2}\log\xi + \sum_{n=1}^{+\infty}\left(\frac{\coth(\pi n\xi) - 1}{n}\right) + \frac{\pi}{6}\xi \quad (9.2)$$

For the case of the square grid $(a = b)$, $F(1) = 0.5275$ and the expression (9.2) reduces to

$$k_p^2 = \frac{2\pi/a^2}{\ln\frac{a}{2\pi r} + 0.5275} \quad (9.3)$$

The expression (9.1) for the permittivity tensor of the wire medium is valid if the wires are relatively thin compared to the lattice periods and if the lattice periods are much smaller than the wavelength, and hence the medium can be homogenized.

In contrast to the usual uniaxial dielectrics that only support two types of modes, namely ordinary and extraordinary. The wire medium supports three different types of modes:

- TE modes (ordinary modes, relative to the orientation of the wires): Waves that polarize in a direction across the wires and do not induce any currents along the wires. In the thin wire approximation, one can regard these as modes as though they are traveling in the free space, without interacting with the wires.
- TM modes (extraordinary modes, relative to the orientation of the wires): Waves that correspond to nonzero currents in the wires and nonzero electric

field along the wires. At the frequencies below the plasma frequency, these waves are evanescent.

- TEM modes (transmission-line modes that are transverse to the orientation of the wires): Waves with nonzero currents in the wires, but with zero electric field along the wires. These modes travel with the speed of light along the wires ($q_x = \pm k$), with their wave vector oriented along the transverse direction. They correspond to the modes of a multiconductor transmission line formed by the wires, and allow one to use the wire medium for canalization.

The wire medium has a strong spatial dispersion. Typically spatial dispersion effects exist in a periodic structures when the wavelength is comparable to the period of the structure [23]. The wire medium is a unique material, in which can be spatially dispersive at any frequency.

9.3 Spatially Dispersive FDTD Formulations

Although a wide variety of complex electromagnetic media have been modeled by using the FDTD method, the list is certainly not all-inclusive. In the past, the focus has been primarily on the frequency dispersion effects; the spatial dispersion effects have been generally ignored. Perhaps the lack of attention to the spatial dispersion effects can be attributed to the following two reasons. First, it is widely assumed that the spatial dispersion effects are usually quite weak, and hence they can be neglected when studying certain electromagnetic crystal structures. However, there is a number of complex materials in which the spatial dispersion effects are strong and play a dominant role in determining their electromagnetic properties. One example such a material is the wire medium. The second possible reason why the spatial dispersion effects have not attracted the attention of many researchers is the complexity in describing the effects. In contrast to the numerous analytical models that deal with the frequency dispersion effects, analytical expressions that describe the spatial dispersion for particular materials are generally unavailable. The wire medium is an exception such that the spatial dispersion effects in this material can be characterized by simple analytical expressions given in (9.1).

Using (9.1) the wire medium can be modeled in the FDTD as a frequency and spatially dispersive dielectric. To account for the dispersive properties of the materials in the FDTD modeling, we introduce the electric flux density in the conventional FDTD updating equations. The x-component of the electric flux density $D_x(\omega, q_x)$ is related to the corresponding component of the electric field intensity $E_x(\omega, q_x)$ in the spectral (frequency-wave vector) domain as follows:

$$D_x(\omega) = \varepsilon(\omega, q_x) E_x(\omega) \tag{9.4}$$

Hence one can obtain the dispersion relationship:

$$\left(k^2 - q_x^2\right) D_x + \left(q_x^2 - k^2 + k_p^2\right) \varepsilon_0 E_x = 0 \tag{9.5}$$

9.3 Spatially Dispersive FDTD Formulations

from which we can obtain the constitutive relations in the space-time domain in the following form:

$$\left(\frac{\partial^2}{\partial x^2} - \frac{1}{c^2}\frac{\partial^2}{\partial t^2}\right)D_x + \left(\frac{1}{c^2}\frac{\partial^2}{\partial t^2} - \frac{\partial^2}{\partial x^2} + k_p^2\right)\varepsilon_0 E_x = 0 \quad (9.6)$$

where we have taken an inverse Fourier transformation and employed the relationships:

$$k^2 \rightarrow -\frac{1}{c^2}\frac{\partial^2}{\partial t^2}, \quad q_x^2 \rightarrow -\frac{\partial^2}{\partial x^2}$$

The above equation only relates the x-components of the electric flux density to the field intensity. The permittivity in both the y- and z-directions is the same as that in the free space since the wires are assumed to be thin and the structure is symmetrical.

The FDTD domain is discretized by using an equally spaced 3-D grid with periods Δx, Δy and Δz along the x-, y-, and z-directions, respectively, and the time step is Δt. For the discretization of (9.6), the central finite difference operators in time δ_t^2 and space δ_x^2 as well as the central averaging operator with respect to time μ_t^2 are used as shown here:

$$\frac{\partial^2}{\partial t^2} \rightarrow \frac{\delta_t^2}{(\Delta t)^2}, \quad \frac{\partial^2}{\partial x^2} \rightarrow \frac{\delta_x^2}{(\Delta x)^2}, \quad k_p^2 \rightarrow k_p^2 \mu_t^2$$

The same central difference and average operators in time δ_t^2 and μ_t^2 were used in Chapter 8 and their definitions are repeated here for convenience. The central difference operator in space δ_x^2 is defined in a similar way as the central difference operator in time, as shown here:

$$\delta_t^2 F|_{m_x,m_y,m_z}^n \equiv F|_{m_x,m_y,m_z}^{n+1} - 2F|_{m_x,m_y,m_z}^n + F|_{m_x,m_y,m_z}^{n-1}$$
$$\delta_x^2 F|_{m_x,m_y,m_z}^n \equiv F|_{m_x+1,m_y,m_z}^n - 2F|_{m_x,m_y,m_z}^n + F|_{m_x-1,m_y,m_z}^n$$
$$\mu_t^2 F|_{m_x,m_y,m_z}^n \equiv \frac{F|_{m_x,m_y,m_z}^{n+1} + 2F|_{m_x,m_y,m_z}^n + F|_{m_x,m_y,m_z}^{n-1}}{4} \quad (9.7)$$

Here F represents the field components; m_x, m_y, m_z are indices corresponding to a specific discretization point in the FDTD domain; and n is the number of time steps. In the discretized (9.6) we get

$$\left[\frac{\delta_x^2}{(\Delta x)^2} - \frac{1}{c^2}\frac{\delta_t^2}{(\Delta t)^2}\right]D_x + \left[\frac{1}{c^2}\frac{\delta_t^2}{(\Delta t)^2} - \frac{\delta_x^2}{(\Delta x)^2} + k_p^2 \mu_t^2\right]\varepsilon_0 E_x = 0 \quad (9.8)$$

Note that in (9.8), the discretization of the term k_p^2 of (9.6) is performed using the central averaging operator μ_t^2 to guarantee stability. In the next section, we analyze the stability of such a discretization scheme as well as that of others that employ a central averaging operator μ_{2t} over a time interval $2\Delta t$, with or without the use of the central averaging operators [24].

Equation (9.8) can be written as:

$$\left[\frac{D_x|^n_{m_x+1,m_y,m_z} - 2D_x|^n_{m_x,m_y,m_z} + D_x|^n_{m_x-1,m_y,m_z}}{(\Delta x)^2}\right.$$

$$-\frac{1}{c^2}\frac{D_x|^{n+1}_{m_x,m_y,m_z} - 2D_x|^n_{m_x,m_y,m_z} + D_x|^{n-1}_{m_x,m_y,m_z}}{(\Delta t)^2}$$

$$+\varepsilon_0\left[\frac{1}{c^2}\frac{E_x|^{n+1}_{m_x,m_y,m_z} - 2E_x|^n_{m_x,m_y,m_z} + E_x|^{n-1}_{m_x,m_y,m_z}}{(\Delta t)^2}\right.$$

$$-\frac{E_x|^n_{m_x+1,m_y,m_z} - 2E_x|^n_{m_x,m_y,m_z} + E_x|^n_{m_x-1,m_y,m_z}}{(\Delta x)^2}$$

$$\left.\left.+k_p^2\frac{E_x|^{n+1}_{m_x,m_y,m_z} + 2E_x|^n_{m_x,m_y,m_z} + E_x|^{n-1}_{m_x,m_y,m_z}}{4}\right]\right] = 0 \quad (9.9)$$

Therefore, the updating equation for E_x in terms of E_x and D_x at previous time steps is as follows:

$$E_x|^{n+1}_{m_x,m_y,m_z} = \frac{1}{a_{1x}}\left[b_{1x}D_x|^{n+1}_{m_x,m_y,m_z} + b_{2x}D_x|^n_{m_x+1,m_y,m_z} + b_{3x}D_x|^n_{m_x,m_y,m_z}\right.$$

$$+ b_{4x}D_x|^n_{m_x-1,m_y,m_z} + b_{5x}D_x|^{n-1}_{m_x,m_y,m_z} - \left(a_{2x}E_x|^n_{m_x+1,m_y,m_z}\right.$$

$$\left.\left.+ a_{3x}E_x|^n_{m_x,m_y,m_z} + a_{4x}E_x|^n_{m_x-1,m_y,m_z} + a_{5x}E_x|^{n-1}_{m_x,m_y,m_z}\right)\right] \quad (9.10)$$

with the coefficients given by

$$a_{1x} = -\frac{\varepsilon_0}{c^2(\Delta t)^2} - \frac{\varepsilon_0 k_p^2}{4}, \qquad b_{1x} = -\frac{1}{c^2(\Delta t)^2}$$

$$a_{2x} = \frac{\varepsilon_0}{(\Delta x)^2}, \qquad b_{2x} = \frac{1}{(\Delta x)^2}$$

$$a_{3x} = \frac{2\varepsilon_0}{c^2(\Delta t)^2} - \frac{2\varepsilon_0}{(\Delta x)^2} - \frac{\varepsilon_0 k_p^2}{2}, \qquad b_{3x} = \frac{2}{c^2(\Delta t)^2} - \frac{2}{(\Delta x)^2}$$

$$a_{4x} = \frac{\varepsilon_0}{(\Delta x)^2}, \qquad b_{4x} = \frac{1}{(\Delta x)^2}$$

$$a_{5x} = -\frac{\varepsilon_0}{c^2(\Delta t)^2} - \frac{\varepsilon_0 k_p^2}{4}, \qquad b_{5x} = -\frac{1}{c^2(\Delta t)^2}$$

During the process of modeling the wire medium by using the spatially dispersive FDTD, we compute **D** from magnetic field intensity **H**, and **H** from **E** are performed by using Yee's conventional FDTD equations [25], while E_x is calculated from D_x using (9.10) and $E_y = \varepsilon_0^{-1}D_y$, $E_z = \varepsilon_0^{-1}D_z$. Note that in (9.10), the central difference approximations in time (for the frequency dispersion) are used for both D_x and E_x, at the position (m_x, m_y, m_z). Central difference approximations are also used in space (for the spatial dispersion) at the time step n in order to update E_x at the time step $n+1$. As a consequence, the storage of D_x and E_x at two previous time steps are required for this procedure.

As discussed in Chapter 8, a field-averaging technique needs to be used for the accurate modeling of the interface between LHM slab and free space. Since the wire medium is uniaxial in nature, the averaging procedure in (9.1) needs to be applied for the electric field component E_x. Furthermore, as pointed out in Chapter 8, the field averaging can be avoided by setting the boundary of the wire medium slab along the y-direction so that it is aligned with the tangential electric field component, as shown in Figure 9.2. Likewise, if the boundary of the wire medium slab along the x-direction is set to be at the same as that of the magnetic field H_z, no additional averaging would be required.

At the boundaries of the wire medium slab along the y-direction, the updating equation in (9.10) includes D_x and E_x at previous time steps in both free space and the wire medium. In the region outside the wire medium, (9.10) reduces to one that relates D_x and E_x in the free space, and the fields E_x are updated locally.

Equation (9.10) incorporates the terms corresponding to both the frequency and spatial dispersion effects. If the terms corresponding to the spatial dispersion that contain Δx are omitted from (9.10), then the rest of the expression reduces to that of a classical updating equation for the Drude material with the collision frequency equal to zero [i.e., $\varepsilon_r(\omega) = 1 - k_p^2/k^2$] [26]:

$$E_x|_{m_x,m_y,m_z}^{n+1} = \left[2 - c^2(\Delta t)^2 k_p^2\right] E_x|_{m_x,m_y,m_z}^{n} - E_x|_{m_x,m_y,m_z}^{n-1}$$
$$+ \varepsilon_0^{-1}\left[D_x|_{m_x,m_y,m_z}^{n+1} - 2D_x|_{m_x,m_y,m_z}^{n} + D_x|_{m_x,m_y,m_z}^{n-1}\right] \qquad (9.11)$$

The updating equation (9.11) is used for modeling uniaxial Drude materials (the frequency-dispersive materials without the spatial dispersion), which can be treated as a conventional but incorrect description of the wire medium [20, 21, 41], in order to demonstrate the significance of taking into account the spatial dispersion effects in modeling the wire medium.

We limit ourselves only to a 1-D wire medium in this study and assume that the lattices of parallel ideally conducting wires are oriented along a single direction as shown in Figure 9.1. The medium can be described [5] in terms of a permittivity

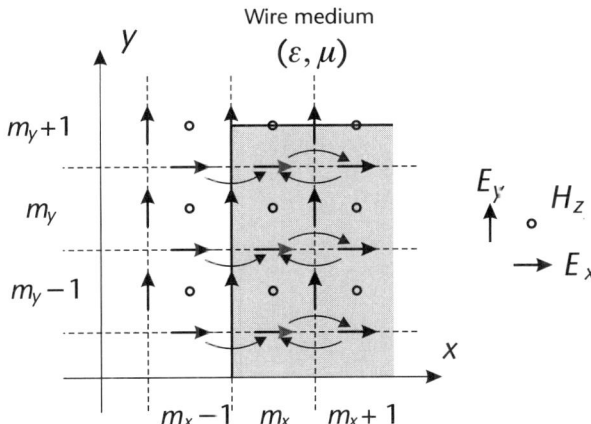

Figure 9.2 Layout of FDTD grid for spatially dispersive FDTD modeling of the wire medium. 2-D FDTD simulation domain for the modeling of the wire medium as a dielectric slab for subwavelength imaging.

tensor given in (9.1), which has one spatially dispersive component. We note, however, there are also 2-D and 3-D wire media [27–29] that consist of two or three orthogonal nonconnected lattices of parallel perfectly conducting wires, which can be described by the permittivity tensors with two or three spatially dispersive components of the same form as in (9.1), respectively. These 2-D and 3-D wire media can also be easily modeled by using the spatially dispersive FDTD method as discussed above. The updating equation (9.10) has to be applied in two or three directions, respectively. The results obtained from both the 2-D and 3-D FDTD simulations of the 1-D wire medium slabs are presented in later sections, while the stability and numerical dispersion analysis of the above developed spatially dispersive FDTD method is presented in Section 9.4.

9.4 Stability and Numerical Dispersion Analysis

Previously, in [30], a stability analysis of the dispersive FDTD algorithm was performed using the von Neumann method and numerical root searching. In this section, we analyze the stability of the proposed spatially dispersive FDTD method by using a technique that combines the von Neumann mand Routh-Hurwitz criterion, shown in [31]. The von Neumann method establishes that, for a finite-difference scheme to be stable, all the roots Z_i of the stability polynomial $S(Z)$ must be located within the unit circle in the Z-plane (i.e., $|Z_i| \leq 1 \ \forall \ i$), where the complex variable Z corresponds to the growth factor of errors and is often called the amplification factor [31].

The wire medium is a uniaxial material where the divergence of the electric field inside the wire medium is nonzero ($\nabla \cdot \mathbf{E} \neq 0$). Therefore, to analyze the numerical stability of the proposed spatially dispersive FDTD method, one must start directly with Maxwell's equations, rather than with the wave equation as in [31] and others for homogeneous materials.

Consider the relation between $\mathbf{D}$ and $\mathbf{E}$ directly expressed from Faraday's and Ampere's laws:

$$\mu_0 \frac{\partial^2 \mathbf{D}}{\partial t^2} + \nabla \times (\nabla \times \mathbf{E}) = 0, \tag{9.12}$$

where μ_0 is the permeability of the free space. Expanding (9.12) in a matrix form, we get

$$\mu_0 \frac{\partial^2}{\partial t^2} \mathbf{D} + \begin{pmatrix} \frac{\partial^2}{\partial x^2} - \Delta & \frac{\partial}{\partial x}\frac{\partial}{\partial y} & \frac{\partial}{\partial x}\frac{\partial}{\partial z} \\ \frac{\partial}{\partial x}\frac{\partial}{\partial y} & \frac{\partial^2}{\partial y^2} - \Delta & \frac{\partial}{\partial y}\frac{\partial}{\partial z} \\ \frac{\partial}{\partial x}\frac{\partial}{\partial z} & \frac{\partial}{\partial y}\frac{\partial}{\partial z} & \frac{\partial^2}{\partial z^2} - \Delta \end{pmatrix} \mathbf{E} = 0 \tag{9.13}$$

where

$$\Delta = \frac{\partial^2}{\partial x^2} + \frac{\partial^2}{\partial y^2} + \frac{\partial^2}{\partial z^2}.$$

9.4 Stability and Numerical Dispersion Analysis

Using the central difference operators, (9.13) can be discretized as

$$\mu_0 \frac{\delta_t^2}{(\Delta t)^2} \mathbf{D} + \begin{bmatrix} \frac{\delta_x^2}{(\Delta x)^2} - \Theta & \frac{\delta_x \delta_y}{\Delta x \Delta y} & \frac{\delta_x \delta_z}{\Delta x \Delta z} \\ \frac{\delta_x \delta_y}{\Delta x \Delta y} & \frac{\delta_y^2}{(\Delta y)^2} - \Theta & \frac{\delta_y \delta_z}{\Delta y \Delta z} \\ \frac{\delta_x \delta_z}{\Delta x \Delta z} & \frac{\delta_y \delta_z}{\Delta y \Delta z} & \frac{\delta_z^2}{(\Delta z)^2} - \Theta \end{bmatrix} \mathbf{E} = 0 \quad (9.14)$$

where

$$\Theta = \frac{\delta_x^2}{(\Delta x)^2} + \frac{\delta_y^2}{(\Delta y)^2} + \frac{\delta_z^2}{(\Delta z)^2}$$

and δ_y and δ_z are defined in the same manner as in [32]. In addition to (9.14), the constitutive relations in the wire medium (9.8) must also be considered and can be written in the matrix form:

$$\begin{bmatrix} \frac{\delta_x^2}{(\Delta x)^2} - \frac{1}{c^2}\frac{\delta_t^2}{(\Delta t)^2} & 0 & 0 \\ 0 & -1 & 0 \\ 0 & 0 & -1 \end{bmatrix} \mathbf{D} + \begin{bmatrix} \frac{1}{c^2}\frac{\delta_t^2}{(\Delta t)^2} - \frac{\delta_x^2}{(\Delta x)^2} + k_0^2 & 0 & 0 \\ 0 & 1 & 0 \\ 0 & 0 & 1 \end{bmatrix} \varepsilon_0 \mathbf{E} = 0 \quad (9.15)$$

For stability analysis in accordance to [31], the following solution is substituted into the discrete equations (9.14) and (9.15),

$$F\big|_{m_x,m_y,m_z}^n = \tilde{F} Z^n e^{j(m_x \Delta x \tilde{q}_x + m_y \Delta y \tilde{q}_y + m_z \Delta z \tilde{q}_z)} \quad (9.16)$$

where $\tilde{F}$ is a complex amplitude, Z is a complex variable that is a measure of the growth of error in a time iteration, and $\tilde{\mathbf{q}} = (\tilde{q}_x, \tilde{q}_y, \tilde{q}_z)^T$ is the numerical wave vector of the discrete mode. After some simple manipulations, we get

$$\frac{\mu}{(\Delta t)^2}(Z-1)^2 \tilde{\mathbf{D}} + 4Z \begin{bmatrix} \Phi - \frac{\sin^2\theta_x}{(\Delta x)^2} & \frac{\sin^2\theta_x}{\Delta x}\frac{\sin^2\theta_y}{\Delta y} & \frac{\sin^2\theta_x}{\Delta x}\frac{\sin^2\theta_z}{\Delta z} \\ \frac{\sin^2\theta_x}{\Delta x}\frac{\sin^2\theta_y}{\Delta y} & \Phi - \frac{\sin^2\theta_y}{(\Delta y)^2} & \frac{\sin^2\theta_y}{\Delta y}\frac{\sin^2\theta_z}{\Delta z} \\ \frac{\sin^2\theta_x}{\Delta x}\frac{\sin^2\theta_z}{\Delta z} & \frac{\sin^2\theta_y}{\Delta y}\frac{\sin^2\theta_z}{\Delta z} & \Phi - \frac{\sin^2\theta_z}{(\Delta z)^2} \end{bmatrix} \tilde{\mathbf{E}} = 0 \quad (9.17)$$

where

$$\Phi = \sum_{\alpha=x,y,z} \frac{\sin^2\theta_\alpha}{(\Delta\alpha)^2}, \quad \theta_\alpha = (\tilde{q}_\alpha \Delta\alpha)/2, \quad \alpha = x,y,z$$

and

$$\left\{ -\frac{1}{c^2(\Delta t)^2} Z^2 + 2\left[\frac{1}{c^2(\Delta t)^2} - \frac{2\sin^2\theta_x}{(\Delta x)^2}\right] Z - \frac{1}{c^2(\Delta t)^2} \right\} \tilde{D}_x + \left\{ \left[\frac{1}{c^2(\Delta t)^2} + \frac{k_p^2}{4}\right] Z^2 \right.$$

$$\left. + \left[-\frac{2}{c^2(\Delta t)^2} + \frac{4\sin^2\theta_x}{(\Delta x)^2} + \frac{k_p^2}{2}\right] Z + \left[\frac{1}{c^2(\Delta t)^2} + \frac{k_p^2}{4}\right] \right\} \varepsilon_0 \tilde{E}_x = 0,$$

$$\tilde{D}_y - \varepsilon_0 \tilde{E}_y = 0, \quad \tilde{D}_z - \varepsilon_0 \tilde{E}_z = 0 \quad (9.18)$$

respectively. The stability criteria is based on the behavior of the determinant of the system of (9.17) and (9.18):

$$S_w(Z) = \left[\frac{1}{c^2(\Delta t)^2} + \frac{k_p^2}{4}\right]Z^4 + 4\left[-\frac{1}{c^2(\Delta t)^2} + \frac{\sin^2\theta_x}{(\Delta x)^2} + \Phi\right]Z^3$$
$$+ \left\{\frac{6}{c^2(\Delta t)^2} - \frac{k_p^2}{2} - \frac{8\sin^2\theta_x}{(\Delta x)^2} - 8\left[1 - \frac{2c^2(\Delta t)^2\sin^2\theta_x}{(\Delta x)^2}\right]\Phi\right.$$
$$\left.+ \frac{4k_p^2 c^2(\Delta t)^2\sin^2\theta_x}{(\Delta x)^2}\right\}Z^2 + 4\left[-\frac{1}{c^2(\Delta t)^2} + \frac{\sin^2\theta_x}{(\Delta x)^2} + \Phi\right]Z$$
$$+ \left[\frac{1}{c^2(\Delta t)^2} + \frac{k_p^2}{4}\right] \quad (9.19)$$

For simplicity, the terms that lead to the stability condition in the free space (i.e., the conventional stability condition) have been omitted from this stability polynomial.

In order to avoid numerical root searching [30] as a way to derive stability conditions, the above stability polynomial can be transformed into the r-plane by using the bilinear transformation

$$Z = \frac{r+1}{r-1} \quad (9.20)$$

The stability polynomial in the r-plane becomes as follows:

$$S_w(r) = \left[\frac{4c^2(\Delta t)^2\sin^2\theta_x}{(\Delta x)^2}\Phi + \frac{k_p^2 c^2(\Delta t)^2\sin^2\theta_x}{(\Delta x)^2}\right]r^4$$
$$+ \left\{k_p^2 + \frac{4\sin^2\theta_x}{(\Delta x)^2} + 4\left[1 - \frac{2c^2(\Delta t)^2\sin^2\theta_x}{(\Delta x)^2}\right]\Phi - \frac{2k_p^2 c^2(\Delta t)^2\sin^2\theta_x}{(\Delta x)^2}\right\}r^2$$
$$+ \left\{\frac{4}{c^2(\Delta t)^2} - \frac{4\sin^2\theta_x}{(\Delta x)^2} - 4\left[1 - \frac{c^2(\Delta t)^2\sin^2\theta_x}{(\Delta x)^2}\right]\Phi + \frac{k_p^2 c^2(\Delta t)^2\sin^2\theta_x}{(\Delta x)^2}\right\} \quad (9.21)$$

Building up the Routh table [31] for the above polynomial, we obtain the following stability conditions:

$$\frac{4\sin^2\theta_x}{(\Delta x)^2}\left[1 - c^2(\Delta t)^2\Phi\right] + 4\left[1 - \frac{c^2(\Delta t)^2\sin^2\theta_x}{(\Delta x)^2}\right]\Phi$$
$$+ k_p^2\left[1 - \frac{2c^2(\Delta t)^2\sin^2\theta_x}{(\Delta x)^2}\right] \geq 0 \quad (9.22)$$

$$4\left[\frac{1}{c^2(\Delta t)^2} - \frac{\sin^2\theta_x}{(\Delta x)^2}\right]\left[1 - c^2(\Delta t)^2\Phi\right] + \frac{k_p^2 c^2(\Delta t)^2\sin^2\theta_x}{(\Delta x)^2} \geq 0 \quad (9.23)$$

9.4 Stability and Numerical Dispersion Analysis

To fulfill these conditions, it is sufficient to fulfill the conventional Courant stability condition [25] given by:

$$\Delta t \leq \frac{1}{c\Phi^{\frac{1}{2}}} \leq \frac{1}{c}\left[\sum_{\alpha=x,y,z}\frac{1}{(\Delta\alpha)^2}\right]^{-\frac{1}{2}} \quad (9.24)$$

Hence the conventional Courant stability condition [25] continues to apply for the modeling of the wire medium and no additional conditions are required.

Note that in the above analysis, the central averaging operator μ_t^2 was used for the discretization of (9.6). The above operator μ_{2t} is defined as

$$\mu_{2t}\mathbf{F}|_{m_x,m_y,m_z}^n = \frac{\mathbf{F}|_{m_x,m_y,m_z}^{n+1} + \mathbf{F}|_{m_x,m_y,m_z}^{n-1}}{2} \quad (9.25)$$

whose use leaves the stability condition (9.22) unchanged; however, now (9.23) becomes

$$4\left[\frac{1}{c^2(\Delta t)^2} - \frac{\sin^2\theta_x}{(\Delta x)^2}\right]\left[1 - c^2(\Delta t)^2\Phi\right] + k_p^2\left[1 + \frac{c^2(\Delta t)^2\sin^2\theta_x}{(\Delta x)^2}\right] \geq 0 \quad (9.26)$$

which indicates that this stability condition is even less restrictive that it is for the conventional FDTD method. However, if no central averaging operator is used when discretizing (9.6), then (9.22) also remains unchanged, but (9.23) is modified to

$$\left[\frac{1}{c^2(\Delta t)^2} - \frac{\sin^2\theta_x}{(\Delta x)^2}\right]\left\{4\left[1 - c^2(\Delta t)^2\Phi\right] - k_p^2 c^2(\Delta t)^2\right\} \geq 0 \quad (9.27)$$

Therefore, this leads to a more restrictive stability condition than the conventional Courant condition, namely:

$$\Delta t \leq \frac{1}{c}\left[\sum_{\alpha=x,y,z}\frac{1}{(\Delta\alpha)^2} + \frac{k_p^2}{4}\right]^{-\frac{1}{2}} \quad (9.28)$$

A comparison of different stability conditions for different averaging schemes is presented in Table 9.1. Figure 9.3 shows a comparison of the magnetic field distributions using different discretization schemes. Using the central averaging operator μ_t^2 and without using the central average operator, it is evident that after 500 time steps, the instability errors begin to appear from inside of the wire medium slab using the latter scheme.

Table 9.1 Comparison of Different Stability Conditions for Different Averaging Schemes

Schemes	Δt	Stability
μ_t	Courant limit	✓
μ_{2t}	Courant limit	✓
No averaging	Courant limit	✗
No averaging	$\leq c^{-1}\left(\Phi + k_p^2/4\right)^{-\frac{1}{2}}$	✓

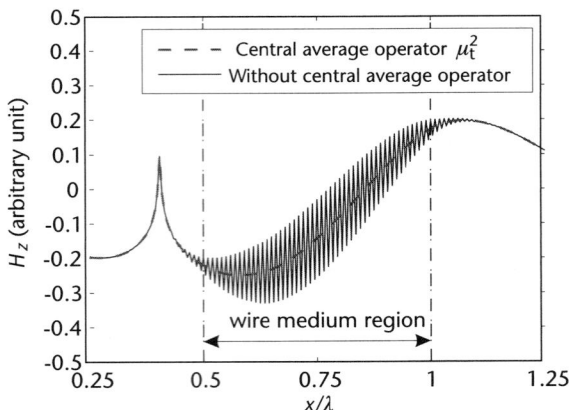

Figure 9.3 Comparison of the internal magnetic field distributions (unit: A/m) for a $0.5\lambda \times \lambda$ wire medium slab ($k_p/k = 4$) excited by a magnetic line source calculated using different discretization schemes. The field is plotted at the time step $n = 570\Delta t$, where Δt is chosen as the Courant limit (i.e., $\Delta t = \Delta x/\sqrt{2}c$, $\Delta x = \lambda/200$).

The numerical dispersion relationship for the wire medium can be found by evaluating the stability polynomial $S_w(Z)$ given by (9.19) on the unit circle of the Z-plane (i.e., when $Z = e^{j\omega\Delta t}$) and equating the results to zero. Substituting the expression

$$Z = e^{j\omega\Delta t} \tag{9.29}$$

into the stability polynomial (9.19), equating the result to zero, and dividing $e^{j2\omega\Delta t}$ on both sides of the equation, one can obtain

$$\left[\frac{1}{c^2(\Delta t)^2} + \frac{k_p^2}{4}\right]e^{j2\omega\Delta t} + 4\left[-\frac{1}{c^2(\Delta t)^2} + \frac{\sin^2\theta_x}{(\Delta x)^2} + \Phi\right]e^{j\omega\Delta t}$$
$$+\frac{6}{c^2(\Delta t)^2} - \frac{k_p^2}{2} - \frac{8\sin^2\theta_x}{(\Delta x)^2} - 8\left[1 - \frac{2c^2(\Delta t)^2\sin^2\theta_x}{(\Delta x)^2}\right]\Phi + \frac{4k_p^2 c^2(\Delta t)^2\sin^2\theta_x}{(\Delta x)^2}$$
$$+4\left[-\frac{1}{c^2(\Delta t)^2} + \frac{\sin^2\theta_x}{(\Delta x)^2} + \Phi\right]e^{-j\omega\Delta t} + \left[\frac{1}{c^2(\Delta t)^2} + \frac{k_p^2}{4}\right]e^{-j2\omega\Delta t} = 0 \tag{9.30}$$

Combining similar terms and according to the fact that

$$e^{j\omega\Delta t} = \cos(\omega\Delta t) + j\sin(\omega\Delta t), \quad e^{-j\omega\Delta t} = \cos(\omega\Delta t) - j\sin(\omega\Delta t)$$

(9.30) can be rewritten as

$$\frac{6}{c^2(\Delta t)^2} - \frac{k_p^2}{2} - \frac{8\sin^2\theta_x}{(\Delta x)^2} - 8\left[1 - \frac{2c^2(\Delta t)^2\sin^2\theta_x}{(\Delta x)^2}\right]\Phi + \frac{4k_p^2 c^2(\Delta t)^2\sin^2\theta_x}{(\Delta x)^2}$$
$$+ \left[\frac{1}{c^2(\Delta t)^2} + \frac{k_p^2}{4}\right]2\cos(2\omega\Delta t) + 4\left[-\frac{1}{c^2(\Delta t)^2} + \frac{\sin^2\theta_x}{(\Delta x)^2} + \Phi\right]2\cos(\omega\Delta t) = 0 \tag{9.31}$$

Applying transforms

$$\cos(\omega\Delta t) = 1 - 2\sin^2\left(\frac{\omega\Delta t}{2}\right), \quad \sin(\omega\Delta t) = 2\sin\left(\frac{\omega\Delta t}{2}\right)\cos\left(\frac{\omega\Delta t}{2}\right)$$

Equation (9.31) is reduced to

$$-\frac{4}{c^2(\Delta t)^2}\sin^2\left(\frac{\omega\Delta t}{2}\right)\cos^2\left(\frac{\omega\Delta t}{2}\right) - k_p^2\sin^2\left(\frac{\omega\Delta t}{2}\right)\cos^2\left(\frac{\omega\Delta t}{2}\right)$$
$$+\frac{4}{c^2(\Delta t)^2}\sin^2\left(\frac{\omega\Delta t}{2}\right) - \frac{4\sin^2\theta_x}{(\Delta x)^2}\sin^2\left(\frac{\omega\Delta t}{2}\right) + \frac{k_p^2 c^2(\Delta t)^2 \sin^2\theta_x}{(\Delta x)^2}$$
$$-4\sin^2\left(\frac{\omega\Delta t}{2}\right)\Phi - \frac{4c^2(\Delta t)^2\sin^2\theta_x}{(\Delta x)^2}\Phi = 0 \qquad (9.32)$$

Using $\cos^2\left(\frac{\omega\Delta t}{2}\right) = 1 - \sin^2\left(\frac{\omega\Delta t}{2}\right)$ and solving for Φ, the numerical dispersion relation for the spatially dispersive FDTD modeling of the wire medium is obtained as:

$$\frac{\left[\frac{4}{c^2(\Delta t)^2} + k_p^2\right]\sin^4\frac{\omega\Delta t}{2} - \left[\frac{4\sin^2\theta_x}{(\Delta x)^2} + k_p^2\right]\sin^2\frac{\omega\Delta t}{2} + \frac{k_p^2 c^2(\Delta t)^2 \sin^2\theta_x}{(\Delta x)^2}}{4\left[\sin^2\frac{\omega\Delta t}{2} - \frac{c^2(\Delta t)^2\sin^2\theta_x}{(\Delta x)^2}\right]} = \Phi \qquad (9.33)$$

If $\Delta\beta \to 0$, where $\beta = x, y, z, t$, then (9.33) reduces to the continuous dispersion relationship for the wire medium [5]:

$$\left(q_x^2 - k^2\right)\left(q_x^2 + q_y^2 + q_z^2 - k^2 + k_p^2\right) = 0 \qquad (9.34)$$

The first and second terms of (9.34) correspond to the transmission-line modes (TEM waves with respect to the orientation of wires) and the extraordinary modes (TM waves), respectively. The ordinary modes (TE waves) did not appear in (9.34) since their contribution was omitted in (9.19) for the sake of simplifying the calculation.

9.5 Perfectly Matched Layer for Wire Medium Slabs

In [33], Berenger's original PML was extended to absorb the electromagnetic waves propagating in anisotropic dielectrics and magnetic media, by introducing the material-independent quantities (electric flux density **D** and magnetic flux density **B**).

PMLs are typically placed at a distance of half-wavelength away from any objects located inside the simulation domain. In order to reduce the time and computer memory requirements for simulations as well as to improve the convergence performance, it is necessary to place PMLs in the close vicinity of the wire medium. Towards this end, we use a similar approach followed in [33] by modifying the Berenger's original PML formulations. In the modified PML (MPML) for the wire medium slab, the quantities **D** and **H** are split, and the updating equation for **D**, for instance, D_{zx} now becomes:

$$D_{zx}|^{n+1}_{m_x,m_y,m_z} = e^{-\sigma'_x \Delta t} D_{zx}|^{n}_{m_x,m_y,m_z}$$
$$+ \frac{\left(1 - e^{-\sigma'_x \Delta t}\right)}{-\sigma'_x \Delta x} \left[H_y|^{n+1/2}_{m_x+1/2,m_y,m_z} - H_y|^{n+1/2}_{m_x-1/2,m_y,m_z} \right] \quad (9.35)$$

However, the updating equations for **H** remain the same as in the Berenger's original PML [34]. The matching conditions are:

$$\sigma'_\alpha = \frac{\sigma^*_\alpha}{\mu_0}, \quad \text{where } \alpha = x,y,z \quad (9.36)$$

where σ'_α and σ^*_α denote the electric conductivity and the magnetic loss inside the MPML, respectively. Comparing the matching conditions for the MPML and with those for Berenger's original PML ($\sigma_\alpha/\varepsilon_0 = \sigma^*_\alpha/\mu_0$), we can readily see that the difference between σ'_α and σ_α is a factor of ε_0. This explains why this factor us missing in the expression for the theoretical reflection coefficient $R(\theta)$ for σ'_α [33],

$$R(\theta) = e^{-2(\cos\theta/c) \int_0^\delta \sigma'_\alpha(\rho) d\rho} \quad (9.37)$$

where δ is the thickness of the MPML. In common with the losses in the Berenger's PML, the losses in MPML should increase gradually with depth ρ, as follows:

$$\sigma'_\alpha(\rho) = \sigma'_{\alpha,\max} \left(\frac{\rho}{\delta}\right)^n \quad (9.38)$$

The optimum values of $\sigma'_{\alpha,\max}$ and n depend on the thickness δ. The thicker the absorber, the larger the optimum values of $\sigma'_{\alpha,\max}$ and n should be. Some numerical experiments pertaining to the selection of $\sigma'_{\alpha,\max}$ and n, for various absorber thicknesses have been reported in [35–38]. Enforcing (9.38) in (9.37) yields the reflection coefficient for the MPML

$$R(\theta) = e^{-2\sigma'_{\alpha,\max} \delta \cos\theta/(n+1)c} \quad (9.39)$$

Typically, n is chosen equal 2 and $\sigma'_{\alpha,\max}$ is calculated to obtain a normal theoretical reflection coefficient $R(0) = 10^{-5}$. In addition to (9.35), the updating relation between **D** and **E** in (9.10) must be extended into the PML in order to realize a match between the wire medium and the MPML.

To evaluate the performance of the MPML for the wire medium, we choose a 2-D computational domain similar to that in Figure 9.10, except that the wire medium slab is now terminated by a 10-cell MPML as shown in Figure 9.4(a). As for the other side, the Berenger's original PML is used to truncate the free space. The source is located close to the edge of the wire medium slab terminated by the MPML and is placed sufficiently far away from the other side of the slab in order

9.5 Perfectly Matched Layer for Wire Medium Slabs

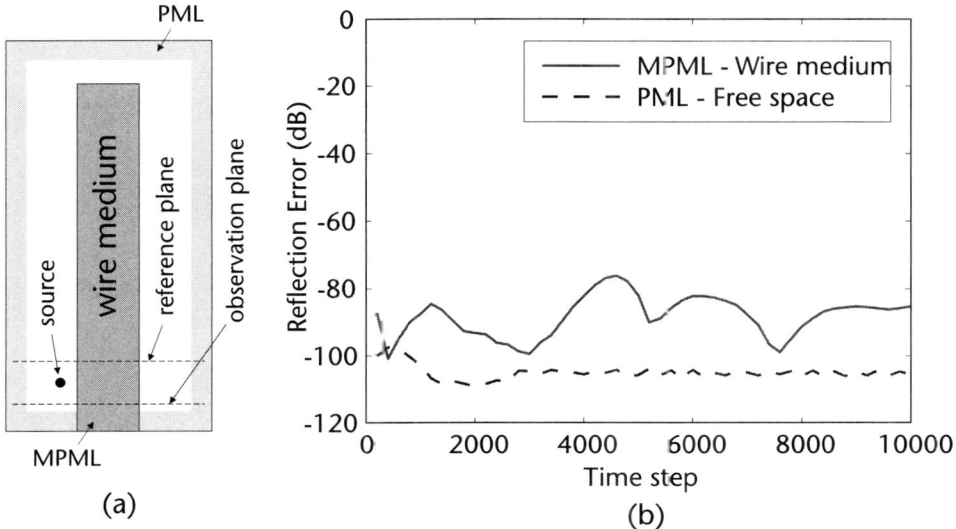

Figure 9.4 (a) Simulation domain and (b) reflection error (in decibels) from the MPML-wire medium and PML-free space interfaces calculated at the observation plane (two cells away from the PMLs) for a wire medium slab excited by a magnetic line source ($d = 0.5\lambda$, $h = 0.1\lambda$ and $k_0/k = 4$) plotted as functions of the time step.

to ensure that the waves reflected back from that side do not reach the reference plane during the simulation period. The observation plane is chosen to be two cells away from the MPML. The reference plane is located at the same distance from the source as the observation plane, but from the opposite side [see Figure 9.4(a)]. The magnetic fields at the observation and reference planes are recorded as H_{PML} and H_{ref}, respectively. The reflection error is defined as

$$\text{reflection error (dB)} = 20 \times \log_{10}\left(\frac{|H_{\text{PML}} - H_{\text{ref}}|}{|H_{\text{max}}|}\right) \qquad (9.40)$$

where $|H_{\text{max}}|$ is the maximum value of the magnetic field at the reference plane. The reflection error, calculated for H_z with this termination is plotted in Figure 9.4. For comparison, the reflection error at the interface of PML-free space is also shown in the same figure. It is found that the level of reflections from the PML is below -70 dB, leading us to conclude that the wire medium is "perfectly" matched.

Figure 9.5 shows the comparison of magnetic field distributions obtained by using: (1) the Berenger's original PML at a $\lambda/2$ distance from the slab and (2) an MPML that truncates the wire medium slab directly. We note that, in comparison with Figure 9.5(a), the simulation domain for the one in Figure 9.5(b) is reduced by 50%, while the convergence is greatly improved, since diffraction from the corners and edges are mitigated when the MPML is used for the termination. For the first case, the iteration error falls below -30 dB only after 1,000 periods (400,000 time steps), while for the latter, the convergence is reached in just 100 periods.

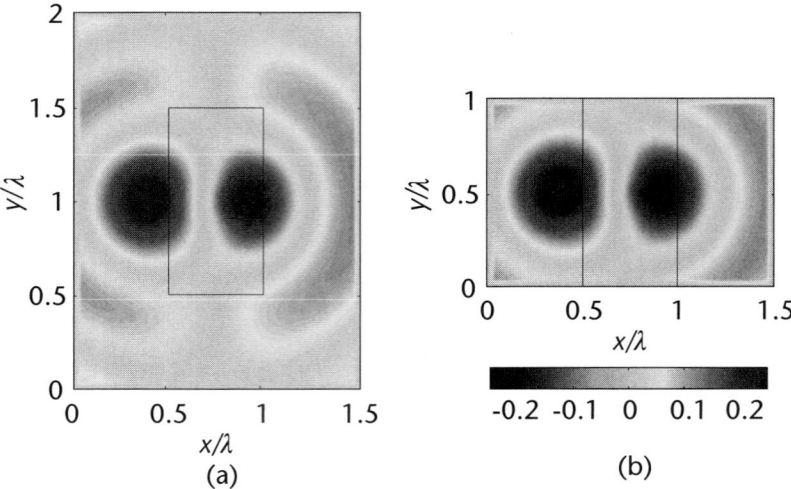

Figure 9.5 (a) Distribution of magnetic field for a finite wire medium slab with dimensions $0.5\lambda \times \lambda$ excited by a magnetic line source ($k_p/k = 4$). (b) The same slab, but terminated by a 10-cell MPML at each side along y-direction. (unit: A/m)

9.6 Numerical Thickness of Wire Medium Slabs

Because of the spatial discretization in the FDTD algorithm, as well as because of the spatial dispersion effect introduced by the wire medium, the *numerical thickness* of the wire medium slab in the FDTD domain may be different from their actual one. It is important to model the thickness correctly for accurate subwavelength imaging by the wire medium slabs, because the transmission characteristics are affected significantly by their thickness. Figure 9.6 shows the simulation domain for calculating the numerical transmission coefficients. The source and image planes are located at the opposite sides of the wire medium slab and are aligned with the locations of the magnetic fields, because the transmission coefficient is calculated using the H_z component. As discussed in the previous sections, the boundaries along the y-direction are aligned with the tangential electric field component E_y to avoid field averaging at material interfaces. The computational domain in the y-direction is truncated by Bloch's PBCs in order to render the wire medium slab effectively infinite along that direction. When updating the field components at the boundaries of the simulation domain in the y-direction, the required field components outside the domain can be calculated by using the PBC and the fields inside the domain. In the x-direction, the computational domain is terminated either by using Berenger's original PML [34] for absorbing propagation waves or the PML proposed in [39], which has been modified for evanescent waves. In common with the procedure used to calculate the transmission coefficient of LHM slabs, we place a soft plane wave source given by Chapter 8 at a distance of $\lambda/10$ away, on the left side of the source plane to avoid the introduction of the singularities in the FDTD simulation. The orientation of the wires in the modeled wire medium is along the x-axis. The polarization of waves is TM with respect to the orientation

9.6 Numerical Thickness of Wire Medium Slabs

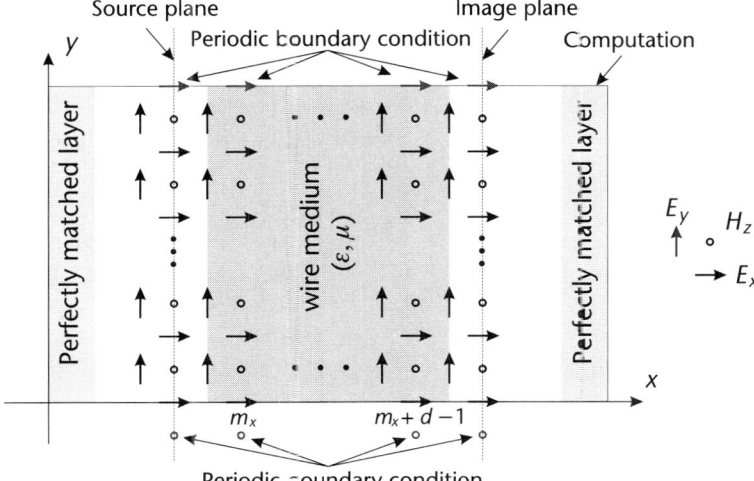

Figure 9.6 Schematic diagram of 2-D TE FDTD simulation domain for the calculation of the numerical transmission coefficient of a wire medium slab with thickness d (normalized by Δx). The arrows and circles indicate the electric and magnetic field components outside the computational domain, to which the PBCs are applied.

of the wires. The electric field is in the x-y plane and the magnetic field is along the z-axis. The thickness of the wire medium slab in the FDTD domain is d_{FDTD} ($d_{\text{FDTD}} = d\Delta x$ where d is the normalized thickness as indicated in Figure 9.6). The FDTD cell size is $\Delta x = \Delta y = \lambda/200$, and the time step is $\Delta t = \Delta x/\sqrt{2}c$, chosen in accordance with the Courant stability condition [25]. The plasma frequency ω_p is inversely proportional to the resolution of the wire medium as a transmission device [40], and this frequency can be chosen as arbitrarily in the numerical simulations. For the present case, we choose the wave number corresponding to the plasma frequency of the wire medium to be four times the wave number of free space (i.e., $k_p = 4k$).

The numerical transmission coefficient is calculated by evaluating the ratio of the transmitted magnetic field at the image plane to that of the incident field at the source plane. The incident field is obtained from simulations carried out in the absence of the wire medium slab (i.e., in free space). The calculated numerical transmission coefficient, and its comparison with the exact analytical solution [40] for the case of a half-wavelength-thick infinite-long wire medium slab, are shown in Figure 9.7. We even see that when the analytical and the numerical thicknesses of the wire medium slab are the same, namely $\lambda/2$, the two results do not agree with each other, because the numerical thickness of the wire medium slab has not been properly chosen in the FDTD domain. However we can vary the thickness parameter so that in the analytical solution, the analytical transmission coefficient agrees with the numerical ones, we then find that the actual analytical thickness of the modeled wire medium slab is $\lambda/2 + \Delta x$. Additionally, we can say that for a slab whose thickness is $\lambda/2$, the correct numerical thickness to use in the FDTD is $d_{\text{FDTD}} = \lambda/2 - \Delta x$, as we also see from Figure 9.7. In fact, according to Figure 9.6,

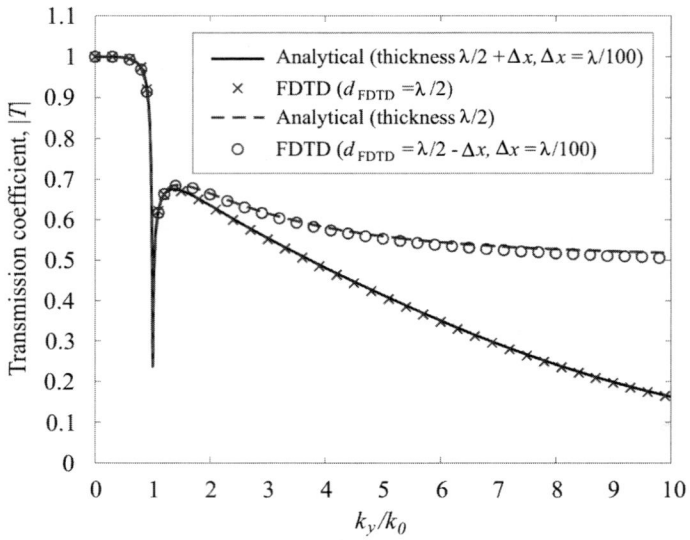

Figure 9.7 Comparison of transmission coefficient for an infinite-long slab of wire medium calculated from FDTD simulations and analytical solution. The thickness of the modeled wire medium slab is $\lambda/2$, and λ is the wavelength at the operating frequency. The numerical thickness of the slab in the FDTD domain is denoted as d_{FDTD}. The FDTD spatial resolution is $\Delta x = \lambda/200$. The analytical solution is given in [40].

the distance between the source and image planes is $d_{\text{FDTD}} + \Delta x$, which is exactly $\lambda/2$, and this serves to explain why the numerical thickness of the wire medium slab is different from the analytical one by Δx.

Next, to investigate the effect of different spatial resolutions used in the FDTD on the numerical transmission coefficient that we calculate, we carry out FDTD simulations with the same configuration as in Figure 9.6, but with different values of Δx. Figure 9.8 shows the calculated numerical transmission coefficient for the two cases, namely $\Delta x = \lambda/100$ and $\lambda/40$, without correcting for the numerical thickness d_{FDTD}. We observe that although the numerical transmission coefficient can always be made to agree with the analytical one the FDTD results do not converge as we vary the spatial resolution in the FDTD, indicating that the thickness of the wire medium slab is not properly modeled. On the other hand, if we properly correct the numerical thickness applied, the FDTD results progressively approach the analytical result as we reduce the FDTD cell size, as evident from the plot in Figure 9.9. Therefore, we conclude, that we must always use the corrected numerical thickness $d_{\text{FDTD}} = \lambda/2 - \Delta x$ in FDTD simulations.

In the following 2-D and 3-D FDTD simulations of the wire medium slabs, the corrected numerical thickness is always used. Note that for modeling finite-sized wire medium slabs, the PBCs in Figure 9.6 are replaced by the PMLs with white space at the slab boundaries along the x-direction. The material interfaces are aligned with the locations of the magnetic field (see Figure 9.2); hence, no additional field averaging is required.

9.6 Numerical Thickness of Wire Medium Slabs

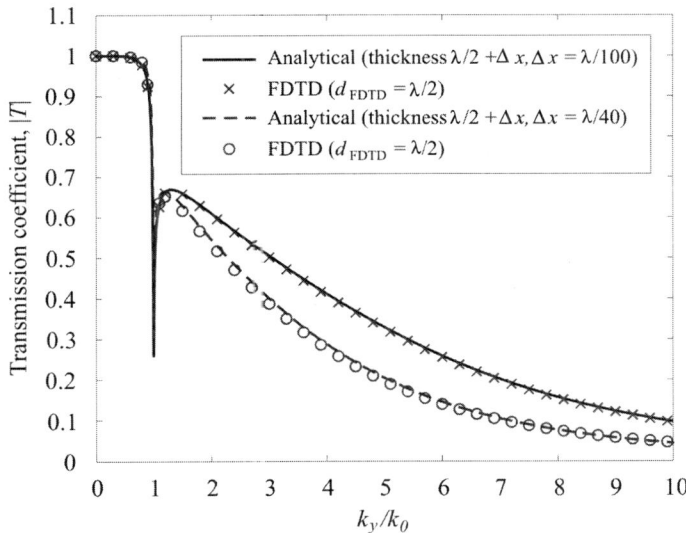

Figure 9.8 Comparison of transmission coefficient for an infinite-long slab of wire medium calculated using analytical solution and from FDTD simulations with different spatial resolutions: $\Delta x = \lambda/100$ and $\Delta x = \lambda/40$ without the correction of numerical thickness. The thickness of the modeled wire medium slab is $\lambda/2$. The numerical thickness of the slab in the FDTD domain in denoted as d_{FDTD}. The analytical solution is given in [40].

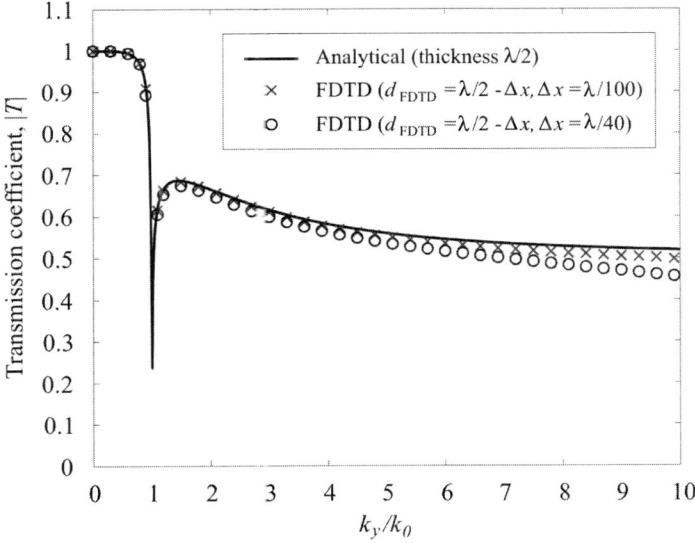

Figure 9.9 Comparison of the transmission coefficient for an infinite-long slab of wire medium calculated using an analytical solution and from FDTD simulations with different spatial resolutions: $\Delta x = \lambda/100$ and $\Delta x = \lambda/40$ with the correction of numerical thickness. The thickness of the modeled wire medium slab is $\lambda/2$. The numerical thickness of the slab in the FDTD domain is denoted as d_{FDTD}. The analytical solution is given in [40].

9.7 Two-Dimensional FDTD Simulations

We begin by implementing the spatially dispersive FDTD method in 2-D simulations with a view to studying wave propagations through a wire medium. The computation domain is infinite in the z-direction and has a rectangular shape in the x-y plane, as shown in Figure 9.10.

The FDTD cell size and the time step are the same as used previously when calculating the transmission coefficient ($\Delta x = \Delta y = \lambda/200$ and $\Delta t = \Delta x/\sqrt{2}c$). The wire medium has a rectangular shape with dimensions $d \times w$ in the $x - y$ plane (see Figure 9.10), and the orientation of the wires is along the x-direction. The wave number corresponding to the plasma frequency of the wire medium is four times that of the wave number of free space ($k_p = 4k$). The thickness of the slab d should be an integer number of half-wavelengths in order to realize the canalization [1, 6]. A magnetic line source radiating sinusoidal signals is placed at a distance of h from the front interface of the wire medium slab. As illustrated in Figure 9.10, the front and back interfaces of the imaging device are treated as the source and image planes, respectively.

For the first simulation, the line source is placed in the close proximity of the front interface of the wire medium slab ($h = \lambda/200$). This setup enables us to illustrate the fact that the waves in the wire medium travel along the wires at the speed of light, in contrast to the free-space case where the waves can travel with the speed of light in any direction. The line source creates a cylindrical wavefront in the free space (see Figure 9.11) since the waves can travel in all directions with the same speed. As soon as the cylindrical wave enters the wire medium, the form of the wavefront changes dramatically and becomes conical (see Figure 9.11).

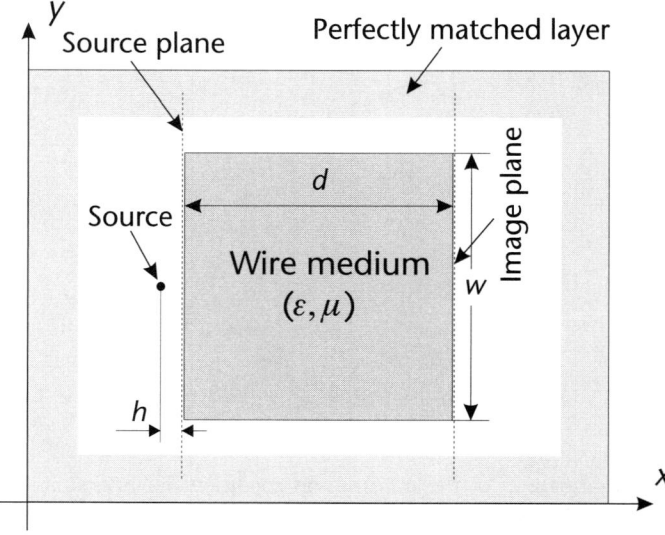

Figure 9.10 The layout of the computational domain for 2-D FDTD simulations. The orientation of wires in the modeled wire medium is along the x-direction.

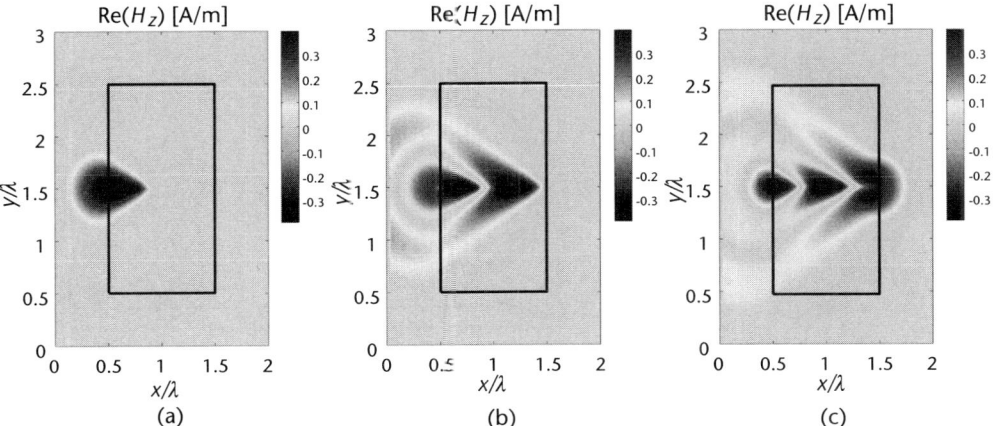

Figure 9.11 Transient propagation of the wave excited by a magnetic sinusoidal line source located at the $\lambda/200$ distance from the front interface of $1\lambda \times 2\lambda$ slab of the wire medium at different time steps: (a) $t = 175\Delta t$, (b) $t = 400\Delta t$, and (c) $t = 530\Delta t$.

The wave travels with the speed of light along the surface of the wire medium in free space, and excites transmission line modes of the wire medium that travel in the orthogonal direction (along the wires) with the same speed; consequently the wavefront becomes conical. Note that the front part of the cone contains the subwavelength information of the source, and the subwavelength image is formed as soon as the cone reaches the back interface of the slab. Therefore, the image is formed with the speed of light and all of the spatial harmonics from the spectrum of the source reach the image plane simultaneously.

In the second simulation, a magnetic line source is placed at $\lambda/10$ away from a wire medium slab with its dimensions of $0.5\lambda \times 1\lambda$. The power flow diagram in the steady state for this case is presented in Figure 9.12. One can see that the power flow changes direction in the vicinity of the interfaces because of the evanescent extraordinary modes of the wire medium and that inside the wire medium, the energy is only transferred along the direction of wires with the help of the transmission line modes. Also, it is noteworthy that no undesirable diffractions from the corners of the slab are observed. This can be explained by the fact that waves inside the wire medium travel and transfer energy only along the x-axis and that no waves travel along the y-axis. That is why the interfaces in the y-direction do not reflect any waves; furthermore, there is no diffraction from the corners. In view of this, it is not necessary to choose the transverse size w of the slab to be significantly larger than the wavelength for the transmission device to function satisfactorily, which is not the case in conventional lenses. In fact, the transverse dimensions of the transmission device can be arbitrary; for example, one or two wavelengths provide good subwavelength imaging performance as our numerical experiments have shown.

In the next simulation, the thickness of the slab is fixed at $\lambda/2$, and the distance between the source and the wire medium slab at $\lambda/10$, though the transverse dimension of the slab is increased to 2λ. Figure 9.13 shows the distributions of

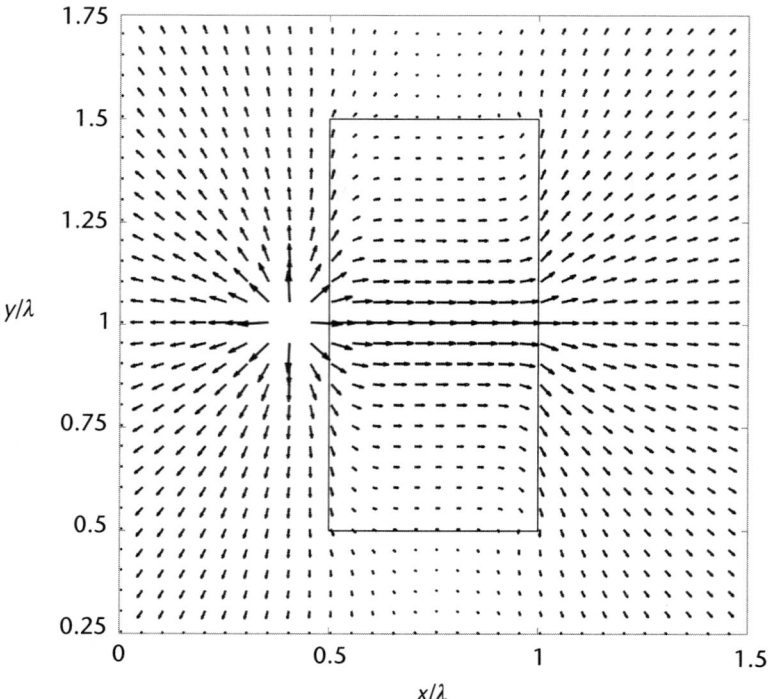

Figure 9.12 Power flow diagram in the steady state for the $0.5\lambda \times 1\lambda$ slab of the wire medium excited by a magnetic line source located at the $\lambda/10$ distance from the front interface.

electric and magnetic fields after the steady state is reached. The absolute values of these fields are presented in Figure 9.14. One can see from Figure 9.13(a) that the nonzero x-component of the electric field is present inside the wire medium slab only in the proximity of the interfaces. This can be easily explained since only the extraordinary modes of the wire medium have a nonzero electric field along the wires, but these modes in the present case are evanescent and decay with distances, and consequently, the x-component of the electric field vanishes at the center of the slab. Only the transmission line modes are present inside the wire medium slab, and hence, the electric and magnetic fields have only y- and z-components, respectively, as may be seen from Figure 9.13(b, c). In accordance to the canalization principle, the transmission line modes transport the information on the fields from the front interface to the back one. This is why the absolute values of the fields are the same at the front and back interfaces, as seen from Figure 9.14. However, the fields in the image plane appear out of phase with respect to those in the source plane since the thickness of the slab is $\lambda/2$.

The images produced by the transmission devices operating in the canalization regime exactly repeat the source distributions if the thickness of the structure is equal to an even number of half-wavelengths and appear out of phase if the thickness is equal to an odd number of half-wavelengths [1]. An example of a case where

9.7 Two-Dimensional FDTD Simulations

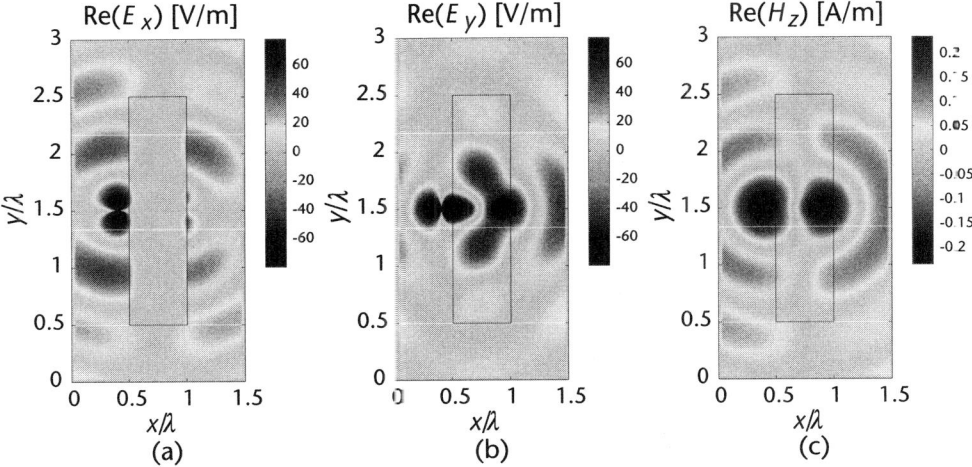

Figure 9.13 (a–c) Distributions of electric and magnetic fields for a $0.5\lambda \times 2\lambda$ slab of the wire medium excited by a magnetic line source located at $\lambda/10$ from the front interface.

the image appears in phase with the source is a wire medium slab with dimensions of $2\lambda \times \lambda$, while the other parameters remain unchanged from the previous case. Figures 9.15(a–c) show the distributions of the electric and magnetic fields in the simulation domain for this case, and it is evident that the fields at the back interface of the slab are very close to the one in the front. Additionally, the distribution of the magnetic field in the plane $y = \lambda$ is plotted in Figure 9.15(d) in order to demonstrate that the field inside the wire medium has harmonic dependence with the period λ along the direction of wires. Using this example, and the fact that the magnetic field is continuous at the interface between free space and the wire

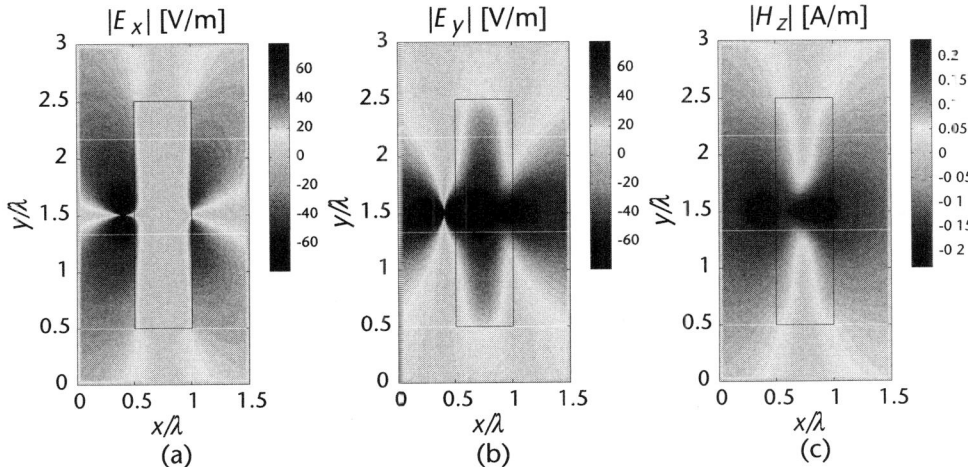

Figure 9.14 (a–c) Absolute values of the fields plotted in Figure 9.13.

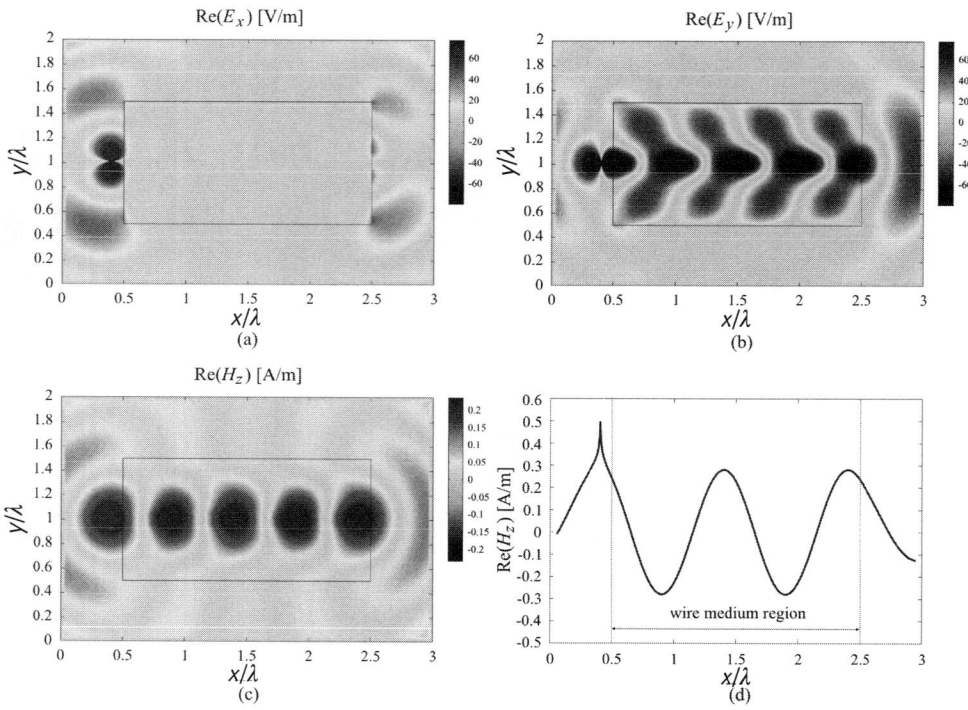

Figure 9.15 (a–d) Distributions of electric and magnetic fields for a $2\lambda \times \lambda$ slab of the wire medium excited by a line source located at $\lambda/10$ from the front interface.

medium, it is possible to explain the canalization principle for the subwavelength imaging: *The field at the back interface of the slab reproduces the distribution at the front interface because the total thickness of the slab is equal to an integral multiple of half-wavelength.*

The wire medium is a unique artificial material that possesses strong spatial dispersion effects. It is not straightforward to compare its properties with media without the spatial dispersion; however, it is possible to reveal certain similarities and differences. For example, if the spatial dispersion effects are neglected in (9.1) then one deals with a uniaxial Drude material with the permittivity tensor given by

$$\bar{\bar{\varepsilon}}_r = \varepsilon_r(\omega)\hat{x}\hat{x} + \hat{y}\hat{y} + \hat{z}\hat{z}, \qquad \varepsilon_r(\omega) = 1 - \frac{k_p^2}{k^2} \tag{9.41}$$

Such a model can be treated as an old and incorrect description of the wire medium [20, 21, 41]. The wire medium in simulations corresponding to Figure 9.13 has been replaced by a local uniaxial Drude material with permittivity tensor (9.41). All other parameters of the structure ($k_p = 4k$, $h = \lambda/10$, $w = 2\lambda$, $d = \lambda/2$—see Figure 9.10) are kept unchanged. The FDTD simulations are performed by using the updating equation given by (9.11). The distributions of the fields in the steady

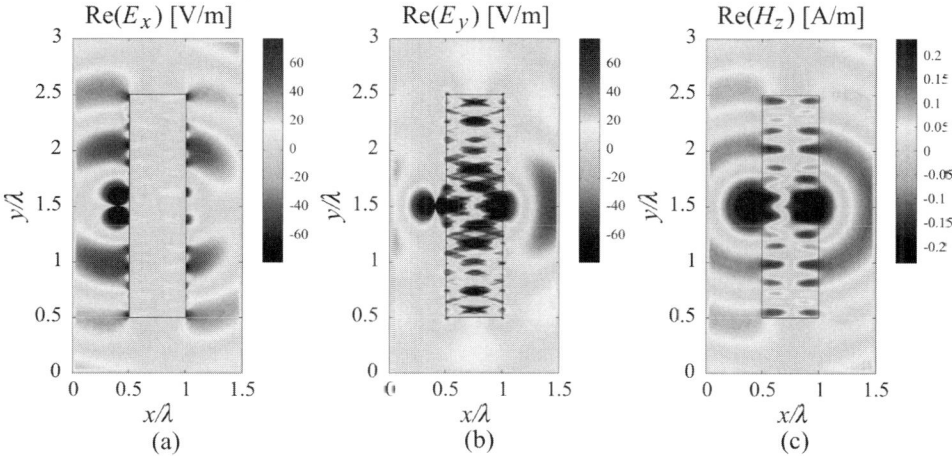

Figure 9.16 (a–c) Distributions of electric and magnetic fields in the steady state for a $0.5\lambda \times 2\lambda$ slab of the uniaxial Drude material (9.41) with $k_p/k = 4$ excited by a line source located at $\lambda/10$ distance from the front interface.

state are presented in Figure 9.16. One can readily observe the strong difference between the field plots in Figures 9.16 and 9.13, which correspond to the uniaxial effective medium model of the wire medium, and the wire medium itself. The waves inside the uniaxial Drude material do not travel precisely along the axis of anisotropy (see Figure 9.17), and hence, they suffer multiple reflections from the edges and corners of the slab. The interference pattern caused by the multiply reflected waves distorts the field distributions in both the source and image planes, as may be seen from Figure 9.18(b). This example shows that it is extremely

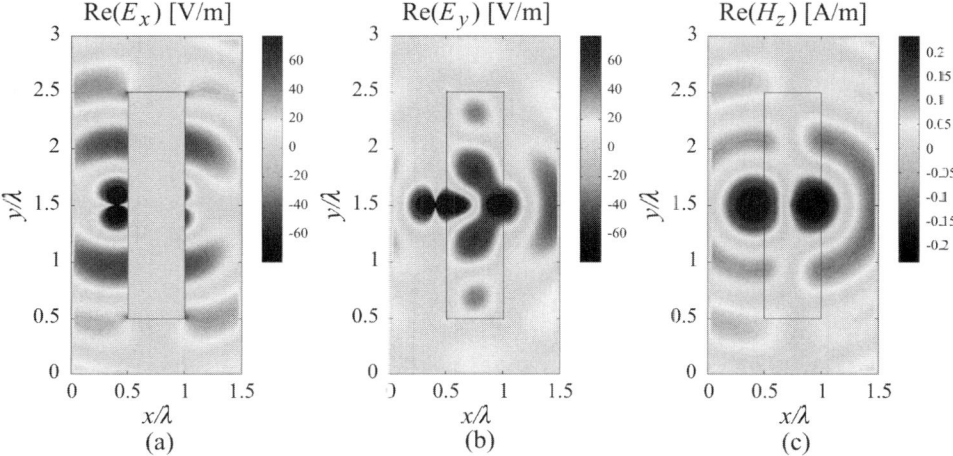

Figure 9.17 (a–c) Distributions of electric and magnetic fields in the steady state for a slab of the uniaxial material with infinite permittivity along anisotropy axis (9.42) excited by a line source located at $\lambda/10$ distance from the front interface.

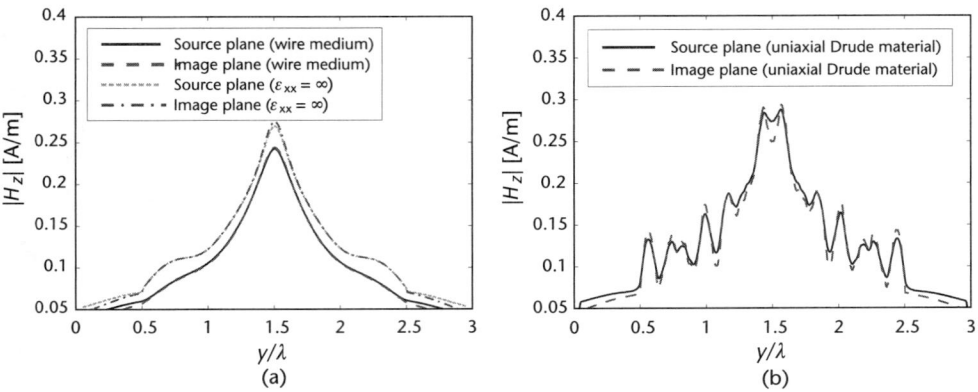

Figure 9.18 Absolute values of magnetic field in the source and image planes of the transmission devices formed by three different materials: (a) wire medium (Figure 9.14), uniaxial material with infinite permittivity along anisotropy axis (Figure 9.17), and (b) uniaxial Drude material (Figure 9.16).

important to take into account of the spatial dispersion while modeling the wire medium.

As a matter of fact, the wire medium behaves similar to a uniaxial material with a permittivity tensor

$$\bar{\bar{\varepsilon}}_r = \infty \hat{x}\hat{x} + \hat{y}\hat{y} + \hat{z}\hat{z} \qquad (9.42)$$

This can be explained by the fact that the component $\varepsilon_r(\omega, q_x)$ of the permittivity tensor of the wire medium (9.1) happens to be infinite for the transmission line modes with $q_x = \pm k$. The FDTD simulations have been performed for the structure, in a manner similar to that depicted in Figure 9.10, but instead of the wire

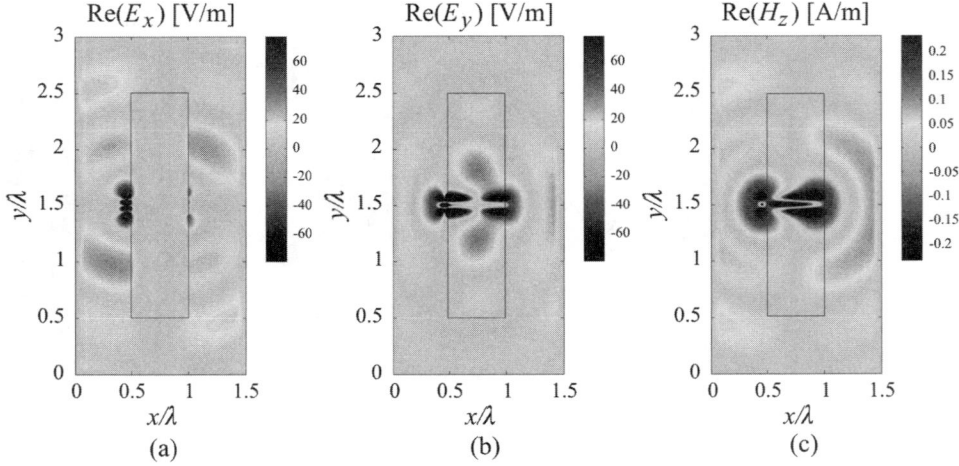

Figure 9.19 (a–c) Distributions of electric and magnetic fields in the steady state for a $0.5\lambda \times 2\lambda$ slab of the wire medium excited by three equally spaced magnetic sources with phase differences equal to $180°$ located at $\lambda/20$ from the front interface.

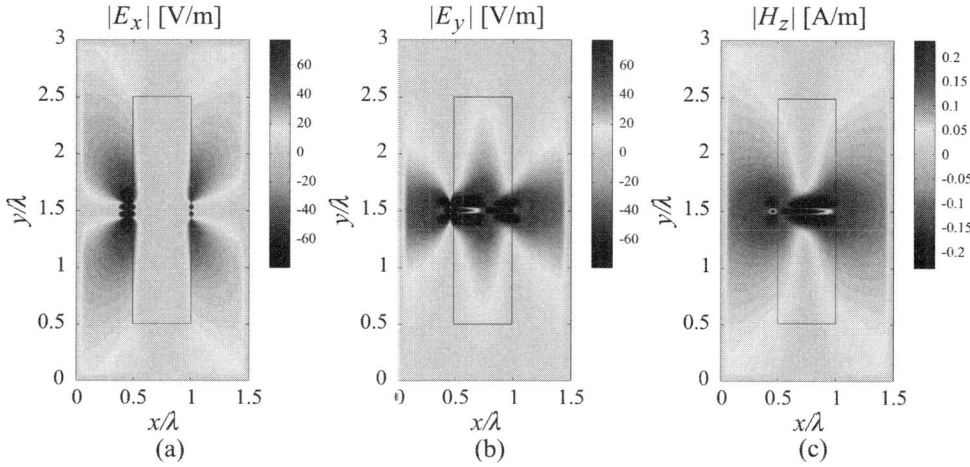

Figure 9.20 (a–c) Absolute values of the fields plotted in Figure 9.19.

medium, the uniaxial material with an infinite permittivity along the anisotropy axis as described in (9.42) has been used. The updating equation for the E_x component inside such a material is relatively simple: $E_x|_{m_x,m_y,m_z}^{n+1} = 0$. The results of the simulation are presented in Figure 9.17. It appears at first sight that Figures 9.13 and 9.17 are identical. However, a more careful comparison immediately reveals the significant differences between the two. Specifically, in the case of the uniaxial material with infinite permittivity along the axis of anisotropy, the x-component

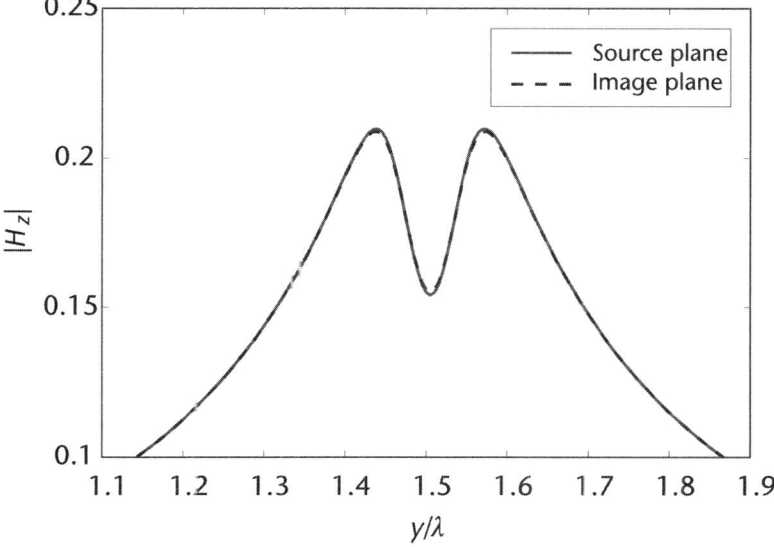

Figure 9.21 Absolute values of magnetic field at the source and image planes for a $0.5\lambda \times 2\lambda$ slab of the wire medium excited by three equally spaced magnetic sources with the phase differences equal to $180°$ located at $\lambda/20$ distance from the front interface.

of the electric field is zero everywhere inside this material; however, for the wire medium there are regions near the front and back interfaces of the slab where this component is nonzero. In other words, the anisotropic model represented in (9.42) allows one to describe the transmission line modes of the wire medium but does not account for the extraordinary modes. The presence of the extraordinary modes in the wire medium renders the distributions of the fields in the source and image planes slightly different from those in the case of the uniaxial material with infinite permittivity along the anisotropy axis, as may be seen from Figure 9.18(a). We see from the plots in Figure 9.18, that both the wire medium slab and the uniaxial material with infinite permittivity along the axis of anisotropy demonstrate a nearly perfect reproduction of the source fields in the image plane as they are almost identical to each other. The difference between the distributions can be explained by the fact that higher-order spatial harmonics experience nonzero reflections from the wire medium slab as shown in [40]. However, as we note from Figure 9.18(b), these reflections are significantly higher for the case of the local uniaxial Drude material, and strong ripples caused by the interference of waves reflected from the edges and corners of the slab adversely affect the imaging operation of the device.

In previous studies, good imaging performance of the wire medium slabs has been demonstrated. However, the subwavelength imaging capabilities of such devices have yet to be confirmed since only single-line source excitations have been used in previous simulations. We will investigate this property of the wire medium by performing additional FDTD simulations using more complex (subwavelength) sources. Three equally spaced magnetic line sources are placed $\lambda/20$ away from the front interface of a $0.5\lambda \times 2\lambda$ wire medium slab. The distance between the sources is $\lambda/20$, and the central source is excited out of phase with respect to the neighboring ones that flank it. The proposed three-source configuration creates a distribution with two strong maxima at the front interface of the wire medium slab, and the distance between these maxima is approximately $\lambda/10$. The results of the spatially dispersive FDTD simulation of the wire medium slab excited by complex sources are presented in Figures 9.19 and 9.20. The size of the computational domain, the operating and plasma frequencies, the FDTD cell size, and the time step remain unchanged from those employed previously. The distribution of the magnetic field in the image plane is presented in Figure 9.21. We see that the two maxima, separated by a distance of $\lambda/10$, are clearly resolved by the device. We conclude, therefore, that the wire medium slab is capable of achieving subwavelength resolution while reproducing the source distribution in the image plane.

9.8 Three-Dimensional FDTD Simulations

The FDTD simulations can be extended directly from the above 2-D case of **H** polarization to the 3-D case. Here a 1-D wire medium is considered where the orientation of the wires is only along the x-direction. As in the previous 2-D case, the wire medium is also modeled using the effective medium method. The FDTD cell size is increased to $\lambda/100$ because of the limitations imposed by computation resources when carrying out the 3-D simulations. The time step is $\Delta t = \Delta x/\sqrt{3}c$ to satisfy the Courant stability condition [25], and the operating frequency is 1 GHz. The plasma frequency is chosen to be six times the operating frequency (i.e., $k_p = 6k$). The simulation domain is shown in Figure 9.22. The MPMLs for the wire medium developed previously, are used in the y- and z-directions to guarantee a fast convergence in simulations of the simulations. The Berenger's original PMLs [34] are used in the x-direction to truncate the free-space region. In order to demonstrate the subwavelength imaging of complex sources, a meander-line antenna is chosen as the source excitation. The meander-line antenna is placed at a distance of $\lambda/20$ to the front interface of the wire medium. The source and image planes are then located at one FDTD cell away from the front and back interfaces of the wire medium slab, respectively. The wire medium slab provides the transportation for the TM-polarized waves [6, 40]; therefore, the electric field components parallel to the orientation of wires at the source plane can be "imaged" with a good resolution. The distributions of the electric field component E_x near the meander-line antenna and at both the source and image planes are compared to demonstrate the subwavelength imaging property of the wire medium slab. Figure 9.23 shows the distribution of the radiated electric field E_x taken at one FDTD cell away from the antenna. It can be seen that the meander-line antenna source creates a field distribution with details at a scale of much less than the wavelength. The comparison of the distributions of E_x at the source and image planes of the wire medium slab is shown in Figures 9.24 and 9.25, respectively.

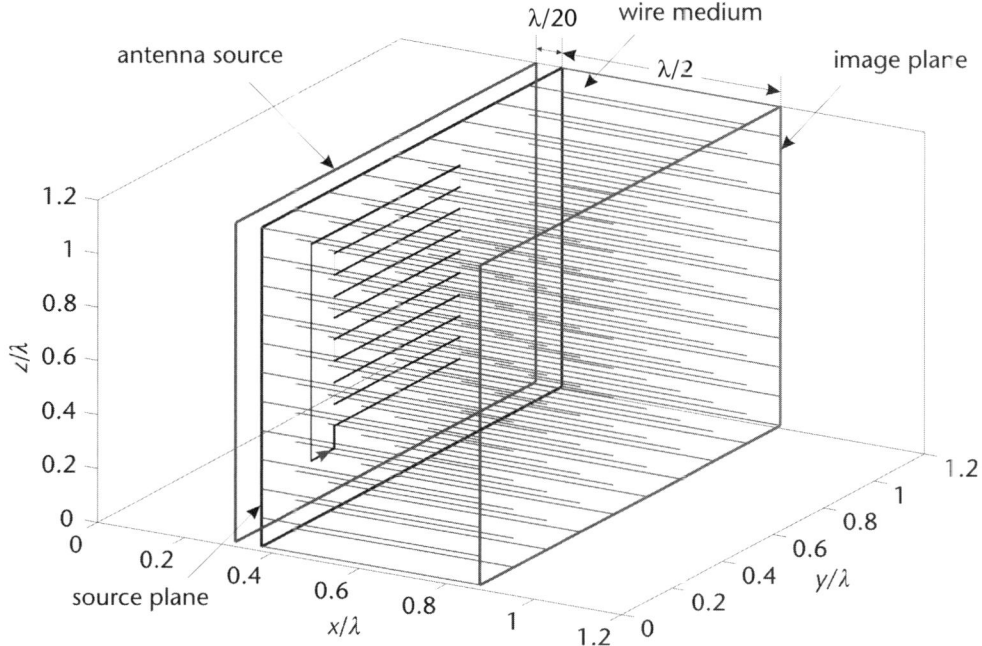

Figure 9.22 3-D FDTD simulation domain for modeling of subwavelength imaging by a 1-D wire medium slab.

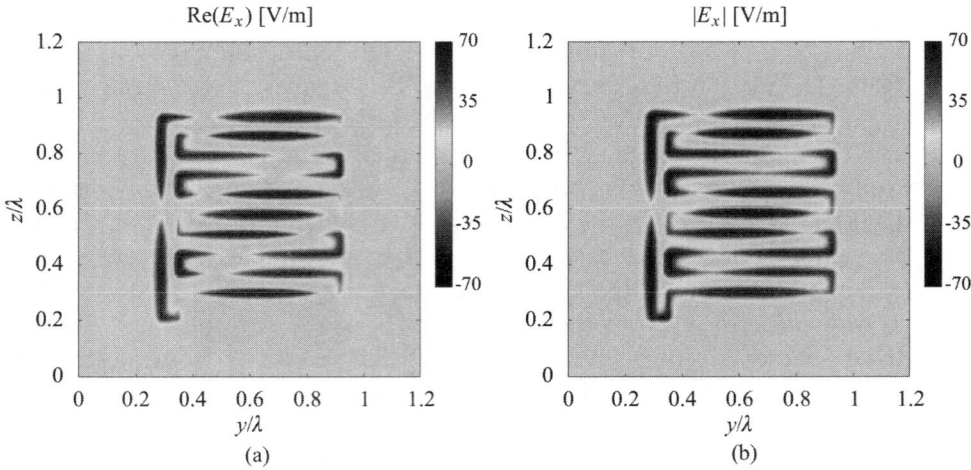

Figure 9.23 Radiated electric field distribution E_x by the meander-line source taken at one FDTD cell away from the source location: (a) real value and (b) absolute value.

The evanescent waves created by the antenna source carry subwavelength details. However, they decay rapidly in the free space and at the source plane of the wire medium slab. We observe from Figures 9.24(a) and 9.25(a) that at a distance of $\lambda/20$ from the meander-line antenna, the field distribution has much lower resolution than that near the source. However through the wire medium slab, the field is transported to the other side and the distribution is repeated with a very good resolution ($< \lambda/10$). The image appears out of phase because the thickness of the wire medium slab is equal to a half-wavelength [see Figure 9.24(b)]. The results in Figures 9.24 and 9.25 confirm that the wire medium slabs possess a good

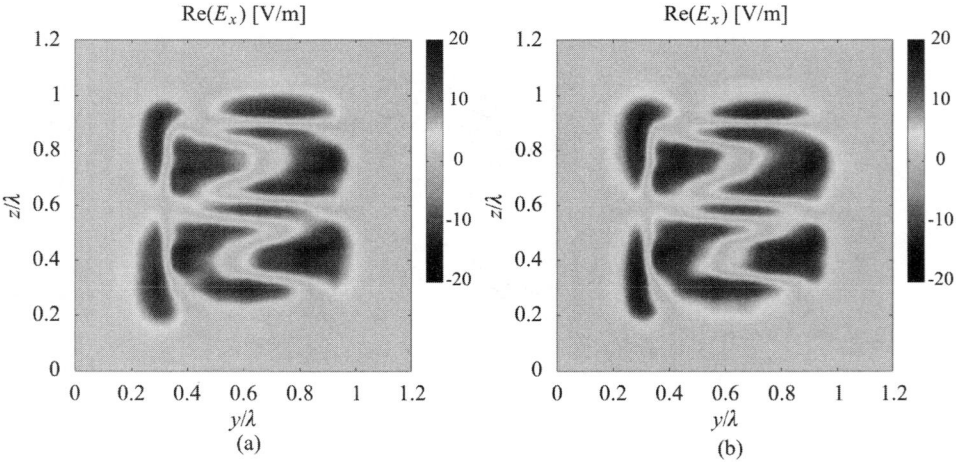

Figure 9.24 Comparison of electric field distributions of E_x at the (a) source plane and (b) image plane of the wire medium slab.

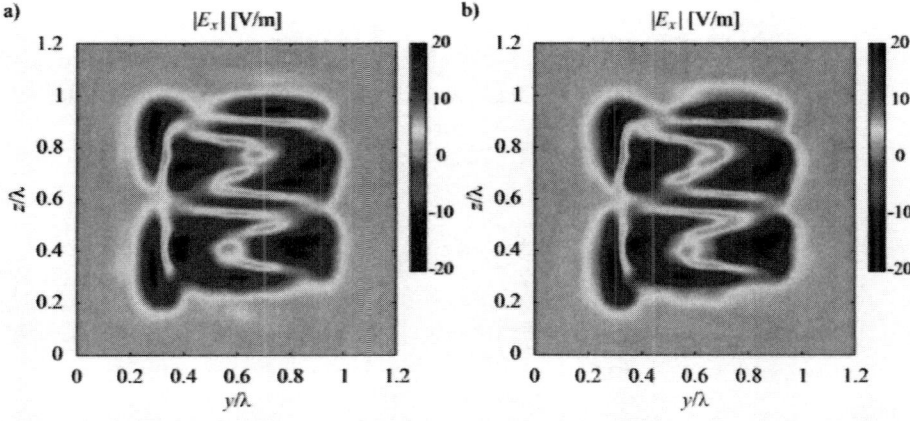

Figure 9.25 The absolute values of the field distributions plotted in Figure 9.24: (a) source plane and (b) image plane.

subwavelength imaging property for practical sources radiated from antennas. The above FDTD simulation results have also been verified through the experimental ones provided in the following section.

9.9 Experimental Verifications

The initial experimental investigation of the subwavelength imaging capability of the wire medium slabs has been performed recently in [6]. The antenna in the form of a letter "P" has been used as a subwavelength source. The clear images of the source are detected at the back interface of the transmission device, and a resolution of $\lambda/15$ is demonstrated for a 18% operation bandwidth. The extensive theoretical studies [40] based on the analysis of transmission and reflection coefficient predicts that the subwavelength imaging should be observed for at least a 4.5% bandwidth for any type of sources. However in practice, for certain sources the imaging bandwidth may be larger. The complexity of the near field produced by the source and the interaction between the source and the transmission device play a crucial role in determining the imaging performance of the system. At the frequencies outside the theoretical minimum band of operation, strong reflections from the wire medium slab are expected in accordance with the theory given in [40], and consequently, the sensitivity of the source with respect to the external fields becomes an issue. If the source is very complex, and contains a substantial amount of subwavelength details then its near-field distribution can be easily distorted by the presence of reflections from two interfaces of the medium, and subwavelength imaging cannot be observed at frequencies outside the theoretical minimum band of operation. However, as observed in [6], if the source is simple and does not contain many subwavelength details then the source is relatively immune from reflections and the image can be successfully transported to the back interface even at some frequencies outside the minimum band of operation.

To investigate the imaging capability of the wire medium slabs in further detail, an experimental study has been carried out using a meander-line antenna printed on a 2-mm-thick slab of duroid whose relative permittivity ε_r is 2.33 [see Figure 9.26(b) for other dimensions]. This source was deliberatively chosen because its near-field distribution is considerably more complex than that of the "P" antenna used in [6]. The return loss (S_{11} parameter) within the frequency band from 840 to 1,060 MHz for the meander-line antenna in the free space is compared with the return loss of the same antenna, but when it is placed close to the front interface of the wire medium slab, see Figure 9.26(a). The results of the comparison are presented in Figure 9.27 and clearly demonstrate that the wire medium slab does not affect the meander-line antenna at the frequency band from 915 to 955 MHz (see the shaded area in Figure 9.27). This implies that within the above frequency range, the meander-line antenna is relatively immune to the presence of the wire medium slab. The slab is "transparent" to the source at these frequencies.

To verify general behavior of the imaging system, a near-field scan has been performed for frequencies in the 840–1,060 MHz frequency band that is significantly wider than the band of 915–955 MHz, where perfect imaging is expected. An mechanical automated near-field scanning device and a 2-mm-long monopole probe made from the central core of a coaxial cable with a 2-mm diameter have been used. The scan area is 24×24 cm^2 with 75 steps in both directions. The probe is oriented normal to the interfaces of both the meander-line antenna and the transmission device and consequently only detects the normal component of

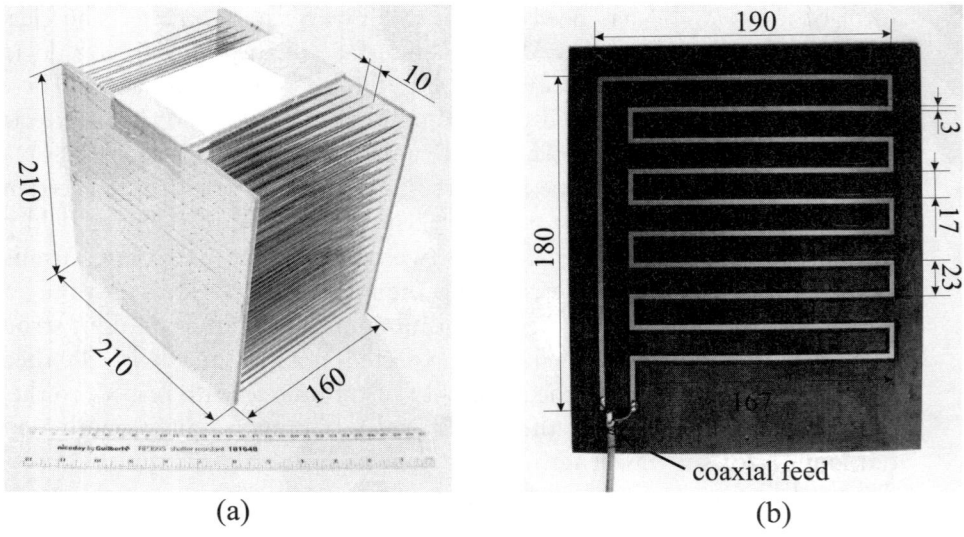

Figure 9.26 (a) The geometries of the transmission device: an 21 × 21 array of wires with 1 mm radii and (b) the near-field source. All dimensions are in millimeters.

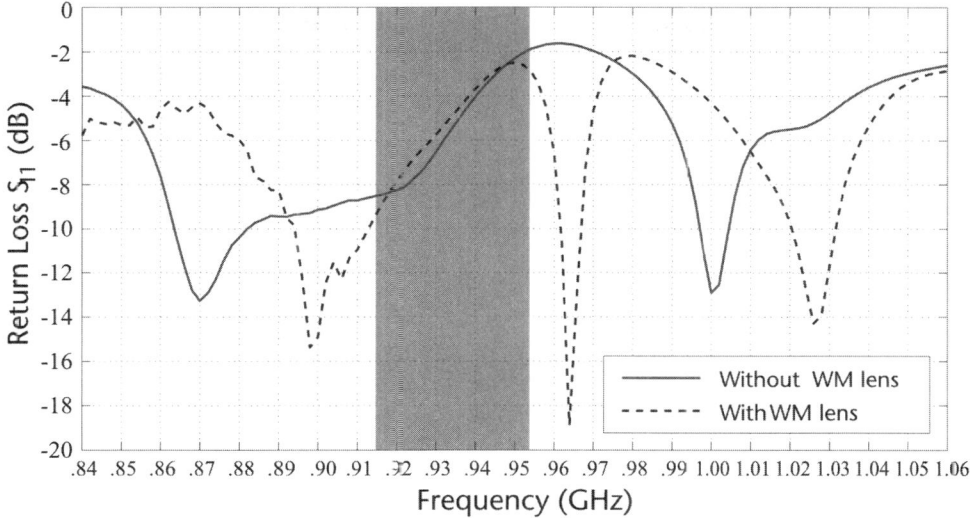

Figure 9.27 The return loss (S_{11} parameter) as a function of frequency for the meander-line antenna in free space and at the interface of the wire medium slab [42].

the electric field. The wire medium slab is capable of imaging only the electromagnetic waves with the TM polarization [6], and only the normal component of electric field is restored at the back interface. The other two components contain the contributions of electromagnetic waves with the TE polarization and are not transported by the wire medium slab. Selected results from the near-field scans at 23 frequencies from 840 to 1,060 MHz with a 10-MHz step are presented in Figures 9.28 and 9.29. At 910–960 MHz the fields at the source plane with and without the presence of the wire medium are nearly identical (see Figure 9.29 for results at 940 MHz). This confirms that the wire medium slab does not introduce significant reflections at these frequencies. At the same time the field in the image plane reproduces the source field with an accuracy of about 2 cm. This confirms that the resolution of the imaging device at this frequency range is about $\lambda/15$.

9.10 Internal Imaging by Wire Medium Slabs

The above 2-D, 3-D FDTD simulation and experimental results demonstrate the potential of the wire medium slabs as transmission devices in which the canalization effect allows the subwavelength information of a source to be transferred to the image plane. In this section, the amplification of evanescent waves because of the resonant excitation of standing waves inside the slab is demonstrated through FDTD simulations; in contrast to the case of LHM slabs, the amplification of evanescent waves is due to the resonant excitation of surface plasmons at the interfaces of the slab.

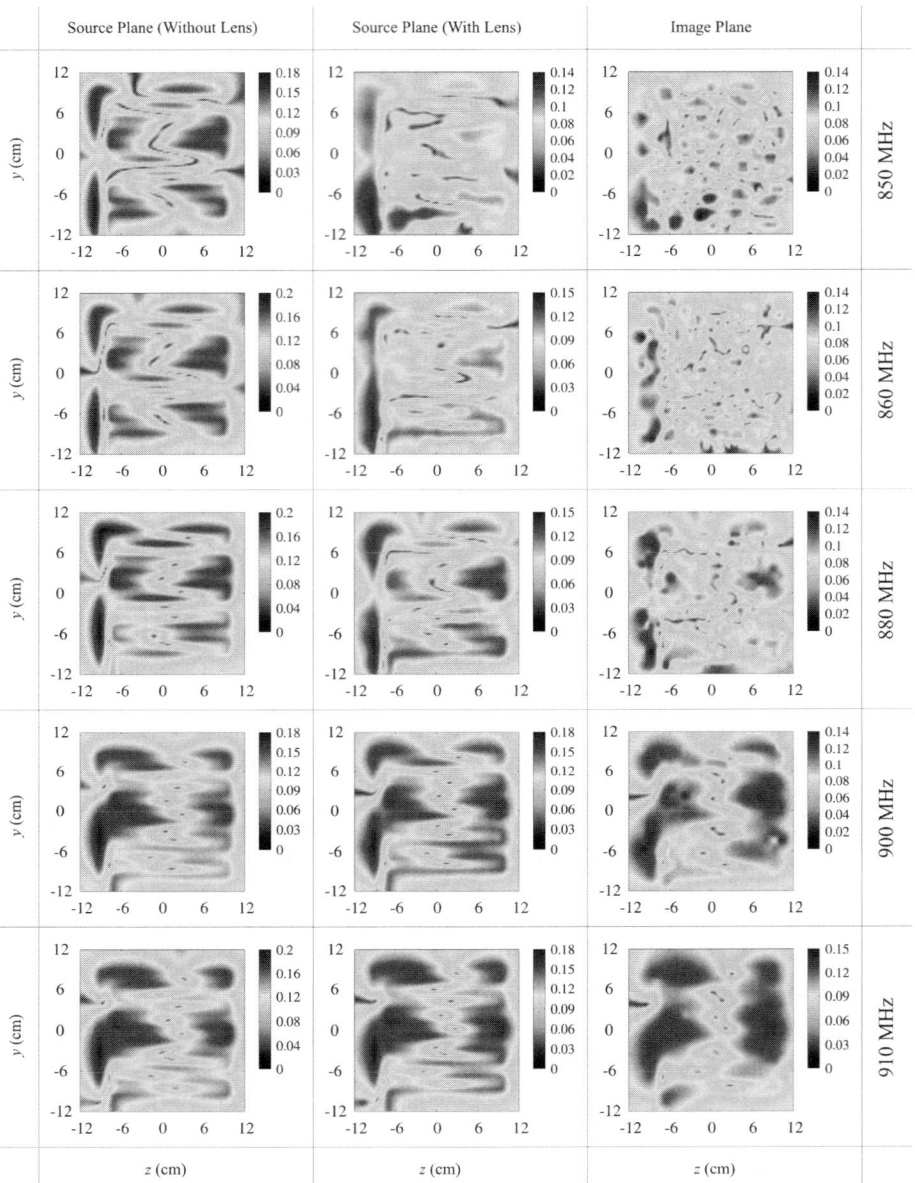

Figure 9.28 Results of the near-field scan at 850 MHz, 860 MHz, 880 MHz, 900 MHz, and 910 MHz (in arbitrary units): the component of electric field normal to the interface at a 2-mm distance from the meander antenna in free space (source plane without wire medium), the same but when the antenna is placed at the front interface of the transmission device (source plane with wire medium) and at a 2-mm distance from the back interface of the wire medium slab (image plane) [42].

Three-dimensional FDTD simulations have been performed to demonstrate internal imaging using the wire medium slab. The simulation domain is the same as the configuration shown in Figure 9.22. The internal image plane is located in the middle of the wire medium slab. Figure 9.30 shows the distributions of the

9.10 Internal Imaging by Wire Medium Slabs

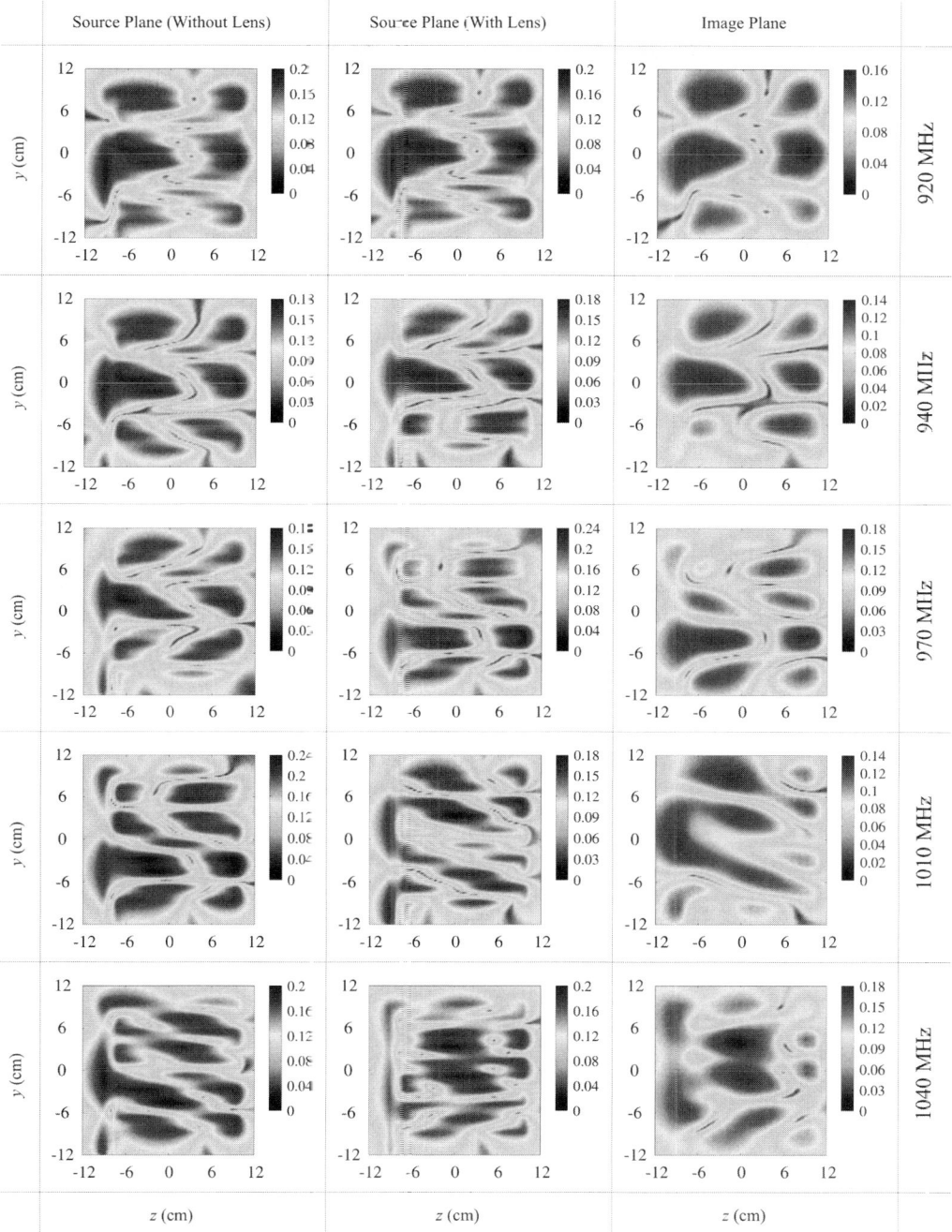

Figure 9.29 Results of the near-field scan at 920 MHz, 940 MHz, 970 MHz, 1,010 MHz, and 1,040 MHz (in arbitrary units): the component of electric field normal to the interface at a 2-mm distance from the meander antenna in the free space (source plane without wire medium), the same but when the antenna is placed at the front interface of the transmission device (source plane with wire medium), and at a 2-mm distance from the back interface of the wire medium slab (image plane) [42].

absolute and real values of the electric displacement components at the source plane, the front interface, and the internal image plane. It can be seen that the source creates a near-field distribution shown in Figure 9.30(a, d) that contains fine subwavelength details at an order of $\lambda/20$ or even smaller. However because evanescent waves carrying subwavelength information decay in free space, the electric field at the front interface of the wire medium slab [Figure 9.30(b, e)] does not contain such fine details, and the amplitude of the field is much lower than that at the source plane. The wire medium operates in the canalization regime, and the same distribution of the electric field as that at the front interface appears at the back one, but this distribution does not provide detailed information about the original source. However, the distribution of electric displacement inside the wire medium [Figure 9.30(c, f)] repeats the original near-field distribution. The intensity of the internal image is even larger than the intensity of the source, which may be due to the overamplification of the certain spectrum of spatial harmonics. The resolution of the internal imaging is about $\lambda/20$.

The internal image that is formed is out of phase with respect to the source, as is evident from Figure 9.30(d, f). Figure 9.19(c) shows that the same effect is present in the 2-D case. This is because the first internal image formed by standing waves inside of the slab is present in the wire medium slab whose thickness is $\lambda/2$. Figure 9.15 shows that for a thicker slab, for example, $d = 2\lambda$, there are four

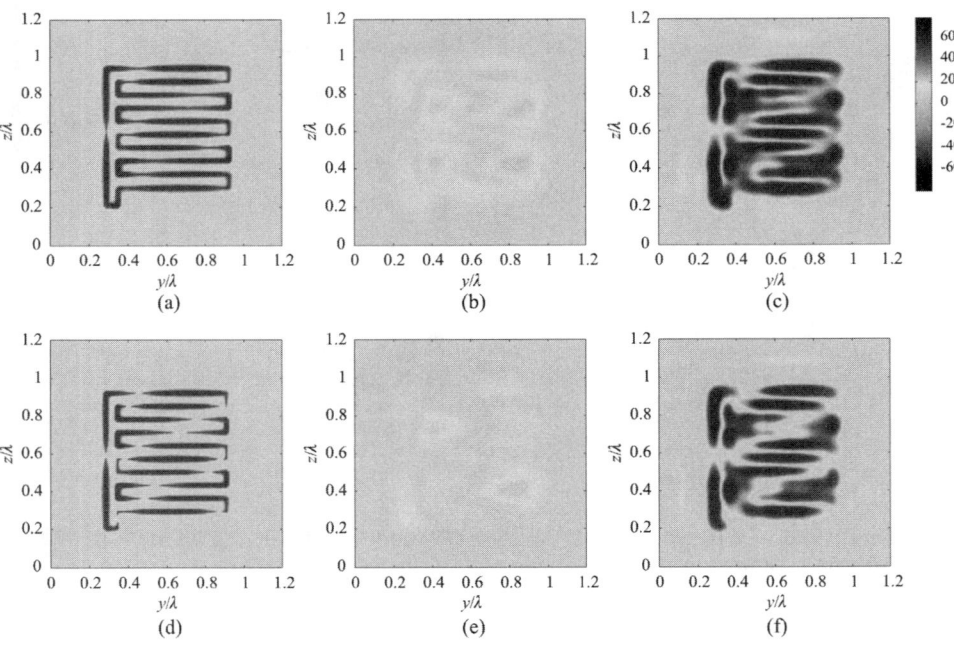

Figure 9.30 Distributions of absolute and real values of electric displacement component along the wires for the slab of wire medium with $\lambda/2$ thickness and $0.7\lambda \times 0.7\lambda$ transverse dimensions excited by a meander-line source located at a $\lambda/20$ distance from the front interface: (a, d) in the source plane; (b, e) at the front interface; and (c, f) inside of the wire medium (middle), respectively.

internal image planes: The first and third ones are out of phase with respect to the source, and the second and fourth ones are in phase.

The preliminary study shows that an image appears near the middle of the $\lambda/2$ thick wire medium slab and does not change significantly beyond the 5–10% of the slab thickness around the internal image plane location. The plane of the image formation cannot be determined exactly from simulations since the distribution of the electric displacement inside the wire medium varies slowly along the direction of the wires.

The field distribution is TEM with respect to the orientation of wires and is formed by the so-called transmission-line modes [5]. The electric field component of such modes along the wires is equal to zero, and the electric displacement inside the wire medium is proportional to the currents flowing along the wires. Therefore in order to detect the internal image, we need to measure the currents along the wires instead of either the electric or the magnetic field; in contrast to this, the canalization regime can be detected using the near-field scanning, where the image is formed at the back interface of the wire medium. Probably the easiest way to capture the internal image is to embed detectors into the metallic wires at the internal image plane and directly measure the current flowing along the wires.

The simulations of different distances between the meander-line source and the wire medium slab have been performed. The internal images with a resolution of approximately $\lambda/20$ have been observed for the distances of $d = \lambda/20$ and $d = \lambda/15$. If the source is placed at a distance of $d = \lambda/10$ from the front interface, the internal image is not as good as in the previous cases and the resolution is less than $\lambda/20$. This leads us to conclude that the internal imaging with a good subwavelength resolution is available only for a limited range of distances between the source and the device. Probably this limitation is attributed to the fact that the dependence of the transfer function for the evanescent spatial harmonics on the transverse wave vector is not exponential as for the case of the LHM slabs.

9.11 Conclusions

A spatially dispersive FDTD method has been developed for the modeling of wave propagation in the wire medium using the effective medium method. The ADE method has been used in order to take into account both the spatial and frequency dispersion effects of the wire medium. The stability analysis shows that the conventional Courant stability limit is preserved when the standard central finite difference approximations and the central average operator are used to discretize the differential equations. Through the use of the MPML, the wire medium can be "perfectly" matched to the absorbing boundaries therefore the convergence performance in simulations is greatly improved since the diffractions from corners and edges of the finite sized wire medium slab are avoided.

The subwavelength transmission devices formed by the wire medium are chosen for the validation of the developed spatially dispersive FDTD formulations. Numerical simulations verify the subwavelength imaging capability of these structures. The results confirm that the wire medium slabs operate in the canalization regime as transmission devices and demonstrate that this regime is not sensitive to the transverse dimensions of these structures. The minimum bandwidth of operation of the wire medium slab as the subwavelength imaging device is confirmed experimentally. The actual bandwidth of operation significantly depends on the complexity and sensitivity of the source to the reflections from the wire medium slab.

The amplification of the evanescent spatial harmonics inside a wire medium slab is reported. It allows to image the sources located at significant distances from the device with a subwavelength resolution. This regime of subwavelength imaging is based on the resonant excitation of standing waves inside the wire medium but is not due to the excitation of surface plasmons as for the case of LHM slabs. The capability of the wire medium slabs operating as internal subwavelength imaging devices is confirmed through 3-D FDTD simulations using the proposed spatially dispersive FDTD method.

References

[1] P. A. Belov, C. R. Simovski, and P. Ikonen, "Canalization of subwavelength images by electromagnetic crystals," *Phys. Rev. B.*, vol. 71, p. 193105, 2005.

[2] P. Ikonen, P. A. Belov, C. R. Simovski, and S. I. Maslovski, "Experimental demonstration of subwavelength field channeling at microwave frequencies using a capacitively loaded wire medium," *Phys. Rev. B*, vol. 73, p. 073102, 2006.

[3] P. A. Belov, Y. Hao, and S. Sudhakaran, "Subwavelength microwave imaging using an array of parallel conducting wires as a lens," *Phys. Rev. B*, vol. 73, p. 033108, 2006.

[4] P. A. Belov and Y. Hao,"Subwavelength imaging at optical frequencies using a transmission device formed by a periodic layered metal-dielectric structure operating in the canalization regime," *Phys. Rev. B*, vol. 73, p. 113110, 2006.

[5] P. A. Belov, R. Marques, S. I. Maslovski, I. S. Nefedov, M. Silverinha, C. R. Simovski, and S. A. Tretyakov, "Strong spatial dispersion in wire media in the very large wavelength limit," *Phys. Rev. B*, vol. 67, p. 113103, 2003.

[6] M. G. Silveirinha, "Nonlocal homogenization model for a periodic array of epsilon-negative rods," *Phys. Rev. E*, vol. 73, p. 046612, 2006.

[7] R. Holland and L. Simpson, "Finite-difference analysis of EMP coupling to thin struts and wires," *IEEE Trans. Electromagn. Compat.*, vol. 23, pp. 88–97, 1981.

[8] G. Ledfelt, "A stable subcell model for arbitrarily oriented thin wires for the FDTD method," *Int. J. Numerical Modeling*, vol. 15, pp. 503–515, 2002.

[9] F. Edelvik, "A new technique for accurate and stable modeling of arbitrarily oriented thin wires in the FDTD method," *IEEE Trans. Electromagn. Compat.*, vol. 45, no. 2, pp. 416–423, 2003.

[10] B. P. Koh, C. J. Railton, and I. J. Craddock, "Wire above ground plane transmission line formulation in FDTD algorithm," *IEE Proceedings—Microwaves, Antennas and Propagation*, vol. 151, no. 3, pp. 249–255, 2004.

[11] K. R. Umashankar, A. Taflove, and B. Beker, "Calculation and experimental validation of induced currents on coupled wires in an arbitrary shaped cavity," *IEEE Trans. Antennas Propagat.*, vol. 35, pp. 1248–1257, 1987.

[12] B. P. Koh, C. J. Railton, and I. J. Craddock, "Wire above ground plane transmission line formulation in FDTD algorithm," *IEE Proceedings—Microwaves, Antennas and Propagation*, vol. 151, no. 3, pp. 249–255, 2004.

[13] M. Bingle, D. B. Davidson, and J. H. Cloete, "Scattering and absorption by thin metal wires in rectangular waveguide—FDTD simulations and physical experiments," *IEEE Trans. Microw. Theory Tech.*, vol. 50, pp. 1621–1627, 2002.

[14] R. M. Makinen, J. S. Juntunen, and M. A. Kivikovski, "An improved thin-wire model for FDTD," *IEEE Trans. Microw. Theory Tech.*, vol. 50, pp. 1245–1255, 2002.

[15] D. J. Riley, "Transient finite-elements for computational electromagnetics: Hybridization with finite differences, modeling thin wires and thin slots," *Proc. 17th Annual Review Progress in Applied Computational Electromagnetics*, Monterey, CA, pp. 128–138, 2001.

[16] D. J. Riley, "Transient finite-elements for computational electromagnetics: Hybridization with finite differences, modeling thin wires and thin slots," *Proc. 17th Annual Review Progress in Applied Computational Electromagnetics*, Monterey, CA, pp. 128–138, 2001.

[17] I. J. Craddock and C. J. Railton, "A new technique for the stable incorporation of static field solutions in the FDTD method for the analysis of thin wires and narrow strips," *IEEE Trans. Microw. Theory Tech.*, vol. 46, pp. 1091–1096, 1998.

[18] C. J. Railton, P. K. Boon, and I. J. Craddock, "The treatment of thin wires in the FDTD method using a weighted residuals approach," *IEEE Trans. Antennas Propagat.*, vol. 52, no. 11, pp. 2941–2949, 2004.

[19] W. Rotman, "Plasma simulations by artificial dielectrics and parallel-plate media," *IRE Trans. Ant. Propag.*, vol. 10, pp. 82–95, 1962.

[20] J. Brown, "Artificial Dielectrics," *Progress in Dielectrics*, vol. 2, pp. 195–225, 1960.

[21] J.B. Pendry, A.J. Holden, W.J. Steward, and I. Youngs, "Extremely low frequency plasmons in metallic mesostructures," *Phys. Rev. Lett.*, vol. 76, no. 25, pp. 4773–4776, 1996.

[22] P. A. Belov, S. A. Tretyakov, and A. J. Viitanen, "Dispersion and reflection properties of artificial media formed by regular lattices of ideally conducting wires," *J. Electromagn. Waves Applic.*, vol. 16, pp. 1153–1170, 2002.

[23] L. D. Landau and E. M. Lifshitz, *Electrodynamics of Continuous Media*, New York: Pergamon Press, 1984.

[24] Y. Zhao, P. A. Belov, and Y. Hao, "Modelling of wave propagation in wire media using spatially dispersive finite-difference time-domain method: numerical aspects," *IEEE Trans. Antennas Propagat.*, vol. 55, 2007.

[25] A. Taflove and S. C. Hagness, *Computational Electrodynamics: The Finite-Difference Time-Domain Method*, 2nd ed., Norwood, MA: Artech House, 2000.

[26] K. P. Prokopidis, E. P. Kosmidou and T. D. Tsiboukis, "An FDTD algorithm for wave propagation in dispersive media using higher-order schemes," *Journal of Electromagn. Waves and Appl.*, vol. 18, no. 9, pp. 1171–1194, 2004.

[27] C. R. Simovski, and P. A. Belov, "Low-frequency spatial dispersion in wire media," *Phys. Rev. E*, vol. 70, p. 046616, 2004.

[28] M. Silveirinha and C. Fernandes, "Homogenization of 3D connected and non-connected wire metamaterials," *IEEE Trans. Microw. Theory Tech.*, vol. 53, no. 4, pp. 1418–1430, 2005.

[29] M. Silveirinha and C. Fernandes, "Homogenization of metamaterial surfaces and slabs: the crossed wire mesh canonical problem," *IEEE Trans. Antennas Propagat.*, vol. 53, no. 1, pp. 59–69, 2005.

[30] P. G. Petropoulos, "Stability and phase error analysis of FDTD in dispersive dielectrics," *IEEE Trans. Antennas Propagat.*, vol. 42, no. 1, pp. 62–69, 1994.

[31] A. Pereda, L. A. Vielva, A. Vegas, and A. Prieto, "Analysing the stability of the FDTD technique by combining the von Neumann method with the Routh-Hurwitz criterion," *IEEE Trans. Microw. Theory Tech.*, vol. 49, no. 2, pp. 377–381, 2001.

[32] F. B. Hildebrand, *Introduction to Numerical Analysis*, New York: McGraw-Hill, 1956.

[33] A. P. Zhao, "Generalized-material-independent PML absorbers used for the FDTD simulation of electromagnetic waves in 3-D arbitrary anisotropic dielectric and magnetic media," *IEEE Trans. Microw. Theory Tech.*, vol. 46, no. 10, pp. 1511–1513, 1998.

[34] J.-P. Berenger, "A perfectly matched layer for the absorption of electromagnetic waves," *J. Computat. Phys.*, vol. 114, no. 2, pp. 185–200, 1994.

[35] Z. Wu and J. Fang, "Numerical implementation and performance of perfectly matched layer boundary condition for waveguide structures," *IEEE Trans. Microw. Theory Tech.*, vol. 43, no. 12, pp. 2676–2683, 1995.

[36] D. S. Katz, C. E. Reuter, E. T. Thiele, R. M. Joseph, and A. Taflove, "Extension of FD-TD simulation capabilities using the Berenger PML," *1995 Dig. USNC/URSI Radio Science Meet.*, Newport Beach, CA, p. 334, 1995.

[37] S. D. Gedney and A. Roden, "Applying Berenger's perfectly matched layer (PML) boundary condition to non-orthogonal FDTD analyzes of planar microwave circuits," *1995 Dig. USNC/URSI Radio Science Meet.*, Newport Beach, CA, p. 333, 1995.

[38] Z. Wu and J. Fang, "Experiments on the perfectly matched layer boundary condition in modeling wave propagation in waveguide components," *1995 Dig. USNC/URSI Radio Science Meet.*, Newport Beach, CA, p. 336, 1995.

[39] J. Fang and Z. Wu, "Generalized perfectly matched layer for the absorption of propagating and evanescent waves in lossless and lossy media," *IEEE Trans. Microw. Theory Tech.*, vol. 44, no. 12, pp. 2216–2222, 1996.

[40] P. A. Belov and M. G. Silveirinha, "Resolution of subwavelength transmission devices formed by a wire medium," *Phys. Rev. E*, vol. 73, p. 056607, 2006.

[41] W. Rotman, "Plasma simulations by artificial dielectrics and parallel-plate media," *IRE Trans. Ant. Propag.*, vol. 10, pp. 82–95, 1962.

[42] P. Belov, Y. Zhao, S. Sudhakaran, A. Alomainy, and Y. Hao, "Experimental study of the subwavelength imaging by a wire medium slab," *Appl. Phys. Lett.*, vol. 89, p. 262109, 2006.

CHAPTER 10
FDTD Modeling of Metamaterials for Optics

10.1 Introduction

So far, we have applied the FDTD method to model metamaterials at microwave frequencies. At optics, there exist some noble metals exhibiting inherent plasmonic resonance similar to those artificial materials at microwaves. Such extraordinary characteristics enable us to obtain negative permittivity directly from materials such as silver or gold without creating resonant particles. Indeed, there has been increasing interest in developing metamaterials at optics based on "surface plasmons" for potential applications in subwavelength imaging and subwavelength optical waveguides. In this chapter, we apply the proposed dispersive FDTD method and its variations to model silver-dielectric layered structures and plasmonic waveguide [1–126]. It is still and perhaps even more problematic to obtain negative permeability at optics, and hence it remains a challenge to design and implement LHMs at optical frequencies. In this chapter, based on effective medium theory, we propose to simulate several structures utilizing optical metamaterials, namely a scanning near-field optical microscopy (SNOM) [12, 14] and electromagnetic cloaking. The latter may even have applications at microwaves in military if such materials can be practically implemented and fabricated with sufficient bandwidth and low loss.

10.2 Dispersive FDTD Modeling of Silver-Dielectric Layered Structures for Subwavelength Imaging

10.2.1 Introduction

The possibility of subwavelength imaging was first proposed by Pendry in 2000 [1]. It was demonstrated in the above work that a slab of LHM [2, 3] (medium with both negative permittivity and permeability) can create images with a nearly unlimited resolution. This idea overcomes the classical restriction on the resolution of imaging systems, the diffraction limit, and became the starting point for the creation of new research area of metamaterials [4], namely development of artificial media possessing extraordinary electromagnetic properties that are usually not available in natural materials. The idea of Pendry's perfect lens is based on such exotic effects observable in LHM as backward waves, negative refraction, and amplification of evanescent waves. The backward waves and negative refraction enable the focusing of the far field of a source. The near field of the source, which contains subwavelength details, is recovered in the image plane by using the amplification of the evanescent modes in the slab. Currently, samples of LHMs have been only created in the microwave region [5]. The creation of LHMs at terahertz frequencies and in the optical range encounters the problems related to getting the required magnetic properties [6, 7], which should be artificially created. In the absence of magnetic properties, the lenses formed by materials with only negative permittivity (for example, silver at optical frequencies) are still capable of creating images with subwavelength resolution, but the operation is restricted to only the p-polarization, and only when the lens is thin compared to the wavelength [1]. This idea was confirmed by recent experimental results [8], which demonstrated the phenomenon of subwavelength imaging obtained by using silver slabs in the optical frequency range. The resolution achieved by such lenses is limited by the losses in silver, but this problem can be resolved by cutting the slab into separate slabs with smaller thicknesses [9, 10], and introducing active materials [15]. Unfortunately, at present there exists no recipe for increasing the thickness of such lenses, except by introducing artificial magnetism at optical frequencies.

Competitive alternatives of LH media at optical frequencies are photonic crystals [16, 17]. The negative refraction effect was reported by Notomi in [18, 19] in photonic crystals at frequencies close to the bandgap edges, and the subwavelength imaging using planar lenses formed by photonic crystals was demonstrated both theoretically [20–24] and experimentally [25, 26]. Unfortunately, the resolution of such lenses is strictly limited by the period of the crystal, as has been proven in [27]. It implies that it is impossible to realize good subwavelength resolution using lenses formed by photonic crystals, not only because they operate in the frequency regime where the wavelength in the crystal is comparable with the period of the lattice, but also because this wavelength can not be shortened too much due to the lack of availability of natural high-contrast materials.

During the studies of negative refraction and imaging in photonic crystals, it was noted that in certain cases subwavelength imaging is realized on the basis of a mechanism other than propagation in LHMs. Actually, the negative refraction in photonic crystals is observed either in the forward wave regime, at frequencies belonging to the first propagation band [20, 24, 27], or in the backward wave regime, which corresponds to the second propagation band [21, 22]. The evidence of nonnegative refraction has been reported by numerous authors [28–32], for crystals operating in the first frequency band. The lenses formed by such crystals indeed operate in the frequency regime that involves neither negative refraction nor an amplification of evanescent waves. This phenomenon was referred to in [32] as *canalization* (as introduced in Chapter 9), wherein the slab of photonic crystals does not operate like a usual lens that focuses radiation into a focal point, but effectively works as a transmission device that delivers subwavelength images from the front interface of the lens to its back. The realization of such phenomenon is possible if the crystal has a flat isofrequency contour [32], and if the thickness of the slab fulfils the Fabry-Perot condition (integer number of half-wavelengths) [32]. The flat isofrequency contour allows the transformation of all spatial harmonics produced by the source, including evanescent modes, into propagating eigenmodes of the crystal. This preserves subwavelength details of the source that usually disappear with distance due to the rapid spatial decay of the evanescent harmonics. These propagating eigenmodes transfer the image across the

10.2 Dispersive FDTD Modeling of Silver-Dielectric Layered Structures

slab from the front to the back interface. Possible reflections from the interfaces are minimized by utilizing the Fabry-Perot resonance phenomenon, which exists for all incidence angles in this case, owing to the flatness of the isofrequency contours.

The lenses operating in the canalization regime have the same restrictions on the resolution provided by the periodicity, as do the lenses working in the LH regime. Specifically, in order to realize subwavelength resolution, it is necessary for the period of the structure to be much smaller than the wavelength. In the microwave region, canalization with $\lambda/6$ resolution [32] has been implemented by using an electromagnetic crystal formed by a lattice of wires periodically loaded by capacitances, a configuration that is also called a capacitively loaded wire medium [33]. Such a crystal has a resonant band-gap at very low frequencies (with wavelength/period ratio $\lambda/a = 14$) and does not contain high-contrast materials. The theoretical and numerical predictions [32] have been confirmed by experimentally [34], and a $\lambda/5$ resolution was demonstrated. A higher resolution can be achieved by using loadings with larger capacitances, but actual devices based on the implementations of these suffer from high losses and exhibit very narrow bandwidths.

It is possibility to realize canalization by utilizing a wire medium, comprised of a lattice of parallel conducting wires [35–38]. Such a device supports a very special type of eigenmodes, the so-called transmission line modes [38], which transport energy strictly along wires with the speed of light and can have arbitrary transverse wave vector components. This implies that such modes correspond to completely flat isofrequency contours, which is the principal requirement for implementing canalization. Detailed analytical, numerical, and experimental studies [39] have shown that flat lenses formed by the wire medium are capable of transmitting subwavelength images with resolution equal to double the period of the structure, which can be made as small as necessary. The lens formed by the wire medium is a unique subwavelength imaging device for microwave frequencies, where metals are ideally conducting. However, such lenses do not function very efficiently at higher frequencies, including those in the visible range, since metals have plasma-like behavior at these frequencies. In the optical range, to achieve properties similar to the wire medium operating at microwave frequencies, it would require the use of a uniaxial optical material, whose permittivity has the form:

$$\bar{\bar{\varepsilon}} = \overline{xx} + \overline{yy} + \infty \, \overline{zz} \tag{10.1}$$

Typically, it is assumed that it would be very difficult, if not impossible, to realize very high values of permittivity. Though this is true for natural materials, the same cannot be said about metamaterials, especially uniaxial ones. The high permittivity can be achieved by employing a layered metal-dielectric structure [40] as shown in Figure 10.1

Such a metamaterial can be described in terms of a permittivity tensor that has the form:

$$\bar{\bar{\varepsilon}} = \varepsilon_\parallel (\overline{xx} + \overline{yy}) + \varepsilon_\perp \overline{zz} \tag{10.2}$$

where

$$\varepsilon_\parallel = \frac{\varepsilon_1 d_1 + \varepsilon_2 d_2}{d_1 + d_2}; \varepsilon_\perp = \left[\frac{\varepsilon_1^{-1} d_1 + \varepsilon_2^{-1} d_2}{d_1 + d_2} \right]^{-1}$$

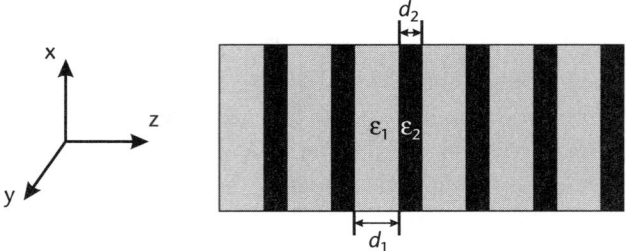

Figure 10.1 Geometry of layered metal-dielectric metamaterial.

In order to realize $\varepsilon_\parallel = 1$ and $\varepsilon_\perp = \infty$, and synthesize a material whose permittivity tensor is given by the form (10.1)—required for the implementation of the canalization regime—it is necessary to choose parameters of the layered material such that $\varepsilon_1/\varepsilon_2 = -d_1/d_2$ and $\varepsilon_1 + \varepsilon_2 = 1$. From the first equation, it is evident that the permittivity of one of the layers should be negative and, hence the structure must be formed by using a combination of dielectric and metallic layers. For example, one can choose $\varepsilon_1 = 2$, $\varepsilon_2 = -1$ and $d_1/d_2 = 2$, or $\varepsilon_1 = 15$, $\varepsilon_2 = -14$ and $d_1/d_2 = 15/14$.

The numerical modeling of subwavelength imaging using the above metal-dielectric structure has been performed in [41] using the commercial simulation package CST Microwave Studio. Here we present the results obtained from the modeling of such a structure by using the FDTD method. Previous studies [44] have shown that the material boundaries have to be carefully treated in FDTD simulations, especially for the case of negative permittivity/permeability materials, because of the evanescent waves involved in the operation of subwavelength imaging. To significantly improve the accuracy of FDTD simulations of LHM for subwavelength imaging, we have proposed an material parameter averaging method along material interfaces [44]. In the following, we first demonstrate that although the averaging of material parameters for the metal-dielectric layered structure is not as crucial as for the case of LHMs, the incorrect modeling of the layered structure leads to a shift in the resonances of the transmission coefficient. With the averaging of material parameters along the interfaces of different materials, we have also examined the subwavelength imaging capability of the metal-dielectric layered structure.

10.2.2 FDTD Modeling of the Silver-Dielectric Layered Structure

We consider here that the layered structure is composed of dielectric material with $\varepsilon_1 = 2$ and lossy isotropic silver slabs with $\varepsilon_2 = -1$. The Drude dispersion model is used for the permittivity of silver:

$$\varepsilon(\omega) = \varepsilon_0 \left(1 - \frac{\omega_p^2}{\omega^2 - j\omega\gamma}\right) \tag{10.3}$$

where ω_p and γ are plasma and collision frequencies, respectively.

The dielectric material can be directly modeled using the conventional FDTD method, while materials with negative permittivity or permeability can be only modeled using the dispersive FDTD method. Although there are various dispersive FDTD methods available for the modeling of LHMs, due to its simplicity and efficiency, we have implemented the ADE method [70], which is discussed in Chapters 6, 8, and 9.

10.2.3 Numerical Results and Discussions

10.2.3.1 Calculation of Numerical Transmission Coefficient

For simplicity, we assume that in these simulations the plasma frequency is $\omega_p = \sqrt{2}\omega$ where ω is the operating frequency, at which the silver slabs are modeled as materials with $\varepsilon_2 = -1$. A small amount of loss is introduced in the simulations by letting $\gamma = 0.005\omega$, which yields a relative permittivity ε_2 given by $\varepsilon_2 = -1 - 0.01j$, to ensure convergence of the simulations. The thickness of the silver layer is assumed to be $\lambda/120$. The permittivity of the dielectric layers is chosen to be $\varepsilon_1 = 2$, and their thickness equals $\lambda/60$. The total number of the modeled silver-dielectric layers is 20, which makes the total thickness of the slab equal one-half wavelength at the operating frequency.

To compute the transmission coefficient, we apply Bloch's PBCs along the y-direction, as shown in Figure 10.2. For periodic structures, the field satisfies the Bloch condition, such that

$$\mathbf{E}(y+L) = \mathbf{E}(y)e^{jk_yL}, \quad \mathbf{H}(y+L) = \mathbf{H}(y)e^{jk_yL} \quad (10.4)$$

where y can be an arbitrary location in the computational domain; k_y is the wave number in the y-direction; and L is the lattice period along the direction of periodicity. When updating the fields at the boundary of the computational domain using the FDTD method, the required fields outside the computational domain can be calculated by using known field values inside the domain via the use of (10.4). Since the period of an infinite structure can be chosen arbitrarily, we use only four FDTD cells in the y-direction ($L = 4\Delta y$) to save computation time. The Berenger's original PML [124] is used along the x-direction, to absorb propagating waves ($k_y < k_0$), and the modified PML [46] is employed to calculate the transmission coefficient for evanescent waves ($k_y > k_0$). We use a soft planewave sinusoidal source, which allows scattered waves to pass through. Phase delays corresponding to different wave numbers is used for the excitations as follows:

$$H_z(i,j_s) = H_z(i,j_s) + s(t)e^{-jk_yi\Delta y} \quad (10.5)$$

where j_s is the location of source along the x-direction; $s(t)$ is a time domain sinusoidal wave function; $i \in [1,I]$ is the index of cell location; and I is the total number of cells in the y-direction ($I = 4$ in our case). Either purely propagating ($k_y < k_0$) or purely evanescent ($k_y > k_0$) waves can be excited by changing the values of the wave number k_y.

The spatial resolution in the FDTD simulations is $\Delta x = \Delta y = \lambda/360$, where λ is the free-space wavelength at the operating frequency. The discretized time step is $\Delta t = \Delta x/\sqrt{2}c$ in accordance with the stability criterion [118], where c is the speed

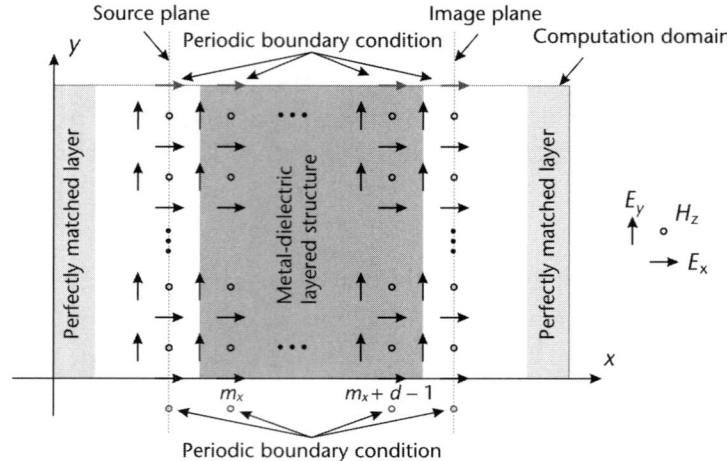

Figure 10.2 Schematic diagram of a 2-D FDTD simulation domain for the calculation of the numerical transmission coefficient.

of light in free space. The source plane is located at a distance of $\lambda/20$ at the front of the interface of the silver-dielectric slab, whose thickness, d, is $\lambda/2$. The front interface of the slab, where the magnetic fields are located, is treated as the source plane, as shown in Figure 10.2. The image plane is located at the back interface of the slab, which is also aligned with the magnetic field. The spatial transmission coefficient is computed as the ratio of the field intensity at the image plane to the source plane for different transverse wave numbers k_y, after the simulations have watched the steady state.

Figure 10.3 shows the transmission coefficient calculated from the FDTD simulations for the silver-dielectric slab using the spatial averaging of permittivity at material interfaces, and its comparison with the result with the spatial averaging is omitted. We note by comparing these results with those for the LHM slabs, previously investigated in [44], that the effect of averaging the layered slab is less significant in the present case. However, the error introduced by omitting spatial averaging causes the resonances in the transmission coefficient to shift, as shown in Figure 10.3. Consequently, to model such a silver-dielectric layered structure correctly and accurately, it is still necessary to apply spatial averaging at the material interfaces. In the following, we will first compare the results, for the simulated finite-sized silver-dielectric layered structures with and without spatial averaging, and then analyze their imaging as well as subwavelength imaging properties.

10.2.3.2 Imaging Property of the Silver-Dielectric Layered Structure

Let us first consider a layered silver-dielectric slab medium that is finite along the y-direction. The number of layers is 20, and the total thickness of each slab is $\lambda/2$. A magnetic point source is excited at a distance of $\lambda/40$ from the front interface of the slab. The simulation is run until a steady state is reached. The distributions of the magnetic field with and without spatial averaging at the material interfaces

10.2 Dispersive FDTD Modeling of Silver-Dielectric Layered Structures

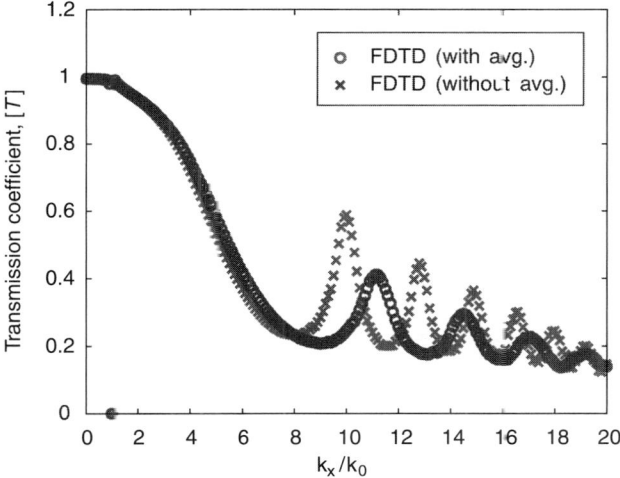

Figure 10.3 Comparison of transmission coefficient of infinite (along the y-direction) silver-dielectric layered slabs calculated from dispersive FDTD simulations with and without averaging of permittivity along the boundaries of different materials.

are plotted in Figure 10.4. From the first sight it appears that Figure 10.4(a, b) are very similar except for the amplitude of the field. They both have ripples near the boundaries along the y-direction of the slab. These ripples are caused by surface waves traveling near the boundary of the slab. These ripples correspond to the resonances in the transmission coefficient shown in Figure 10.3 and are primarily associated with the first resonance. This can be gleaned by examining the period of the ripples in Figure 10.4(a, b), which appear to agree well with each other, if we realize that the first resonance shifts to a higher location, without the spatial averaging, and, hence, the period of the ripples in Figure 10.4(b) is seen to be smaller than that in Figure 10.4(a). The comparison also demonstrates that spatial averaging does affect the FDTD simulations of layered structures, although it is less pronounced than in the case of LHM slabs [44], where a numerical (nonphysical) amplification in transmission coefficient occurs because of an incorrect modeling of the material interfaces.

We also plot the field distribution along the y-direction at both the source and image planes of the layered silver-dielectric slab, computed by using a spatial averaging in the FDTD simulations, as shown in Figure 10.5. We observe that the central part of the source distribution is not transferred accurately to the image plane. This part of source contains deep subwavelength information but cannot be delivered to the image plane because the transmission coefficient is small in the higher-order wave vector region (see Figure 10.3). However, we can argue that the image still contains some subwavelength information, since the half-power width of the distribution is less than 0.1λ. Also, we note that the image shows good correspondence with the source distribution outside the central region.

Since the layered silver-dielectric medium operates in the canalization regime, which is insensitive to the transverse dimension of the device, it is also interesting

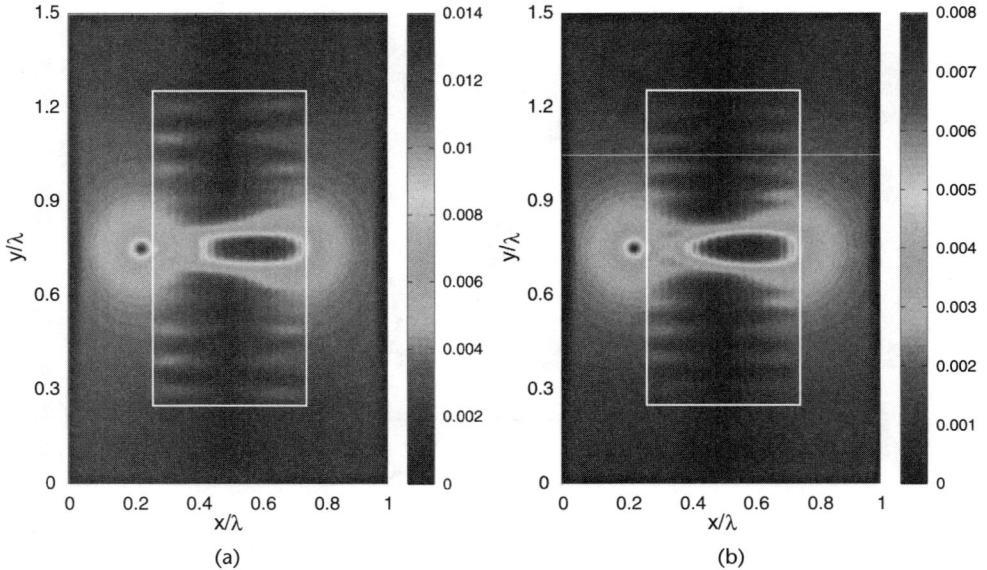

Figure 10.4 Distributions of magnetic field for the silver-dielectric layered structure for the case of (a) with and (b) without spatial averaging at material interfaces from FDTD simulations.

to investigate such an effect in the numerical simulations. Therefore we reduce the transverse dimension of the layered structure from λ to $\lambda/3$ while keeping the rest of the dimensions of the slab and the simulations parameters unchanged. The steady-state distribution of the magnetic field for such a reduce-sized structure is shown in Figure 10.6(a), and a comparison of the magnetic field distributions in

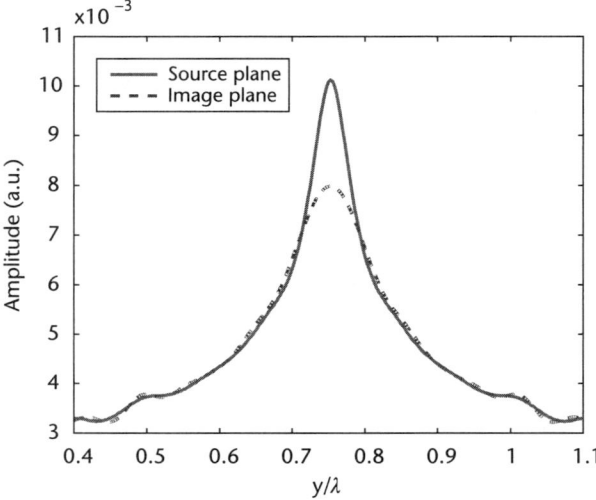

Figure 10.5 Comparison of magnetic field distributions at the source and image planes of the silver-dielectric layered structure with transverse dimension of λ. A point magnetic source is used for the excitation.

10.2 Dispersive FDTD Modeling of Silver-Dielectric Layered Structures

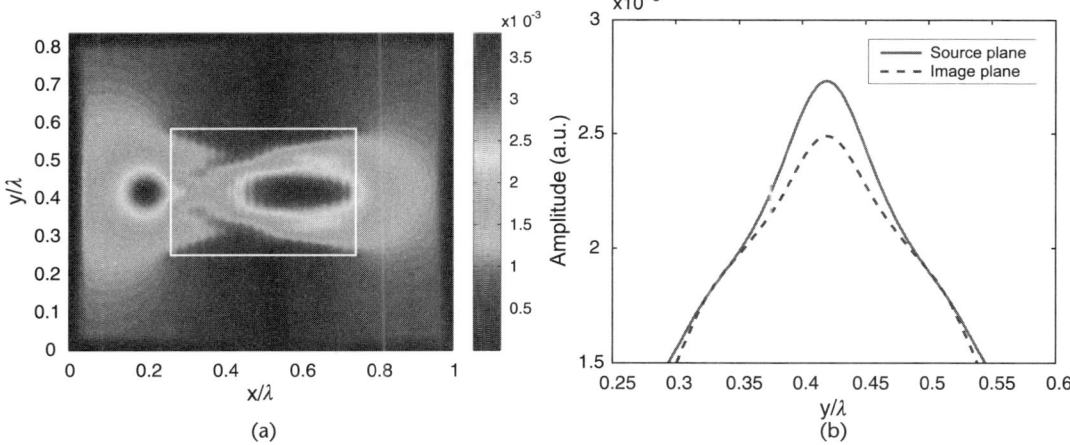

Figure 10.6 (a, b) Comparison of magnetic field distributions at the source and image planes of the layered silver-dielectric medium with a transverse dimension of $\lambda/3$. A point magnetic source is used for the excitation.

the source and image planes is plotted in Figure 10.6(b). It can be seen that the ripples are still present inside the slab, and they have a pattern similar to the one depicted in Figure 10.4(a). The field distribution outside the slab shows a smooth distribution, as evident from Figure 10.6(b). The difference between the source and image distributions is attributable to the fact that the deviation of transmission coefficient deviates from unity for such a structure, limiting its imaging capability. We note, by comparing the plots in Figures 10.6(a) and 10.4(a), that the imaging property of the layered silver-dielectric medium is indeed insensitive to its transverse dimension.

10.2.3.3 Subwavelength Imaging Property of the Layered Silver-Dielectric Medium

To demonstrate the subwavelength imaging capability of the layered silver-dielectric medium, we replace the single magnetic point source by three magnetic point sources that are separated by a distance of $\lambda/20$ and have an initial phase difference between them that equals $180°$. The sources are placed at a distance of $\lambda/40$ from the slab. The field distribution in this configuration has two distinct maxima at the front interface of the slab (the source plane). Next, we increase the transverse dimension of the slab to 0.6λ. The steady-state distribution of the magnetic field is shown in Figure 10.7(a). Note that the ripples are still present inside the slab and that their patterns are similar. It is also evident that the field propagates across the slab to deliver the image to the other side. We plot the distribution of the magnetic field at both the source and image planes in Figure 10.7(b). It can be seen that the distance between the two maxima created by the three-source configuration is $\lambda/10$ and that they are resolved in the image plane. This, in turn, demonstrates the subwavelength imaging capability of the layered silver-dielectric medium.

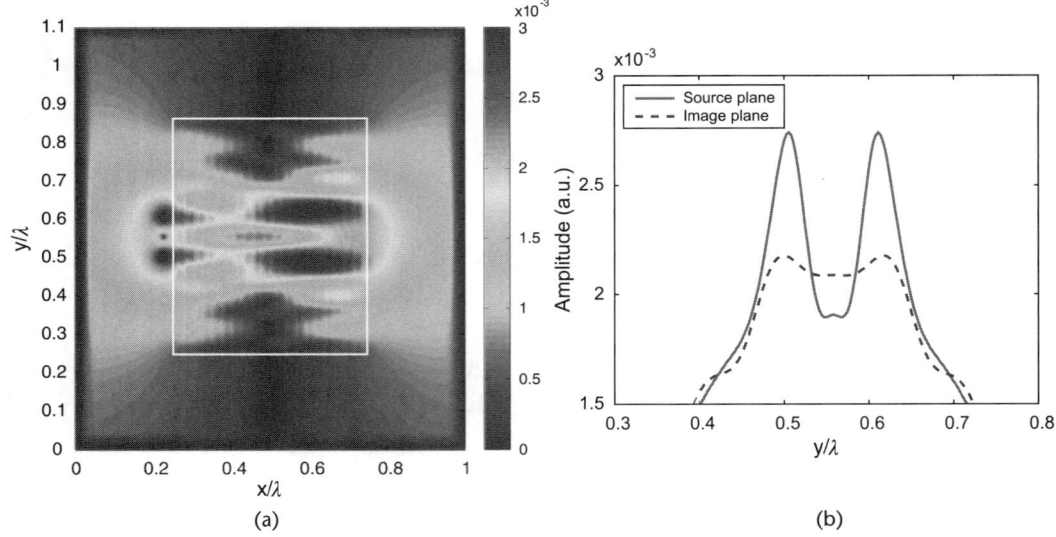

Figure 10.7 (a, b) Comparison of magnetic field distributions at the source and image planes of the layered silver-dielectric medium with a transverse dimension of 0.6λ. Three point magnetic sources are used for the excitation with the phase difference between adjacent sources equal to $180°$.

In summary, we have modeled the layered silver-dielectric medium by using the ADEs in the dispersive FDTD method. In addition to the dispersive FDTD method, we have also applied our previously proposed spatial averaging scheme at the interfaces between different materials. An incorrect modeling of the material interfaces, omitting the spatial averaging, leads to a shift of the resonances in the computed transmission coefficient. We argue that it is necessary to apply the field averaging for accurate modeling of layered structures. In addition we have demonstrated the subwavelength imaging capability of the above structure via numerical simulations and showed that this property does not depend strongly on its transverse dimension.

10.3 A Metamaterial Scanning Near-Field Optical Microscope

10.3.1 Introduction

Generally, when a metamaterial slab is used as a "perfect lens" to produce subwavelength images, it is by assuming that there exist two virtual planes, on the object and the image sides, respectively. Pendry [1] has shown that a metamaterial slab could be used as a superlens that focuses evanescent modes and resolves objects only a few nanometers wide in the optical domain. Fang et al. [8] have reported the results of an experimental, subdiffraction-limited, optical imaging system comprising of a silver superlens. They have pointed out that two stringent criteria must be met when using this device [3], namely that the surface of the film must be extremely smooth and that its thickness must be optimized. It would be interesting to find an alternate approach retrieving the image that imposes less stringent conditions: Alu et al. [111] have shown how metamaterial layers placed at the entrance and on the exit face of a subwavelength aperture in an flat and opaque screen could be used to enhance wave transmission through this aperture. In this chapter, we propose an alternative implementation of the perfect lens by using the standard methods of SNOM [12], which entails the insertion of a metamaterial slab between the object and the probe, rather than an air pocket or a dielectric slab.

10.3.2 Theory

The diffracted spectrum of an object, whose dimensions are much smaller than the wavelengths, is comprised of fields that have either an exponential (growing or decaying), or a trigonometric (propagating) behaviors. Information on the geometry of this object is transported via the exponential components of the field. For the dielectric slab case, the probe dimensions should be much smaller than the wavelength, and it should be placed as close as possible to the object. Thus, the slab width should be much smaller than the wavelength, so that the exponential field should decay as little as possible as it recedes from the object. On the other hand, this is less relevant in a hypothetical metamaterial slab, in which the exponential fields grow as they recede from the object, although this behavior is obviously affected by unavoidable losses. According to [13], a small object lit by a propagating field generates a diffracted field, which is partly exponential. Conversely, it can be argued by applying the reciprocity theorem that a small probe placed in an exponential field converts part of this field into a propagating field, which could be detected in the far region.

Let us suppose that the source is a plane wave of wavelength λ_0, approximately obtained through an array of 10 linear sources situated close to an object consisting of two infinitely long slits in the x-direction, each with a very small width $w \ll \lambda_0$. Their separation distance s is very small (i.e., $s \ll \lambda_0$). The width of the metamaterial slab is denoted by d. The probe is an infinitely long slit in the x-direction, whose width $W \ll \lambda_0$, and which may be displaced up and down along the slab in the y-direction to achieve a point-by-point scanning of the object (see Figure 10.8). This type of probe is advantageous to use when we wish to scan only part of the object. Moreover, we could select the probe that is best suited to our needs from a number of different available probes. For instance, we found that if the level of intensity detected with a given probe is too weak, we could use a vibrating metallic conical probe [13] instead to enhance the intensity.

10.3.3 Simulation

Let us assume that the center of the probe is located on the axis of symmetry of the imaging device and let us choose it in the middle of the metamaterial slab. We will use a dispersive FDTD type of simulation, but other numerical simulations [14] can also be used. We have assumed, for this simulation, that $\lambda_0 = 0.5$ μm, $s = 0.0775$ μm, $w = 0.115$ μm, $d = 0.08$ um, $\omega_{pe} = \omega_{pm} = \sqrt{2}\omega_0$, and $v_e = v_m = 0$. Also, the metallic plates are 0.01-m-thick and are assumed to be perfectly conducting. The FDTD cell size is $\lambda_0/200$, and the computational region

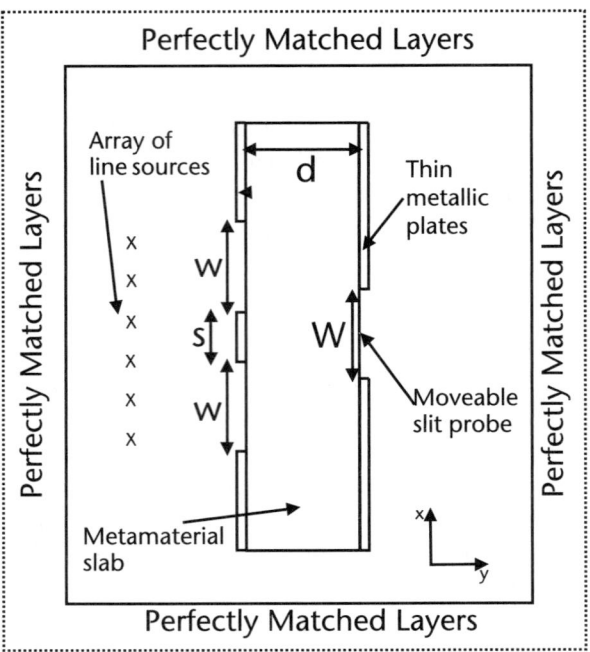

Figure 10.8 The geometry of the metamaterial SNOM.

is 2000 × 600 cells. An 8-layer uniaxial perfectly matched layer (UPML) is used as an absorbing boundary, and only the TM polarization is considered in the simulation. The computed field intensity contours in the y-z plane outside the device are shown in Figure 10.9. The field intensities are plotted in Figure 10.10 as functions of y at fixed distances z where they attain highest values, namely, at $z = 2.75\lambda_0$ and $z = 3.8\lambda_0$. The resolution (1.46) defined as the ratio of maximum to minimum intensities is good at $z = 3.8\lambda_0$. The simulation has been repeated without the probe, and the results are plotted in Figure 10.11. It can be seen from this plot that the intensity level is lower in this case than it was in the previous case by about a factor of two. The resolution is also lower (1.26), and it occurs only at $z = 4\lambda_0$. There is no resolution at $z = 2.95\lambda_0$, showing that the focal region is narrower. Using a thin slab ($d = 0.04$m) instead of the previous thick one ($d = 0.08$m), it can be shown that the intensity is higher (about eightfold) in the absence of the probe, that the resolution is lower (1.3 as opposed to 1.9), and that the focal region is narrower. If we remove the metamaterial slab, leaving air between the object and the probe, we observe that the image is around 200 times weaker, showing the decisive role of the metamaterial slab in realizing high-intensity images provided, of course, that we can obtain a slab of real material that possesses the characteristics we have assumed for the metamaterial slab.

In summary, a metamaterial scanning near-field optical microscope has been proposed. The object, consisting of a pair of long subwavelength slits, is separated

10.3 A Metamaterial Scanning Near-Field Optical Microscope

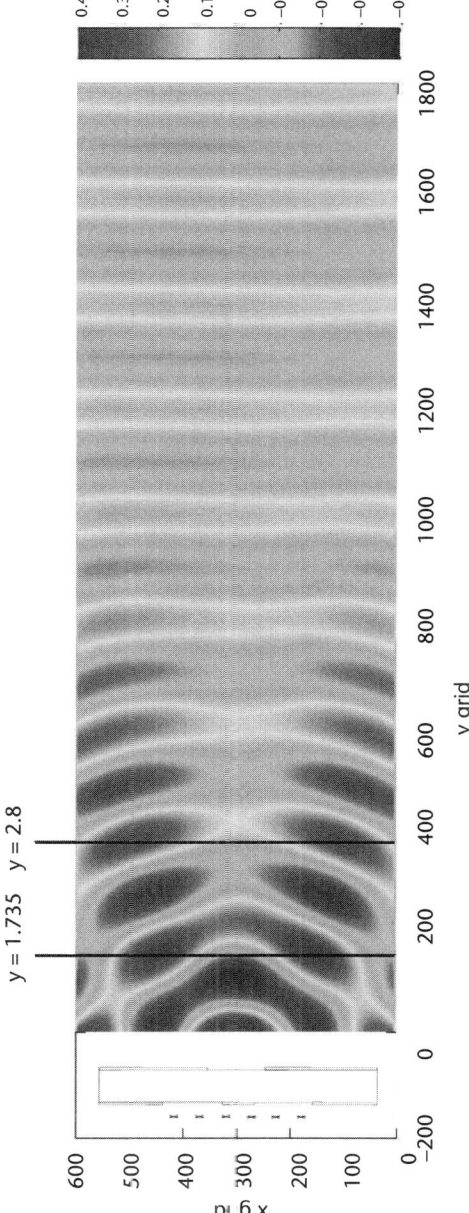

Figure 10.9 The electric field intensity (V/m) contours in the y-z plane (100 cells corresponding to 1 μm). y=0 is located 0.75 μm under the middle of the probe while z=0 is located 0.25 μm to the right of the middle of the metamaterial slab.

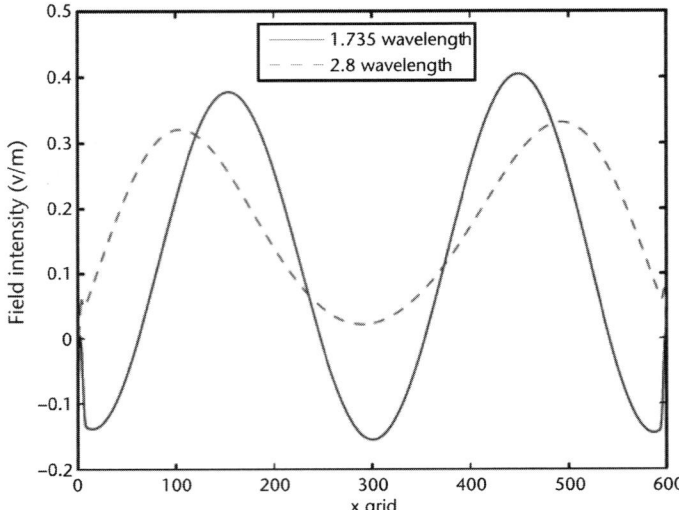

Figure 10.10 The resolution for an object consisting of two slits, probe present (400 cells corresponding to 1 μm).

from the probe, which is a single long subwavelength slit, by a metamaterial slab, and the probe slides along this slab. It has been found that by comparing the results for the cases with and without probe cases, the resolution is better when the probe is present. Also, it is advantageous to use a probe when we wish to scan only a

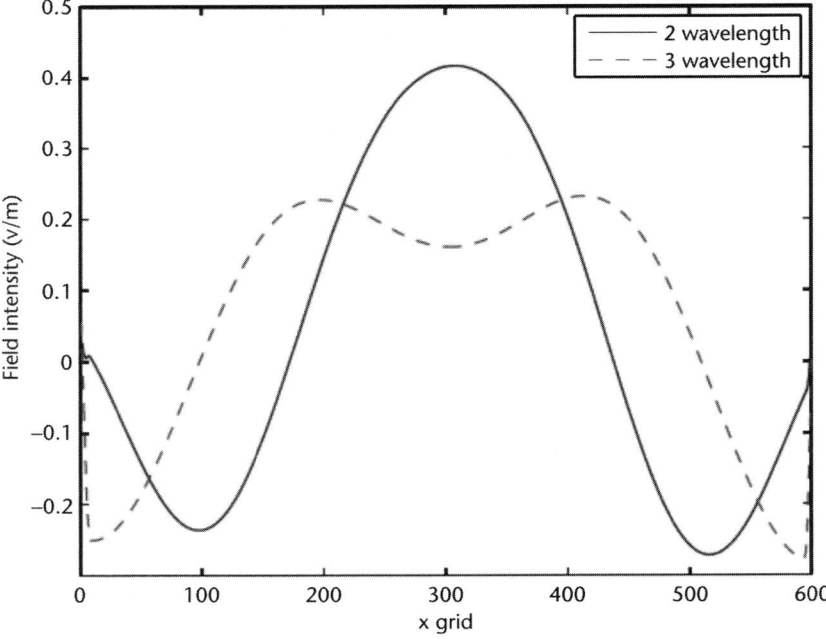

Figure 10.11 The resolution for an object consisting of two slits, probe missing (400 cells corresponding to 1 μm).

part of the object. Moreover, since a number of different probes is available, we could select the probe best suited to our needs. The various parameters and the components involved in the microscope can be optimized to obtain the strongest intensity with the highest resolution.

10.4 FDTD Study of Guided Modes in Nanoplasmonic Waveguides

It is well known that photonic crystals (PCs) offer unique opportunities to control the flow of light [47] and that we can design periodic dielectric structures that have a bandgap in a particular frequency range. Periodic dielectric rods, from which we have removed one or more rows of elements, can be used as waveguiding devices when operating at bandgap frequencies. Researchers have invested a considerable effort to obtain a bandgap that is complete and wide. It has been shown that a triangular lattice of air holes in a dielectric background has a complete bandgap for the TE mode, while a square lattice of dielectric rods in air has a bandgap for the TM mode [48]. The devices operating in the bandgap frequencies are not the only options available to us for guiding the flow of light. Another waveguiding mechanism is the total internal reflection (TIR) in 1-D periodic dielectric rods [49]. It is shown in [49] that a single row of either dielectric rods, or air holes, supports waveguiding modes and therefore can also be used as waveguide. In [50], the design of such waveguides consisting of several rows of dielectric rods with various spacings is proposed.

Recently, a new method for guiding electromagnetic waves in structures whose dimensions are below the diffraction limit has been proposed. The structures are termed "plasmonic waveguides," and their operating principle is based on near-field interactions between closely spaced noble metal nanoparticles (spacing $\ll \lambda$), which can be efficiently excited at their surface plasmon frequency. The guiding principle relies on coupled plasmon modes set up by the near-field dipole interactions that lead to coherent propagation of energy along the array. Analogous structures, which serve as waveguides in the microwave regime, include periodic metallic cylinders that support propagating waves [51], arrays of flat dipoles that support guided waves [52], and Yagi-Uda antennas [53, 54]. Although these structures can be scaled to optical frequencies with appropriate material properties, their dimensions are limited by the so-called diffraction limit $\lambda/(2n)$. On the other hand, plasmonic waveguides employ the localization of electromagnetic fields near metal surfaces to confine and guide light in regions much smaller than the free-space wavelength, and then can effectively overcome the diffraction limit. Previous analysis of plasmonic structures include the plasmon propagation along metal stripes, wires, or grooves in metal [55–60], and the coupling between plasmons on metal particles in order to guide energy [61, 62]. Such subwavelength structures can also find their applications as efficient absorbers and as electrically small receiving antennas at microwave frequencies. Recently, composite materials containing randomly distributed electrically conductive material and nonelectrically conductive material have been designed [63]. They are noted to exhibit a plasma-type response at frequencies well below the plasma frequencies of the bulk material.

As mentioned earlier, the FDTD method [118] is one of the most widely used numerical techniques because of its flexibility in handling material dispersive media, as well as arbitrary shaped inclusions. In [65], the optical pulse propagation below the diffraction limit is studied by using the FDTD method. The method has also been used to investigate waveguides formed by several rows of silver nanorods arranged in a hexagonal lattice [66]. Even though the FDTD method has been employed for plasmonic structures, the accuracy of such modeling has yet to be proven. When modeling curved structures, one must either use extremely fine mesh to mitigate the staircasing errors in the conventional FDTD or modify the algorithm to improve the numerical accuracy, by using special treatments at the interfaces between dissimilar materials [67], or by working with the improved conformal algorithms [68] for curved surfaces.

In addition to the modifications at material interfaces, the material frequency dispersion must also be accounted for in the FDTD modeling [69–71]. However, modeling dispersive materials with curved surfaces still remains a challenging topic, because the algorithm is complex and suffers from numerical instabilities. An alternative way to solve this problem is to utilize the concept of effective permittivities (EPs) [72, 74, 123] in the Cartesian coordinate system, and modify the dispersive FDTD scheme accordingly, without compromising the stability of the algorithm. In this section, we first propose a novel conformal dispersive FDTD algorithm combining the EPs with an ADE method [118], then apply the developed method to the modeling of plasmonic waveguides formed by an array of circular or elliptical silver cylinders at optical frequencies. We also verify the numerical FDTD simulation results by comparing them with those obtained by using a frequency domain embedding method [75].

10.4.1 Conformal Dispersive FDTD Method Using Effective Permittivities (EPs)

A staircasing approximation is used to model curved electromagnetic structures in the conventional FDTD algorithm on a Cartesian grid. Figure 10.12(a) shows an example layout of an infinite-long cylinder in free space represented in a 2-D Cartesian FDTD domain. The staircasing approximation of the shape introduces spurious numerical resonant modes that do not exist in the actual structure. On the other hand, using the concept of the filling factor, which is defined as the ratio of the area of material ε_2 to the area of the partially filled FDTD cell, the curvature can be properly accounted in the FDTD as shown in Figure 10.12(b), where different levels of darkness indicate different filling factors of material ε_2. The accuracy of modeling can be significantly improved above than in the staircased approximation, as will be shown in a later section.

According to [74], in general the EP is given by

$$\varepsilon_{\text{eff}} = \varepsilon_{\|}(1 - n^2) + \varepsilon_{\perp} n^2 \tag{10.6}$$

where n is the projection of the unit normal vector **n** along the field component as shown in Figure 10.13 and $\varepsilon_{\|}$ and $\varepsilon_{\perp}$ are parallel and perpendicular permittivities to the material interface, respectively. They are defined as:

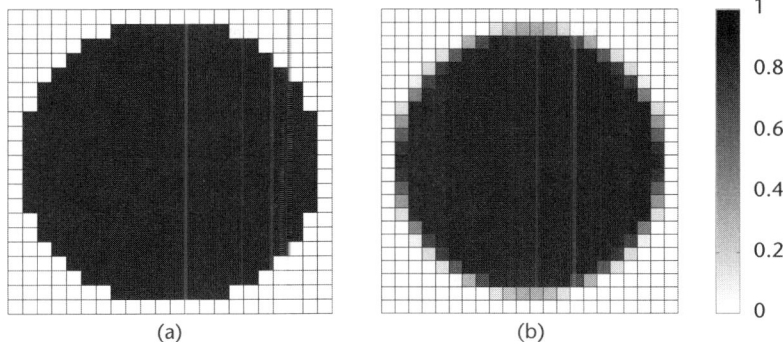

Figure 10.12 Comparison of the filling ratio for E_y component in FDTD modeling of a circular cylinder using (a) staircase approximations and (b) a conformal scheme. The radius of circular cylinder is 10 cells.

$$\varepsilon_\| = f\varepsilon_2 + (1-f)\varepsilon_1 \tag{10.7}$$

$$\varepsilon_\perp = [f/\varepsilon_2 + (1-f)/\varepsilon_1]^{-1} \tag{10.8}$$

where f is the filling factor of the material ε_2 in a given FDTD cell.

In this book, we consider the inclusions as silver cylinders that can be modeled at optical frequencies by using the Drude dispersion model

$$\varepsilon_2(\omega) = \varepsilon_0 \left(1 - \frac{\omega_p^2}{\omega^2 - j\omega\gamma}\right) \tag{10.9}$$

where ω_p and γ are the plasma and the collision frequencies, respectively. At frequencies below the plasma frequency, the real part of the permittivity is negative. In this book, we assume that the silver cylinders are embedded in free space ($\varepsilon_1 = \varepsilon_0$).

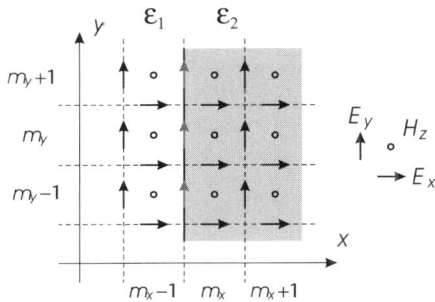

Figure 10.13 Layout of a quarter circular inclusion in orthogonal FDTD grid for E_y component. The radius of circular cylinder is three cells.

In order to account for the frequency dispersion of the material, we introduce the electric flux density **D** into the conventional FDTD updating equations. **D** is updated directly at each time step from **H** and **E**, which can in turn, be calculated from **D** by using the following steps. Substitute (10.7) and (10.8) into (10.6) and using the expressions for ε_1 and ε_2 (10.9), we obtain the constitutive relation in the frequency domain, which reads

$$\{\omega^4 - 2\gamma j\omega^3 - [\gamma^2 + (1-f)\omega_p^2]\omega^2 + \gamma(1-f)\omega_p^2 j\omega\}\mathbf{D}$$
$$= [\omega^4 - 2\gamma j\omega^3 - (\gamma^2 + \omega_p^2)\omega^2 + \gamma\omega_p^2 j\omega + f(1-f)(1-n^2)\omega_p^4]\varepsilon_0 \mathbf{E} \quad (10.10)$$

Using the inverse Fourier transformation (i.e., replacing $j\omega$ with $\partial/\partial t$), we obtain the following constitutive relation in the time domain

$$\left\{\frac{\partial^4}{\partial t^4} + 2\gamma\frac{\partial^3}{\partial t^3} + [\gamma^2 + (1-f)\omega_p^2]\frac{\partial^2}{\partial t^2} + \gamma(1-f)\omega_p^2\frac{\partial}{\partial t}\right\}\mathbf{D}$$
$$= \left[\frac{\partial^4}{\partial t^4} + 2\gamma\frac{\partial^3}{\partial t^3} + (\gamma^2 + \omega_p^2)\frac{\partial^2}{\partial t^2} + \gamma\omega_p^2\frac{\partial}{\partial t} + f(1-f)(1-n^2)\omega_p^4\right]\varepsilon_0\mathbf{E} \quad (10.11)$$

The FDTD simulation domain is represented by an equally spaced 3-D grid, with periods Δx, Δy and Δz along the x-, y-, and z-directions, respectively. For discretization of (10.11), we use the central finite difference operators in time (δ_t), and the central average operator with respect to time (μ_t):

$$\frac{\partial^4}{\partial t^4} \rightarrow \frac{\delta_t^4}{(\Delta t)^4}, \quad \frac{\partial^3}{\partial t^3} \rightarrow \frac{\delta_t^3}{(\Delta t)^3}\mu_t, \quad \frac{\partial^2}{\partial t^2} \rightarrow \frac{\delta_t^2}{(\Delta t)^2}\mu_t^2, \quad \frac{\partial}{\partial t} \rightarrow \frac{\delta_t}{\Delta t}\mu_t^3$$
$$1 \rightarrow \mu_t^4 \quad (10.12)$$

where the time step is Δt and the operators δ_t and μ_t are defined as in [122]:

$$\delta_t \mathbf{F}|_{m_x,m_y,m_z}^n \equiv \mathbf{F}|_{m_x,m_y,m_z}^{n+\frac{1}{2}} - \mathbf{F}|_{m_x,m_y,m_z}^{n-\frac{1}{2}} \quad (10.13)$$

$$\mu_t \mathbf{F}|_{m_x,m_y,m_z}^n \equiv \frac{\mathbf{F}|_{m_x,m_y,m_z}^{n+\frac{1}{2}} + \mathbf{F}|_{m_x,m_y,m_z}^{n-\frac{1}{2}}}{2} \quad (10.14)$$

Here **F** represents the field components and m_x, m_y, m_z are indices corresponding to a certain discretization point in the FDTD domain. The discretized version of (10.11) reads

$$\left\{\frac{\delta_t^4}{(\Delta t)^4} + 2\gamma\frac{\delta_t^3}{(\Delta t)^3}\mu_t + [\gamma^2 + (1-f)\omega_p^2]\frac{\delta_t^2}{(\Delta t)^2}\mu_t^2 + \gamma(1-f)\omega_p^2\frac{\delta_t}{\Delta t}\mu_t^3\right\}\mathbf{D}$$
$$= \left[\frac{\delta_t^4}{(\Delta t)^4} + 2\gamma\frac{\delta_t^3}{(\Delta t)^3}\mu_t + (\gamma^2 + \omega_p^2)\frac{\delta_t^2}{(\Delta t)^2}\mu_t^2 + \gamma\omega_p^2\frac{\delta_t}{\Delta t}\mu_t^3\right.$$
$$\left. + f(1-f)(1-n^2)\omega_p^4\mu_t^4\right]\varepsilon_0\mathbf{E} \quad (10.15)$$

10.4 FDTD Study of Guided Modes in Nanoplasmonic Waveguides

Note that we have retained all the fourth-order terms in the above equations to be to guarantee the numerical stability. Equation (10.15) can be written as

$$\frac{D^{n+1} - 4D^n + 6D^{n-1} - 4D^{n-2} + D^{n-3}}{(\Delta t)^4} + \gamma \frac{D^{n+1} - 2D^n + 2D^{n-2} - D^{n-3}}{(\Delta t)^3}$$

$$+ \left[\gamma^2 + (1-f)\omega_p^2\right] \frac{D^{n+1} - 2D^{n-1} + D^{n-3}}{4(\Delta t)^2} + \gamma(1-f)\omega_p^2 \frac{D^{n+1} + 2D^n - 2D^{n-2} - D^{n-3}}{8\Delta t}$$

$$= \varepsilon_0 \frac{E^{n+1} - 4E^n + 6E^{n-1} - 4E^{n-2} + E^{n-3}}{(\Delta t)^4} + \varepsilon_0 \gamma \frac{E^{n+1} - 2E^n + 2E^{n-2} - E^{n-3}}{(\Delta t)^3}$$

$$+ \varepsilon_0 \left(\gamma^2 + \omega_p^2\right) \frac{E^{n+1} - 2E^{n-1} + E^{n-3}}{4(\Delta t)^2} + \varepsilon_0 \gamma \omega_p^2 \frac{E^{n+1} + 2E^n - 2E^{n-2} - E^{n-3}}{8\Delta t}$$

$$+ \frac{\varepsilon_0 f(1-f)(1-n^2)\omega_p^4}{16}\left(E^{n+1} + 4E^n + 6E^{n-1} 4E^{n-2} + E^{n-3}\right) \quad (10.16)$$

The indices m_x, m_y and m_z are omitted from (10.16) since E and D are colocated. We solve for E^{n+1}, and obtain the following updating equation for E in the FDTD:

$$E^{n+1} = [b_0 D^{n+1} + b_1 D^n + b_2 D^{n-1} + b_3 D^{n-2} + b_4 D^{n-3} \\ - (a_1 E^n + a_2 E^{n-1} + a_3 E^{n-2} + a_4 E^{n-3})]/a_0 \quad (10.17)$$

with the coefficients given by

$$a_0 = \varepsilon_0 \left[\frac{1}{(\Delta t)^4} + \frac{\gamma}{(\Delta t)^3} + \frac{\gamma^2 + \omega_p^2}{4(\Delta t)^2} + \frac{\gamma \omega_p^2}{8\Delta t} + \frac{f(1-f)(1-n^2)\omega_p^4}{16}\right]$$

$$a_1 = \varepsilon_0 \left[-\frac{4}{(\Delta t)^4} - \frac{2\gamma}{(\Delta t)^3} + \frac{\gamma \omega_p^2}{4\Delta t} + \frac{f(1-f)(1-n^2)\omega_p^4}{4}\right]$$

$$a_2 = \varepsilon_0 \left[\frac{6}{(\Delta t)^4} - \frac{\gamma^2 + \omega_p^2}{2(\Delta t)^2} - \frac{3f(1-f)(1-n^2)\omega_p^4}{8}\right]$$

$$a_3 = \varepsilon_0 \left[-\frac{4}{(\Delta t)^4} + \frac{2\gamma}{(\Delta t)^3} - \frac{\gamma \omega_p^2}{4\Delta t} + \frac{f(1-f)(1-n^2)\omega_p^4}{4}\right]$$

$$a_4 = \varepsilon_0 \left[\frac{1}{(\Delta t)^4} - \frac{\gamma}{(\Delta t)^3} + \frac{\gamma^2 + \omega_p^2}{4(\Delta t)^2} - \frac{\gamma \omega_p^2}{8\Delta t} + \frac{f(1-f)(1-n^2)\omega_p^4}{16}\right]$$

$$b_0 = \frac{1}{(\Delta t)^4} + \frac{\gamma}{(\Delta t)^3} + \frac{\gamma^2 + (1-f)\omega_p^2}{4(\Delta t)^2} + \frac{\gamma(1-f)\omega_p^2}{8\Delta t}$$

$$b_1 = -\frac{4}{(\Delta t)^4} - \frac{2\gamma}{(\Delta t)^3} + \frac{\gamma(1-f)\omega_p^2}{4\Delta t}$$

$$b_2 = \frac{6}{(\Delta t)^4} - \frac{\gamma^2 + (1-f)\omega_p^2}{2(\Delta t)^2}$$

$$b_3 = -\frac{4}{(\Delta t)^4} + \frac{2\gamma}{(\Delta t)^3} - \frac{\gamma(1-f)\omega_p^2}{4\Delta t}$$

$$b_4 = \frac{1}{(\Delta t)^4} - \frac{\gamma}{(\Delta t)^3} + \frac{\gamma^2 + (1-f)\omega_p^2}{4(\Delta t)^2} - \frac{\gamma(1-f)\omega_p^2}{8\Delta t} \quad (10.18)$$

The computations of **H** and **D** are performed by using Yee's standard updating equations in free space. Note that if the plasma frequency is equal to zero ($\omega_p = 0$), then (10.17) reduces to the update equation in the free space (i.e., $\mathbf{E} = \mathbf{D}/\varepsilon_0$).

10.5 FDTD Calculation of Dispersion Diagrams

Bloch's PBCs [77–82] can be used with the FDTD method to model periodic structures and compute their dispersion diagrams [83, 84]. For all periodic structures, the fields must satisfy the Bloch condition, such that

$$\mathbf{E}(d+\mathbf{a}) = \mathbf{E}(d)e^{j\mathbf{k}\mathbf{a}}, \quad \mathbf{H}(d+\mathbf{a}) = \mathbf{H}(d)e^{j\mathbf{k}\mathbf{a}} \quad (10.19)$$

where d is an arbitrary point the computational domain, **k** is the wave vector, and **a** is the lattice vector along the direction of periodicity. When updating the fields at the boundary of the computation domain using FDTD, the required fields outside the computation domain can be calculated by using known field values inside the domain via the use of (10.19).

First, we apply the conformal dispersive FDTD method we have developed to calculate the dispersion diagram for 1-D plasmonic waveguides formed by an array of periodic, infinite-long (along the z-direction), circular silver cylinders. Let us consider the TE modes in the 2-D simulation domain (x-y) for which the only nonzero fields are E_x, E_y, and H_z. The domain, as shown in Figure 10.14, is truncated by using Bloch's PBCs in the x-direction and Berenger's PMLs [124] in the y-direction. Berenger's PML performs well when absorbing propagating waves [124]. However, the same is not true for evanescent waves, for which the field grows inside the PML. Since the waves radiated by a point or line sources consist of both the propagating and evanescent components, we add some extra space—typically a quarter of a wavelength at the frequency of interest—between the PMLs and the circular inclusion to allow for the evanescent waves to decay before reaching the PMLs.

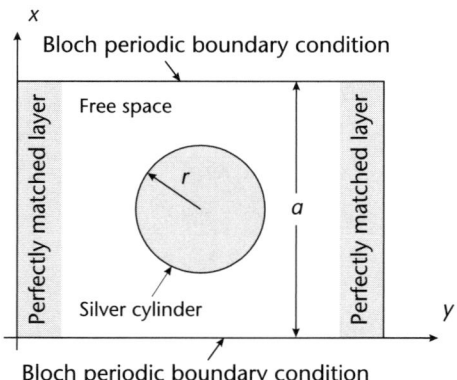

Figure 10.14 The layout of the 2-D FDTD computation domain for calculating the dispersion diagram for 1-D periodic structures. The inclusion has a circular cross-section with radius r, and the period of the 1-D infinite structure is a.

10.5 FDTD Calculation of Dispersion Diagrams

The radius of the silver cylinders is $r = 2.5 \times 10^{-8}$ m, and the period is $a = 7.5 \times 10^{-8}$ m. The plasma and collision frequencies are $\omega_p = 9.39 \times 10^{15}$ rad/s and $\gamma = 3.14 \times 10^{13}$ Hz, respectively, chosen to closely match the bulk dielectric function of silver [86] (see Figure 10.14). The FDTD cell size is $\Delta x = \Delta y = 2.5 \times 10^{-9}$ m, and the time step, chosen in accordance with the Courant stability criterion [118], is $\Delta t = \Delta x/(\sqrt{2}c)$ s, where c is the speed of light in the free space. The stability condition for a higher-order FDTD method is typically more stringent than the conventional one. However, we have found no evidence of any instability once we have applied the average operator μ_t even after 40,000 time steps, in all of the simulations.

A wideband magnetic line source is placed at an arbitrary location in the free-space region of the 2-D simulation domain in order to excite all resonant modes of the structure within the frequency range of interest (normalized frequency $\bar{f} = \omega a/(2\pi c) \in [0 \sim 0.5]$):

$$g(t) = e^{-\left(\frac{t-t_0}{\tau}\right)^2} \cdot e^{j\omega t} \qquad (10.20)$$

where t_0 is the initial time delay, τ defines the pulse width and ω is the center frequency of the pulse ($\bar{f} = 0.25$). The magnetic fields at 100 random locations in the free-space region are recorded during simulations, transformed into the frequency domain, and combined to extract individual resonant mode corresponding to each local maximum. For each wave vector, 40,000 time steps are used in our simulations to obtain enough accurate frequency domain results.

In order to demonstrate the advantage of EPs and validate the proposed conformal dispersive FDTD method, we have also performed simulations using staircase approximations for the circular cylinder, as shown in Figure 10.12(a). Figure 10.15 compares the first resonant frequency (transverse mode) corresponding to the wave vector $k_x = \pi/a$ of the plasmonic waveguide, calculated by using the FDTD method with staircase approximations, with the FDTD method with EPs, and the frequency-domain embedding method [87]. With the same FDTD spatial resolutions, the model using EP shows excellent agreement with the results from the frequency domain embedding method. However, in contrast to this, the staircase approximation not only leads to a shift in the main resonant frequency, but also introduces a spurious numerical resonant mode that does not exist in actual structures. The same effect has also been found for nondispersive dielectric cylinders [88]. It is also shown in Figure 10.15 that although one may correct the main resonant frequency using finer meshes, the spurious resonant mode still remains.

The problem of frequency shift and spurious modes becomes even more severe when calculating the higher guided modes near the "flat-band" region (i.e., the region where waves travel at a very low phase velocity). Even with a refined spatial resolution, the staircase approximation fails to provide correct results (not shown). On the other hand, using the proposed conformal dispersive FDTD scheme, all resonant modes are correctly captured in the FDTD simulations, as demonstrated by the comparison with the embedding method, shown in Figure 10.16.

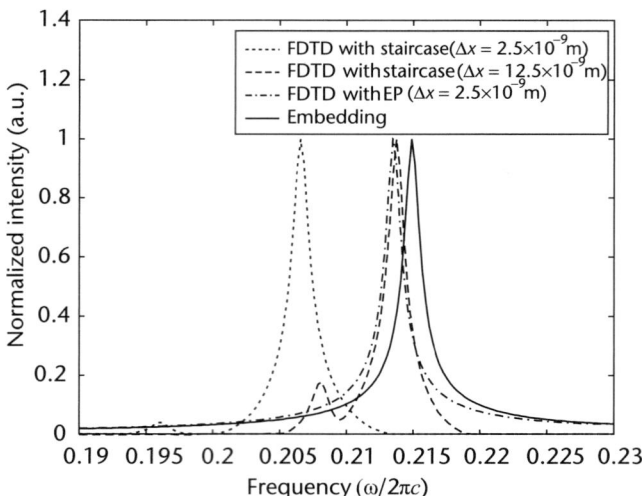

Figure 10.15 Comparison of the first resonant frequency (transverse mode) at wave vector $k_x = \pi/a$ calculated using the FDTD method with staircase approximations, the FDTD method with EPs, and the frequency domain embedding method.

According to previous analysis using the frequency-embedding method, the fundamental mode in the modelled plasmonic waveguide is the transverse mode, and the second guided mode is longitudinal [87], which is also shown by the distribution of the electric field intensities in Figure 10.17 from our FDTD simulations. The higher guided modes are referred to as the "plasmon modes." To demonstrate the field symmetries of the TE mode considered in our simulations, we have plotted

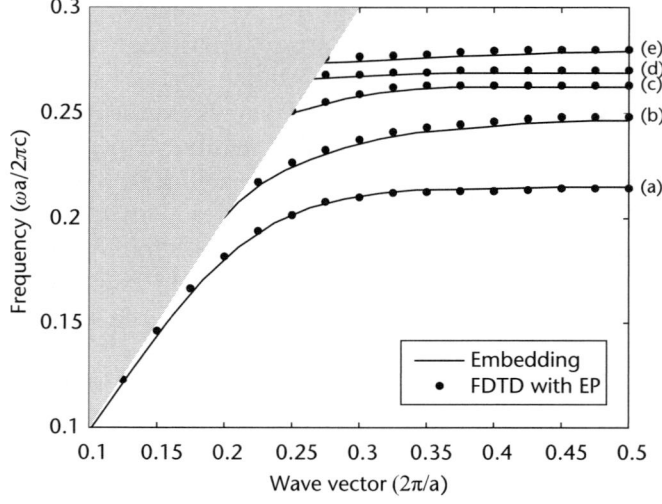

Figure 10.16 Comparison of dispersion diagrams for an array of infinite-long (along the z-direction) circular silver cylinders calculated using the FDTD method with EPs and the frequency domain embedding method.

10.5 FDTD Calculation of Dispersion Diagrams

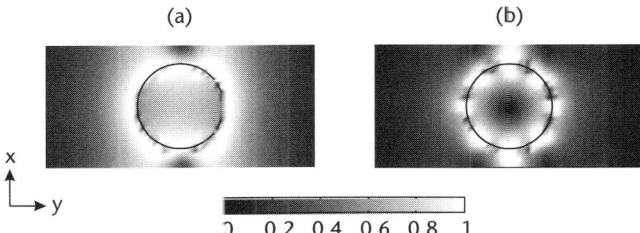

Figure 10.17 Normalized total electric field intensities corresponding to (a) transverse and (b) longitudinal modes [87] at wave number $k_x = \pi/a$ as marked in Figure 10.16. The structure is infinite along the x-direction.

the distributions of the magnetic field corresponding to different resonant modes at wave number $k_x = \pi/a$ as marked in Figure 10.16 and shown in Figure 10.18.

Sinusoidal sources are used for the excitation of various single modes, and the sources are placed at different locations corresponding to different symmetries of the field patterns. All field patterns are plotted after the steady state has been reached in simulations. The modes depicted in Figure 10.18(a, c, e) are even modes [relative to the direction of periodicity of the waveguide (i.e., x-axis)], while those displayed in Figure 10.18(b, e) are considered as odd modes.

The above comparison of the simulation results calculated by using the conformal dispersive FDTD method and the embedding method clearly demonstrates the effectiveness of applying the EPs in FDTD modeling. Furthermore, in contrast to the embedding method, the main advantage of the FDTD method is that arbitrary shaped geometries can be easily modeled. We have applied the conformal dispersive FDTD method to study the effect of different inclusions on the

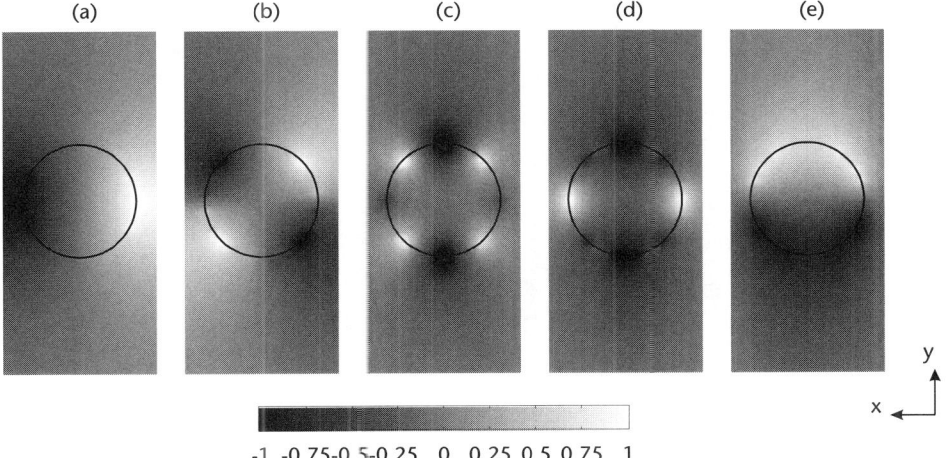

Figure 10.18 Normalized distributions of magnetic fields corresponding to different resonant modes at wave number $k_x = \pi/a$ as marked in Figure 10.16 (a, c, d): even modes, and (b, e): odd modes. The structure is infinite along the x-direction. (Note that the coordinate has been rotated 90° anticlockwise from Figure 10.14 for better presentation of the figure.)

dispersion diagrams of 1-D plasmonic waveguides. The geometries considered are two rows of periodic infinite-long (along the z-direction) circular silver cylinders arranged in square lattice and a single row of periodic infinite-long (along the z-direction) elliptically shaped silver cylinders. The elliptical cylinder has a ratio of semimajor-to-semiminor axis 2:1, where the semiminor axis is equal to the radius of the circular element (25.0 nm). For the two rows of circular nanorods, the spacing between the two rows (center-to-center distance) is 75 nm. The dispersion diagrams for these structures are plotted in Figures 10.19 and 10.21.

Comparing the dispersion diagrams for a single circular element in Figure 10.16 with that for the two circular elements shown in Figure 10.19, we can see that the dispersion diagram has been modified due to the change of inclusions. The strong coupling between the two elements introduces additional guided modes to appear in the dispersion diagram. Such a phenomenon has also been studied previously for dielectric (nondispersive) nanorods [50]. The distributions of the magnetic field for selected guided modes, as marked in Figure 10.19, are plotted in Figure 10.20. The modes in Figure 10.20(a, c, d) are even modes while Figure 10.20(b, e) are odd modes.

The dispersion diagram for a periodic structure comprised of single elliptical elements as inclusions is shown in Figure 10.21. We observe that an increased number of guided modes appears in this case than appeared earlier for circularly shaped inclusions. This is attributable to the fact that the geometrical shape of the inclusions has been changed from circular to elliptical. For instance, the frequency corresponding to the lowest mode is now lower because of an increase in the volume of the inclusion. The distributions of magnetic fields are plotted in Figure 10.22 for selected guided modes. The modes plotted in Figure 10.22(a, d) are even, while the plots for the odd modes appear in Figure 10.22(b, c, e).

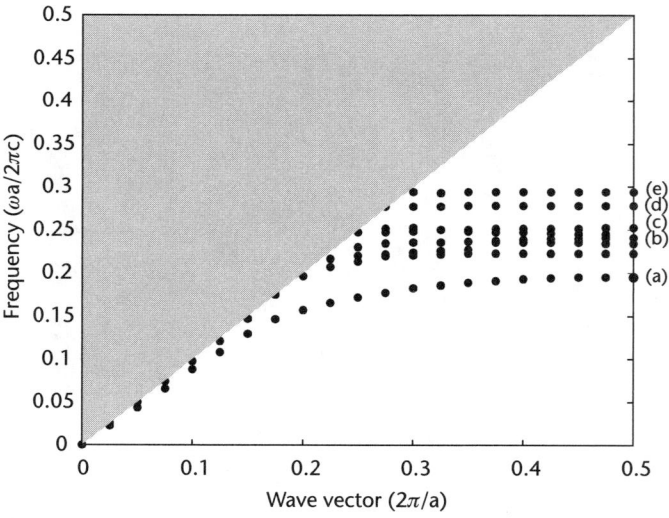

Figure 10.19 (a–e) Dispersion diagram for two rows of periodic infinite-long (along the z-direction) circular silver cylinders arranged in square lattice calculated from conformal dispersive FDTD simulations.

10.5 FDTD Calculation of Dispersion Diagrams

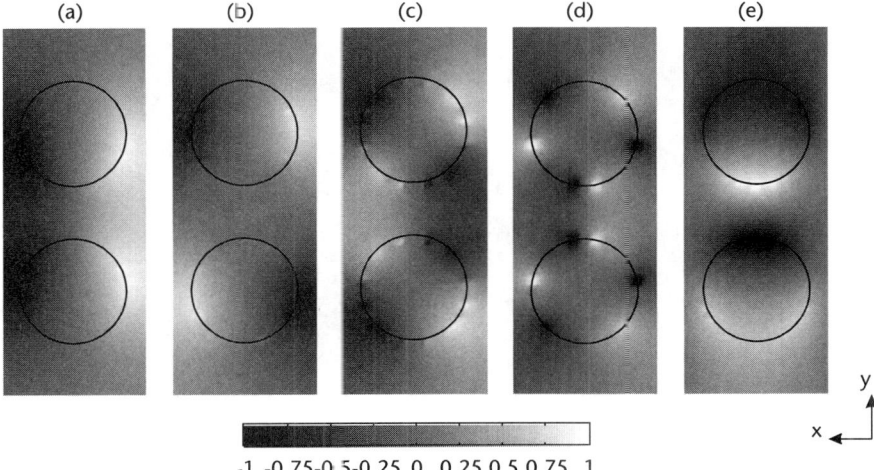

Figure 10.20 Normalized distributions of magnetic fields corresponding to different guided modes as marked in Figure 10.19: (a, c, d) are even modes, and (b, e) are odd modes. The structure is infinite along the x-direction. (Note that the coordinate has been rotated 90° anticlockwise from Figure 10.14 for better presentation of the figure.)

10.5.1 Wave Propagation in Plasmonic Waveguides Formed by Finite Number of Elements

In order to study wave propagations in plasmonic waveguides formed by a finite number of silver nanorods, we have replaced the PBCs in the x-direction with PMLs and have increased the number of cells in the free-space region of the simulation domain. The number of nanorods in this study is seven. The spacing

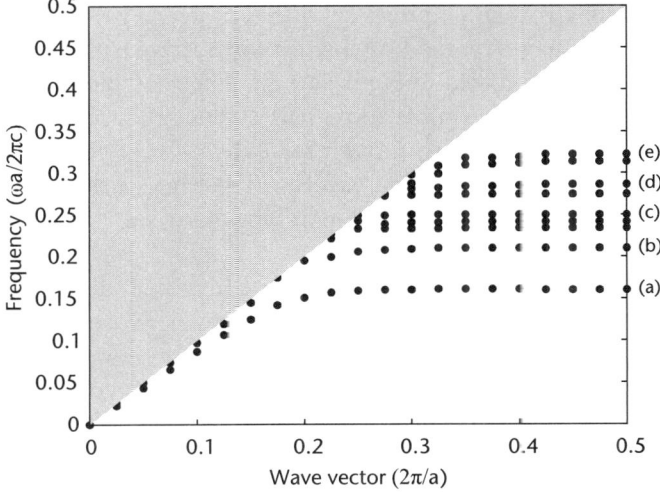

Figure 10.21 (a–e) Dispersion diagram for a single row of periodic infinite-long (along the z-direction) elliptical silver cylinders calculated from conformal dispersive FDTD simulations.

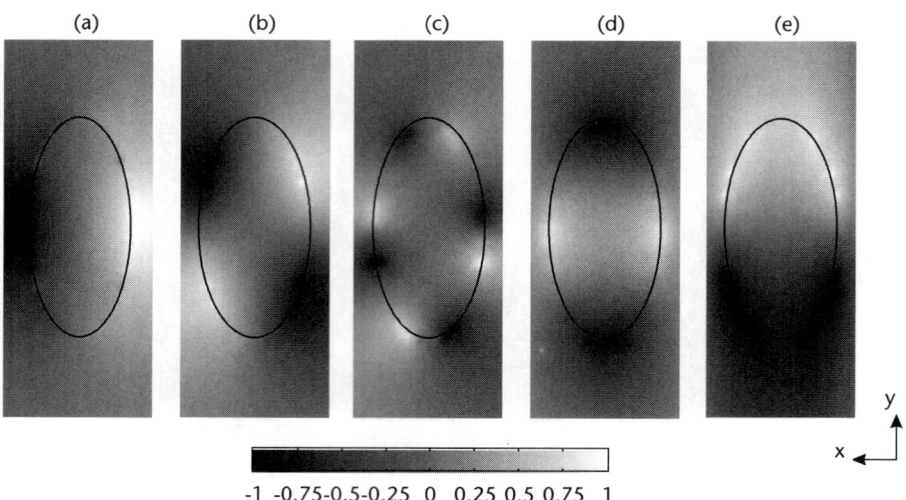

Figure 10.22 Normalized distributions of magnetic fields corresponding to different guided modes as marked in Figure 10.21: (a, d) are even modes, and (b, c, e) are odd modes. The structure is infinite along the x-direction. (Note that the coordinate has been rotated 90° anticlockwise from Figure 10.14 for better presentation of the figure.)

(pseudoperiod) between adjacent elements remains the same as that for infinite structures considered in the previous section. For a single-mode excitation, we choose the frequency of the corresponding mode from the dispersion diagram and excite with sinusoidal sources at one end of the waveguides at different locations, depending on the symmetry of different guided modes.

For the plasmonic waveguides formed by different types of inclusions, we have chosen certain types of eigenmodes for: (1) a single row of circular cylinders, shown in Figure 10.18(a); (2) two rows of circular cylinders, as shown in Figure 10.20(d); and (3) a single row of elliptical cylinders, appearing in Figure 10.22(e). The distributions of the magnetic field intensities for different waveguides operating in these guided modes are plotted in Figure 10.23. The field plots are taken after the steady state is reached in the simulations. It is evident that single guided modes are coupled into these waveguides, though the excitation of certain modes depends highly on the symmetry of the field patterns. The energy that can be coupled into the waveguides also depends on the matching between the source and the plasmonic waveguide.

In summary, we have developed a conformal dispersive FDTD method for the modeling of plasmonic waveguides formed by an array of periodic, infinite-long, silver cylinders at optical frequencies. The conformal scheme is based on effective permittivities, and its main advantage is that it introduces no numerical instabilities because only a conventional orthogonal FDTD grid is used for the simulations. The material frequency dispersion is accounted for by using an auxiliary differential equation method. A comparison of dispersion diagrams for 1-D periodic silver cylinders, computed by using these different ways—the conformal dispersive FDTD method, the conventional dispersive FDTD method with staircase

10.6 FDTD Modeling of Electromagnetic Cloaking Structures

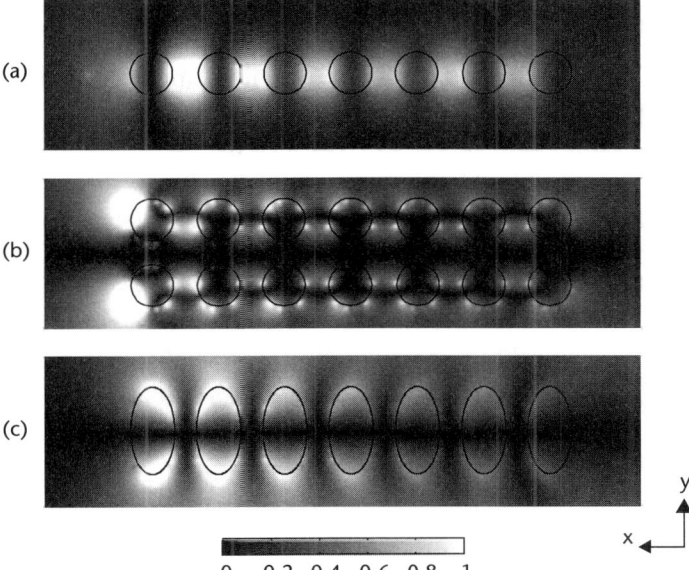

Figure 10.23 Normalized distributions of magnetic field intensity corresponding to different guided modes for seven-element plasmonic waveguides formed by (a) a single row of circular nanorods [the corresponding eigenmode is shown in Figure 10.18(a)]; (b) two rows of circular nanorods arranged in square lattice [the corresponding eigenmode is shown in Figure 10.20(c)]; and (c) a single row of elliptical nanorods [the corresponding eigenmode is shown in Figure 10.22(e)]. (Note that the coordinate has been rotated 90° anticlockwise from Figure 10.14 for better presentation of the figure.)

approximations, and the frequency domain embedding method—demonstrates the accuracy of the proposed approach. It is shown that by adding additional elements, or changing the geometry of inclusions, the corresponding dispersion diagram can be modified. Numerical simulations of plasmonic waveguides formed by seven elements show that while the eigenmodes in infinite structures can be excited, they depend highly on the symmetry of field patterns of certain modes. Further work includes the investigation of the effects of different numbers of elements in the plasmonic waveguides on the guided modes and the calculation of group velocity of different modes propagating in these waveguides. Although the results presented in this book focus on optical frequencies, it is anticipated that novel applications can be found in the designs of small antenna and efficient absorbers with future advances in microwave plasmonic materials.

10.6 FDTD Modeling of Electromagnetic Cloaking Structures

The widespread interest in the invisibility of objects has led to the recent development in electromagnetic cloaking structures. Pendry et al. have proposed an electromagnetic material through which electromagnetic fields can be controlled and manipulated to propagate around its interior region like the flow of water [89];

hence, objects placed inside would become "invisible" to external electromagnetic fields. The proposed cloaking structure in [89] requires the use of inhomogeneous and anisotropic media, with both the permittivity and the permeability being independently controlled and radially dependent. The magnitudes of the relative permittivity and permeability of the perfect cloak are less than one; therefore, such a cloak cannot be constructed by using naturally existing materials, However, recent development of metamaterials [90] (artificially engineered structures with unussal electromagnetic properties that cannot be obtained naturally) makes it possible for us to construct such cloaking structures. However, in common with the negative-index metamaterials [90], the cloaking materials are inevitably dispersive and, therefore, band-limited. Furthermore, the complete set of material parameters proposed in [89] requires the control of all the components of permittivity and permeability of the material, which makes its practical realization difficult. This has led to the use of reduced sets of material parameters for both TE [91] and TM [92] cases. Such reduced parameters for the TM case eliminates the dependence on the magnetic properties of the material, and this is especially important for the realization of cloaking in the optical frequency range, because of the absence of optical magnetism in nature. However, considerable reflections occur because of the impedance mismatch at the outer boundary of such a simplified cloak. Under the assumption of the geometric optics, a higher-order transformation has been proposed in [93] to improve the performance and minimize the scattering introduced by the cloak.

The development of Pendry's cloak is based on the coordinate transformation technique [89, 94], which has also evoked other research topics such as the design of magnifying perfect and super lenses [95]; the transformation media that rotate electromagnetic fields [96]; the design of reflectionless complex media for shifting and splitting optical beams [97]; and the design of conformal antennas [98]. The spatial transformation technique has also been applied to analyze eccentric elliptical cloaks in [99] and for acoustic cloaking in [100, 101]. Other theoretical studies of the cloaking structure include the assessment of the sensitivity of an ideal cloak to small perturbations [102]; performance of the cylindrical cloaks comprised of simplified material parameters [103–105]; realization of cloaking using a concentric layered structure of homogeneous isotropic materials [106]; improvement of the cloaking performance using soft-and-hard surface lining [107]; and broadband cloaking using sensors and active sources near the surface of a region [108]. The experimental demonstration of a simplified cloak consisting of SRRs has been reported at microwave frequencies [109]. For the optical frequency range, the cloak can be constructed either by embedding silver wires in a dielectric medium [92] or by using a gold-dielectric concentric layered structure [110].

It is worth mentioning that there exist different approaches to rendering objects invisible, for example, by canceling the dipolar scattering using plasmonic coatings [111, 112] and by using a LHM coating [113]. Among these two, the plasmonic coating approach is limited to objects with the subwavelength scale, and the coating depends on the geometry and material parameters of the object to be cloaked. The performance realized by using LHM coating is also affected by the objects placed inside, whose dimensions are on the order of the wavelength. In contrast to these two approaches, Pendry's cloaking technique is more general and can be applied to objects with arbitrary dimensions and for arbitrary wavelengths.

Pendry's invisible cloak has been modeled by using both the analytical and numerical methods. Besides the widely used coordinate transformation technique [89, 94, 95–101, 104, 114], a cylindrical wave expansion technique [102] and a method based on the full-wave Mie scattering model [115] have also been applied. In addition, the full-wave finite element method (FEM)–based commercial simulation package COMSOL Multiphysics has been extensively used to model different cloaks and has validated theoretical predictions [91–93, 99, 104], because the code can deal with anisotropic as well as radially dependent material parameters. While most of the numerical simulations have been performed in the frequency domain, little attention has been paid to the time-domain analysis of the cloaks. Frequency-domain techniques such as the FEM can be inefficient when we design wideband solutions. To date the only time-domain analysis of the cloak has been presented in [116] by using a time-dependent scattering theory. In this section, we propose a dispersive FDTD method to deal with both the frequency and radial dependent permittivity and permeability for the analysis, design, and optimization of cloaking structures.

10.6.1 Dispersive FDTD Modeling of the Cloaking Structure

A complete set of material parameters of the ideal cloak is given by [89]:

$$\varepsilon_r = \mu_r = \frac{r - R_1}{r}, \quad \varepsilon_\phi = \mu_\phi = \frac{r}{r - R_1},$$

$$\varepsilon_z = \mu_z = \left(\frac{R_2}{R_2 - R_1}\right)^2 \frac{r - R_1}{r} \quad (10.21)$$

where R_1 and R_2 are the inner and outer radii of the cloak, respectively. It can be easily verified from (10.21) that the ranges of the permittivity and the permeability within the cloak are $\varepsilon_r, \mu_r \in [0, (R_2 - R_1)/R_2]$, $\varepsilon_\phi, \mu_\phi \in [R_2/(R_2 - R_1), \infty]$ and $\varepsilon_z, \mu_z \in [0, R_2/(R_2 - R_1)]$. Since the values of ε_r, μ_r, ε_z and μ_z are less than unity, the cloak cannot be modeled directly using the conventional FDTD method, which deals with material parameters that are constant at any particular location r. However, one can map the material parameters of the cloak by using a dispersive material models, for example, a Drude model for the ε_r

$$\varepsilon_r(\omega) = 1 - \frac{\omega_p^2}{\omega^2 - j\omega\gamma} \quad (10.22)$$

where ω_p and γ are the plasma and collision frequencies of the material, respectively. The radial dependent material parameters (10.21) can be achieved by varying the plasma frequency. Note that in practice, the plasma frequency of the material depends on the periodicity of the SRRs [109], as well as of the wires [92] and varies along the radial direction. Furthermore, different dispersion models (e.g., Debye and Lorentz) can be also considered for the modeling of the cloak, which will lead to slightly different FDTD formulas from the following ones.

Since the conventional FDTD method [117, 118] is able to deal with frequency-independent materials, the frequency-dependent FDTD method will be referred to as the dispersive FDTD method [119–121]. For simplicity, we will implement the dispersive FDTD method for a 2-D TE case, for which only three field components are nonzero: E_x, E_y, and H_z. Hence the cloak we model is cylindric and infinitely long in the z-direction [$\mu_r = \mu_\phi = \varepsilon_z = 0$ in (10.21)], though an extension to a 3-D FDTD method that models a 3-D cloak [89] is relatively straightforward. There exist dispersive FDTD methods based on different approaches that can deal with frequency-dependent material parameters. These include: the recursive convolution (RC) method [119], the ADE method [120], and the Z-transform method [121]. In view of its simplicity, we have chosen the ADE method for modeling the cloak.

The ADE dispersive FDTD method is based on Faraday's and Ampere's laws, which are written as:

$$\nabla \times \mathbf{E} = -\frac{\partial \mathbf{B}}{\partial t} \tag{10.23}$$

$$\nabla \times \mathbf{H} = \frac{\partial \mathbf{D}}{\partial t} \tag{10.24}$$

It also utilizes the constitutive relations $\mathbf{D} = \varepsilon \mathbf{E}$ and $\mathbf{B} = \mu \mathbf{H}$, where ε and μ are expressed by (10.21). Equations (10.23) and (10.24) can be discretized by following a normal procedure [117, 118] that leads to the conventional FDTD updating equations:

$$\mathbf{B}^{n+1} = \mathbf{B}^n - \Delta t \cdot \tilde{\nabla} \times \mathbf{E}^{n+\frac{1}{2}} \tag{10.25}$$

$$\mathbf{D}^{n+1} = \mathbf{D}^n + \Delta t \cdot \tilde{\nabla} \times \mathbf{H}^{n+\frac{1}{2}} \tag{10.26}$$

where $\tilde{\nabla}$ is the discretized curl operator, Δt is the FDTD time step, and n is the number of time steps.

In addition, we need to include ADEs that can be discredited through the following steps. Note that the FDTD formulas are only given for the electric fields, and the update equation for the magnetic field can be obtained in the same way. Since the material parameters given in (10.21) are in cylindrical coordinates, the coordinate transformation

$$\begin{bmatrix} \varepsilon_{xx} & \varepsilon_{xy} \\ \varepsilon_{yx} & \varepsilon_{yy} \end{bmatrix} = \begin{bmatrix} \varepsilon_r \cos^2\phi + \varepsilon_\phi \sin^2\phi & (\varepsilon_r - \varepsilon_\phi)\sin\phi\cos\phi \\ (\varepsilon_r - \varepsilon_\phi)\sin\phi\cos\phi & \varepsilon_r \sin^2\phi + \varepsilon_\phi \cos^2\phi \end{bmatrix} \tag{10.27}$$

for the conventional Cartesian FDTD mesh is used. The tensor form of the constitutive relation is given by

$$\varepsilon_0 \begin{bmatrix} \varepsilon_{xx} & \varepsilon_{xy} \\ \varepsilon_{yx} & \varepsilon_{yy} \end{bmatrix} \begin{bmatrix} E_x \\ E_y \end{bmatrix} = \begin{bmatrix} D_x \\ D_y \end{bmatrix} \Leftrightarrow \varepsilon_0 \begin{bmatrix} E_x \\ E_y \end{bmatrix} = \begin{bmatrix} \varepsilon_{xx} & \varepsilon_{xy} \\ \varepsilon_{yx} & \varepsilon_{yy} \end{bmatrix}^{-1} \begin{bmatrix} D_x \\ D_y \end{bmatrix} \tag{10.28}$$

where

$$\begin{bmatrix} \varepsilon_{xx} & \varepsilon_{xy} \\ \varepsilon_{yx} & \varepsilon_{yy} \end{bmatrix}^{-1} = \frac{1}{\varepsilon_r \varepsilon_\phi} \begin{bmatrix} \varepsilon_r \sin^2\phi + \varepsilon_\phi \cos^2\phi & (\varepsilon_\phi - \varepsilon_r)\sin\phi\cos\phi \\ (\varepsilon_\phi - \varepsilon_r)\sin\phi\cos\phi & \varepsilon_r \cos^2\phi + \varepsilon_\phi \sin^2\phi \end{bmatrix} \tag{10.29}$$

10.6 FDTD Modeling of Electromagnetic Cloaking Structures

Note that the inverse of the permittivity tensor matrix exists only when $\varepsilon_r \neq 0$ and $\varepsilon_\phi \neq 0$, which is not the case for the inner boundary of the cloak. In our FDTD simulations, we place a perfect electric conductor (PEC) cylinder, with a radius equal to R_1 inside the cloak to guarantee the validity of (10.29).

Substituting (10.29) into (10.28) yields

$$\begin{cases} \varepsilon_r \varepsilon_\phi \varepsilon_0 E_x = \left(\varepsilon_r \sin^2\phi + \varepsilon_\phi \cos^2\phi\right) D_x + \left(\varepsilon_\phi - \varepsilon_r\right) \sin\phi \cos\phi D_y \\ \varepsilon_r \varepsilon_\phi \varepsilon_0 E_y = \left(\varepsilon_r \cos^2\phi + \varepsilon_\varrho \sin^2\phi\right) D_y + \left(\varepsilon_\phi - \varepsilon_r\right) \sin\phi \cos\phi D_x \end{cases} \quad (10.30)$$

Expressing ε_r in the Drude form of (10.21), (10.30) can be written as

$$\begin{cases} \varepsilon_0 \varepsilon_\phi \left(\omega^2 - j\omega\gamma - \omega_p^2\right) E_x = \left[\left(\omega^2 - j\omega\gamma - \omega_p^2\right) \sin^2\phi + \varepsilon_\phi \left(\omega^2 - j\omega\gamma\right) \cos^2\phi\right] D_x \\ \qquad + \left[\varepsilon_\phi \left(\omega^2 - j\omega\gamma\right) - \left(\omega^2 - j\omega\gamma - \omega_p^2\right)\right] \sin\phi \cos\phi D_y, \\ \varepsilon_0 \varepsilon_\phi \left(\omega^2 - j\omega\gamma - \omega_p^2\right) E_y = \left[\left(\omega^2 - j\omega\gamma - \omega_p^2\right) \cos^2\phi + \varepsilon_\phi \left(\omega^2 - j\omega\gamma\right) \sin^2\phi\right] D_y \\ \qquad + \left[\varepsilon_\phi \left(\omega^2 - j\omega\gamma\right) - \left(\omega^2 - j\omega\gamma - \omega_p^2\right)\right] \sin\phi \cos\phi D_x \end{cases} \quad (10.31)$$

Notice that ε_ϕ is retained in (10.31), because its value is always greater than one (except at the inner surface of the cloak) and can be directly used in conventional FDTD update equations [117, 118]. Using an inverse Fourier transform and the following rules:

$$j\omega \to \frac{\partial}{\partial t}, \quad \omega^2 \to -\frac{\partial^2}{\partial t^2}, \quad (10.32)$$

the first equation of (10.31) can be rewritten in the time domain as

$$\varepsilon_0 \varepsilon_\phi \left(\frac{\partial^2}{\partial t^2} + \gamma \frac{\partial}{\partial t} + \omega_p^2\right) E_x = \left[\left(\frac{\partial^2}{\partial t^2} + \gamma \frac{\partial}{\partial t} + \omega_p^2\right) \sin^2\phi + \varepsilon_\phi \left(\frac{\partial^2}{\partial t^2} + \gamma \frac{\partial}{\partial t}\right) \cos^2\phi\right] D_x$$
$$+ \left[\varepsilon_\phi \left(\frac{\partial^2}{\partial t^2} + \gamma \frac{\partial}{\partial t}\right) - \left(\frac{\partial^2}{\partial t^2} + \gamma \frac{\partial}{\partial t} + \omega_p^2\right)\right] \sin\phi \cos\phi D_y \quad (10.33)$$

The FDTD simulation domain is represented by an equally spaced 3-D grid, whose periods are Δx, Δy, and Δz along the x-, y-, and z-directions, respectively. To discretize (10.33), we use central finite difference operators in time (δ_t and δ_t^2) and central averaging operators with respect to time (μ_t and μ_t^2), that is, we let:

$$\frac{\partial^2}{\partial t^2} \to \frac{\delta_t^2}{(\Delta t)^2}, \quad \frac{\partial}{\partial t} \to \frac{\delta_t}{\Delta t}\mu_t, \quad \omega_p^2 \to \omega_p^2 \mu_t^2$$

where the operators δ_t, δ_t^2, μ_t, and μ_t^2 are defined as in [122]:

$$\delta_t \mathbf{F}|^n_{m_x,m_y,m_z} \equiv \mathbf{F}|^{n+\frac{1}{2}}_{m_x,m_y,m_z} - \mathbf{F}|^{n-\frac{1}{2}}_{m_x,m_y,m_z}$$

$$\delta_t^2 \mathbf{F}|^n_{m_x,m_y,m_z} \equiv \mathbf{F}|^{n+1}_{m_x,m_y,m_z} - 2\mathbf{F}|^n_{m_x,m_y,m_z} + \mathbf{F}|^{n-1}_{m_x,m_y,m_z}$$

$$\mu_t \mathbf{F}|^n_{m_x,m_y,m_z} \equiv \frac{\mathbf{F}|^{n+\frac{1}{2}}_{m_x,m_y,m_z} + \mathbf{F}|^{n-\frac{1}{2}}_{m_x,m_y,m_z}}{2}$$

$$\mu_t^2 \mathbf{F}|^n_{m_x,m_y,m_z} \equiv \frac{\mathbf{F}|^{n+1}_{m_x,m_y,m_z} + 2\mathbf{F}|^n_{m_x,m_y,m_z} + \mathbf{F}|^{n-1}_{m_x,m_y,m_z}}{4} \quad (10.34)$$

In (10.34), $\mathbf{F}$ represents the field components, and m_x, m_y, m_z are the indices corresponding to a certain discretization point in the FDTD domain. The discretized (10.33) reads:

$$\varepsilon_0 \varepsilon_\phi \left[\frac{\delta_t^2}{(\Delta t)^2} + \gamma \frac{\delta_t}{\Delta t} \mu_t + \omega_p^2 \mu_t^2 \right] E_x = \left\{ \left[\frac{\delta_t^2}{(\Delta t)^2} + \gamma \frac{\delta_t}{\Delta t} \mu_t + \omega_p^2 \mu_t^2 \right] \sin^2 \phi \right.$$
$$+ \varepsilon_\phi \left[\frac{\delta_t^2}{(\Delta t)^2} + \gamma \frac{\delta_t}{\Delta t} \mu_t \right] \cos^2 \phi \right\} D_x + \left\{ \varepsilon_\phi \left[\frac{\delta_t^2}{(\Delta t)^2} + \gamma \frac{\delta_t}{\Delta t} \mu_t \right] \right.$$
$$\left. - \left[\frac{\delta_t^2}{(\Delta t)^2} + \gamma \frac{\delta_t}{\Delta t} \mu_t + \omega_p^2 \mu_t^2 \right] \right\} \sin \phi \cos \phi \, D_y \quad (10.35)$$

Note that, in (10.35), the discretization of the term ω_p^2 of (10.33) is performed by using the central averaging operator μ_t^2, to ensure improved stability; the central averaging operator μ_t is used for the term containing γ to preserve the second-order feature of the equation. Equation (10.35) can be written as

$$\varepsilon_0 \varepsilon_\phi \left[\frac{E_x^{n+1} - 2E_x^n + E_x^{n-1}}{(\Delta t)^2} + \gamma \frac{E_x^{n+1} - E_x^{n-1}}{2\Delta t} + \omega_p^2 \frac{E_x^{n+1} + 2E_x^n + E_x^{n-1}}{4} \right]$$
$$= \sin^2 \phi \left[\frac{D_x^{n+1} - 2D_x^n + D_x^{n-1}}{(\Delta t)^2} + \gamma \frac{D_x^{n+1} - D_x^{n-1}}{2\Delta t} + \omega_p^2 \frac{D_x^{n+1} + 2D_x^n + D_x^{n-1}}{4} \right]$$
$$+ \varepsilon_\phi \cos^2 \phi \left[\frac{D_x^{n+1} - 2D_x^n + D_x^{n-1}}{(\Delta t)^2} + \gamma \frac{D_x^{n+1} - D_x^{n-1}}{2\Delta t} \right]$$
$$+ \sin \phi \cos \phi \left\{ \varepsilon_\phi \left[\frac{D_y^{n+1} - 2D_y^n + D_y^{n-1}}{(\Delta t)^2} + \gamma \frac{D_y^{n+1} - D_y^{n-1}}{2\Delta t} \right] \right.$$
$$\left. - \left[\frac{D_y^{n+1} - 2D_y^n + D_y^{n-1}}{(\Delta t)^2} + \gamma \frac{D_y^{n+1} - D_y^{n-1}}{2\Delta t} + \omega_p^2 \frac{D_y^{n+1} + 2D_y^n + D_y^{n-1}}{4} \right] \right\}$$
$$\quad (10.36)$$

Therefore the update equation for E_x can be obtained as

$$E_x^{n+1} = \left[a D_x^{n+1} + b D_x^n + c D_x^{n-1} + d \overline{D_y}^{n+1} + e \overline{D_y}^n + f \overline{D_y}^{n-1} - \left(g E_x^n + h E_x^{n-1} \right) \right] / l$$
$$\quad (10.37)$$

10.6 FDTD Modeling of Electromagnetic Cloaking Structures

where the coefficients a to l are given by

$$a = \sin^2\phi \left[\frac{1}{(\Delta t)^2} + \frac{\gamma}{2\Delta t} + \frac{\omega_p^2}{4} \right] + \varepsilon_\phi \cos^2\phi \left[\frac{1}{(\Delta t)^2} + \frac{\gamma}{2\Delta t} \right]$$

$$b = \sin^2\phi \left[-\frac{2}{(\Delta t)^2} + \frac{\omega_p^2}{2} \right] - \varepsilon_\phi \cos^2\phi \frac{2}{(\Delta t)^2}$$

$$c = \sin^2\phi \left[\frac{1}{(\Delta t)^2} - \frac{\gamma}{2\Delta t} + \frac{\omega_p^2}{4} \right] + \varepsilon_\phi \cos^2\phi \left[\frac{1}{(\Delta t)^2} - \frac{\gamma}{2\Delta t} \right]$$

$$d = \left\{ \varepsilon_\phi \left[\frac{1}{(\Delta t)^2} + \frac{\gamma}{2\Delta t} \right] - \left[\frac{1}{(\Delta t)^2} + \frac{\gamma}{2\Delta t} + \frac{\omega_p^2}{4} \right] \right\} \sin\phi \cos\phi$$

$$e = \left\{ \varepsilon_\phi \left[-\frac{2}{(\Delta t)^2} \right] - \left[-\frac{2}{(\Delta t)^2} + \frac{\omega_p^2}{2} \right] \right\} \sin\phi \cos\phi,$$

$$f = \left\{ \varepsilon_\phi \left[\frac{1}{(\Delta t)^2} - \frac{\gamma}{2\Delta t} \right] - \left[\frac{1}{(\Delta t)^2} - \frac{\gamma}{2\Delta t} + \frac{\omega_p^2}{4} \right] \right\} \sin\phi \cos\phi$$

$$g = \varepsilon_0 \varepsilon_\phi \left[-\frac{2}{(\Delta t)^2} + \frac{\omega_p^2}{2} \right], \quad h = \varepsilon_0 \varepsilon_\phi \left[\frac{1}{(\Delta t)^2} - \frac{\gamma}{2\Delta t} + \frac{\omega_p^2}{4} \right], \quad l = \varepsilon_0 \varepsilon_\phi \left[\frac{1}{(\Delta t)^2} + \frac{\gamma}{2\Delta t} + \frac{\omega_p^2}{4} \right]$$

Note that the field quantities $\overline{D_y}$ in (10.37) are locally averaged values of D_y since the x- and y-components of the field reside at different locations in the FDTD domain. The averaged value can be calculated by using [123]

$$\overline{D_y}(i,j) = \frac{D_y(i,j) + D_y(i+1,j) + D_y(i,j-1) + D_y(i+1,j-1)}{4} \tag{10.38}$$

where (i,j) are the coordinates of the location of the field component. Following the same procedure, we can obtain the update equations for the second equation of (10.30) and for the magnetic field component H_z. Equations (10.25)–(10.37) and the equations for E_y and H_z form the FDTD updating equation set in the context of the well-known leap-frog scheme [117].

Since the FDTD method is inherently a numerical technique, the spatial and time discretizations have important effects on the accuracy of simulation results. Also, since the permittivity is frequency-dependent, one can expect a slight difference between the analytical and numerical material parameters because of the discrete time step employed in the FDTD algorithm. In general, a spatial resolution (FDTD cell size) of $\Delta x < \lambda/10$ is required [118] to model conventional dielectrics with the relative permittivity/permeability greater than unity. However, we know from our previous analysis [125] that for metamaterials, especially for the case of LHMs, the numerical errors introduce an nonphysical resonance in the transmission coefficient because of time discretization. To mitigate this problem, a requirement of $\Delta x < \lambda/80$ is proposed. For the case of the cloak, we follow the same approach as in [125], and substitute the plane-wave solution

$$\mathbf{E}^n = \mathbf{E} e^{jn\omega\Delta t}, \quad \mathbf{D}^n = \mathbf{D} e^{jn\omega\Delta t} \tag{10.39}$$

into (10.36). We then compare the resulting equation with the first equation of (10.30) and find that $\widetilde{\varepsilon}_\phi$ has the exact analytical value, while $\widetilde{\varepsilon}_r$ takes the following form:

$$\widetilde{\varepsilon}_r = \varepsilon_0 \left[1 - \frac{\omega_p^2 (\Delta t)^2 \cos^2 \frac{\omega \Delta t}{2}}{2 \sin \frac{\omega \Delta t}{2} \left(2 \sin \frac{\omega \Delta t}{2} - j\gamma \Delta t \cos \frac{\omega \Delta t}{2} \right)} \right] \quad (10.40)$$

Note that (10.40) simplifies to the Drude dispersion model (10.22) when $\Delta t \to 0$. In Figure 10.24 we compare the analytical (10.22) and numerical relative permittivities (10.40) for the case of $\varepsilon_r = 0.1$ (lossless). It is apparent that the conventional criteria for the FDTD spatial resolution does not guarantee accuracy for the modeling of the cloaks. We note that the numerical error is still around 2% even when we choose $\Delta x = \lambda / 40$.

By using the expression of the numerical permittivity (10.40), we can correct the errors introduced by the discrepancy between the numerical and analytical material parameters. For example, if the required permittivity is $\varepsilon_r = \varepsilon_r' + j\varepsilon_r''$, we can calculate the corrected plasma and collision frequencies as

$$\widetilde{\omega}_p^2 = \frac{2 \sin \frac{\omega \Delta t}{2} \left[-2(\varepsilon_r' - 1) \sin \frac{\omega \Delta t}{2} - \varepsilon_r'' \gamma \Delta t \cos \frac{\omega \Delta t}{2} \right]}{(\Delta t)^2 \cos^2 \frac{\omega \Delta t}{2}}, \quad \widetilde{\gamma} = \frac{2 \varepsilon_r'' \sin \frac{\omega \Delta t}{2}}{(\varepsilon_r' - 1) \Delta t \cos \frac{\omega \Delta t}{2}}$$
(10.41)

after simple derivations. Our FDTD simulations show that the simulation becomes unstable before reaching the steady state if we use $\Delta x = \lambda / 35$ without correcting the numerical material parameters. The cause of such an instability remains an open question at present, though the correction of material parameters guarantees stable FDTD simulations. Consequently we will use the corrected material parameters (10.41) in the examples presented next.

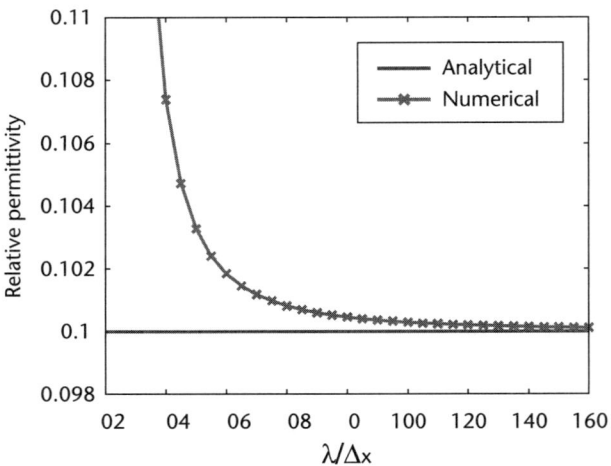

Figure 10.24 The comparison between the analytical (10.22) and numerical material parameters (10.40) for different FDTD spatial resolutions for the case of $\varepsilon_r = 0.1$ (lossless).

10.6.2 Numerical Results and Discussion

The dispersive FDTD method technique described above has been implemented for a 2-D TE case. The computation domain is shown in Figure 10.25. The following parameters are used in the simulations: FDTD cell size $\Delta x = \Delta y = \lambda/150$, where λ is the wavelength at the operating frequency $f = 2.0$ GHz, and the time step $\Delta t = \Delta x / \sqrt{2}c$, chosen according to the Courant stability criterion [118]. We assume an ideal lossless case [i.e., that the collision frequency in (10.22) is equal to zero ($\gamma = 0$)]. The radial dependent plasma frequency can be computed from (10.41) with a given value of ε_r obtained from (10.21). The radii of the cloak are: $R_1 = 0.2$m and $R_2 = 0.4$m. The computational domain is truncated by using Berenger's PML [124] in the y-direction to absorb the waves outgoing from the computation domain without introducing reflections. We also use the PBCs in the x-direction to model a plane-wave source. The source is implemented by specifying a complete column of FDTD cells using a certain wave function (sinusoidal source in our case), as shown in Figure 10.25.

First, we consider the ideal cloak, whose material parameters are given by (10.21), where $\mu_r = \mu_\phi = \varepsilon_z = 0$. Figure 10.26 shows the distributions of the electric and magnetic field components calculated from the dispersive FDTD simulation of the ideal cloak. Note that only the central part of the simulation domain is shown and that the actual computation domain is larger. We see from Figure 10.26(b, c, d) that the distribution is almost identical to the published results [91] calculated by using the analytical and frequency domain methods, except that there are small ripples in the magnetic field. In fact, the field distribution presented in [91] shows even stronger ripples, which may be due to the inadequate spatial resolution used in the calculation. These ripples are purely numerical and disappear when a fine mesh is used. In our case, the slight ripples are also caused by the staircasing approximation of the circular surface of the cloak, since a Cartesian mesh is used. There are also nonzero scattered fields in the x-component of the electric field outside the cloak [see Figure 10.26(b)], which is also a sign of numerical errors because an ideal cloak does not introduce any scattering outside the cloak. The staircasing approximation only causes a very small amount of numerical errors due

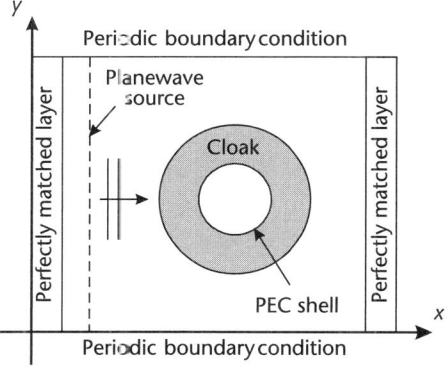

Figure 10.25 A 2-D FDTD simulation domain for the case of plane-wave incidence on the cloak.

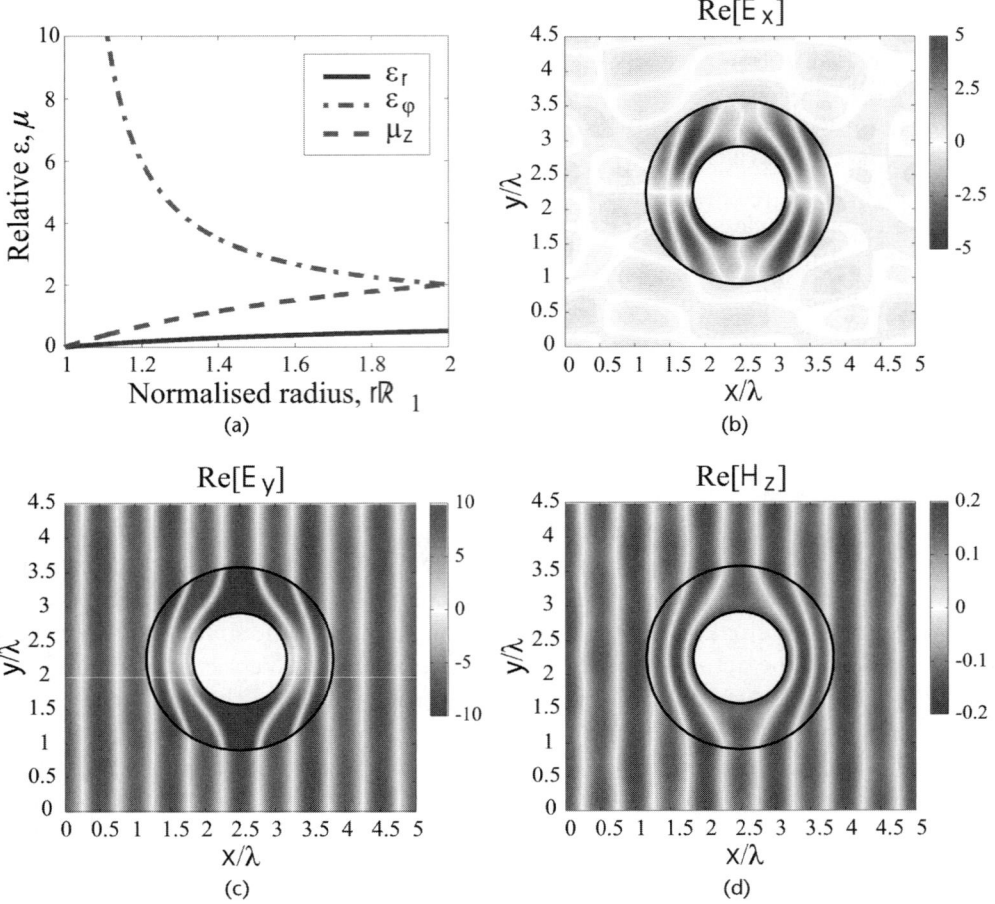

Figure 10.26 (a) Material parameters for an infinite ideal cylindrical cloak [91] where all ε_r, ε_ϕ, and μ_z are radial dependent. (b, c, d) Field distributions from dispersive FDTD simulations of the cloak: (b) x-component of the electric field, (c) y-component of the electric field, and (d) the magnetic field.

to the fine mesh used in our simulations. However, this problem can be further reduced and the accuracy of simulations can be improved by using a conformal scheme in conjunction with the dispersive FDTD method. This entails using an effective permittivity at material boundaries, as it was done for the case of isotropic dispersive materials at planar [125] and curved boundaries [126]. However, the complexity stems from the anisotropy of the cloaking material, which leads to an eighth-order differential equation to be discretized. Recall that the order of the differential equation for the case of isotropic dispersive materials is four [126].

As mentioned previously, it is difficult to realize the ideal cloak that requires all the components of the permittivity and permeability to be radial-dependent. Therefore, it has been proposed in [92] that a reduced set of material parameters be used, while keeping the same wave trajectory, to construct a simplified cloak.

In [92], the reduced set of material parameters is given by:

$$\varepsilon_r = \left(\frac{R_2}{R_2 - R_1}\right)^2 \left(\frac{r - R_1}{r}\right)^2, \qquad \varepsilon_\phi = \left(\frac{R_2}{R_2 - R_1}\right)^2, \qquad \mu_z = 1 \qquad (10.42)$$

Following the same procedure, we have also modified the dispersive FDTD method proposed above to model the reduced set of material parameters and analyzed its cloaking performance. The dimensions of the simplified cloak remain the same as those of the ideal one. Figure 10.27 shows the steady-state field distributions computed by using the procedure described above. Such a cloak significantly reduces the complexity of practical realization since only ε_r is radially dependent, as shown in Figure 10.27(a). However, considerable reflections occur because of the impedance mismatch at the outer boundary of the simplified cloak, as may be seen from Figure 10.27(b, d). Interestingly, the y-component of the electric field is only affected slightly by the scattered field. Note that here we only consider the simplified nonmagnetic cloak [92]; however, the simplified cloak proposed in [91] can be modeled in a similar way.

Scattering from the simplified cloak (10.42), introduced by the impedance mismatch can be reduced by using an improved cloak based on a higher-order transformation [93]. The material parameters are given by [93]

$$\varepsilon_r = \left(\frac{r'}{r}\right)^2, \qquad \varepsilon_\phi = \left[\frac{\partial g(r')}{\partial r'}\right]^{-2}, \qquad \mu_z = 1 \qquad (10.43)$$

where $r = g(r') = [(R_1/R_2)(r'/R_2 - 2) + 1]r' + R_1$. We have also modeled such a cloak and plotted the field distributions in Figure 10.28. Its dimensions are kept the same as in the previous two cases. In fact the dimensions of this cloak are at its limit, since it is required to satisfy $R_1/R_2 < 0.5$, to guarantee a monotonic transformation [93]. The improved cloak imposes an additional dependency of the permittivity on the radius, as shown in Figure 10.28(a). It is evident that indeed a cloak designed by using a higher-order transformation exhibits an improved impedance at its outer boundary, and hence, reduces the scattered field considerably. Notice that the wavefront only starts to bend near the inner surface of the cloak, which is different from the case of the ideal cloak shown in Figure 10.26. This is due to the slow variation of the impedance as we traverse inward from the outer boundary of the cloak.

For the sake of demonstration, we have plotted in Figure 10.29 the power flow diagrams as well as the scattering patterns for the above ideal cloak; the simplified cloak based on a linear transformation; and the simplified cloak based on the higher-order transformation. The power flow diagrams show that for the ideal case [Figure 10.29(a)], the wavefront enters the cloak smoothly, bends around the central region, and returns to its original pattern after leaving the cloak. The cloak based on the higher-order transformation [Figure 10.29(c)] shows a similar pattern with a smooth bending of the wavefront near the central region of the cloak. However, the power flow is disturbed before entering the cloak, because of reflections, while the wavefront leaving the cloak has a relatively smooth distribution. For the case of the cloak based on the linear transformation, as shown in Figure 10.29(b), the waves do not strictly follow the trend of the bending inside the cloak and

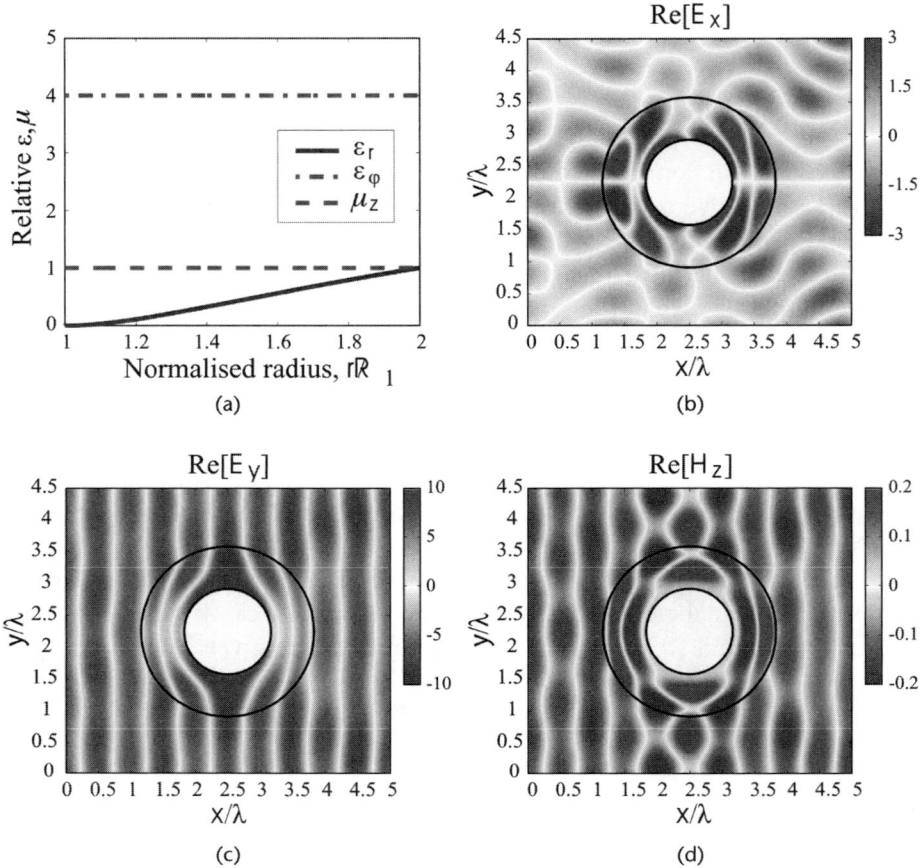

Figure 10.27 (a) Material parameters for an infinite simplified cylindrical cloak using a linear transformation [92] where only ε_r is radial dependent. (b, c, d) Field distributions from dispersive FDTD simulations of the cloak: (b) x-component of the electric field, (c) y-component of the electric field, and (d) the magnetic field.

propagate in arbitrary directions, and the external field is significantly disturbed as a consequence. This can be clearly identified from the scattering patterns that are referred to the free-space case with absence of the cloak, and then normalized to the scattering pattern of a PEC cylinder without the cloak, as shown in Figure 10.29(d). For all the cloaks, the scattering at the back of the cloak (relative to the direction of wave incidence) is dramatically reduced. However, the level of the scattered field for the linear cloak is almost the same as that of a PEC cylinder, leading to the conclusion that the object placed inside this simplified cloak can be detected from its front side, similar to the conclusion drawn in [103]. For the cloaks based on higher-order transformation, scattering is reduced by around four times as compared to that in the case of linear transformation. Theoretically the ideal cloak has a zero-scattered field, though the nonzero but small values, as appeared in the plot in Figure 10.29(d) are caused mainly by the staircasing approximation in the FDTD simulations, and they will tend to zero when either an extremely fine mesh is used or a conformal scheme is employed, as mentioned earlier.

10.6 FDTD Modeling of Electromagnetic Cloaking Structures

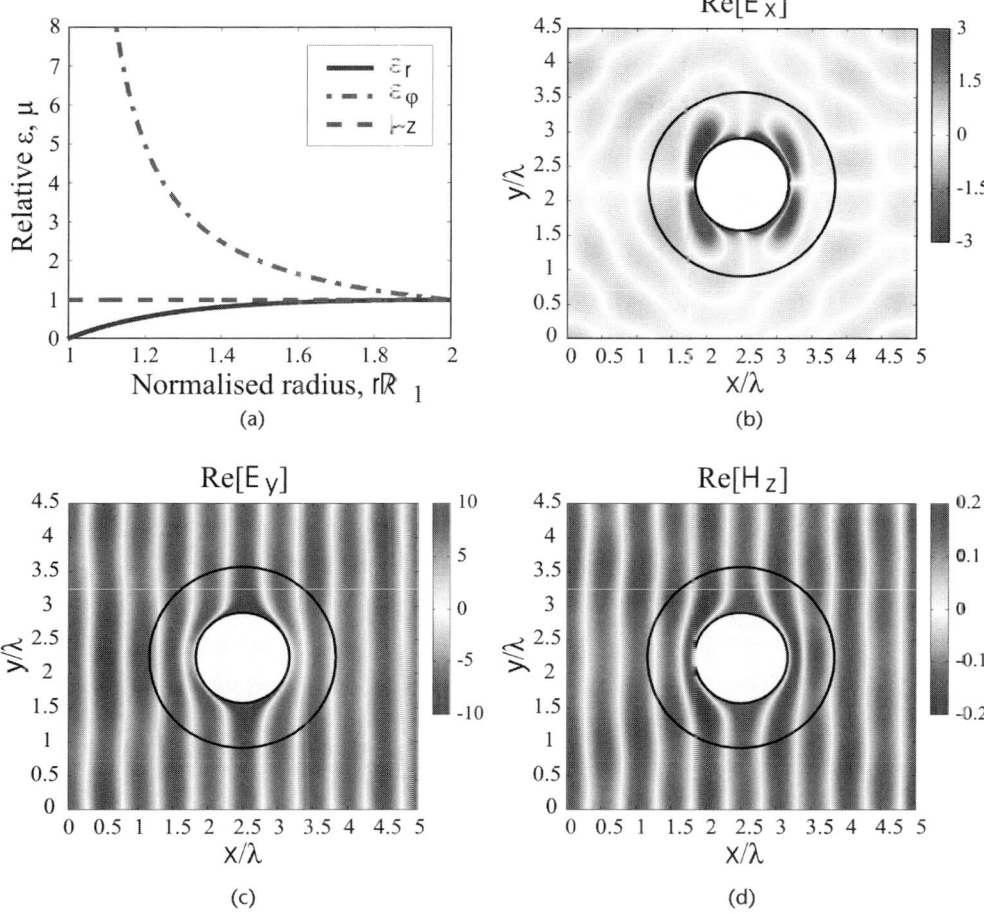

Figure 10.28 (a) Material parameters for an infinite simplified cylindrical cloak using a high-order transformation [93] where only ε_r and ε_ϕ are radial dependent. (b, c, d) Field distributions from dispersive FDTD simulations of the cloak. (b) x-component of the electric field, (c) y-component of the electric field, and (d) the magnetic field.

In summary, we have proposed a dispersive FDTD scheme for modeling cloaking structures. The unusual material parameters (the relative magnitudes of the permittivity and permeability are less than one) are mapped to the Drude dispersion model, which is then taken into account in the FDTD simulations by using a method based on ADEs. The proposed method has been implemented in a two-dimensional case, and three different cylindrical cloaks have been considered in our simulations: the ideal cloak, the linear transformation–based cloak, and the higher-order transformation–based cloak. It is found from the simulations that cloaks based on the linear transformations introduce a level of back-scattering similar to the one of a PEC cylinder without the cloak, causing the possibility of the object being detected. Such scattering can be significantly reduced by using a cloaking based on the higher-order transformation. In this book, we have only considered lossless cloaks. The "ideal" cloak with a material loss of $\tan \delta = 0.1$ has been modeled in [91] using the finite element method, and the case for $\tan \delta = 0.01$,

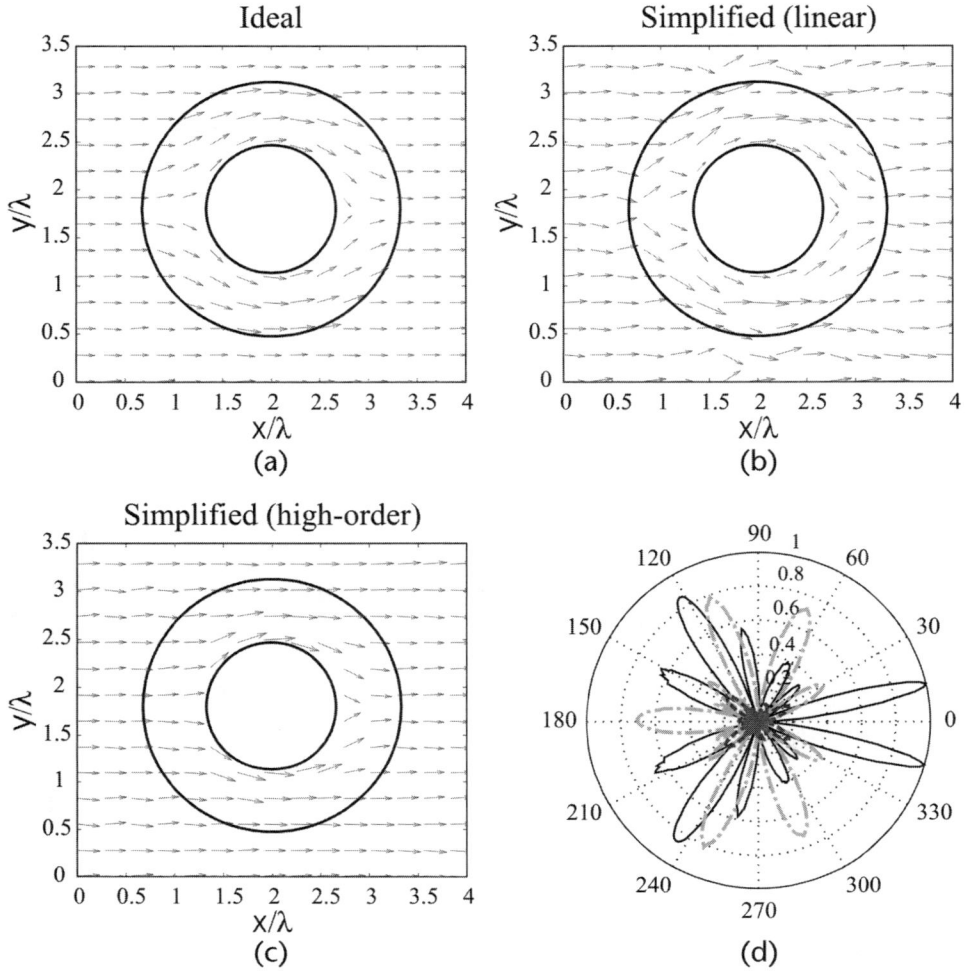

Figure 10.29 Power flow diagrams for (a) the ideal cloak [91], (b) the simplified cloak based on the linear transformation [92], and (c) the simplified cloak based on a high-order transformation [93]. (d) Comparison of the scattering patterns for different cloaks and for the case of the PEC cylinder without cloak.

0.1 and 1 have been modeled in [115] by using a full-wave Mie scattering theory. Lossy cloaks can also be directly modeled using the above proposed dispersive FDTD method by specifying a certain value for the collision frequency for ε_r in the Drude model, and by defining a dielectric loss for ε_ϕ.

References

[1]   J. Pendry, "Negative refraction index makes perfect lens," *Phys. Rev. Lett.*, vol. 85, p. 3966, 2000.

[2] V. Veselago, "The electrodynamics of substances with simultaneously negative values of ε and μ," *Sov. Phys. Usp.*, vol 10, p. 509, 1968.

[3] D. R. Smith, "How to Build a Superlens," *Science*, Vol. 308, April 22, 2005, pp. 502-503.

[4] D. Smith, J. Pendry, and M. Wiltshire, "Metamaterials and negative refractive index," *Science*, vol. 305, pp. 788-792, 2004.

[5] R. A. Shelby, D. R. Smith, and S. Schultz, "Experimental verification of a negative index of refraction," *Science*, vol. 292, pp. 77-79, 2001.

[6] T. Yen, W. Padilla, N. Fang, D. Vier, D. Smith, J. Pendry, D. Basov, and Z. Zhang, "Terahertz magnetic response from artificial materials," *Science*, vol. 303, pp. 1494-1496, 2004.

[7] S. Linden, C. Enkrich, M. Wegener, J. Zhou, T. Kochny, and C. Soukoulis, "Magnetic response of metamaterials at 100 tetrahertz," *Science*, vol. 306, pp. 1351-1353, 2004.

[8] N. Fang, H. Lee, C. Sun, and X. Zhang, "Subdiffraction-limited optical imaging with a silver superlens," *Science*, vol. 308, pp. 534-537, 2005.

[9] E. Shamonina, V. Kalinin, K. Ringhofer, and L. Solymar, "Imaging, compression and poynting vector streamlines for negative permittivity materials," *Electron. Lett.*, vol. 37, pp. 1243-1244, 2001.

[10] S. A. Ramakrishna, J. B. Pendry, M. C. K. Wiltshire, and W. J. Stewart, "Imaging the near field," *Journal of Modern Optics*, vol. 50, pp. 1419-1430, 2003.

[11] Alu, A., F. Bilotti, N. Engheta and L. Vegni, "How Metamaterials May Significantly Affect the Wave Transmission Through a Sub-Wavelength Hole in a Flat Perfectly Conducting Screen," *I.E.E. Workshop*, London, U.K., November 24, 2003, pp. 1-6.

[12] Vigoureux, J.M., F. Depasse, and C. Girard, "Superresolution of Near-Field Optical Microscopy Defined from Properties of Confined Electromagnetic Waves," *Applied Optics*, Vol. 31, No. 16, June 1, 1992, pp. 3036-3045.

[13] H. Cory, A.C. Boccara, J.C. Rivoal, and A. Lahrech, "Electric Field Intensity Variation in the Vicinity of a Perfectly Conducting Conical Probe: Application to Near-Field Microscopy," *Microw. and Opt. Techn. Let.*, Vol. 18, No. 2, 1998, pp. 120-124.

[14] Leviatan, Y., "Electromagnetic Coupling between Two Half-Space Regions Separated by Two Slot-Perforated Parallel Conducting Screens," *IEEE Trans. Microw. Theo. and Techn.*, Vol. 36, No. 1, 1988, pp. 44-52.

[15] S. A. Ramakrishna and J. B. Pendry, "Removal of absorption and increase in resolution in a near-field lens via optical gain," *Phys. Rev. B*, vol. 67, p. 201101, 2003.

[16] J. Joannopoulos, R. Mead, and J. Winn, *Photonic Crystals: Molding the Flow of Light*, Princeton, NJ: Princeton University Press, 1995.

[17] K. Sakoda, *Optical Properties of Photonic Crystals*, Berlin: Springer-Verlag, 2001.

[18] M. Notomi, "Theory of light propagation in strongly modulated photonic crystals: refractionlike behavior in the vicinity of the photonic band gap," *Phys. Rev. B*, vol. 62, no. 16, pp. 10696-10705, 2000.

[19] M. Notomi, "Negative refraction in photonic crystals," *Optical and Quantum Electronics*, vol. 34, pp. 133-143, 2002.

[20] C. Luo, S. G. Johnson, J. D. Joannopoulos, and J. B. Pendry, "All-angle negative refraction without negative eective index," *Phys. Rev. B*, vol. 65, p. 201104, 2002.

[21] C. Luo, S. G. Johnson, and J. D. Joannopoulos, "All-angle negative refraction in three-dimentionally periodic photonic crystal," *Appl. Phys. Lett.*, vol. 81, no. 13, pp. 2352-2354, 2002.

[22] C. Luo, S. G. Johnson, and J. D. Joannopoulos, "Negative refraction without negative index in metallic photonic crystals," *Optics Express*, vol. 11, no. 7, pp. 746-754, 2003.

[23] X. Zhang, "Absolute negative refraction and imaging of unpolarized electromagnetic waves by two-dimensional photonic crystals," *Phys. Rev. B*, vol. 70, p. 205102, 2004.

[24] X. Zhang, "Image resolutio depending on slab thickness and object distance in a two-dimentional photonic-crystal-based superlens," *Phys. Rev. B*, vol. 70, p. 195110, 2004.

[25] P. V. Parimi, W. T. Lu, P. Vodo, and S. Sridhar, "Imaging by at lens using negative refraction," *Nature*, vol. 426, p. 404, 2003.

[26] A. Berrier, M. Mulot, M. Swillo, M. Qiu, L. Thylen, A. Talneau, and S. Anand, "Negative refraction at infrared wavelengths in a two-dimentional photonic crystal," *Phys. Rev. Lett.*, vol. 93, no. 7, p. 073902, 2004.

[27] C. Luo, S. G. Johnson, J. D. Joannopoulos, and J. B. Pendry, "Sub-wavelength imaging in photonic crystals," *Phys. Rev. B*, vol. 68, p. 045115, 2003.

[28] H.-T. Chien, H.-T. Tang, C.-H. Kuo, C.-C. Chen, and Z. Ye, "Directed diffraction without negative refraction," *Phys. Rev. B*, vol. 70, p. 113101, 2004.

[29] Z.-Y. Li and L.-L. Lin, "Evaluation of lensing in photonic crystal slabs exhibiting negative refraction," *Phys. Rev. B*, vol. 68, p. 245110, 2003.

[30] C.-H. Kuo and Z. Ye, "Optical transmission of photonic crystal structures formed by dielectric cylinders: Evidence for non-negative refraction," *Phys. Rev. E*, vol. 70, p. 056608, 2004.

[31] D. N. Chigrin, S. Enoch, C. M. S. Torres, and G. Tayeb, "Self-guiding in two-dimentional photonic crystals," *Optics Express*, vol. 11, no. 10, pp. 1203–1211, 2003.

[32] P. Belov, C. Simovski, and P. Ikonen, "Canalization of subwavelength images by electromagnetic crystals," *Phys. Rev. B*, vol. 71, p. 193105, 2005.

[33] P. Belov, C. Simovski, and S. Tretyakov, "Two-dimensional electromagnetic crystals formed by reactively loaded wires," *Phys. Rev. E*, vol. 66, p. 036610, 2002.

[34] P. Ikonen, P. Belov, C. Simovski, and S. Maslovski, "Experimental demonstration of sub-wavelength image channeling using capacitively loaded wire medium," *Phys. Rev. B*, vol. 73, p. 073102, 2006.

[35] W. Rotman, "Plasma simulations by artificial dielectrics and parallel-plate media," *IRE Trans. Ant. Propag.*, vol. 10, p. 82, 1962.

[36] J. Brown, "Artificial Dielectrics," *Progress in Dielectrics*, vol. 2, pp. 195–225, 1960.

[37] J. Pendry, A. Holden, W. Steward, and I. Youngs, "Extremely low frequency plasmons in metallic mesostructures," *Phys. Rev. Lett.*, vol. 76, p. 4773, 1996.

[38] P. Belov, R. Marques, S. Maslovski, I. Nefedov, M. Silverinha, C. Simovski, and S. Tretyakov, "Strong spatial dispersion in wire media in the very large wavelength limit," *Phys. Rev. B*, vol. 67, p. 113103, 2003.

[39] P. A. Belov, Y. Hao, and S. Sudhakaran, "Subwavelength imaging by wire media," *Phys. Rev. Lett.*, 2005.

[40] J. T. Shen, P. B. Catrysse, and S. Fan, "Mechanism for designing metallic metamaterials with a high index of refraction," *Phys. Rev. Lett.*, vol. 94, p. 197401, 2005.

[41] P. A. Belov and Y. Hao, "Subwavelength imaging at optical frequencies using a transmission device formed by a periodic layered metal- dielectric structure operating in the canalization regime," *Physical Review B*, vol. 73, p. 113110, 2006.

[42] Y. Zhao, P. A. Belov, and Y. Hao, "Accurate modeling of left-handed metamaterials using finite-difference time-domain method with spatial averaging at the boundaries," *Journal of Optics A: Pure and Applied Optics*, vol. 9, pp. 468–475, 2007.

[43] O. P. Gandhi, B.-Q. Gao, and J.-Y. Chen, "A frequency-dependent finite-difference time-domain formulation for general dispersive media," *IEEE Trans. Microwave Theory Tech.*, vol. 41, pp. 658–664, 1993.

[44] A. Taflove and S. C. Hagness, *Computational Electrodynamics: The Finite-Difference Time-Domain Method*, 2nd ed., Norwood, MA: Artech House, 2000.

[45] J.-P. Berenger, "A perfectly matched layer for the absorption of electromagnetic waves," *J. Computat. Phys.*, vol. 114, no. 2, pp. 185–200, 1994.

[46] J. Fang and Z. Wu, "Generalized perfectly matched layer for the absorption of propagating and evanescent waves in lossless and lossy media," *IEEE Trans. Microw. Theory Tech.*, vol. 44, no. 12, pp. 2216-2222, 1996.

[47] J. D. Joannopoulos, R. D. Meade, and J. N. Winn, *Photonic Crystals: Molding the Flow of Light*, Princeton, NJ: Princeton University Press, 1995.

[48] G. Qiu, F. Lin, and Y. Li, "Complete two-dimensional bandgap of photonic crystals of a rectangular Bravais lattice," *Opt. Commun.*, vol. 219, pp. 285-288, 2003.

[49] S. Fan, J. Winn, A. Devenyi, J. C. Chen, R. D. Meade, and J. D. Joannopoulos, "Guided and defect modes in periodic dielectric waveguides," *J. Opt. Soc. Am. B*, vol. 12, pp. 1267-1272, 1995.

[50] D. Chigrin, A. Lavrinenko, and C. Sotomayor Torres, "Nanopillars photonic crystal waveguides," *Opt. Express*, vol. 12, pp. 617-622, 2004.

[51] J. Shefer, "Periodic cylinder arrays as transmission lines," *IEEE Trans. Microwave Theory Tech.*, vol. 11, pp. 55-61, January 1963.

[52] B. A. Munk, D. S. Janning, J. B. Pryor, and R. J. Marhefka, "Scattering from surface waves on finite FSS," *IEEE Trans. Antennas Propagat.*, vol. 49, pp. 1782-1793, December 2001.

[53] R. J. Mailloux, "Antenna and wave theories of infinite Yagi-Uda arrays," *IEEE Trans. Antennas Propagat.*, vol. 13, pp. 499-506, July 1965.

[54] A. D. Yaghjian, "Scattering-matrix analysis of linear periodic arrays," *IEEE Trans. Antennas Propagat.*, vol. 50, pp. 1050-1064, August 2002.

[55] J. C. Weeber, A. Dereux, C. Girard, J. R. Krenn and J. P. Goudonnet, "Plasmon polaritons of metallic nanowires for controlling submicron propagation of light," *Phys. Rev. B*, vol. 60, pp. 9061-9068, 1999.

[56] B. Lamprecht, J. R. Krenn, G. Schider, H. Ditlbacher, M. Salerno, N. Felidj, A. Leitner, F. R. Aussenegg, and J. C. Weeber, "Surface plasmon propagation in microscale metal stripes," *Appl. Phys. Lett.*, vol. 79, pp. 51-53, 2001.

[57] T. Yatsui, M. Kourogi, and M. Ohtsu, "Plasmon waveguide for optical far/near-field conversion," *Appl. Phys. Lett.*, vol. 79, pp. 4583-4585, 2001.

[58] R. Zia, M. D. Selker, P. B. Catrysse, and M. L. Brongersma, "Geometries and materials for subwavelength surface plasmon modes," *J. Opt. Soc. Am. A*, vol. 21, pp. 2442-2446, 2004.

[59] R. Charbonneau, N. Lahoud, G. Mattiussi, and P. Berini, "Demonstration of integrated optics elements based on long-ranging surface plasmon polaritons," *Opt. Express*, vol. 13, pp. 977-984, 2005.

[60] D. F. P. Pile and D. K. Gramotnev, "Channel plasmon-polariton in a triangular groove on a metal surface," *Opt. Lett.*, vol. 29, pp. 1069-1071, 2004.

[61] M. Quinten, A. Leitner, J. R. Krenn, and F. R. Aussenegg, "Electromagnetic energy transport via linear chains of silver nanoparticles," *Opt. Lett.*, vol. 23, pp. 1331-1333, 1998.

[62] M. L. Brongersma, J. W. Hartman, and H. A. Atwater, "Electromagnetic energy transfer and switching in nanoparticle chain arrays below the diffraction limit," *Phys. Rev. B*, vol. 62, pp. 16356-16359, 2000.

[63] T. J. Shepherd, C. R. Brewitt-Taylor, P. Dimond, G. Fixter, A. Laight, P. Lederer, P. J. Roberts, P. R. Tapster, and I. J. Youngs, "3D microwave photonic crystals: Novel fabrication and structures," *Electron. Lett.*, vol. 34, pp. 787-789, 1998.

[64] A. Taflove, *Computational Electrodynamics: The Finite Difference Time Domain Method*, Norwood, MA: Artech House, 1995.

[65] S. A. Maier, P. G. Kik, and H. A. Atwater, "Optical pulse propagation in metal nanoparticle chain waveguides," *Phys. Rev. B*, vol. 67, p. 205402, 2003.

[66] W. M. Saj, "FDTD simulations of 2D plasmon waveguide on silver nanorods in hexagonal lattice," *Opt. Express*, vol. 13, no. 13, pp. 4818–4827, June 2005.

[67] K.-P. Hwang and A. C. Cangellaris, "Effective permittivities for second-order accurate FDTD equations at dielectric interfaces," *IEEE Microwave Wirel. Compon. Lett.*, vol. 11, pp. 158–160, 2001.

[68] Y. Hao and C. J. Railton, "Analyzing electromagnetic structures with curved boundaries on Cartesian FDTD meshes," *IEEE Trans. Microwave Theory Tech.*, vol. 46, pp. 82–88, January 1998.

[69] R. Luebbers, F. P. Hunsberger, K. Kunz, R. Standler, and M. Schneider, "A frequency-dependent finite-difference time-domain formulation for dispersive materials," *IEEE Trans. Electromagn. Compat.*, vol. 32, pp. 222–227, August 1990.

[70] O. P. Gandhi, B.-Q. Gao, and J.-Y. Chen, "A frequency-dependent finite-difference time-domain formulation for general dispersive media," *IEEE Trans. Microwave Theory Tech.*, vol. 41, pp. 658–664, April 1993.

[71] D. M. Sullivan, "Frequency-dependent FDTD methods using Z transforms," *IEEE Trans. Antennas Propagat.*, vol. 40, pp. 1223–1230, October 1992.

[72] N. Kaneda, B. Houshmand, and T. Itoh, "FDTD analysis of dielectric resonators with curved surfaces," *IEEE Trans. Microwave Theory Tech.*, vol. 45, pp. 1645–1649, September 1997.

[73] J.-Y Lee and N.-H Myung, "Locally tensor conformal FDTD method for modeling arbitrary dielectric surfaces," *Microw. Opt. Tech. Lett.*, vol. 23, pp. 245–249, November 1999.

[74] A. Mohammadi and M. Agio, "Contour-path effective permittivities for the two-dimensional finite-difference time-domain method," *Opt. Express*, vol. 13, pp. 10367–10381, 2005.

[75] J. E. Inglesfield, "A method of embedding," *J. Phys. C: Solid State Phys.*, vol. 14, pp. 3795–3806, 1981.

[76] F. B. Hildebrand, *Introduction to Numerical Analysis*, New York: McGraw-Hill, 1956.

[77] C. T. Chan, Q. L. Yu, and K. M. Ho, "Order-N spectral method for electromagnetic waves," *Phys. Rev. B*, vol. 51, pp. 16635–16642, 1995.

[78] H. Holter and H. Steyskal, "Infinite phased-array analysis using FDTD periodic boundary conditions-pulse scanning in oblique directions," *IEEE Trans. Antennas Propagat.*, vol. 47, pp. 1508–1514, 1999.

[79] M. Turner and C. Christodoulou, "FDTD analysis of phased array antennas," *IEEE Trans. Antennas Propagat.*, vol. 47, pp. 661–667, 1999.

[80] D. T. Prescott and N. V. Shuley, "Extensions to the. FDTD method for the analysis of innitely periodic arrays," *IEEE Microwaves and Guided Waves Letters*, vol. 4, pp. 352–354, October 1994.

[81] J. R. Ren, O. P. Gandhi, L. R. Walker, J. Fraschilla, and C. R. Boerman, "Floquet-based FDTD analysis of two-dimensional phased array antennas," *IEEE Microwave and Guided Wave Letters*, vol. 4, pp. 109–111, 1994.

[82] J. A. Roden, S. D. Gedney, M. P. Kesler, J. G. Maloney, and P. H. Harms, "Time-domain analysis of periodic structures at oblique incidence: Orthogonal and nonorthogonal FDTD implementations," *IEEE Trans. Microwave Theory and Techniques*, vol. 46, pp. 420–427, 1998.

[83] S. Fan, P. R. Villeneuve, and J. D. Joannopoulos, "Large omnidirectional band gaps in metallodielectric photonic crystals," *Phys. Rev. B*, vol. 54, pp. 11245–11251, 1996.

[84] M. Qiu and S. He, "A nonorthogonal finite-difference time-domain method for computing the band structure of a two-dimensional photonic crystal with dielectric and metallic inclusions," *J. Appl. Phys.*, vol. 87, pp. 8268–8275, 2000.

[85] J. R. Berenger, "A perfectly matched layer for the absorption of electromagnetic waves," *J. Computat. Phys.*, vol. 114, pp. 185-200, October 1994.

[86] P. B. Johnson and R. W. Christy, "Optical constants of the noble metals," *Phys. Rev. B*, vol. 6, pp. 4370-4379, 1972.

[87] N. Giannakis, J. Inglesfield, P. Belov, Y. Zhao, and Y. Hao, "Dispersion properties of subwavelength waveguide formed by silver nanorods," *Photon06*, September 3-7, 2006, Manchester, U.K.

[88] W. Song, Y. Hao, and C. Parini, "Comparison of nonorthogonal and Yee's FDTD schemes in modelling photonic crystals," *Opt. Express*, 2006.

[89] J. B. Pendry, D. Schurig, and D. R. Smith, "Controlling electromagnetic fields," *Science* 312, 1780-1782 (2006).

[90] V. G. Veselago, "The electrodynamics of substances with simultaneously negative value of ε and μ," *Sov. Phys. Usp.* 10, 509 (1968).

[91] S. A. Cummer, B.-I. Popa, D. Schurig, and D. R. Smith, "Full-wave simulations of electromagnetic cloaking structures," *Phys. Rev. E*, 74, 036621 (2006).

[92] W. Cai, U. K. Chettiar, A. V. Kildishev, and V. M. Shalaev, "Optical cloaking with metamaterials," *Nat. Photonics* 1, 224-227 (2007).

[93] W. Cai, U. K. Chettiar, A. V. Kildishev, and V. M. Shalaev, "Nonmagnetic cloak with minimized scattering," *Appl. Phys. Lett.*, 91, 111105 (2007).

[94] U. Leonhardt, "Optical conformal mapping," *Science* 312, 1777-1780 (2006).

[95] M. Tsang and D. Psaltis, "Magnifying perfect lens and superlens design by coordinate transformation," *Phys. Rev. B*, 77, 035122 (2008).

[96] H. Chen and C. T. Chan, "Transformation media that rotate electromagnetic fields," *Appl. Phys. Lett.*, 90, 241105 (2007).

[97] M. Rahm, S. A. Cummer, D. Schurig, J. B. Pendry, and D. R. Smith, "Optical design of reflectionless complex media by finite embedded coordinate transformations," *Phys. Rev. Lett.*, 100, 063903 (2008).

[98] Y. Luo, J. Zhang, L. Ran, H. Chen, and J. A. Kong, "Controlling the emission of electromagnetic sources by coordinate transformation." *ArXiv.org:0712.3776v1*, (2007).

[99] D.-H. Kwon and D. H. Werner, "Two-dimensional eccentric elliptic electromagnetic cloaks," *Appl. Phys. Lett.* 92, 013505 (2008).

[100] S. A Cummer, and D. Schurig, "One path to acoustic cloaking," *New Jour. Physics*, 9, 45 (2007).

[101] H. Chen, and C. T. Chan, "Acoustic cloaking in three dimensions using acoustic metamaterials," *Appl. Phys. Lett.*, 91 183518 (2007).

[102] Z. Ruan, M. Yan, C. W. Neff, and M. Qiu, "Ideal cylindrical cloak: perfect but sensitive to tiny perturbations," *Phys. Rev. Lett.*, 99 113903 (2007).

[103] M. Yan, Z. Ruan, and M. Qiu, "Cylindrical invisibility cloak with simplified material parameters is inherently visible," *Phys. Rev. Lett.*, 99, 233901 (2007).

[104] G. Isic, R. Gajic, B. Novakovic, Z. V. Popovic, and K. Hingerl, "Radiation and scattering from imperfect cylindrical electromagnetic cloaks," *Opt. Express*, 16, 1413-1422 (2008).

[105] B. Zhang, H. Chen, B.-I. Wu, Y. Luo, L. Ran, and J. A. Kong, "Response of a cylindrical invisibility cloak to electromagnetic waves," *Phys. Rev. B*, 76, 121101 (2007).

[106] Y. Huang, Y. Feng, T. Jiang, "Electromagnetic cloaking by layered structure of homogeneous isotropic materials," *Opt. Express*, 15, 11133-11141 (2007).

[107] A. Greenleaf, Y. Kurylev, M. Lassas, and G. Uhlmann, "Improvement of cylindrical cloaking with the SHS lining," *Opt. Express*, 15, 12717-12734 (2007).

[108] D. A. B. Miller, "On perfect cloaking," *Opt. Express*, 14, 12457-12466 (2006).

[109] D. Schurig, J. J. Mock, B. J. Justice, S. A. Cummer, J. B. Pendry, A. F. Starr, and D. R. Smith, "Metamaterial electromagnetic cloak at microwave frequencies," *Science*, 314, 977–980 (2006).

[110] I. I. Smolyaninov, Y. J. Hung, and C. C. Davis, "Electromagnetic cloaking in the visible frequency range," *ArXiv.org:0709.2862v2*, (2007).

[111] A. Alu and N. Engheta, "Achieving transparency with plasmonic and metamaterial coatings," *Phys. Rev. E*, 72, 016623 (2005).

[112] M. G. Silveirinha, A. Alu, and N. Engheta, "Parallel-plate metamaterials for cloaking structures," *Phys. Rev. E*, 75, 036603 (2007).

[113] G. W. Milton and N. P. Nicorovici, "On the cloaking effects associated with anomalous localized resonance," *Proc. R. Soc. A*, 462, 3027–3059 (2006).

[114] D. Schurig, J. B. Pendry, and D. R. Smith, "Calculation of material properties and ray tracing in transformation media," *Opt. Express*, 14, 9794–9804 (2006).

[115] H. Chen, B.-I. Wu, B. Zhang, and J. A. Kong, "Electromagnetic wave interactions with a metamaterial cloak," *Phys. Rev. Lett*, 99, 063903 (2007).

[116] R. Weder, "A rigorous time-domain analysis of full-wave electromagnetic cloaking (invisibility)," *ArXiv.org:0704.0248v4*, (2007).

[117] K. S. Yee, "Numerical solution of initial boundary value problems involving Maxwell's equations in isotropic media," *IEEE Trans. Antennas Propgat.*, 14, 302–307 (1966).

[118] A. Taflove and S. Hagness *Computational Electrodynamics: The Finite-Difference Time-Domain Method*, 2nd ed., Norwood, MA: Artech House, 2000.

[119] R. Luebbers, F. P. Hunsberger, K. Kunz, R. Standler, and M. Schneider, "A frequency-dependent finite-difference time-domain formulation for dispersive materials," *IEEE Trans. Electromagn. Compat.*, 32, 222–227 (1990).

[120] O. P. Gandhi, B.-Q. Gao, and J.-Y. Chen, "A frequency-dependent finite-difference time-domain formulation for general dispersive media," *IEEE Trans. Microwave Theory Tech.*, 41, 658–664 (1993).

[121] D. M. Sullivan, "Frequency-dependent FDTD methods using Z transforms," *IEEE Trans. Antennas Propagat.*, 40, 1223–1230 (1992).

[122] F. B. Hildebrand, *Introduction to Numerical Analysis*, New York: McGraw-Hill, 1956.

[123] J.-Y. Lee and N.-H. Myung, "Locally tensor conformal FDTD method for modelling arbitrary dielectric surfaces," *Microw. Opt. Tech. Lett.*, 23, 245–249 (1999).

[124] J. R. Berenger, "A perfectly matched layer for the absorption of electromagnetic waves," *J. Computat. Phys.*, 114, 185 (1994).

[125] Y. Zhao, P. A. Belov, and Y. Hao, "Accurate modelling of left-handed metamaterials using a finite-difference time-domain method with spatial averaging at the boundaries," *Jour. Opt. A: Pure Appl. Opt.*, 9, 468–475 (2007).

[126] Y. Zhao and Y. Hao, "Finite-difference time-domain study of guided modes in nano-plasmonic waveguides," *IEEE Trans. Antennas Propag.*, 55, 3070–3077 (2007).

CHAPTER 11

Overviews and Final Remarks

11.1 Introduction

One of the principal objectives of research and development in metamaterials is to explore new opportunities in science and technology. To ensure that the newly gained knowledge and findings of metamaterial research in academia is transferred to the industry, it is highly desirable to perform an efficient and accurate characterization of the material properties of the medium via numerical modeling. Recent advances in computer techniques for numerical simulations have revolutionized the way in which electromagnetic problems are now analyzed and how new materials are designed for novel applications. Recent years have witnessed a steady growth of new computational techniques, enhancements to existing techniques, new implementations of existing techniques, and new applications of numerical modeling. Sorting through this wealth of information, and then choosing the numerical technique that is best suited for a particular application, can be a daunting task. It is our hope that the reader will benefit from this book, especially through the detailed discussion of the FDTD method, when dealing with the problem of designing metamaterials.

In this chapter, we will try to summarize the contents of the book and discuss future challenges both in the area of FDTD and metamaterials research [1–11].

11.2 Overview of Advantages and Disadvantages of the FDTD Method in Modeling Metamaterials

Since the FDTD method was first introduced by Yee in 1966 [1], it has been growing steadily, aided by the fact that the computing costs continue to drop with advances in digital devices and computing architectures alike. It has found successfull applications in a number of disciplines such as electromagnetics, acoustics [2], optics [3], and biology [4]. The fundamental concepts of the FDTD method are described in Chapters 2 and 3, particularly from the point of view of modeling metamaterials, to set the stage for the rest of the book.

One of the major reasons for the rapid growth of the FDTD method is that it has many strengths, some of which are summarized as follows:

- It is a very versatile modeling technique. It is also very intuitive, so it does not require an extensive amount of advanced preparation on the part of users.
- It is a time-domain technique, which yields wideband results via the use of FFT applied to the time-domain response, generated by a single simulation.

- It is a grid-based numerical technique, which enables us to model objects with arbitrary material fillings.
- It provides transient electromagnetic fields directly with a flexible choice of excitations, and the field solutions can be visualized by using computer-graphic tools.
- It can be easily parallelized for ultra-high-performance computing, because it works well on distributed processors.

In common with other numerical techniques, the FDTD method also has a few shortcomings. First of all, it requires the entire computational domain to be discretized and gridded, and the cell must be a small fraction of the smallest wavelength, as well as small enough to capture the smallest feature of the object being modeled. The former imposes a heavy computational burden, and the latter makes it difficult to model extra long and thin structures (e.g., the imaging lenses described in [6]). The exception is the case of a periodic structure, which can still be manageable, despite its fine features, when thin wire metamaterials are modeled by using the unit cell approach, which was discussed in Chapter 4. When modeling metamaterials consisting of thin wires and layers, one can either use the effective medium representation (presented in Chapters 6, 8, and 9) or the parallel FDTD code to model the physical structures as detailed in Chapter 7. Although the latter approach is computationally intensive, it provides rigorous solutions that the simplified models may fail to generate.

In addition, since conventional FDTD cells are cubical in nature, objects with curved surfaces must be staircased. For configurations with sharp edges, it may be necessary to use a sufficiently fine grid to adequately represent these edges using a staircased approximation and, consequently, to use a small time step [5]. In this respect, the finite-difference method is not as flexible as the FEM method when dealing with arbitrarily shaped object with curved surfaces and edges. In Chapter 4, we used the nonorthogonal FDTD algorithm to model EBGs consisting of an array of cylindrical metallic rods. An alternative way to solve this problem is to utilize the concept of effective permittivities (EPs) in the Cartesian coordinate system. We have combined the EP concept with the dispersive FDTD scheme, presented in Chapter 10, and we have applied it to the modeling of plasmonic waveguides.

11.3 Overview of Metamaterial Applications and Final Remarks

In the last few chapters, we have presented extensive discussions of various FDTD-based techniques for modeling metamaterials in general, and periodic structures in particular, that are typical ingredients of artificial dielectrics exhibiting metamaterial characteristics. Some of the important attributes of metamaterials, which have made them attractive for several novel applications, are described as follows:

- *Performance enhancement of small antennas using ENG.* It has been suggested that one way we can utilize an ε negative (ENG) medium, for instance, is to take a wire antenna, whose dimensions are very small compared

11.3 Overview of Metamaterial Applications and Final Remarks

to the wavelength, and enclose it with a shroud of an ENG medium. It has been argued that this will enable us to compensate for the positive reactive energy of the small-wire antenna, which is known to be a high-Q structure with a large capacitive reactance, with the reactive energy proportional to $-\varepsilon|E|^2$, where the minus sign is associated with the ENG medium. Then we can mitigate the problem of inefficiency of a small antenna, as well as the narrow bandwidth (high Q) problems associated with the same, if we can strike a good balance between the positive reactive energy of the antenna with the negative one of the ENG shroud to match it. We will discuss below the feasibility of achieving this cancelation.

- *Achieving superlensing by using DNG media that magnify evanescent waves.* Perhaps no topic has drawn more attention of metamaterial researchers as that of fabricating a superlens by using a DNG slab. The figure that has appeared more than any other in the metamaterial literature is the picture of refracting rays that originate from a point source, undergo negative refraction through the DNG slab, and then focus on the other side to form the image in a manner predicted by Vesalago back in 1968. The important question we raise is: Does the effective medium approach to characterizing a DNG-type slab, which is physically realized by employing periodic inclusions such as split rings and dipoles (see Figure 8.1) in a background medium, accurately describe the refraction of wave in the medium, or are the real-life behaviors of the propagating fields substantially different from those predicted by using the ε_{eff} and μ_{eff} parameters to describe the DNG medium?

- *Performance enhancement of microstrip patch antennas using a metamaterial superstrate.* A large number of recent publications have proposed performance enhancement of conformal antennas [e.g., microstrip patch antennas (MPAs)], by covering them with a DNG superstrate, whose function is to take the energy emanating from the MPA and focus it at infinity (see Figure 11.1).

The question we ask is: Should we restrict ourselves solely to the DNG-type superstrates, so that we can utilize their lensing properties, or should

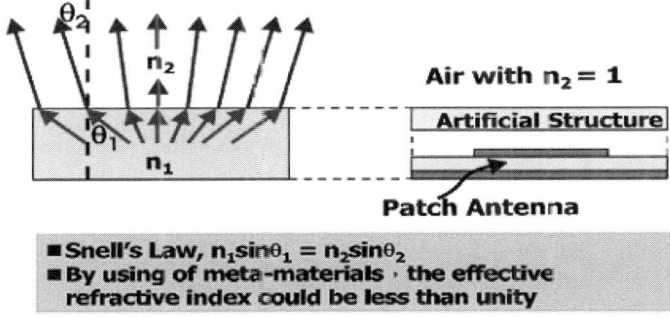

Figure 11.1 Concept of the high-gain patch antenna achieved by the use of a metamaterial superstrate.

we also explore some other types of media, such as an ENG or a DPS? Since we know from effective medium theory that a medium with a negative ε_{eff} and positive μ_{eff}, or vice versa, would have an imaginary $n(=\sqrt{\varepsilon_{eff}\mu_{eff}})$, we may intuitively argue that we should not be using a medium with an imaginary "n" because the field radiated by the antenna will not penetrate the superstrate. Nor should we consider planar and homogeneous DPS superstrates, which have no focusing properties. So the question we may ask is: Is our best approach to designing the superstrates to go directly to a DNG medium and summarily discard other types such as ENG and DPS media right from the start?

- *Electromagnetic cloaks.* Another topic that is currently drawing considerable attention from the metamaterial community is "cloaking." The objective of cloaking is to coat a target with metamaterials such that it becomes invisible to the interrogating wave. The issue here is the realizability of the desired profile for the effective ε and μ that would provide the designed shielding characteristics. This is especially true when we wish the cloak to be effective for arbitrary polarizations and angles of incidence and to be effective over a wide bandwidth.

In the following four sections we will discuss the questions raised above in connection with the four applications we have listed. A common thread that runs through the discussion presented below is that we will seek the answers to the questions, posed above, by examining the field solutions of the problems at hand that we generate via rigorous numerical simulations. We follow this strategy to examine the validity of the effective medium theories, and of the plethora of interesting characteristics of the metamaterials predicted by the above theories that form the underlying foundations for the design of a variety of devices, such as lenses and highly directive antennas. Of course, our ultimate goal is to make a strong case to the reader that rigorous numerical analysis is the only reliable way to predict the performance of metamaterial-based devices and that all designs based on effective medium theories must be carefully examined using rigorous simulations before devices incorporating these designs are fabricated. Before proceeding with the proposed examination of the effective medium theories, we would like to refresh the memory of the reader and recall how the effective medium parameters are typically determined. Towards this end, we refer the reader to Chapters 6, 8, and 9, where this topic has been discussed in great detail.

Recall that the first step is to interrogate the metamaterial slab with a normally incident plane wave and compute or measure the reflection and transmission coefficients, or, equivalently, the S-parameters. We then back out the refractive index "n" and follow this by extracting the effective material parameters utilizing (7.1)–(7.8). From then on, the typical procedure followed by most researchers in the field is to use the effective medium descriptions of the homogenized version of the original periodic structure and to predict the performance of the devices that utilize the metamaterials. We add, parenthetically, that we have done the same in this book on numerous occasions, namely made use of the effective medium approach, as is undoubtedly evident to the reader who has gone through the previous chapters. Whether or not this approach yields reliable results that we can trust is the key question we wish to address in the following.

11.3.1 Small Antennas Enclosed by an ENG Shell

It is in [7] that by using an ε_r negative (ENG) shell, the matching of a small dipole antenna can be significantly improved. For instance, the improvement in gain compared with that of the original unmatched antenna can be over 50 dB. Figure 11.2 shows the antenna configuration simulated using Ansoft HFSS.

The radius of the dipole is 2.5 mm, and the total height is 10 mm. The inner and outer radii of the ENG shell are 10 mm and 19.51 mm, respectively. The dipole antenna enclosed by the ENG shell is designed to have a resonant frequency around 350 MHz. The material properties of the shell are assumed to be $\varepsilon_r = -3$ and $\mu_r = 1$. A small loss is also assumed, for the convergence of the simulations, by setting $\tan\delta = 0.001$. Since an ENG shell would have negative stored energy (electric), such a shell would significantly improve the matching of the dipole antenna. The simulated return loss plotted in Figure 11.2(b) comfirms this. We see that the gain of the dipole encased in an ENG shell has a dramatic improvement, by about 50 dB compared with the dipole antenna without the shell. In the above simulations we have used a small amount of losses $\tan\delta = 0.001$. However for practical applications, such small losses may not be achievable, and therefore it is important to investigate the effect of large losses on antenna performance. Chapter 7 models the real physical structure of resonant particles based on the unit cell approach and followed (7.1)–(7.8) to extract the material properties. If we neglect both dielectric and conductive losses, a loss tangent of 0.01 and fractional bandwidth of 1.2% can be achieved for a metamaterial made of particles with their electrical sizes close to 0.06 wavelengths. Therefore, we have resimulated the dipole enclosed in an ENG shell, using more realistic material parameters, for instance, $\tan\delta = 0.01$ and 0.1. The calculated return losses for these two cases are plotted in Figure 11.3. It can be seen that the matching of the antenna becomes worse due to the change of material properties of the ENG shell. A comparison of antenna Q with the Chu-Harrington limit is shown in Figure 11.3(b) which demonstrates the significance of losses in metamaterials and their impact in antenna performance.

In order to compare with the ENG shell enclosed dipole antenna, we have also simulated a small biconical antenna with similar effective dimensions of the ENG dipole; the length of one arm is 20 mm. The geometry of antenna is shown in Figure 11.4(a). Such a small antenna has very high impedance and the resonance of the antenna is supposed to be at a very high frequency. We obtain the antenna input impedance at 350 MHz from simulations (i.e., $Z = 0.26 - j242.84$) and normalize the impedance in order to compare with the dipole antenna enclosed in the ENG shell. The return loss of the antenna using normalized impedance (equivalent to impedance matching) is plotted in Figure 11.4(b) and the gain in Figure 11.4(c).

Despite the difficulties in matching the antenna in practice, the gain of the small biconical antenna is only -30 dB, which is 25 dB lower than the ENG dipole antenna with $\tan\delta = 0.001$. However, with a loss tangent of $\tan\delta = 0.1$, the metamaterial-loaded antenna demonstrates no better performance (gain -28.9 dB

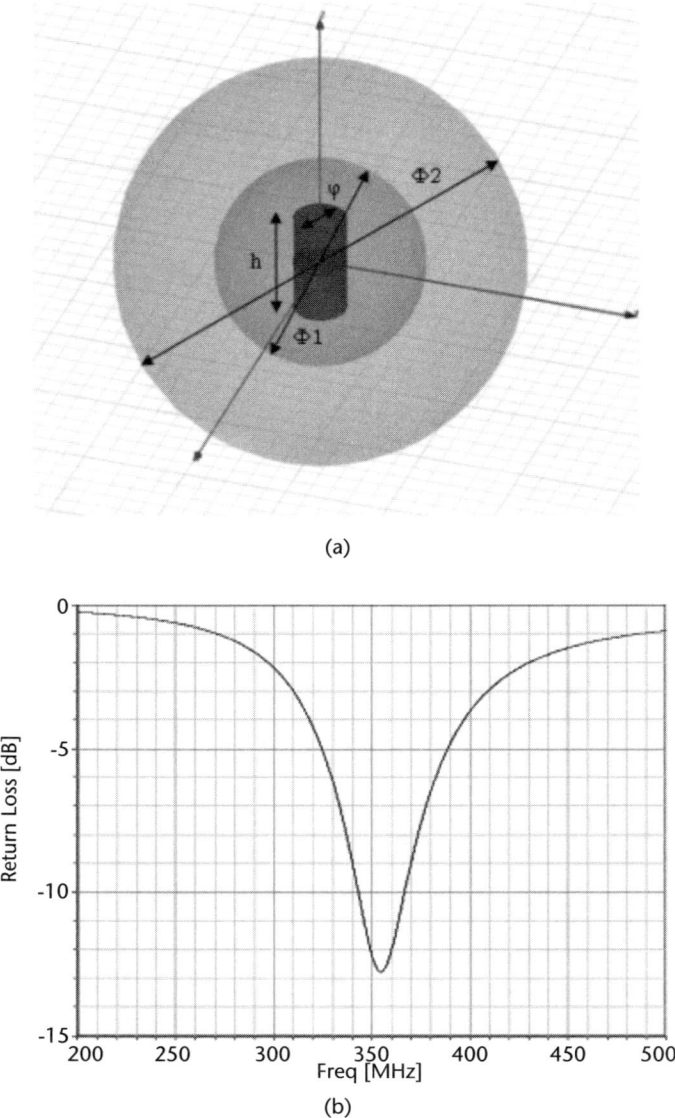

Figure 11.2 (a) Geometry of a dipole enclosed by an ENG shell with $\varepsilon_r = -3$ and $\mu_r = 1$. The dimensions are h = 10 mm, Φ = 5 mm, Φ_1 = 10 mm, Φ_2 = 19.51 mm, and λ = 845.1 mm. (b) The return loss of the ENG dipole.

with $\tan\delta = 0.1$) than the conventional biconical antenna. This indicates that the advantage of using metamaterial loading for small antennas will only be realized if the metamaterial possesses extremely low losses and sufficient bandwidth. It should be evident, therefore, that accurate parameter extraction is needed by modeling physical structures before the effective medium approach is adopted.

We might think, at this point, that we can solve the problem at hand simply by synthesizing an ENG medium with low losses, but this would both be misleading and incorrect for two reasons. First, if we use a slab of wire medium, such as

11.3 Overview of Metamaterial Applications and Final Remarks

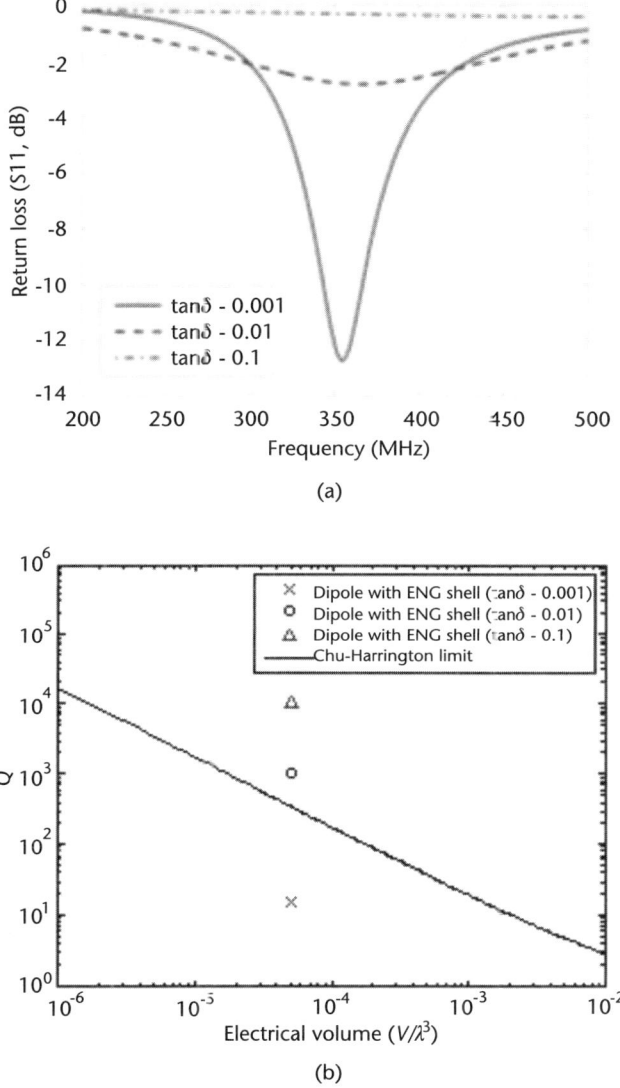

Figure 11.3 (a) Comparison of return loss (S_{11}, in dB) of a dipole antenna with the ENG shell with different amount of losses tan δ = 0.01 and tan δ = 0.1. (b) Comparison of antenna Q for dipole enclosed within ENG shell with different losses with the theoretical Chu-Harrington limit.

the one shown in Chapter 8, to realize an ε negative medium, on the basis of the reflection and transmission characteristics of the slab when illuminated by a normally incident plane wave, we would not necessarily have the same behavior when we use a spherical shell configuration, which is obviously nonplanar and whose radius is small, and therefore, which can only accommodate small wire segments. The effective medium approach would not be valid unless we could demonstrate that such an ENG shell responds equally well for all angles of incidence, since the dipole generates a spectrum of waves, encompassing both visible and invisible ranges.

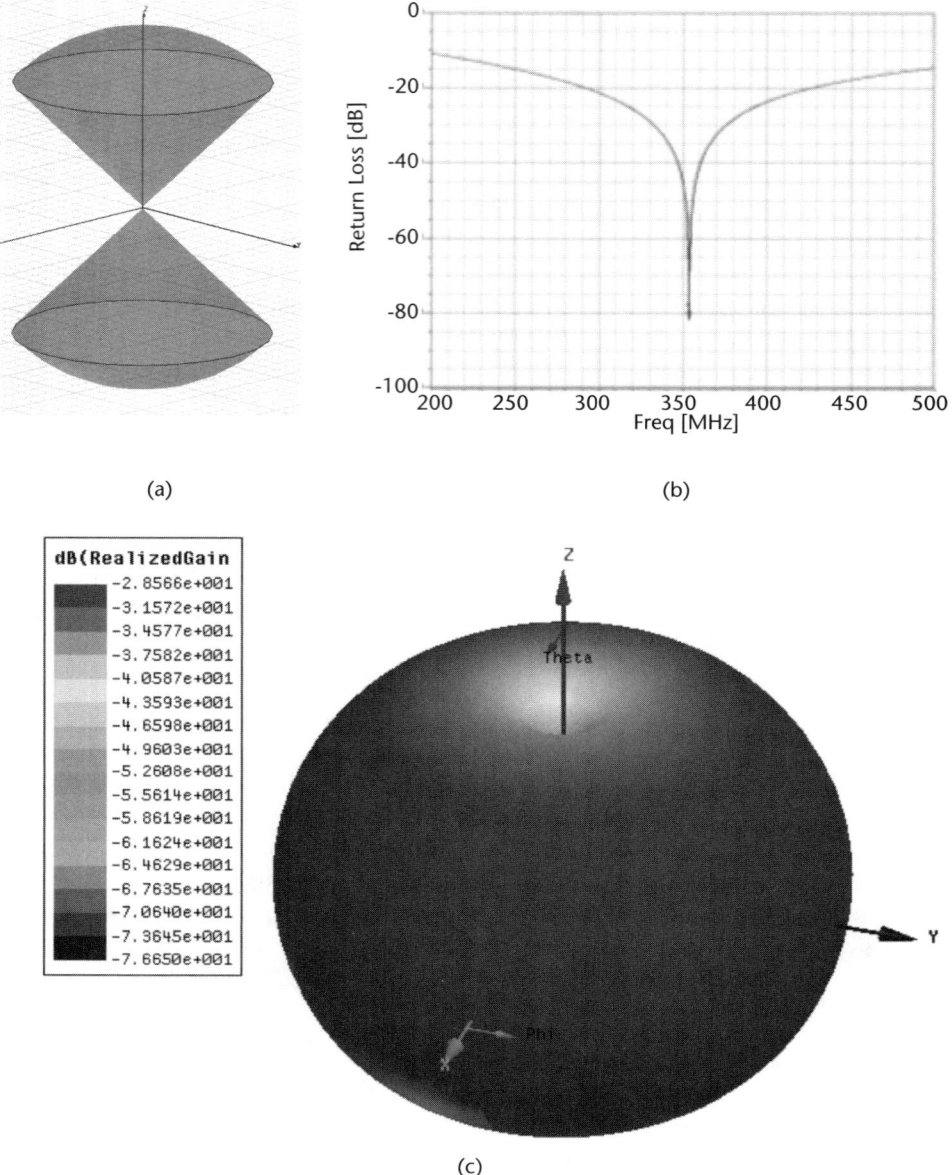

Figure 11.4 (a) The geometry of a biconical antenna. (b) The return loss for the biconical antenna. (c) The gain pattern for the biconical antenna.

The second issue is even more serious. Let us say we rigorously analyze the shell structure with embedded wires and find that on the basis of its S-parameter characteristics its effective ε is negative. This, in turn, would imply that the electric energy stored in this medium would be negative. We could then go on to argue that we could cancel the positive reactive energy of the dipole by covering it with the ENG shell, and thereby realize an impedance match for the antenna. However, if we rigorously analyze the shell configuration by computing the electric field inside the shell region, and then use it to evaluate the energy, we will find that the stored electric energy is, in fact, positive in contradiction to what was predicted by the effective medium approach. The reason we maintain that the stored energy must be positive is that the true ε of the medium inside the shell, in which the wires are embedded, is obviously positive and, hence, $\varepsilon|E|^2$ must be positive everywhere in the shell region. (Note that the E-field is identically zero inside the wires, and the part of the volume within the shell that is occupied by the wires does not contribute to the stored energy.)

For this reason, a dipole enclosed by a shell with embedded wires with small lengths would neither achieve the desired matching, nor the directivity enhancement, that we expected the ENG metamaterial shell to deliver.

11.3.2 Focusing and Superlensing Effects

The cornerstone of the focusing phenomenon associated with DNG slab is the negative refraction, in the absence of which we would never be able to achieve focusing with a *planar* homogeneous slab. So before proceeding very far, we should first take a close and scrutinizing look at the negative refraction phenomenon itself.

The negative refraction phenomenon is based on the application of Snell's law to the medium under consideration. The basis for this law is relatively straightforward because it just requires us to match the projection of the $\vec{k}$ vectors in the half-space problem, shown in the figure above, along the interface of the two media. Obviously, this would immediately lead us to insist that the ray incident from the left upon the interface from the medium above (free space in the picture) at an oblique angle must bend toward the left, in order to satisfy the well-known Snell's Law, assuming, of course that the half space below is "homogeneous." Figure 11.5(b) also implies that the Poynting vector associated with the ray follows its direction (i.e., is directed toward the left of the normal to the surface) for this incident angle.

At this point it is rather important to recognize that, to date, no homogeneous medium with DNG characteristics has been discovered and that the only media

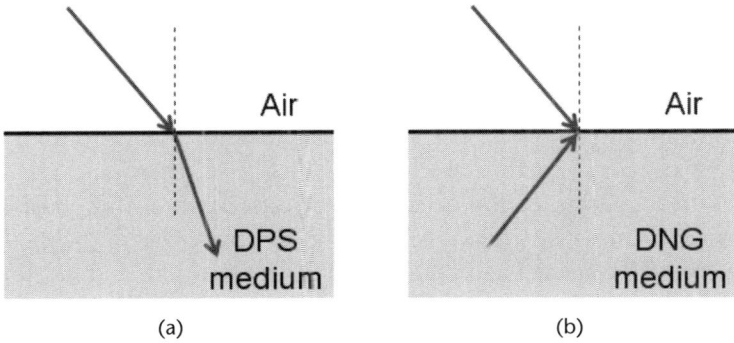

Figure 11.5 (a) Positive and (b) negative refraction phenomena.

that have been claimed to be DNG are *periodic* structures—artificially synthesized at microwaves or naturally occurring in crystalline structures at optics.

Thus it behooves us to ask the question whether the propagation characteristics within a periodic structure (doubly periodic along the transverse directions and truncated-periodic in the longitudinal direction) display the same behavior as that depicted by the ray picture in Figure 11.5(b).

Perhaps the most widely used DNG material at microwaves is the SRR and wire combination, which we have discussed previously on numerous occasions (see Chapter 6, for instance). In order to mimic a ray, let us excite a slab of the above DNG structure with a Gaussian beam incident from above.

We begin by establishing the DNG behavior of the slab by interrogating it with a plane wave (at normal incidence) and computing the S-parameters. We then back out the refractive index n as well as the effective parameters ε and μ, and plot them in Figures 11.6 and 11.7 as functions of frequency. It is evident that the medium indeed displays DNG characteristics for this normal incidence case, albeit in a narrow frequency range around 15.3 GHz.

We now illuminate the slab with a Gaussian beam and follow it as it propagates through the slab and eventually exits to free space on the other side of the slab in the region below. A careful examination of the field distribution, both within the slab as well as in the air region outside, reveals that the behavior of the field is highly granular, and not as "clean" as the ray picture would suggest. More importantly, we find that a dominant "backward" wave is excited within the slab, though there are other harmonics besides the β_{-1} (backward Floquet mode), namely β_0 and β_1 whose presence can also be detected in the field distribution. In light of this, it is important to recognize that the ray picture, which implies refraction in a "single" direction does not adequately describe the wave propagation in the periodic structure. What is even more important is the fact that the propagation inside the artificial dielectric slab, which is comprised of periodic inclusions, can be highly dependent upon the direction of propagation. For instance, referring to the SRR-type DNG structure in Figure 8.1, we notice immediately that the propagation characteristics in this medium are bound to be very different in the x-direction (transverse) when compared to that in the z-direction (longitudinal). This is because the H-field component of the incident wave interacts with the loops when the wave is normally incident, since it is orthogonal to the plane of the loop, but the interaction would be minimal when the incident angles is close to 90°. Since the propagation inside the periodic structure must satisfy the phase-matching condition at the interface, the behavior of the wave number in the transverse direction plays a key role, since it must exactly match the projected wave number of the incoming wave in the transverse direction in order to satisfy the continuity of the fields at the interface. Thus, if and only if the slab supports a backward wave in the "transverse" direction, then, and only then, the energy can be propagating in the negative-x-direction while the phase propagates in the opposite (positive-x) direction, to be in synchronization with the incident wave. Since the SRRs in the example geometry we are discussing do not interact with the H-field of the incident wave when the propagation is in the x-direction, we do not expect a backward wave to be excited that has a component in that direction.

11.3 Overview of Metamaterial Applications and Final Remarks

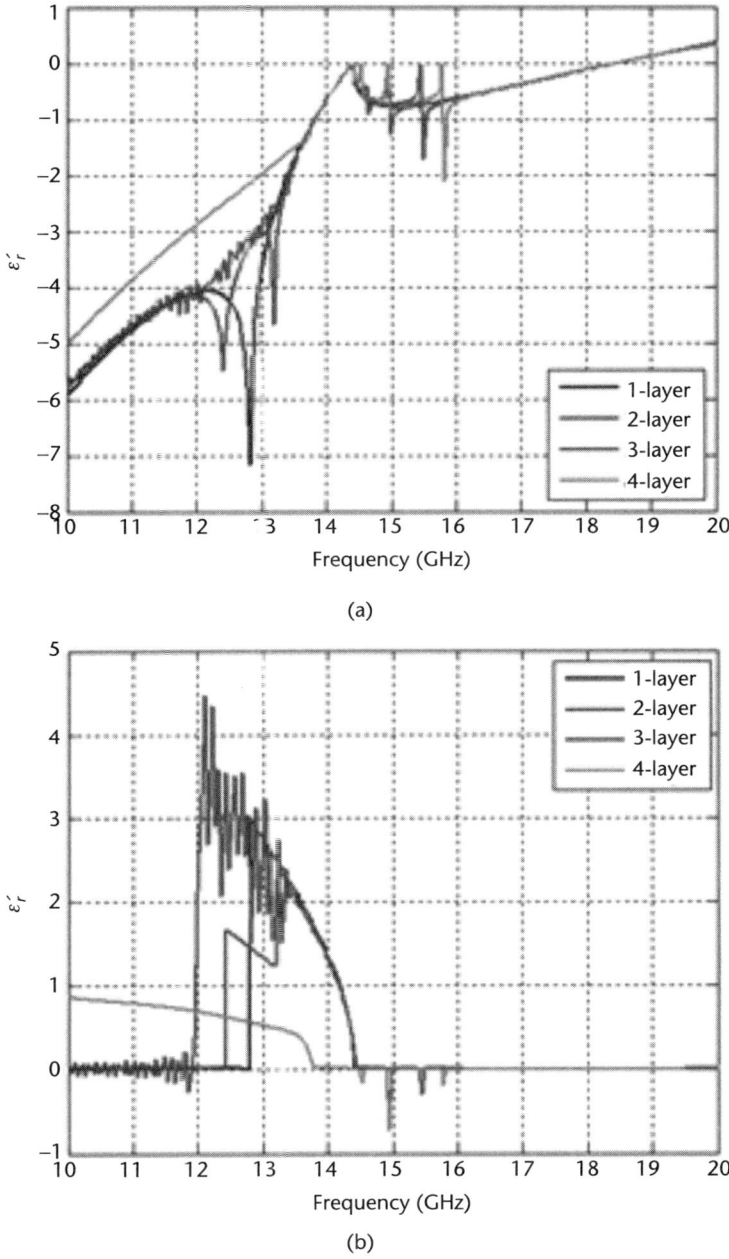

Figure 11.6 The effective electric permittivity extracted for the (SRR+wire) metamaterial slab, starting from one and up to four layers: (a) real and (b) imaginary parts.

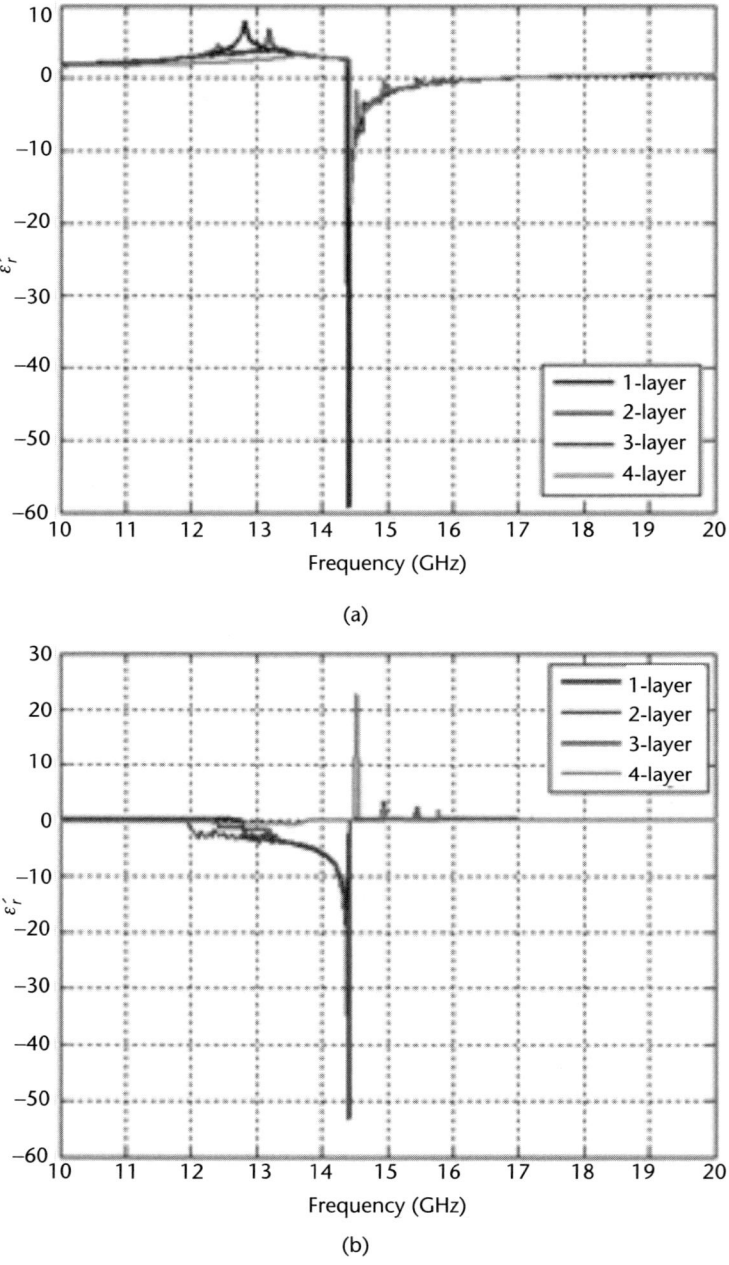

Figure 11.7 The effective magnetic permeability extracted for the (SRR+wire) metamaterial slab, starting from one and up to four layers: (a) real and (b) imaginary parts.

11.3 Overview of Metamaterial Applications and Final Remarks

And yet, in the homogeneous and isotropic type of effective medium model, the excitation of a backward wave in the longitudinal direction also implies a similar type of propagation characteristic in the transverse direction. This serves to explain the difference between the effective medium and physical models, insofar as the field propagation inside the slab is concerned, especially because there is a fundamental and phenomenological difference between the refraction in a homogeneous medium and Floquet-mode diffraction in a periodic structure. It is also important to reiterate the fact that almost always the extraction of the effective parameters is based on the reflection and transmission properties at a "normal" incidence, and that it is a common practice to predict the behavior of the slab on the basis of these parameters for "arbitrary" angles of incidence. This is the case, as for instance, when the illuminating field is a spectrum of plane waves radiated by a small dipole (source in a focusing problem), or emanates from a patch antenna (metamaterial enhancement of antenna problem).

We now turn to the prism problem, shown in Figure 5.42 (also discussed in Chapter 5). The "refraction" in these structures, which again comprise SRRs and wire segments, has been touted as evidence of negative refraction in a DNG material. However, a study of the wave propagation inside the prism reveals, once again, the presence of Floquet harmonics that determine the field propagation in the interior of the prism, and it again confirms that there is no "clean" refracted field as suggested by the ray propagation picture in the context of a homogenized and isotropic effective medium that replace the quasiperiodic structure. Thus, what is typically interpreted as the negative refractive phenomenon is nothing but the propagation of one of the several Floquet harmonics excited in the prism, which contains a periodic array of inclusions (truncated, of course). This serves to explain the results of a recent experimental study of such a prism [8] in which it was reported that the level of the so-called negatively refracted wave was down by about 10 dB below what was expected, and this low level could not be accounted for by adding up all the conduction losses. Consequently, the gap of about 30% of the power in the loss budget could not be explained by using the effective medium approach. Of course, we would have no difficulty in clearing up this discrepancy if we were to add up the powers of all the Floquet harmonics, rather than look at just the "refracted" field alone, which is really associated with just one of the harmonics.

Next, let us turn to the focusing problem in which we study the behavior of the fields propagating through a DNG slab, whose geometry the same as in Figure 8.1. The simulation result of the physical structure are presented in Figures 11.8 and 11.9.

By examining these results, we can come up with the following observations:

1. The field propagation within the slab is governed, once again, by the Floquet harmonics corresponding to the longitudinal direction. (Note that this information is completely lost when we homogenize the problem and replace it by an effective medium, and that the same is also true for the anisotropic property of the medium, as we have pointed out before.)
2. There is no "focusing" in the longitudinal distinction, but only a monotonic decay of the amplitude as we move away from the exit plane into the image

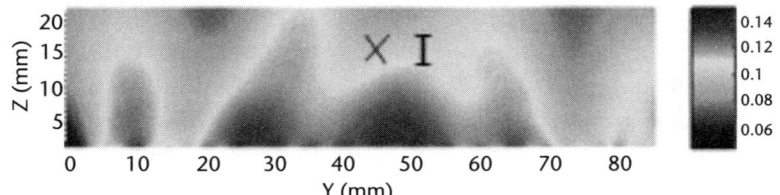

Figure 11.8 The magnitude of Ey at 15.3 GHz on the YZ plane (E-plane) in the free-space region when the (SRR + wire) metamaterial slab is excited by a small dipole on the other side. The expected image position predicted by effective medium theory is marked as I.

region. There is some focusing type of behavior in the transverse direction in the expected "image" plane, but by no means it is a "clean" image. Diffractions of other Floquet harmonics also contribute and introduce a multitude of sidelobes.

3. There is no crossing of rays inside the slab as predicted by the ray picture (see Figure 11.10) for the DNG slab. However, the presence of backward waves excited within the slab can be used to explain the phase behavior in the exit aperture and the focusing type of characteristics they engender in the output region.

These observations lead us to conclude that the behavior of the field propagation is far different in this DNG slab than that predicted by the effective medium theory, and the same is true for its focusing characteristics. Once again, we have run into this conundrum because of the oversimplification introduced in the use of the effective medium approach, which ignores the physics associated with the periodic nature of the medium, and, consequently, the behaviors of the Floquet harmonics associated with such a medium. At this point it is worthwhile for us to take a look at some of the geometries of DNG materials that are purported to exhibit negative refraction. These include the crystalline structures investigated

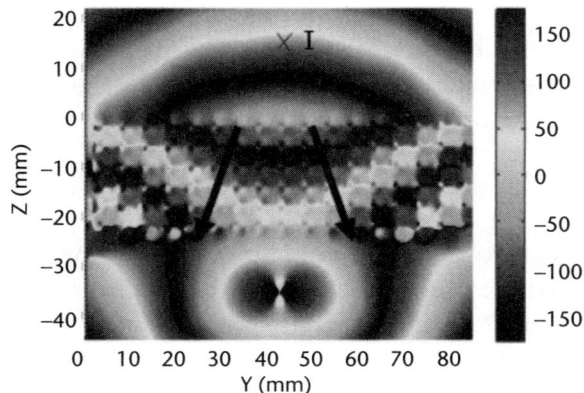

Figure 11.9 The phase of Ey at 15.3 GHz on the YZ plane (E-plane) when the (SRR + wire) metamaterial slab, located between $z = -23.75$ mm and 0 mm, is excited by a small dipole.

11.3 Overview of Metamaterial Applications and Final Remarks 367

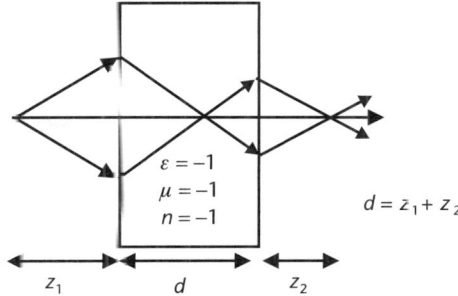

Figure 11.10 Superlensing effect by a DNG slab.

by [9], and the so-called mushroom configuration studied by a number of workers. We notice right away that, unlike the (SRR + dipole) medium, the mushroom structure does have the same type of propagation characteristics in the transverse and longitudinal distinctions by virtue of its geometry, and, hence, it is able to support backward waves both in the transverse and longitudinal directions. Thus this structure should be able to support negative-refraction-like behavior, although once again, it is really the propagation of the Floquet harmonics associated with the structure that accurately describes the nature of the field propagation. This can be verified by referring to the field propagation inside the mushroom region which is flanked by parallel plates at both ends (see Figure 11.11). The field plots, presented in Figures 11.12 and 11.13 show, once again, the presence of Floquet harmonics, in one of which a backward wave field is dominant in a certain frequency range. It shows backward propagation, both in the longitudinal and transverse directions, and hence a "negative-refraction-type" of behavior. However, a closer examination reveals that the Floquet harmonic behavior is not identical to that of negative refraction and that there are several subtle but important differences between the two.

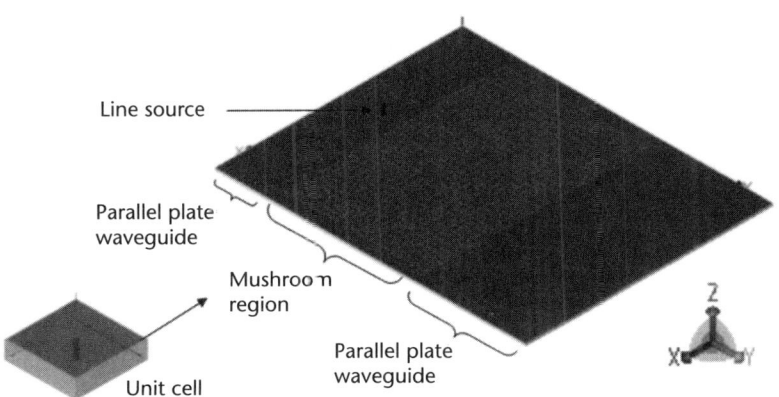

Figure 11.11 A 2D superlens based on mushroom structures.

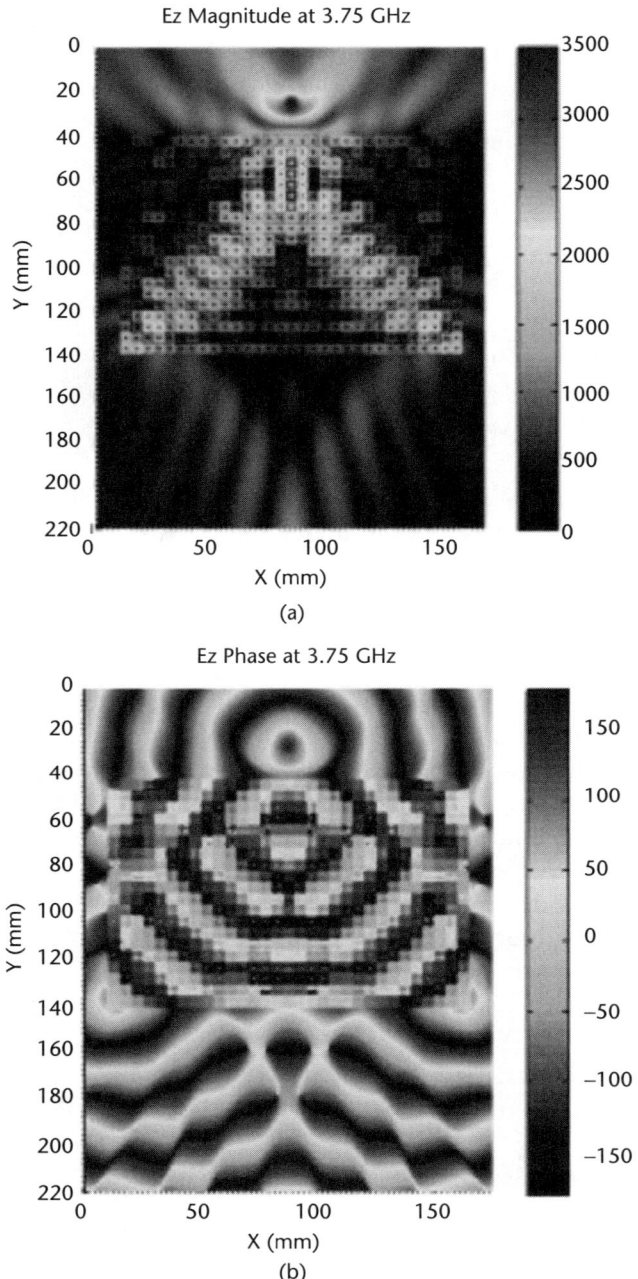

Figure 11.12 (a) Magnitude and (b) phase of Ez measured just above the ground plane at 3.75 GHz.

11.3 Overview of Metamaterial Applications and Final Remarks

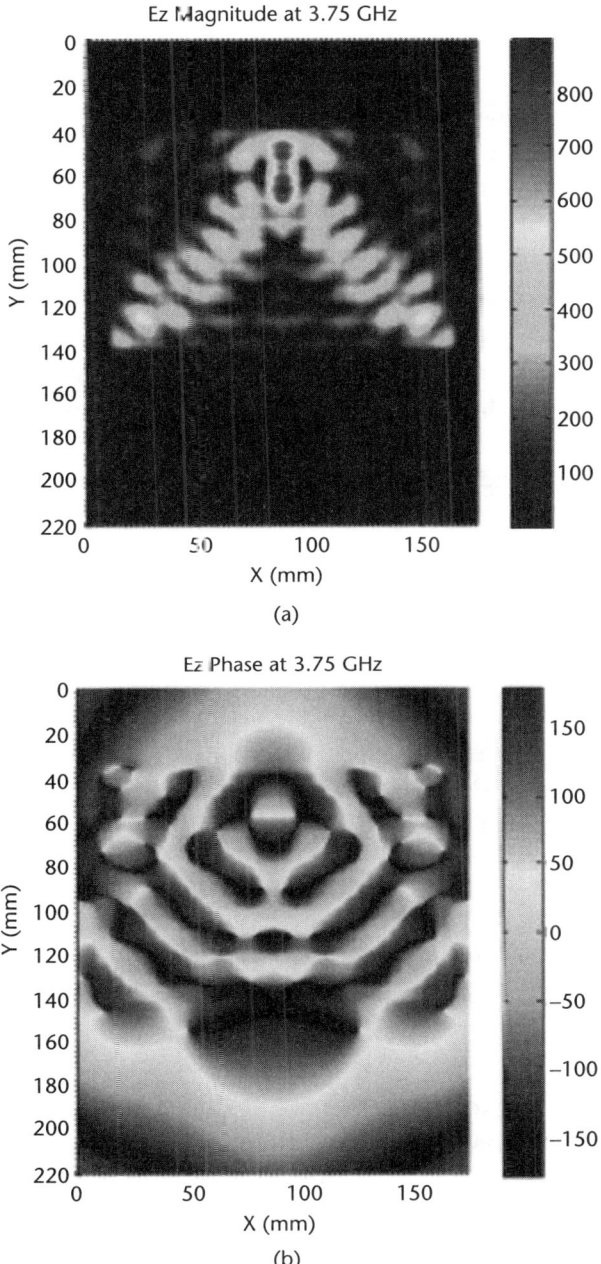

Figure 11.13 (a) Magnitude and (b) phase of Ez measured 3.5 mm above the top of mushroom at 3.75 GHz.

The "focusing" behavior can again be attributed to the backward wave launched inside the slab which, in turn, creates a phase lead as we move away from the center in the exit aperture, as opposed to a phase lag in a diverging wave. However, the amplitude distribution is considerably different from that of a converging spherical wave. Furthermore, the phase behavior shows the presence of other Floquet harmonics, which contribute substantially in the output region and manifest themselves in the way of sidelobes in that region that flanks the main beam.

11.3.3 Performance Enhancement of Planar Antennas

Let us now move to the next topic in our list, namely the performance enhancement of planar antennas by using a metamaterial superstrate. As mentioned earlier, most workers in the field just go directly to designing a DNG superstrate for the microstrip patch, arguing that this would provide the focusing of the rays emanating from the antenna, and they stay clear of either the DPS or the ENG superstrate, because the former would not have any focusing properties, and the latter would block the transmission through it. However, it has been adequately demonstrated that, in general, superstrate designs utilizing the DPS and/or ENG superstrates perform considerably better, insofar as the directivity enhancement of a microstrip patch antenna is concerned. This is because the directivity enhancement is achieved not by focusing, as pictured in Figure 11.1, but by utilizing the characteristics of the Fabry-Perot resonator that is formed by the ground plane underneath and the superstrate above, which acts like a partially reflecting surface (PRS), with the antenna playing the role of an exciter of the Fabry-Perot cavity. Since the superstrate is a PRS, and not a Vesalago lens type of slab, it works fine when DPS or ENG slabs are used for the cover. In any case, the effective medium approach, though it simplifies the analysis, leads to less-than-optimal designs. Thus the DNG-based designs cannot compete with the DPS and ENG type of designs that are eliminated from consideration at the outset, when one follows, exclusively, the simplified effective medium type of approach to design the patch antenna/metamaterial composites.

11.3.4 Electromagnetic Cloaks

The prospect of designing a shroud or a cloak using metamaterials is intriguing, and the physics of cloaking is usually based on bending of the rays through the cloak in a way such that they do not "see" the obstacle being shrouded. However, the task of synthesizing a radially varying dielectric is not at all straightforward, and once again we might resort to the artificial medium approach to describing the shroud. As long as we do not have anisotropic inclusions that preclude the possibility of describing the medium with a homogenized and isotropic medium, we may be able to describe the cloak by using such a representation.

However, the main sticking point in this type of cloak is not only the frequency dependence of the performance of the cloak (reduction of RCS), but also similar dependence on the angle of incidence and polarization of the incident wave. This, in turn, can limit the performance severely in a practical situation in which the enemy radar cannot be expected to cooperate to provide us the best-case scenario that suits us.

11.3 Overview of Metamaterial Applications and Final Remarks

The point of this discussion, however, is that whatever physical structure is chosen for the shroud, we must simulate that configuration before we can trust the results. To be sure this is not an easy task, particularly in the context of ray techniques, because no robust asymptotic methods exist for simulating PEC structures covered by inhomogeneous dielectrics. Thus we must resort to rigorous numerical simulations, using, say, the FDTD, to design the cloak and to see if we can develop a "magic" shroud that is insensitive to the variations in the frequency over a wide band, as well as the angle of incidence and polarization, over a wide frequency range, using real structures but not their effective media representations.

Before closing, we mention that the rigorous simulations, mentioned herein, have been reported in [10], and that they have been generated by using the conformal and parallelized FDTD solver GEMS [11], which is capable of handling problems with upward of $10E + 9$ unknowns, often required to simulate the problems discussed in this chapter. Interested readers are referred to [10, 11] for further details.

References

[1] K. S. Yee, "Numerical solution of initial boundary value problems involving Maxwell's equations in isotropic media," *IEEE Trans. on Antennas and Propagations*, vol. 14, no. 3, pp. 302–307, 1966.

[2] W. C. Chew and Q. H. Liu, "Using perfectly matched layers for elastodynamics," *IEEE Antennas and Propagat. Soc. Int. Symp.*, vol. 1, pp. 366–369, Baltimore, MD, July 1996.

[3] J. Yamauchi, H. Kanbara, and H. Nakano, "Analysis of optical waveguides with high-reflection coatings using the FD-TD method," *IEEE Photonics Technol. Lett.*, vol. 10, no. 1, pp. 111–113, January 1998.

[4] D. M. Sullivan, O. P. Gandhi, and A. Taflove, "Use of the finite-difference time-domain method in calculating EM absorption in man models," *IEEE Trans. Biomed. Eng.*, vol. 35, no. 3, pp. 179–186, 1988.

[5] R. Holland, "Pitfalls of staircase meshing," *IEEE Trans. Electromagn. Compat.*, vol. 35, no. 4, pp. 434–439, 1993.

[6] P. A. Belov, Y. Hao, and S. Sudhakaran, "Subwavelength microwave imaging using an array of parallel conducting wires as a lens," *Phys. Rev. B*, vol. 73, p. 033108, 2006.

[7] R. W. Ziolkowski and A. Erentok, "Metamaterial-based efficient electrically small antennas," *IEEE Trans. on Antennas and Propagation*, vol. 54, no. 7, pp. 2113–2130, 2006.

[8] C. Brewitt-Taylor, "Theoretical and experimental studies of metamaterials for radar absorption and for low-profile antennas," *Loughborough Antennas & Propagation Conference*, Loughborough, U.K., March 2008.

[9] H. Kosaka, T. Kawashima, A. Tomita, M. Notomi, T. Tamamura, T. Sato, and S. Kawakami, "Superprism phenomena in photonic crystals," *Phys. Rev. B*, vol. 58, no. 16, pp. 10096R–10098R, October 1998.

[10] L.-C. Ma, "Implementation of parallelized finite difference time domain (FDTD) algorithm and its application to the modeling of metamaterials," Ph.D. dissertation, Pennsylvania State Univ., University Park, PA, 2008.

[11] GEMS: http://www.2comu.com/.

List of Abbreviations

2-D	two-dimensional
3-D	three-dimensional
ABC	absorbing boundary condition
ADE	auxiliary differential equation
ADI	alternating direction implicit (method)
BW	backward wave
CAD	computer-aided design
CLWM	capacitively loaded wire medium
CPU	central processing unit
CST	computer simulation technology
DNG	double negative (material)
EBG	electromagnetic bandgap
EM	electromagnetic
FDM	finite difference method
FDPM	frequency-domain prony method
FDTD	finite-difference time-domain
FEM	finite element method
FFT	fast Fourier transform
FSS	frequency-selective surface
HFSS	high-frequency structure simulator
KKR	Korringa-Kohn-Rostoker
LD-NFDTD	local distorted nonorthogonal FDTD
LHM	left-handed metamaterial
MD-EBG	metallo-dielectric electromagnetic bandgap
MoM	method of moments
NFDTD	nonorthogonal FDTD
PC	photonic crystal
PBC	periodic boundary condition
PDE	partial differential equation
PEC	perfect electric conductor
PLRC	piecewise linear recursive convolution
PMC	perfect magnetic conductor
PML	perfectly matched layer
PS	point source
PWE	plane wave expansion
PSTD	pseudospectral time domain
RC	recursive convolution
SPP	surface plasmon polariton
SRR	split-ring resonator

TE	transverse electric
TEM	transverse electromagnetic
TM	transverse magnetic
TMM	transfer-matrix method
UPML	uniaxial perfectly matched layer
UC-EBG	uniplanar compact electromagnetic bandgap
WM	wire medium

About the Authors

Yang Hao received a Ph.D. from the Centre for Communications Research (CCR) at the University of Bristol, United Kingdom, in 1998. From 1998 to 2000, he was a postdoctorate research fellow with the School of Electrical and Electronic Engineering, University of Birmingham, United Kingdom. In 2000, he joined the Antenna Engineering Group, Queen Mary College, University of London, London, United Kingdom, first as a lecturer and then as a reader in 2005 and as a professor in 2007. Professor Hao is active in a number of areas, including computational electromagnetics, electromagnetic bandgap structures and microwave metamaterials, antennas and radio propagation for body-centric wireless networks, active antennas for millimeter/submillimeter applications, and photonic integrated antennas. He is a coeditor of *Antennas and Radio Propagation for Body-Centric Wireless Communications* (Artech House, 2006).

Professor Hao is an associate editor for *IEEE Antennas and Wireless Propagation Letters* and a guest editor and an associate editor for *IEEE Transactions on Antennas and Propagation*. He is a Senior Member of the IEEE and a member of the Technical Advisory Panel of the IET Antennas and Propagation Professional Network. He is also a member of the IEEE AP-S New Technology Directions Committee and was elected as a Fellow of ERA Foundation in 2007. He has served as an invited and keynote speaker, a conference general chair, a session chair, and a short course organizer at many international conferences.

Raj Mittra is a professor in the Electrical Engineering Department at Pennsylvania State University. He is also the director of the Electromagnetic Communication Laboratory, which is affiliated with the Communication and Space Sciences Laboratory of the EE Department. Prior to joining Penn State, he was a professor in electrical and computer engineering at the University of Illinois at Urbana–Champaign. He is a Life Fellow of the IEEE and a past president of the AP-S, and he has served as the editor of the *Transactions of the Antennas and Propagation Society*. Professor Mittra has been awarded the Guggenheim Fellowship, the IEEE Centennial and Millennium Medals, the IEEE/AP-S Distinguished Achievement Award, the AP-S Chen-To Tai Distinguished Educator Award, and the Electromagnetics Award of the IEEE. He has about 1,000 publications to his credit that include 31 books or book chapters on electromagnetics, antennas, microwaves, and electronic packaging. He has supervised the completion of 100 Ph.D. theses and an equal number of M.S. theses, and has mentored over 50 postdoctoral candidates.

Professor Mittra is also the president of RM Associates, a consulting organization that provides services to industrial and governmental organizations, both in the United States and abroad.

Index

A

Absorbing boundary condition, 15, 78, 81, 160
ADI-FDTD, 40
Antenna array, 2, 13, 40, 140, 141
Antenna patterns, 120
Artificial dielectrics, 2, 4, 11, 354
Artificial magnetic materials, 7, 8
Auxiliary differential equation (ADE), 17, 40, 147, 156, 242, 332

B

Backward waves, 2, 3, 4, 9, 123, 307, 366
Bianisotropy, 8
Bloch state, 28, 29
Bloch-Floquet method, 16
Bloch's Theorem, 25, 27, 28, 32
Branch line coupler, 130, 132
Brillouin Zones, 25
Broadband Balun, 137

C

Capacitively loaded wire medium, 59, 309
Cartesian coordinate, 17, 69, 76, 78, 80, 322, 354
CFL: Courant-Friedrich-Levy, 70, 71, 72
Collision frequencies, 124, 149, 164, 244, 245, 257, 310, 323, 327, 335, 340
Composite Right/Left Handed Transmission Line (CRLH TL), 17, 129, 130, 132, 133, 134, 135, 137, 139, 140
COMSOL Multiphysics, 335

Coordinate transformation, 334, 335, 336
Coplanar waveguide (CPW), 132, 137
Couplers, 41, 47, 49, 132, 133, 134
CST microwave studio, 7, 268, 310

D

Debye media, 147
Defect modes, 46, 54, 55, 56, 58
Delay line, 137, 138, 141
Diffraction Limits, 9, 13, 58, 59, 217, 239, 240, 307, 316, 321, 322
Directivity, 2, 41, 47, 48, 115, 116, 117, 120, 121, 143, 361
Dispersion diagram, 17, 25, 27, 30, 31, 32, 33, 40, 48, 49, 50, 84, 85, 88, 91, 94, 98, 99, 100, 101, 109, 110, 112, 113
Distributed Bragg reflector (DBR), 2

E

EBG resonator antenna, 47, 115, 117, 119, 120
EBG Superstrates, 47
EBG Tunable Filters, 47, 53
EBG Waveguide, 41, 47, 49, 53
Effective medium theory (EMT), 16, 18, 123, 156, 182, 189, 211, 241, 307, 356, 366
Electromagnetic bandgap structure (EBG), 25–66, 91–101
Electromagnetic crystals, 2, 4, 11, 13, 16, 54
Electromagnetic scattering, 34, 142, 144
Evanescent waves, 16, 127, 128, 161, 169, 217, 239–41, 251–53, 255, 256, 258, 260, 264, 267, 282, 296, 299, 302, 307, 308, 310, 326, 355

377

F

Fabry-Perot cavity, 2, 370
Far field, 120, 143, 165, 166, 239, 241, 307
Fast Fourier transformation, 93
Fat dipoles, 14
Ferromagnetic materials, 3
Figure of merit (FoM), 18, 173–238
Finite element method (FEM), 335
Floquet mode, 28, 362
Frequency-selective surface, 47
Fundamental translation vector, 26, 33

G

Generalized Rayleigh's Identity Method, 33, 34
Guided wave, 8, 132, 321

H

Helmholtz equation, 33, 34
High frequency structure simulator (HFSS), 113, 114, 232, 262, 357
High impedance surfaces (HIS), 2, 7, 13

I

Indefinite media, 9, 10
Indium tin oxide (ITO), 55
Inversion approach, 174, 182–213, 233
Invisible cloak, 335

K

Korringa-Kohn-Rostoker (KKR), 33

L

Leaky wave antennas (LWA), 2
Left-handed materials, 1, 148–55
Liquid crystal, 54–58
Local distorted NFDTD (LD-NFDTD), 76

M

Maxwell's equations, 4, 17, 32, 35, 36, 39, 67, 75, 80, 274

Method of Moments (MoM), 16
Microwave filters, 2
Mur's Absorbing Boundary Conditions, 78

N

Near field, 10, 13, 14, 18, 59, 61, 160, 161, 241, 297–303, 307–19
Negative refraction, 3, 4, 9, 10, 16, 18, 141, 143, 144, 148, 151, 174, 180, 209, 219, 221, 222, 223, 227, 236, 239–41, 253, 267, 307, 308, 355, 361, 365–67

P

Parameter extraction, 18, 358
Perfect lens, 9, 18, 40, 59, 239–64, 307, 316, 317
Perfectly matched layer (PML), 78, 279–81, 318
Periodic boundary condition, 25, 27, 78, 81
Phase compensation, 160
Phase shifters, 137, 138
Photonic bandgap structure, 11, 105
Plasmonic waveguide, 13, 17, 18, 307–33, 354
Pseudo-spectral time-domain, 147
Plane wave expansion (PWE), 17, 33, 35, 109

R

Radar absorbing materials, 9
Radomes, 10, 148, 161, 163
Recursive convolution (RC) method, 147, 156

S

Small antennas, 2, 15, 134, 354, 357, 358
Spatial dispersion, 40, 148, 269–92
Spatial harmonics, 2, 16, 17, 136, 287, 294, 302–4, 308
Split ring resonator (SRR), 7–10, 124, 144, 148, 174, 229
Super-cell approach, 85

Surface plasmon polariton (SPR), 241
Surface waves, 13, 44–46, 152, 239, 313

T

Transfer-Matrix Method (TMM), 36, 102, 108, 110
Translational Symmetry, 25, 26, 28, 29
Traveling wave tubes (TWT), 2, 25

U

Uniaxial perfectly matched layer, 318
Uniplanar compact electromagnetic bandgap, 42

Z

Zero phase delay, 141, 148
Zeroth order resonator, 134
Z-transform method, 147, 336

Recently Related Titles from Artech House

Analytical Modeling in Applied Electromagnetics, Sergei Tretyakov

Applications of Neural Networks in Electromagnetics, Christos Christodoulou and Michael Georgiopoulos

Computational Electrodynamics: The Finite-Difference Time Domain Method, Third Edition, Allen Taflove and Susan C. Hagness

Electromagnetics, Microwave Circuit, and Antenna Design for Communications Engineering, Second Edition, Peter Russer

Grid Computing for Electromagnetics, Luciano Tarricone and Alessandra Esposito

Numerical Analysis for Electromagnetic Integral Equations, Karl F. Warnick

Parallel Finite-Difference Time-Domain Method, Wenhua Yu, Raj Mittra, Tao Su, Yongjun Liu, and Xiaoling Yang

Problem Solving in Electromagnetics, Microwave Circuit, and Antenna Design for Communications Engineering, Karl F. Warnick and Peter Russer

Wavelet Applications in Engineering Electromagnetics, Tapan K. Sarkar, Magdalena Salazar-Palma, Michael C. Wicks

WIPL-D Microwave: Circuit and 3D EM Simulation for RF & Microwave Applications—Software and User's Manual, Branko M. Kolundzija, Jovan S. Ognjanovic, and Tapan K. Sarkar

For further information on these and other Artech House titles, including previously considered out-of-print books now available through our In-Print-Forever® (IPF®) program, contact:

Artech House	Artech House
685 Canton Street	46 Gillingham Street
Norwood, MA 02062	London SW1V 1AH UK
Phone: 781-769-9750	Phone: +44 (0)20 7596-8750
Fax: 781-769-6334	Fax: +44 (0)20 7630-0166
e-mail: artech@artechhouse.com	e-mail: artech-uk@artechhouse.com

Find us on the World Wide Web at: www.artechhouse.com